# Soil Fertility and Fertilizers

Edited by Kye Young

New York

Published by Syrawood Publishing House,
750 Third Avenue, 9th Floor,
New York, NY 10017, USA
www.syrawoodpublishinghouse.com

**Soil Fertility and Fertilizers**
Edited by Kye Young



International Standard Book Number: 978-1-68286-387-9 (Hardback)



**Cataloging-in-publication Data**

Soil fertility and fertilizers / edited by Kye Young.
p. cm.
Includes bibliographical references and index.
ISBN 978-1-68286-387-9
1. Soil fertility. 2. Fertilizers. 3. Soil science. 4. Soil biology. I. Young, Kye.
S633 .S65 2017
631.8--dc23

Printed in the United States of America.

# TABLE OF CONTENTS

# PREFACE

Soil fertility is a primary aspect of soil science. It refers to the study of techniques and methods used for enhancement of fertility in soil. It also includes soil structure, soil moisture retention and soil conservation among other practices. It is used in agriculture mainly to reduce soil erosion and soil degradation. The most common fertilizers are natural fertilizers like manure, peat, etc. and artificial fertilizers like ammonium nitrate, etc. While understanding the long-term perspectives of the topics, this book makes an effort in highlighting their impact as modern tool for the growth of soil fertility. It aims to shed light on some of the unexplored aspects and the recent researches in this field. Students, researchers, experts and all associated with this are will benefit from this text.

This book aims to highlight the current researches and provides a platform to further the scope of innovations in this area. This book is a product of the combined efforts of many researchers and scientists from different parts of the world. The objective of this book is to provide the readers with the latest information in the field.

I would like to express my sincere thanks to the authors for their dedicated efforts in the completion of this book. I acknowledge the efforts of the publisher for providing constant support. Lastly, I would like to thank my family for their support in all academic endeavors.

**Editor**

# 1

# Determination of Critical Nitrogen Dilution Curve Based on Stem Dry Matter in Rice

**Syed Tahir Ata-Ul-Karim, Xia Yao, Xiaojun Liu, Weixing Cao, Yan Zhu***

National Engineering and Technology Center for Information Agriculture, Jiangsu Key Laboratory for Information Agriculture, Nanjing Agricultural University, Nanjing, Jiangsu, P. R. China

## Abstract

Plant analysis is a very promising diagnostic tool for assessment of crop nitrogen (N) requirements in perspectives of cost effective and environment friendly agriculture. Diagnosing N nutritional status of rice crop through plant analysis will give insights into optimizing N requirements of future crops. The present study was aimed to develop a new methodology for determining the critical nitrogen ($N_c$) dilution curve based on stem dry matter ($S_{DM}$) and to assess its suitability to estimate the level of N nutrition for rice (*Oryza sativa* L.) in east China. Three field experiments with varied N rates (0–360 kg N $ha^{-1}$) using three Japonica rice hybrids, Lingxiangyou-18, Wuxiangjing-14 and Wuyunjing were conducted in Jiangsu province of east China. $S_{DM}$ and stem N concentration (SNC) were determined during vegetative stage for growth analysis. A $N_c$ dilution curve based on $S_{DM}$ was described by the equation ($N_c = 2.17W^{-0.27}$with W being $S_{DM}$ in t $ha^{-1}$), when $S_{DM}$ ranged from 0.88 to 7.94 t $ha^{-1}$. However, for $S_{DM} < 0.88$ t $ha^{-1}$, the constant critical value $N_c = 1.76\%$ $S_{DM}$ was applied. The curve was dually validated for N-limiting and non-N-limiting growth conditions. The N nutrition index (NNI) and accumulated N deficit ($N_{and}$) of stem ranged from 0.57 to 1.06 and 51.1 to −7.07 kg N $ha^{-1}$, respectively, during key growth stages under varied N rates in 2010 and 2011. The values of $\Delta N$ derived from either NNI or $N_{and}$ could be used as references for N dressing management during rice growth. Our results demonstrated that the present curve well differentiated the conditions of limiting and non-limiting N nutrition in rice crop. The $S_{DM}$ based $N_c$ dilution curve can be adopted as an alternate and novel approach for evaluating plant N status to support N fertilization decision during the vegetative growth of Japonica rice in east China.

**Editor:** Guoping Zhang, Zhejiang University, China

**Funding:** This work was supported by grants from the National High-Tech Research and Development Program of China (863 Program) (2011AA100703), Special Program for Agriculture Science and Technology from Ministry of Agriculture in China (201303109), Priority Academic Program Development of Jiangsu Higher Education Institutions (PAPD), and Science and Technology Support Plan of Jiangsu Province (BE2011351, BE2012302). The funders had no role in study design, data collection and analysis, decision to publish, or preparation of the manuscript.

**Competing Interests:** The authors have declared that no competing interests exist.

* Email: yanzhu@njau.edu.cn

## Introduction

Estimating nitrogen (N) nutritional status is a key to investigating, monitoring, and managing cropping systems [1]. Conventional farming has led to extensive use of N as a tool for ensuring profitability in the soils with uncertain fertility levels, which has raised the concerns about environmental sustainability. A reliable diagnosis of crop N requirement and nutritional status give insight into optimization of qualitative and quantitative aspects of crop production. It also improve N use efficiency and add to environmental protection [2]. Soil and plant-based strategies are two principle approaches, extensively used to derive information about the N nutrition status of crops, for satisfying their demand for N and to minimize N losses [3]. The former rarely describes the intensity of N release over a longer period, so the latter are widely accepted and adopted. Therefore, the present study investigates a plant-based strategy for an in-season assessment of N nutrition status for rice crop.

In plant-based approaches, the N nutrition status is generally monitored to determine the requirement for top dressing in crops [3]. For this purpose, several plant-based diagnostic tools, such as critical N concentration ($N_c$) approach, chlorophyll meter, hyper-spectral reflectance and remote sensing, have been successfully used for in-season N management [4]. They differ in scope, in context of reference spatial scale, in terms of monetary and time resources, as well as skills and expertise required for their implementation at field [5]. Despite being simple, chlorophyll meter readings are affected by leaf thickness, abiotic stress and nutrient variability [6]. Canopy reflectance method's accuracy is affected by solar illumination, soil background effects and sensor viewing geometry [4]. However, the concept of $N_c$ can be used as a potential alternate to these techniques, and it can give insight into relative N status of a crop. The present study utilizes this concept for an in-season N fertilizer management in rice crop.

The concept of $N_c$ is crop specific, precise, simple and biologically sound, because it is based on actual crop growth. Whole plant dry matter based $N_c$ approach was successfully applied for N management in winter wheat [7,8], corn [9] and spring wheat [10]. This approach was successfully applied for a Indica rice in tropics and Japonica rice in subtropical temperate region [11,12]. Dry matter partitioning among different plant organs affects the weight/N concentration relationship, and changes the shape of the dilution curve, thus limits its acceptance as a reliable method [13,14]. The concept of $N_c$ for specific plant

**Table 1.** Changes of stem dry matter ($S_{DM}$) with time (days after transplantation) under different N rates in two rice cultivars in experiments conducted during 2010 and 2011.

| Year | Cultivar | DAT | Sampling date | Stem dry matter/Applied N (kg ha$^{-1}$) | | | | | F prob. | LSD |
|---|---|---|---|---|---|---|---|---|---|---|
| | | | | 0 | 80 | 160 | 240 | 320 | | |
| 2010 | LXY-18 | 16 | 07-Jul | 0.23 | 0.27 | 0.38 | 0.48 | 0.49 | * | 0.028 |
| | LXY-18 | 26 | 17-Jul | 0.63 | 0.78 | 0.95 | 1.11 | 1.12 | * | 0.055 |
| | LXY-18 | 36 | 27-Jul | 1.04 | 1.28 | 1.55 | 1.77 | 1.81 | * | 0.075 |
| | LXY-18 | 48 | 08-Aug | 2.23 | 2.73 | 3.25 | 3.61 | 3.51 | * | 0.226 |
| | LXY-18 | 60 | 20-Aug | 3.87 | 4.47 | 4.94 | 5.23 | 5.29 | * | 0.146 |
| | LXY-18 | 70 | 30-Aug | 4.72 | 5.56 | 6.7 | 7.01 | 7.22 | * | 0.279 |
| | WXJ-14 | 16 | 07-Jul | 0.22 | 0.27 | 0.32 | 0.36 | 0.35 | * | 0.019 |
| | WXJ-14 | 26 | 17-Jul | 0.39 | 0.54 | 0.73 | 0.9 | 0.91 | * | 0.045 |
| | WXJ-14 | 36 | 27-Aug | 0.55 | 0.8 | 1.13 | 1.38 | 1.46 | * | 0.063 |
| | WXJ-14 | 48 | 08-Aug | 1.22 | 1.65 | 1.99 | 2.18 | 2.23 | * | 0.11 |
| | WXJ-14 | 60 | 20-Aug | 2.77 | 3.46 | 3.72 | 4.19 | 3.97 | * | 0.233 |
| | WXJ-14 | 70 | 30-Sep | 3.69 | 4.4 | 5.04 | 5.81 | 5.7 | * | 0.205 |
| | | | | Stem dry matter/Applied N (kg ha$^{-1}$) | | | | | | |
| | | | | 0 | 90 | 180 | 270 | 360 | | |
| 2011 | LXY-18 | 18 | 09-Jul | 0.22 | 0.33 | 0.4 | 0.56 | 0.59 | * | 0.042 |
| | LXY-18 | 30 | 21-Jul | 0.67 | 0.76 | 0.92 | 1.19 | 1.19 | * | 0.053 |
| | LXY-18 | 42 | 02-Aug | 1.12 | 1.28 | 1.42 | 1.78 | 1.76 | * | 0.132 |
| | LXY-18 | 54 | 15-Aug | 2.24 | 2.41 | 2.76 | 3.43 | 3.48 | * | 0.127 |
| | LXY-18 | 64 | 25-Aug | 3.59 | 4.02 | 4.37 | 4.75 | 4.85 | * | 0.164 |
| | LXY-18 | 74 | 04-Sep | 5.68 | 5.91 | 6.41 | 7.84 | 8.04 | * | 0.172 |
| | WXJ-14 | 18 | 09-Jul | 0.19 | 0.28 | 0.32 | 0.34 | 0.36 | * | 0.02 |
| | WXJ-14 | 30 | 21-Jul | 0.37 | 0.6 | 0.71 | 0.86 | 0.9 | * | 0.06 |
| | WXJ-14 | 42 | 02-Aug | 0.54 | 0.96 | 1.2 | 1.37 | 1.47 | * | 0.128 |
| | WXJ-14 | 54 | 15-Aug | 1.51 | 1.84 | 2.24 | 2.52 | 2.68 | * | 0.137 |
| | WXJ-14 | 64 | 25-Aug | 2.49 | 3.07 | 3.36 | 4.06 | 4.04 | * | 0.169 |
| | WXJ-14 | 74 | 04-Sep | 4.41 | 5.04 | 5.75 | 6.2 | 6.27 | * | 0.178 |

*: F statistic significant at 0.01 probability level.

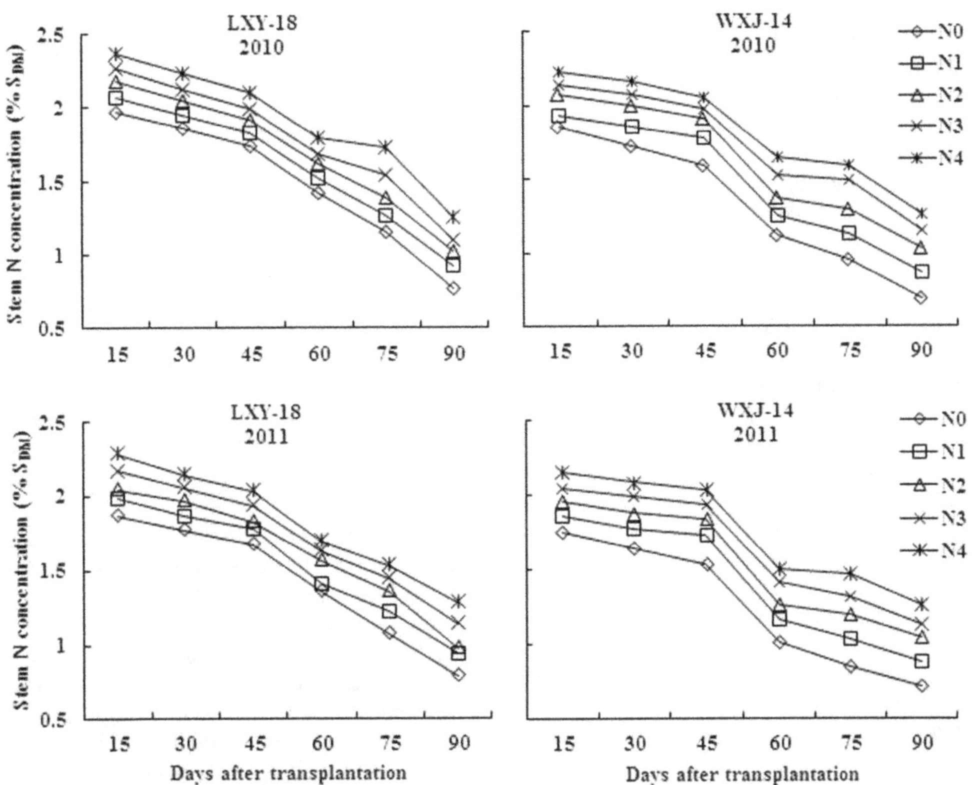


**Figure 1. Changes of stem nitrogen concentration (% $S_{DM}$) with time (days after transplantation) for rice under different N rates in experiments conducted during 2010 and 2011.**

organs (e.g., leaves and stem) is similar to that on whole plant basis. Leaf based diagnosis of N status in crops is affected by progressive shading by newer leaves, decline of leaf N concentration due to aging, pest attack, abiotic stresses and increase in the proportion of structural tissues [15]. Stem sap nitrate concentration is influenced by phenological phase, cultivar, temperature and solar radiation [16]. During vegetative phase, the contribution of stem dry matter ($S_{DM}$) towards total plant dry matter is significantly higher than that of leaf dry matter ($L_{DM}$), hence it is the most determining factor for N dilution of the whole plant [17]. Thus, the idea of using $N_c$ curve based on $S_{DM}$ over whole plant dry matter and $L_{DM}$ based methods, can be used as an alternate approach for determination of $N_c$ dilution curve.

The objectives of this work were to develop a $N_c$ dilution curve based on $S_{DM}$ and to assess the plausibility of this curve to estimate N nutrition status of Japonica rice. The estimation based on this approach will be more reliable than existing methods due to consistency at different growth stages.

## Materials and Methods

### Ethics statement

The experiments land is owned and managed by Nanjing Agricultural University, Nanjing, China. Nanjing Agricultural University permits and approvals obtained for the work and study. The field studies did not involve wildlife or any endangered or protected species.

### Experimental details

Three field experiments with multiple N rates (0–360 kg N $ha^{-1}$) were conducted using three contrasting Japonica rice hybrids, Lingxiangyou-18 (LXY-18), Wuxiangjing-14 (WXJ-14) and Wuyunjing (WYJ), at Yizheng (32°16′N, 119°10′E) and Jiangning (31°56′N, 118°59′E) located in lower Yangtze River Reaches of east China. The soil was clay loam and was classified as Ultisoles. The rice-wheat cropping system is practiced in the region. The applied N rates varied significantly among different farmers. The average rate of N fertilizer reached 387 kg $ha^{-1}$ during the period of 2004–2008 [18].

The whole experimental area was ploughed and subsequently harrowed before transplanting. All bunds were compacted to prevent seepage into and from adjacent plots. A plastic lining was installed to a depth of 40 cm between drain and the bund of each plot to minimize seepage across the bunds towards the drains. To further minimize seepage of water from control plot ($N_0$), double bunds were constructed separating them and the adjacent plots. Experiments were arranged in a randomized complete block design with three replications. The size of each experimental plot was 8 m by 4.5 m, with planting density of approximately 22.2 hills per $m^2$. At site 1, soil pH, organic matter, total N, available phosphorous (P), and available potassium (K) were 6.2, 17.5 g $kg^{-1}$, 1.6 g $kg^{-1}$ 43 mg $kg^{-1}$, 90 mg $kg^{-1}$, and 6.4, 15.5 g $kg^{-1}$, 1.3 g $kg^{-1}$ 38 mg $kg^{-1}$, and 85 mg $kg^{-1}$ in 2010 and 2011, respectively. The corresponding soil properties were 6.5, 13.5 g $kg^{-1}$, 1.13 g $kg^{-1}$ 45 mg $kg^{-1}$, 91 mg $kg^{-1}$ in 2007 at site 2. For experiments conducted at site 1 in 2010 and 2011, treatment consisted of five N rates as 0, 80, 160, 240, and 320 kg N $ha^{-1}$, and 0, 90, 180, 270, and 360 kg N $ha^{-1}$, respectively, while for experiment conducted at site 2 in 2007, treatment consisted of three N rates as 110, 220, and 330 kg N $ha^{-1}$. N in all experiments was distributed as 50% at pre planting, 10% at tillering, 20% at jointing, and 20% at booting, with urea as the N source. Aside from N fertilizer, phosphorus (135 kg $ha^{-1}$) and potassium (190 kg $ha^{-1}$) fertilizers were basally incorporated at the last harrowing and leveling in all plots before transplanting as monocalcium phosphate $Ca(H_2PO_4)_2$ and potassium chloride (KCl). Rice seedlings at five leaves stage were transplanted in experimental fields on June 20 (site 1) in 2010 and 2011, and on 29 June (site 2) in 2007, respectively. Pre-emergence herbicides were used to control weeds at early growth stages. Also plots were regularly hand-weeded until canopy was closed to prevent weed damage. Insecticides were used to prevent insect damage. All other agronomic practices were used according to local recommendations to avoid yield loss.

## Sample collection and measurement

Rice plants were sampled from each plot at the intervals of 10–12 days from 0.23 $m^2$ area (5 hills) at active tillering, mid tillering, stem elongation, panicle initiation, booting and heading stages during the period of each experiment for growth analysis. The plants were manually severed at ground level on each sampling date. Fresh plants were divided into green leaf blades and culm plus sheath. Samples were oven-dried at 105°C for half an hour to rapidly stop metabolism and then at 70°C until constant weight to obtain stem dry matter ($S_{DM}$, t $ha^{-1}$). The dried stem samples were ground and analyzed for total stem N concentration (SNC, %) by Kjeldahl method. Stem N accumulation (SNA, kg N $ha^{-1}$) was obtained as summed product of the $S_{DM}$ by the SNC. The SNC of whole-plant stem was calculated as SNA divided by $S_{DM}$.

## Statistical analysis

The $S_{DM}$ and SNC data for each sampling date, year and variety was separated and subjected to analysis of variance (ANOVA) using GLM procedures in SPSS-16 software package (SPSS Inc., Chicago. IL, USA). The differences among treatment means were measured by using the least significant difference (LSD) test at 90% level of significance, instead of classically used 95% in order to reduce the occurrence of Type II errors that could be high in such field experiments. For each measurement date, year and variety, the variation in the SNC versus $S_{DM}$ across the different N levels was combined into a bilinear relation composed of a linear regression representing the joint increase in SNC and $S_{DM}$ and a vertical line corresponding to an increase in SNC without significant variation in $S_{DM}$. The theoretical $N_c$ points corresponds to the ordinate of the breakout of the bilinear regression. Regression analysis was performed using Microsoft Excel (Microsoft Cooperation, Redmond, WA, USA).

## Construction and validation of critical, maximum and minimum N dilution curves

For determination of $N_c$ dilution curve it is necessary to determine the N concentration that did not limit the $S_{DM}$ production either by its excess or deficiency. The data used to construct the $N_c$ dilution curve came from two experiments conducted in 2010 and 2011 by distinguishing the data points for N-limiting and non-N-limiting growth. The N-limiting growth treatment is defined as a treatment for which an additional N application leads to a significant increase in $S_{DM}$. The non-N-

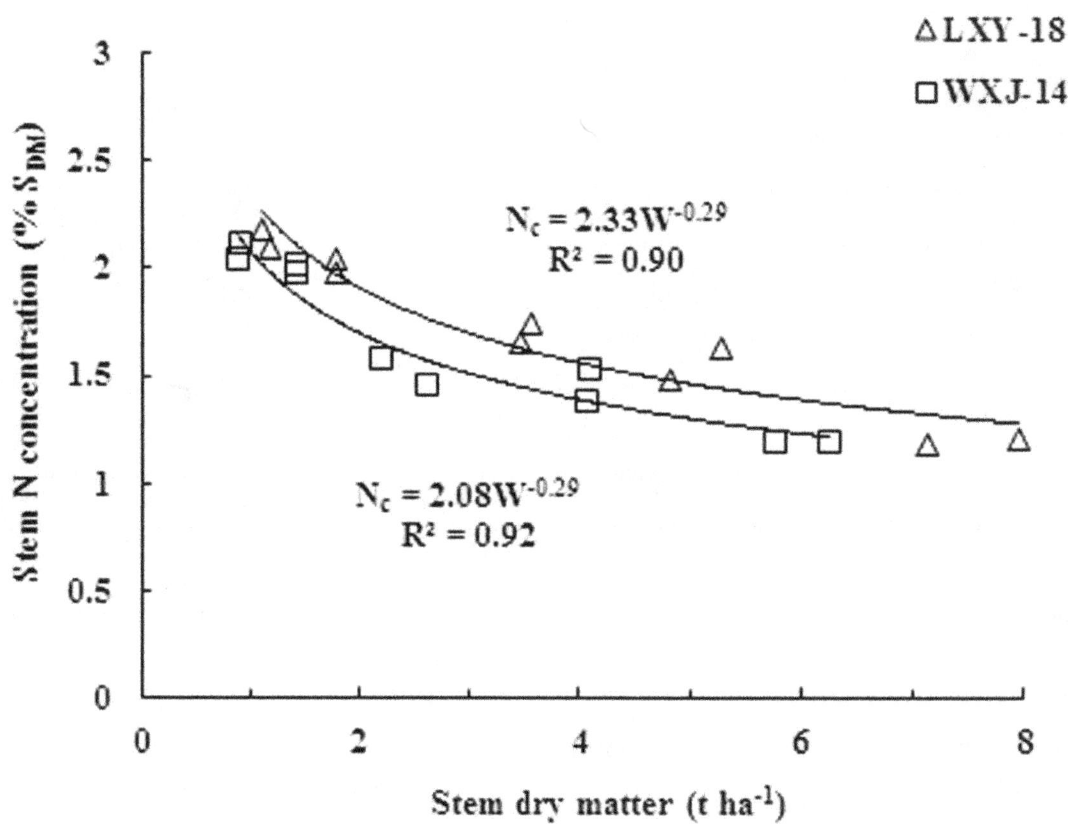


**Figure 2. Critical nitrogen data points and $N_c$ dilution curves in stem obtained by non-linear fitting for two rice cultivars (LXY-18, $N_c = 2.33W^{-0.29}$ and WXJ-14, $N_c = 2.08W^{-0.29}$) under different N rates in experiments conducted during 2010 and 2011.**

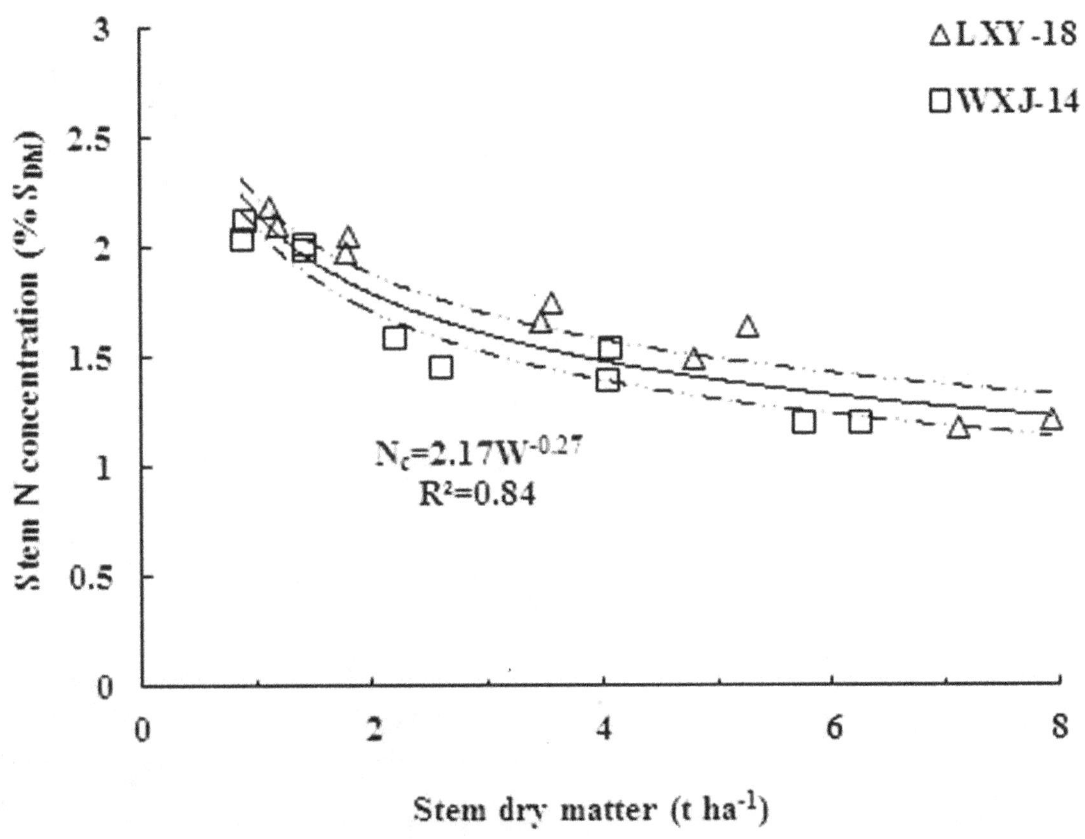


**Figure 3. Critical nitrogen data points used to define the $N_c$ dilution curve when data were pooled over for two rice cultivars (LXY-18 and WXJ-14).** The solid line represents the $N_c$ dilution curve ($N_c = 2.17W^{-0.27}$; $R^2 = 0.84$) describing the relationship between the $N_c$ and stem dry matter of rice. The dotted lines represent the confidence band ($P = 0.95$).

limiting growth treatment is defined as a treatment, for which a supplement of N application does not lead to an increase in $S_{DM}$ and, at the same time, exhibits a significant increase in SNC. If at the same measurement date, statistical analysis distinguished at least one set of N-limiting and non-N limiting data point, these data points were used either for construction of the $N_c$ dilution curve or to validate it [7]. Consistent with earlier studies, an allometric function based on power regression (Freundlich model) was used to determine the relationship between the observed decreases in $N_c$ with increasing $S_{DM}$. The $N_c$ dilution curve was validated first by using the data points not retained for establishing the parameters of the allometric function in 2010 and 2011, and then with independent data set from experiment conducted in 2007.

The data points ($n = 13$) from most plethoric N treatments ($N_4$ plots) was assumed to represent the maximum N dilution curve ($N_{max}$) while the minimum N dilution curve ($N_{min}$) was determined by using the data points ($n = 13$) from the most N-limiting treatments for which N application was zero ($N_0$ check plots).

### Calculation of critical N dilution curve based diagnostic tools

To identify the N status in the $S_{DM}$ of rice during vegetative growth, the nitrogen nutrition index (NNI) and accumulated nitrogen deficit ($N_{and}$) were established for each sampling date, experiment and variety. The NNI value was obtained by dividing the total N concentration of $S_{DM}$ by $N_c$ value determined by critical dilution curve, [9]. The $N_{and}$ value for rice crop on each sampling date was obtained by subtracting the N accumulation under the $N_c$ condition ($N_{cna}$) from actual N accumulation ($N_{na}$) under different N rates [12]. For in-season recommendation of supplemental N application, the difference value of NNI (ΔNNI), $N_{and}$ ($\Delta N_{and}$) and difference value of N application rate (ΔN between different N treatments was calculated according to the method proposed by Ata-Ul-Karim et al. [12].

## Results

### Stem dry matter and nitrogen concentration

The $S_{DM}$ production was significantly affected by N fertilization during the growth period of rice. The increase in $S_{DM}$ followed a continuous increasing trend along with sampling dates for both the varieties during each year with increasing N rates from $N_0$ to $N_4$; however, there was no significant difference between $N_3$ and $N_4$ in all the cases (Table 1). This increase in the $S_{DM}$ production with N fertilization may be linked to a higher absorption of N fertilizer. $S_{DM}$ ranged from minimum 0.22 t $ha^{-1}$ and 0.19 t $ha^{-1}$ ($N_0$) in WXJ-14 to a maximum of 7.22 t $ha^{-1}$ and 8.04 t $ha^{-1}$ ($N_4$) in LXY-18 during 2010 and 2011, respectively. The results showed that there was no positive correlation between $S_{DM}$ and N rates, as the $S_{DM}$ tend to decrease when N rate exceeded a critical level. During each experimental year, $S_{DM}$ conferred with the following inequality under different N ratess.

$$S_{DM0} < S_{DM1} < S_{DM2} < S_{DM3} = S_{DM4} \quad (1)$$

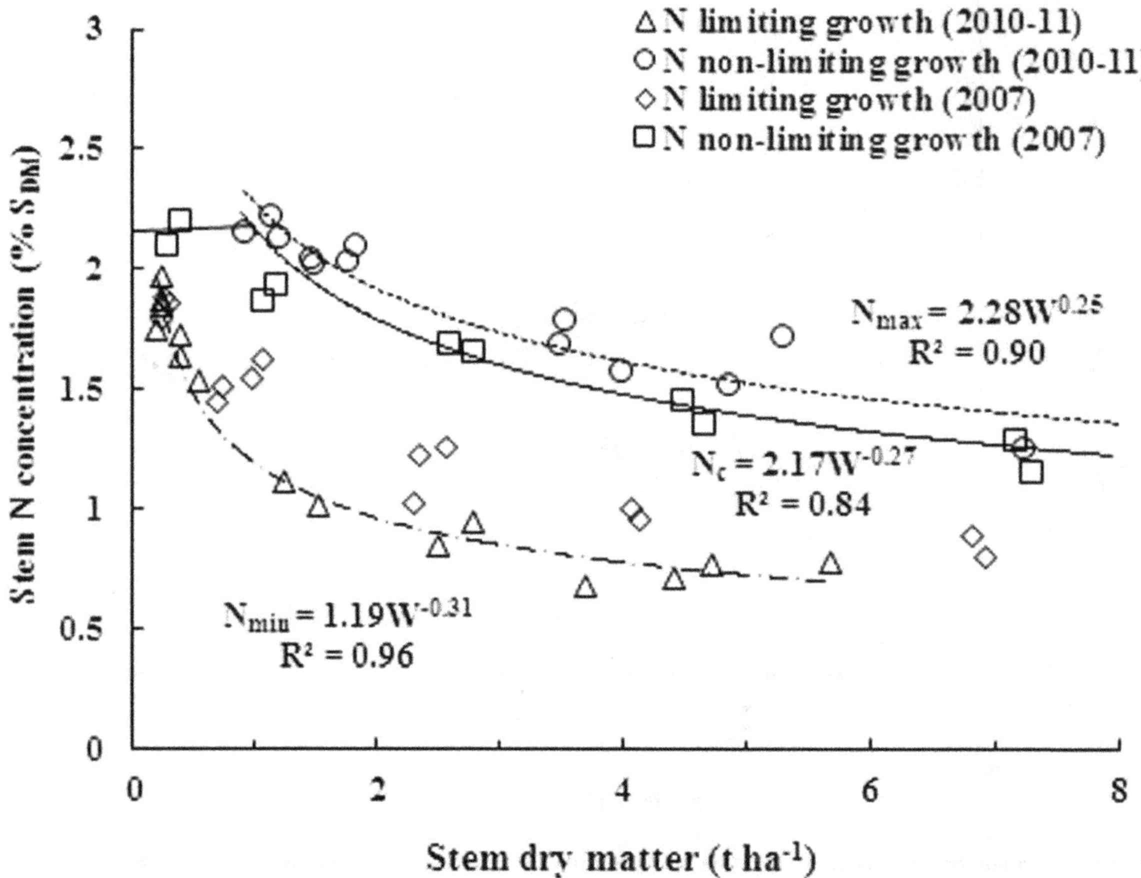


**Figure 4. Comprehensive validation of $N_c$ dilution curve using independent data set from experiment conducted in 2007.** Data points (◇) represent N limiting growth conditions, while (□) represent N non-limiting conditions. The solid line in the middle represents the $N_c$ curve ($N_c = 2.17W^{-0.27}$) describing the relationship between the $N_c$ and stem dry matter of rice. The data points (Δ) and (○) not engaged for establishing the parameters of allometric function (2010 and 2011) were used to develop two boundary curves, (-●-●-●) minimum limit curve ($N_{min} = 1.19\ W^{-0.31}$) and (------) maximum limit curve ($N_{max} = 2.27W^{-0.25}$).

where $S_{DM0}$, $S_{DM1}$, $S_{DM2}$, $S_{DM3}$ and $S_{DM4}$ stands for $S_{DM}$ of $N_0$, $N_1$, $N_2$, $N_3$ and $N_4$, respectively.

Stem N concentration response to N fertilizer rates was usually linear and a higher rate of N mostly resulted in a higher SNC, hitherto a decline in SNC was observed with increasing $S_{DM}$ from active tillering to heading. Maximum variation in SNC of both cultivars was observed on 16 and 18 DAT, while minimum on 70 and 74 DAT, in years 2010 and 2011, respectively. The SNC ranged from 2.28 to 0.78 for LXY-18 and 2.16 to 0.71 for WXJ-14 during 2010, while 2.36 to 0.77 for LXY-18 and 2.23 to 0.68 for WXJ-14 during 2011 (Fig. 1).

## Determination of critical nitrogen dilution curves based on stem dry matter

A set of twenty theoretical data points for both cultivar, obtained from two experiments (10 data points for each cultivar) from active tillering to heading were used to calculate the $N_c$ for a given level of $S_{DM}$. The $S_{DM}$ data that fit the statistical criteria for establishing $N_c$ dilution curve varied from 0.88 t ha$^{-1}$ to 7.94 t ha$^{-1}$. A power functions were fitted to the calculated $N_c$ points as equations (2) and (3), the coefficient for which were 0.90 and 0.92 for LXY-18 and WXJ-14, respectively (Fig. 2).

$$N_c = 2.33W^{-0.29} \qquad (W \geq 0.88\ t\ ha^{-1}, R^2 = 0.90, n = 10) \quad (2)$$

$$N_c = 2.08W^{-0.29} \qquad (W \geq 0.88\ t\ ha^{-1}, R^2 = 0.92, n = 10) \quad (3)$$

where W is the $S_{DM}$ expressed in t ha$^{-1}$; $N_c$ is the critical N concentration in stem expressed in % $S_{DM}$; *a* and *b* are estimated parameters. The parameter *a* represents the N concentration in the $S_{DM}$ when W = 1 t ha$^{-1}$, and *b* represents the coefficient of dilution describing the relationship between N concentration and $S_{DM}$.

The F-value (0.72) of two curves was less than the critical value of $F_{(1-18)} = 4.41$ at 5% probability level, showing non-significant difference between the curves [19], thus the data for the two varietal groups were united, and a unified dilution curve was determined as equation 4.

$$N_c = 2.17W^{-0.27} \qquad (W \geq 0.88\ t\ ha^{-1}, R^2 = 0.84, n = 20) \quad (4)$$

The model accounted for 84% of the total variance. At early growth stages of rice crop, the $N_c$ varied between 2.24% $S_{DM}$ to 2.10% $S_{DM}$ (95% confidence interval) for a $S_{DM}$ of 0.88 t ha$^{-1}$ at the lower end while 7.94 t ha$^{-1}$ at the higher end, respectively (Fig. 3).

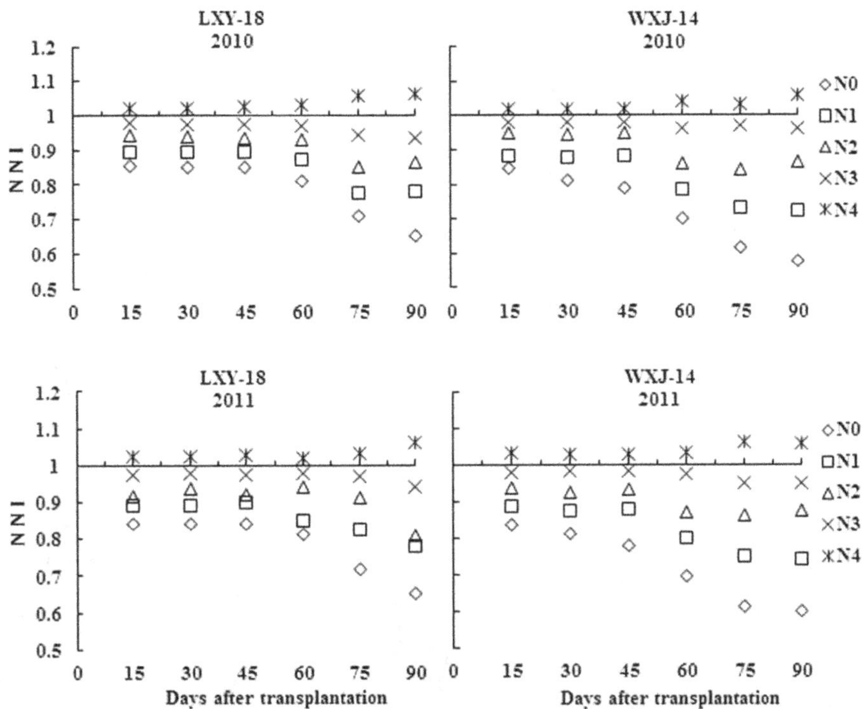


**Figure 5. Changes of nitrogen nutrition index (NNI) with time (days after transplantation) for rice stem under different N rates in experiments conducted during 2010 and 2011.**

For the $S_{DM}$ range of 0.1 to 0.88 t $ha^{-1}$, corresponding to early growth stages, increasing N rates at sowing did not significantly affect $S_{DM}$, because N requirement is relatively low during these early stages. Therefore, the $N_c$ dilution curve cannot be applied to the low $S_{DM}$<0.88 t $ha^{-1}$ at early growth stages due to relatively smaller decline of $N_c$ with increasing $S_{DM}$. For these $S_{DM}$, the $N_c$ could not been determined by the same statistical method because the very high slope of the linear regression resulted in a highly variable estimate [7]. Hence, for the data points of $S_{DM}$ ranging from 0.37 to 0.88 t $ha^{-1}$ a constant $N_c$ (1.76% $S_{DM}$) was calculated as the mean value between the minimum N concentration of non-limiting N points (2.26% $S_{DM}$) and the maximum N concentration of limiting N points (1.25% $S_{DM}$), based on extrapolation of equation 4.

The above $S_{DM}$ based $N_c$ dilution curve was dually validated for N-limiting and non-N-limiting situations within the range for which it was developed. First, the curve was partially validated by combining the data points not engaged for establishing the parameters of the allometric function. In addition, the comprehensive validation of the curve was performed by using the data points from an independent experiment conducted in 2007. The results revealed that the N concentration data that led to the highest significant yields in $S_{DM}$ were positioned close to or above the $N_c$ dilution curve and considered to be non-N-limiting concentrations, whereas the data for the lowest significant $S_{DM}$ yields, were positioned close to or under the $N_c$ dilution curve and classified as N limiting values (Fig. 4). To determine $N_{max}$, data points were selected only from non-N-limiting treatments (n = 13), and for $N_{min}$, data points were selected from the treatment without N application (n = 13). Thus, the present $N_c$ dilution curve could well discriminate the N limiting and non-N-limiting growing conditions in this study

## Changes of NNI and $N_{and}$

Nitrogen nutrition index and $N_{and}$ are helpful in determining the crop nutrition status i.e. deficient, optimal or excess of N nutrition. N nutrition is considered as optimum when NNI = 1 and $N_{and}$ = 0, while NNI>1 and $N_{and}$<0 indicates luxury consumption of N nutrition, values of NNI<1 and $N_{and}$>0 represents N shortage. NNI and $N_{and}$ can be used to quantify the intensity of the N stress after the onset of N deficiency. Our results of significant differences in NNI and $N_{and}$ across the growing seasons, N rates, and phenological stages in rice are in agreement with earlier reports for maize and wheat [10]. As seen in Figure 5 and 6, during 2010 and 2011 the NNI ranged from 0.65 to 1.06 for LXY-18 and 0.57 to 1.06 for WXJ-14, while the $N_{and}$ ranged from 51.1 kg $ha^{-1}$ to −7.07 kg $ha^{-1}$ for LXY-18 and 43.3 kg $ha^{-1}$ to −4.5 kg $ha^{-1}$ for WXJ-14. The results showed that NNI amplified while $N_{and}$ declined with increasing N rates, while both intensified steadily with growth of rice crop and reached to peaks at heading stage for $N_0$, $N_1$, $N_2$ and $N_3$ (N limiting treatments), nevertheless, for $N_3$ this intensification was minor. In contrast, surplus N nutrition existed till heading stage for $N_4$ (non-N-limiting treatment). The estimates based on NNI and $N_{and}$ can be used to identify the N nutritional status at any stage of rice growth, allowing us to assess whether the N fertilizer dosage was ample enough to obtain higher yield in practice. These results confirmed the plausibility of using NNI and $N_{and}$ to assess the status of N nutrition in rice plants growing under various conditions and stages.

Figure 7 and 8 showed that ΔN had a positive correlation with ΔNNI and $\Delta N_{and}$. The simple linear regression equation showed non-significant differences between two varieties, although noticeable differences were observed among different phenological stages. Therefore, ΔN during growth period for both varieties could be derived from ΔNNI and $\Delta N_{and}$, respectively, according

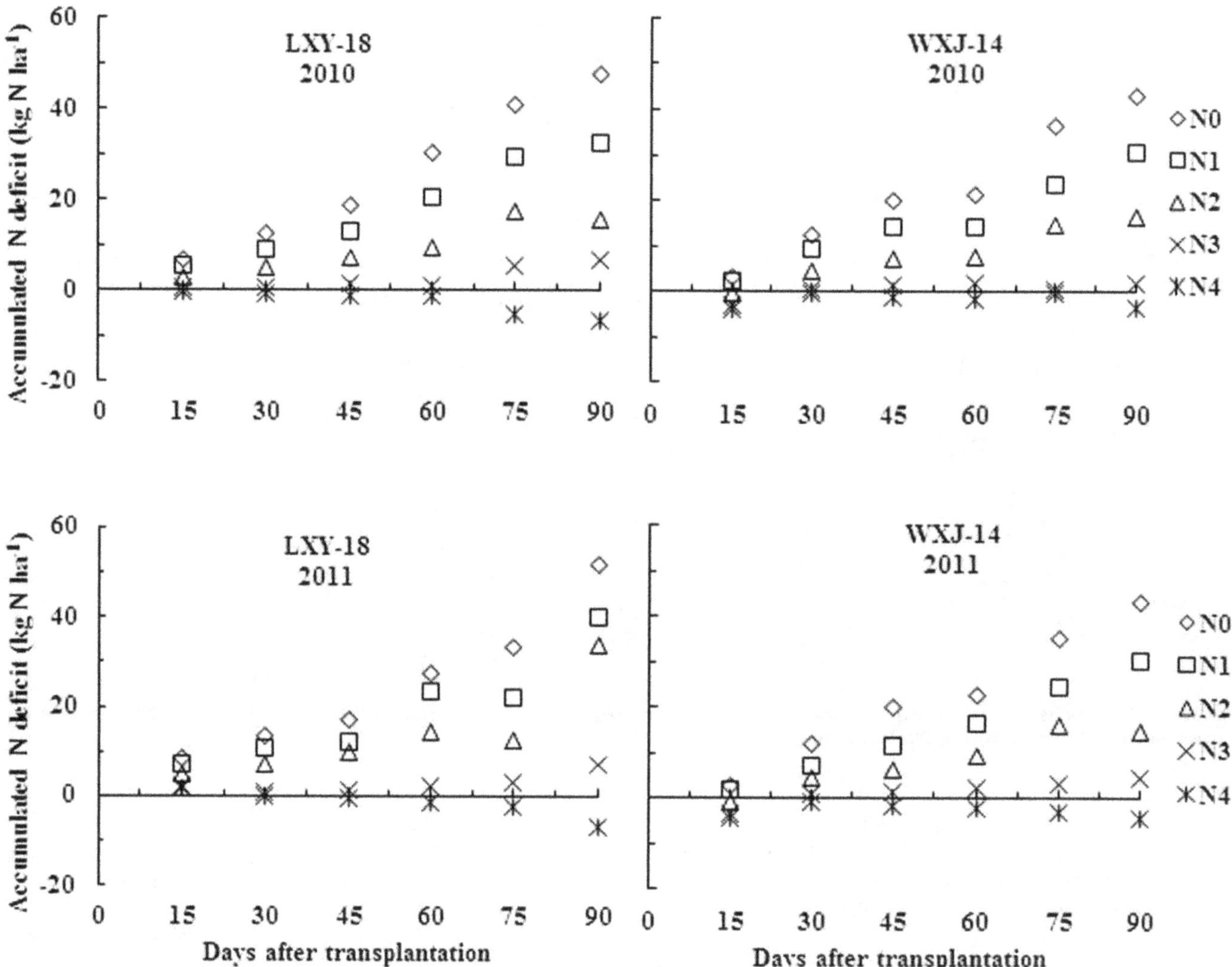


**Figure 6. Changes of accumulated N deficit ($N_{and}$) with time (days after transplantation) for rice stem under different N rates in experiments conducted during 2010 and 2011.**

to the equations 5 & 6 as follows:

$$\Delta N = A \times \Delta NNI + B \quad (5)$$

$$\Delta N = C \times \Delta N_{and} + D \quad (6)$$

The parameters A, B, C, and D could be calculated from days after transplanting (DAT) using the equations as:

$$A = -16.60 \times DAT + 2101 \quad (R^2 = 0.95) \quad (7)$$

$$B = -0.024 \times DAT^2 + 2.57 \times DAT - 40.07 \quad (R^2 = 0.62) \quad (8)$$

$$C = 18.97 ln(DAT) - 89.22 \quad (R^2 = 0.98) \quad (9)$$

$$D = -9.98 ln(DAT) + 52.76 \quad (R^2 = 0.19) \quad (10)$$

The ΔN obtained on the basis of relationship between ΔNNI, $\Delta N_{and}$ and ΔN, allowed us to make corrective decisions of N dressing recommendation for the precise N management during or even before the period of highest demand of the rice crop.

## Discussion

Application of N fertilizer for crop production is an economically viable option in terms of low cost as compared to the value of the marketable agricultural products themselves; however, N usage cannot assure a significant increase in crop productivity due to diminishing returns after certain levels. There is an increasing demand by strategy makers for simple-to-use, technically established and economically viable N indicators, which may allow monitoring and assessment of policy measures and offer tools for farm N management. With the advent of technology, more emphasis should be put on plant-based indicators, which simultaneously reflect the interactions between the plant and the

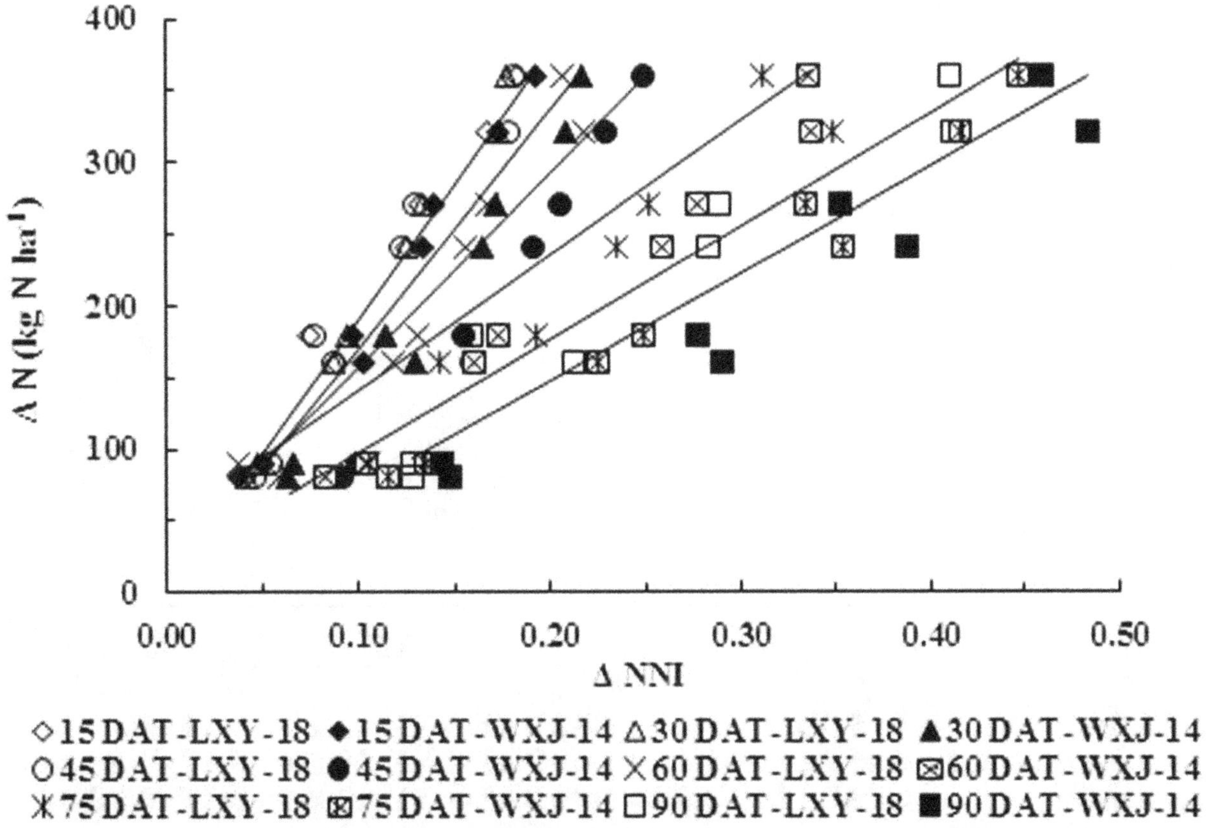


**Figure 7. Relationship between changes of nitrogen nutrition index (ΔNNI) and changes of nitrogen application rates (ΔN, kg N ha$^{-1}$) at different growth stages in experiments conducted during 2010 and 2011.** The open symbols represent different growth stages for LXY-18 while filled symbols represent different growth stages for WXJ-14. ($\Delta N = A \times \Delta NNI + B$; $A = -16.60 \times DAT + 2101$, $R^2 = 0.95$; $B = -0.024 \times DAT^2 + 2.57 \times DAT - 40.07$, $R^2 = 0.62$).

soil. So far, there have been several reports on estimating the $N_c$ concentration on whole plant dry matter basis in various crops, including rice [11,12], and on $L_{DM}$ basis in rice [20], yet no attempt was made to determine the $N_c$ dilution curve on $S_{DM}$ basis for any crop including rice. The current study has developed a $S_{DM}$ based $N_c$ dilution curve for rice in east China, thus providing a new approach for diagnosing and regulating N in crop species.

## Minimum and maximum nitrogen dilution curves

An obvious variability in SNC for a given range of $S_{DM}$ was observed when all the data from three year experiments were analyzed for interpretation. This variability in SNC towards maturity of rice crop in present study was in agreement with earlier studies on winter wheat [7] and Japonica rice [12], and this variability could be attributed to a decline in the fraction of total plant N associated with photosynthesis [21], change in leaf/shoot ratio and self-shading of leaves [8].

Two boundary curves for N maximum ($N_{max}$) and minimum ($N_{min}$) have been determined by using maximum and minimum N concentration in $S_{DM}$ and can be represented as equations:

$$N_{max} = 2.28\,W^{-0.25} \tag{11}$$

$$N_{min} = 1.19\,W^{-0.31} \tag{12}$$

The $N_{max}$ curve corresponding to the maximum N uptake in the $S_{DM}$ without interfering with productivity and it can be considered as the first assessment of a maximum N dilution on $S_{DM}$ basis in crops, and can be obtained with increasing N rates for maximum growth and N accumulation. This curve is an estimate of the maximum N accumulation capacity of stem which is regulated by mechanism associated with the growth and availability of soil N directly or indirectly via N metabolism [22]. The $N_{max}$ curve in the present study shows a luxury consumption of N under $N_4$ treatment, when N concentration exceeds $N_c$ dilution curve and $S_{DM}$ does not increase with increasing N rate. In contrast, the $N_{min}$ curve is considered as a lower limit at which the N metabolism would soon stop to function. It corresponds to the minimum N taken up by rice plants under $N_0$ treatment in present study. Thus, the $N_{min}$ were used as the threshold concentration for proper metabolic functionality of the plant.

Moreover, the value of parameter $b$ for the $N_{max}$ was not significantly different from that of $N_c$ dilution curve, which indicate that the partitioning of dry matter remains relatively constant when N uptake exceeds the $N_c$ dilution curve. This is consistent with the concept of $N_c$ dilution curve, which represents the lowest N at which maximum dry matter accumulation occurs. This implies that under luxury consumption of N, when N exceeds $N_c$ dilution curve, dry matter accumulations does not increase with N and hence, dry matter partitioning will have similar value of parameter $b$. In contrast, for $N_{min}$ curve under N stress, the value for parameter $b$ tended to be slightly lower than the dilution curve.

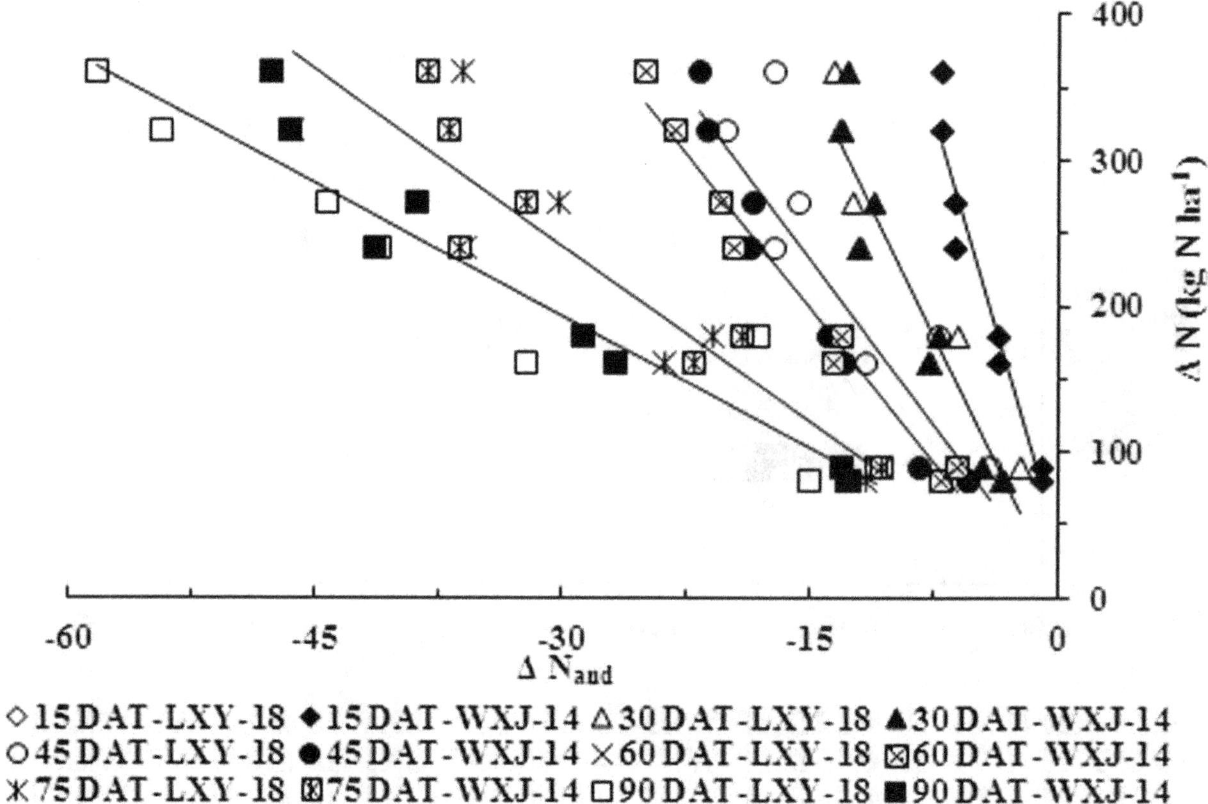


**Figure 8. Relationship between changes of accumulated N deficit ($\Delta N_{and}$) and changes of nitrogen application rates ($\Delta N$, kg N ha$^{-1}$) at different growth stages in experiments conducted during 2010 and 2011.** The open symbols represent different growth stages for LXY-18 while filled symbols represent different growth stages for WXJ-14 ($\Delta N = C \times \Delta N_{and} + D$; $C = 18.97 \ln(DAT) - 89.22$, $R^2 = 0.98$ and $D = -9.98 \ln(DAT) + 52.76$, $R^2 = 0.19$), respectively.

The relatively low value for *b* was associated with a change in dry matter partitioning.

## Comparison with other critical nitrogen dilution curves

The concept of $N_c$ dilution curve on whole plant dry matter and $L_{DM}$ basis have already been successfully implicated for several crops including rice, yet no attempt was made to construct a $S_{DM}$ based $N_c$ dilution curve in any crop including rice. Figure 9 showed that the parameter *a* of $N_c$ dilution curve on $S_{DM}$ basis with Japonica rice developed in present study (2.17) was lower than the reference curve on whole plant dry matter basis of Indica rice in tropics (5.20) by Sheehy et al. [11] as well as lower than the curves developed with Japonica rice on whole plant dry matter basis (3.53) by Ata-Ul-Karim et al. [12], and $L_{DM}$ basis (3.76) by Yao et al. [20].

The differences observed between the parameter *a* of dilution curve developed in present study and the curves on whole plant dry matter basis [11,12] were due to morphological aggregation of structural components, which relates to the weight/N concentration in the whole plant [13]. Stress responses may cause differences in the partitioning of dry matter among various plant organs, and thereby affect the shape of the dilution curves. Moreover, dissimilarities in climatic conditions and genetic differences of Indica and Japonica rice contributed to the differences between the curves. The ability of Indica to hold higher plant N content and total N uptake [23–25] and faster growth rate [26], compared with those of Japonica rice, also lead to the differences between $N_c$ dilution curve of Sheehy et al. [11] and that described in the present study. The differences of $S_{DM}$ based curve with that of $L_{DM}$ based one [20] are mainly attributed to leaf/stem ratio, because decrease in stem N during vegetative phase is related to decline in the metabolic biomass with high N contents, and increase in proportion of structural and non-photosynthetic biomass with low N contents [8]. Thus, higher proportion of structural biomass in stem than in leaves is responsible for the differences between the $L_{DM}$ and $S_{DM}$ based curves of Japonica rice.

The parameter *b* of the dilution curve indicates the dilution intensity of N during growth and the higher values of *b* indicate lower N dilutions [17].The coefficients *b* were (−0.50, −0.28, −0.22 and −0.274) for $N_c$ dilution curve of Indica rice and for Japonica rice based on whole plant dry matter, $L_{DM}$, and on $S_{DM}$, respectively. The observed differences between the coefficients *b* of Indica rice and current $S_{DM}$ based dilution curve might be explained by the differences in duration of vegetative phase in tropical and subtropical climates, while the differences between coefficients *b* of the curves of Japonica rice based on $L_{DM}$ and $S_{DM}$ were directly related to the distribution of dry matter between green leaves and the stem [17]. In contrast, the differences between coefficients *b* of the curves of Japonica rice on whole plant dry matter basis compared with that of $S_{DM}$ basis, are negligible due to the reason that stem have a dilution effect on the N in the above ground tissues, because of their higher weight percentage in the total dry matter [27]. Therefore, the $S_{DM}$ based dilution curve can be used as a potential alternative for in-season estimation of

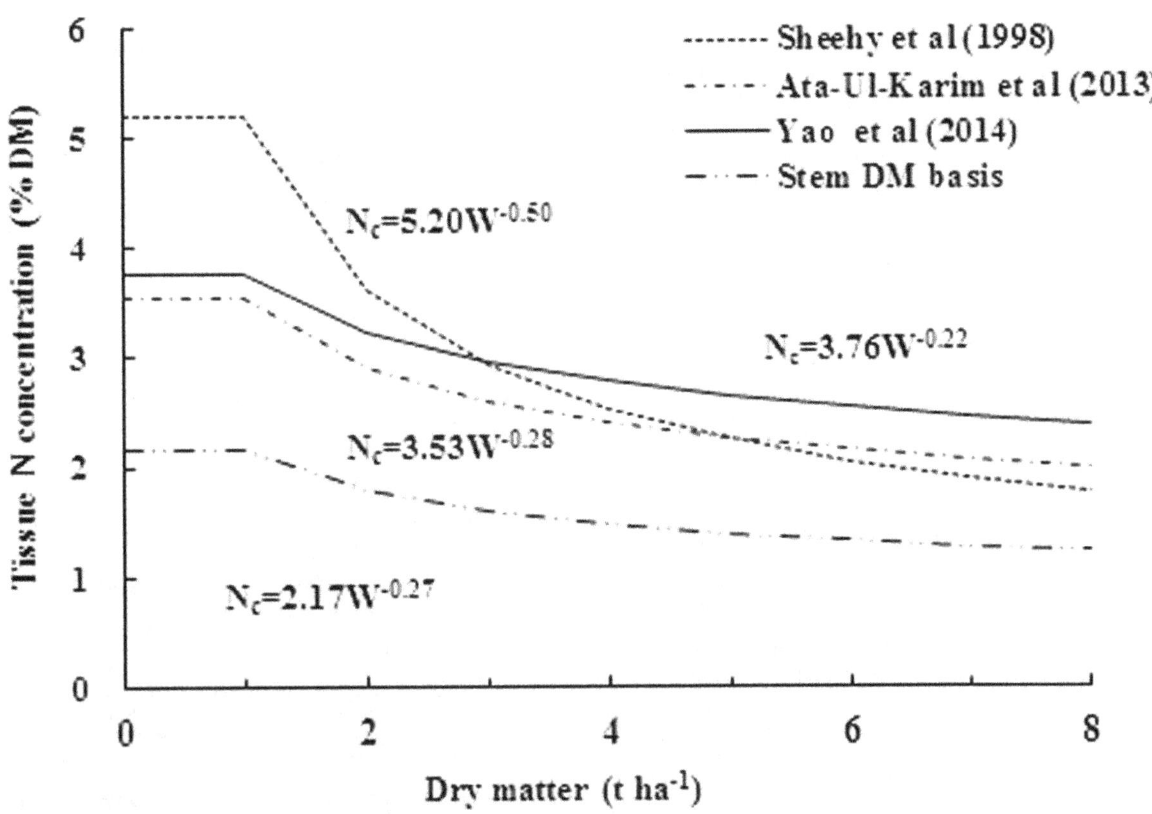


**Figure 9. Comparison of different $N_c$ dilution curves.** The (------) represents the $N_c$ dilution curve of Sheehy et al. (1998) ($N_c = 5.20W^{-0.50}$) on plant dry matter basis in Indica rice under tropic environment. The (-•-•-•) represents the $N_c$ dilution curve of Ata-Ul-Karim et al. (2013) ($N_c = 3.53W^{-0.28}$) on plant dry matter basis in Japonica rice in Yangtze River Reaches. The (——) line represents $N_c$ dilution curve of Yao et al. (2014) ($N_c = 3.76W^{-0.22}$) on leaf dry matter basis in Japonica rice in Yangtze River Reaches, and the (-••-••-) line represents $N_c$ dilution curve on stem dry matter basis in present study ($N_c = 2.17W^{-0.27}$).

plant N nutrition status, instead of existing whole plant dry matter and $L_{DM}$ based approaches.

## Implication for nitrogen diagnosis

The application of the present $N_c$ dilution curve as a diagnostic tool for accurate N management to make corrective decisions of N dressing recommendation during rice production is very interesting. The $N_c$ dilution curve can be used for a priori analysis intended to optimize fertilizer N management or for a posteriori diagnosis intended to detect N limiting nutrition for rice within experimental trials or fields in production. The a priori diagnosis of plant N status consists of timely detection of plant N deficiency during the crop growth cycle to determine the necessity of applying additional N fertilizer. Present study showed that the $N_c$ dilution curve, resulting NNI and $N_{and}$ effectively distinguished conditions of deficient, optimal and surplus N nutrition in rice. The values of ΔN in present study obtained on the basis of relationship between ΔNNI, $\Delta N_{and}$ and ΔN, permitted us to make corrective decisions of N dressing recommendation for precise N management during or even before the period of peak demand of the rice crop. The main limitation in using the present NNI and $N_{and}$ directly as diagnostic tools is the need to determine the actual dry matter and N concentration, which can be monitored by the non-destructive means including remote sensing [28–30]. Moreover, a good correlation between these analytical tools and chlorophyll meter readings was previously reported by [9]. These indirect methods could possibly be a substitute for assessing NNI and $N_{and}$ and portray crops and environments in conditions where they cannot be measured directly [31]. Thus, the models of NNI and $N_{and}$, based on $N_c$ dilution curve in relation to actual growth status, can be exploited directly for the estimation of crop N status to recommend the necessities of further N application during plant growth. These novel algorithms can also be combined into crop growth and management models to forecast crop N status and quantify N dressing plan. Although, NNI and $N_{and}$ calculated in present study distinguished well the N-limiting and non-N-limiting growth conditions, a more comprehensive validation using different N management practices, N availabilities and cultivars is mandatory to robustly confirm the reliability of NNI and $N_{and}$ usage as an investigative indicators for different ecological regions and rice production systems.

## Conclusions

In conclusion, we found that N fertilization endorses increase in the $S_{DM}$, which was influenced by variations in SNC. A higher rate of N fertilizer generally increased SNC in Japonica rice; however, towards advancing maturity this increase followed a declining trend under different N levels, sampling dates and growing seasons. $S_{DM}$ during vegetative growth period ranged from minimum value of 0.19 ($N_0$) in WXJ-14 to a maximum value of 8.04 ($N_4$) in LXY-18, whereas SNC varied from 0.68% in WXJ-14 to 2.36% in LXY-18 on $S_{DM}$ basis under different N rates and growth stages. A new $N_c$ dilution curve on $S_{DM}$ basis for Japonica rice grown in east China was developed and can be described by equation, $N_c = 2.17W^{-0.274}$, when $S_{DM}$ ranges from 0.88 and 7.94 t $ha^{-1}$, however for $S_{DM}$<0.88 t $ha^{-1}$, the constant critical value $N_c = 1.76\%$ $S_{DM}$ was applied, which was independent of $S_{DM}$. Additionally, the values of NNI and $N_{and}$ at different sampling dates for N limiting condition were generally <1 and >0, while > 1 and <0, respectively for non-N-limiting supply. The values of ΔN derived on the basis of relationship between ΔNNI, $\Delta N_{and}$ and ΔN, can be used to make corrective decisions of N dressing recommendation for precise N management, prior to or on the onset of the period of highest demand of the rice crop. We conclude that the $S_{DM}$ based dilution curve developed in the present study offers a new vision into plant N status and can possibly be adopted as an alternate practical tool for reliable diagnosis of plant N status to correct N fertilization decision during the vegetative growth of rice in east China.

## Author Contributions

Conceived and designed the experiments: ST AUK XY XL WC YZ. Performed the experiments: ST AUK XY XL. Analyzed the data: ST AUK XY YZ. Wrote the paper: ST AUK YZ.

## References

1. Jaggard K, Qi A, Armstrong M (2009) A meta-analysis of sugarbeet yield responses to nitrogen fertilizer measured in England since 1980. J Agric Sci-(Camb) 147: 287–301.
2. Ghosh M, Mandal B, Mandal B, Lodh S, Dash A (2004) The effect of planting date and nitrogen management on yield and quality of aromatic rice (*Oryza sativa*). J Agric Sci-(Camb) 142: 183–191.
3. Cabangon R, Castillo E, Tuong T (2011) Chlorophyll meter-based nitrogen management of rice grown under alternate wetting and drying irrigation. Field Crops Res 121: 136–146.
4. Lin FF, Qiu LF, Deng JS, Shi YY, Chen LS, et al. (2010) Investigation of SPAD meter-based indices for estimating rice nitrogen status. Compu Electron Agric 71: S60–S65.
5. Confalonieri R, Debellini C, Pirondini M, Possenti P, Bergamini L, et al. (2011) A new approach for determining rice critical nitrogen concentration. J Agric Sci-(Camb) 149: 633–638.
6. Smeal D, Zhang H (1994) Chlorophyll meter evaluation for nitrogen management in corn. Commun Soil Sci Plant Anal 25: 1495–1503.
7. Justes E, Mary B, Meynard JM, Machet JM, Thelier-Huche L (1994) Determination of a critical nitrogen dilution curve for winter wheat crops. Ann Bot 74: 397–407.
8. Yue S, Meng Q, Zhao R, Li F, Chen X, et al. (2012) Critical nitrogen dilution curve for optimizing nitrogen management of winter wheat production in the North China Plain. Agron J 104: 523–529.
9. Ziadi N, Brassard M, Bélanger G, Cambouris AN, Tremblay N, et al. (2008) Critical nitrogen curve and nitrogen nutrition index for corn in eastern Canada. Agron J 100: 271–276.
10. Ziadi N, Belanger G, Claessens A, Lefebvre L, Cambouris AN, et al. (2010) Determination of a critical nitrogen dilution curve for spring wheat. Agron J 102: 241–250.
11. Sheehy JE, Dionora MJA, Mitchell PL, Peng S, Cassman KG, et al. (1998) Critical nitrogen concentrations: implications for high-yielding rice (*Oryza sativa* L.) cultivars in the tropics. Field Crops Res 59: 31–41.
12. Ata-Ul-Karim ST, Yao X, Liu X, Cao W, Zhu Y (2013) Development of critical nitrogen dilution curve of Japonica rice in Yangtze River Reaches. Field Crops Res 149: 149–158.
13. Kage H, Alt C, Stützel H (2002) Nitrogen concentration of cauliflower organs as determined by organ size, N supply, and radiation environment. Plant Soil 246: 201–209.
14. Vouillot MO, Huet P, Boissard P (1998) Early detection of N deficiency in a wheat crop using physiological and radiometric methods. Agronomie 18: 117–130.
15. Ziadi N, Bélanger G, Gastal F, Claessens A, Lemaire G, et al. (2009) Leaf nitrogen concentration as an indicator of corn nitrogen status. Agron J 101: 947–957.
16. Lemaire G, Jeuffroy MH, Gastal F (2008) Diagnosis tool for plant and crop N status in vegetative stage: Theory and practices for crop N management. Eur J Agron 28: 614–624.
17. Oliveira ECAd, de Castro Gava GJ, Trivelin PCO, Otto R, Franco HCJ (2013) Determining a critical nitrogen dilution curve for sugarcane. J Plant Nutr Soil Sci 176: 712–723.
18. Chen J, Huang Y, Tang Y (2011) Quantifying economically and ecologically optimum nitrogen rates for rice production in south-eastern China. Agric Ecosyst Environ 142: 195–204.
19. Hahn WS (1997) Statistical Methods for Agriculture and Life Science. Seol: Free Academy Publishing Co. 747 p.
20. Yao X, Ata-Ul-Karim ST, Zhu Y, Tian Y, Liu X, et al. (2014) Development of critical nitrogen dilution curve in rice based on leaf dry matter. Eur J Agron 55: 20–28.
21. Bélanger G, Richards JE (2000) Dynamics of biomass and N accumulation of alfalfa under three N fertilization rates. Plant Soil 219: 177–185.
22. Gayler S, Wang E, Priesack E, Schaaf T, Maidl FX (2002) Modeling biomass growth, N-uptake and phenological development of potato crop. Geoderma 105: 367–383.
23. Islam M, Islam M, Sarker A (2008) Effect of phosphorus on nutrient uptake of Japonica and Indica rice. J Agric Rural Dev 6: 7–12.
24. Shan Y, Wang Y, Yamamoto Y, Huang J, Yang L, et al. (2001) Study on the differences of nitrogen uptake and use efficiency in different types of rice. J Yangzhou Univ (Nat Sci Ed) 4: 42.
25. Yoshida H, Horie T, Shiraiwa T (2006) A model explaining genotypic and environmental variation of rice spikelet number per unit area measured by cross-locational experiments in Asia. Field Crops Res 97: 337–343.
26. Ying J, Peng S, He Q, Yang H, Yang C, et al. (1998) Comparison of high-yield rice in tropical and subtropical environments: I. Determinants of grain and dry matter yields. Field Crops Res 57: 71–84.
27. Oliveira ECAd, Freire FJ, Oliveira RId, Freire M, Simoes Neto DE, et al. (2010) Extração e exportação de nutrientes por variedades de cana-de-açúcar cultivadas sob irrigação plena. Rev Bras de Ciênc Solo 34: 1343–1352.
28. Wang W, Yao X, Tian Y, Liu X, Ni J, et al. (2012) Common spectral bands and optimum vegetation indices for monitoring leaf nitrogen accumulation in rice and wheat. J Integr Agric 11: 2001–2012.
29. Zhao B, Yao X, Tian Y, Liu X, Ata-Ul-Karim ST, et al. (2014) New critical nitrogen curve based on leaf area index for winter wheat. Agron J 106: 379–389.
30. Ata-Ul-Karim ST, Zhu Y, Yao X, Cao W (2014) Determination of critical nitrogen dilution curve based on leaf area index in rice. Field Crops Res: In press.
31. Debaeke P, Rouet P, Justes E (2006) Relationship between the normalized SPAD index and the nitrogen nutrition index: Application to durum wheat. J Plant Nutr 29: 75–92.

2

# Reconciling Pesticide Reduction with Economic and Environmental Sustainability in Arable Farming

**Martin Lechenet[1], Vincent Bretagnolle[2], Christian Bockstaller[3,4], François Boissinot[5], Marie-Sophie Petit[6], Sandrine Petit[1], Nicolas M. Munier-Jolain[1]***

**1** Institut National de la Recherche Agronomique, Unité Mixte de Recherche 1347 Agroécologie, Dijon, Côte d'Or, France, **2** Centre d'Etudes Biologiques de Chizé - Centre National de Recherche Scientifique, Beauvoir sur Niort, Deux-Sèvres, France, **3** Institut National de la Recherche Agronomique, Unité de Recherche 1121 Agronomie et Environnement, Colmar, Haut-Rhin, France, **4** Université de Lorraine, Vandœuvre-lès-Nancy, Meurthe-et-Moselle, France, **5** Chambre d'Agriculture des Pays de la Loire, Angers, Maine-et-Loire, France, **6** Chambre Régionale d'Agriculture de Bourgogne, Quetigny, Côte d'Or, France

## Abstract

Reducing pesticide use is one of the high-priority targets in the quest for a sustainable agriculture. Until now, most studies dealing with pesticide use reduction have compared a limited number of experimental prototypes. Here we assessed the sustainability of 48 arable cropping systems from two major agricultural regions of France, including conventional, integrated and organic systems, with a wide range of pesticide use intensities and management (crop rotation, soil tillage, cultivars, fertilization, etc.). We assessed cropping system sustainability using a set of economic, environmental and social indicators. We failed to detect any positive correlation between pesticide use intensity and both productivity (when organic farms were excluded) and profitability. In addition, there was no relationship between pesticide use and workload. We found that crop rotation diversity was higher in cropping systems with low pesticide use, which would support the important role of crop rotation diversity in integrated and organic strategies. In comparison to conventional systems, integrated strategies showed a decrease in the use of both pesticides and nitrogen fertilizers, they consumed less energy and were frequently more energy efficient. Integrated systems therefore appeared as the best compromise in sustainability trade-offs. Our results could be used to re-design current cropping systems, by promoting diversified crop rotations and the combination of a wide range of available techniques contributing to pest management.

**Editor:** Raul Narciso Carvalho Guedes, Federal University of Viçosa, Brazil

**Funding:** Funding for the study was provided by the French National Research Agency ANR (STRA-08-02 Advherb project) and from Région Bourgogne. The Burgundy farms network was developed in the framework of the Réseau Mixte Technologique "Systèmes de Culture Innovants" and the project "Plus d'Agronomie, Moins d'intrants" initiated by Région Bourgogne. The long term experiment at Dijon-Epoisses was partly funded by the European Network of Excellence ENDURE. The funders had no role in study design, data collection and analysis, decision to publish, or preparation of the manuscript.

**Competing Interests:** The authors have declared that no competing interests exist.

* E-mail: nicolas.munier-jolain@dijon.inra.fr

## Introduction

Reconciling agricultural productivity with other components of sustainability remains one of the greatest challenges for agriculture [1]. A key issue will be to achieve substantial reductions in the level of pesticide use for environmental and health reasons [2,3]. Agriculture in temperate climates is widely dominated by conventional intensive farming systems, with highly specialized crop productions and a heavy reliance on pesticides and mineral fertilizers [4]. However, increasing environmental concerns about intensive farming practices has contributed to the emergence of innovative farming systems, such as organic and integrated farming, typically presented as alternative paths to reduce pesticide use as compared to current conventional systems [5,6,7]. Whether these systems better meet sustainability criteria has been a matter of debate [8,9]. Integrated farming, recently promoted in Europe through the 2009/128/EC European directive [10], is defined as a crop protection management based on Integrated Pest Management (IPM) principles, which emphasizes physical and biological regulation strategies to control pests while reducing the reliance on pesticides [11]. It can be regarded as an intermediate between conventional farming, with high levels of inputs, and organic farming, which prohibits the use of synthetic pesticides and fertilizers. Organic and integrated farming have in common the combined use of management approaches to replace, at least in part, synthetic inputs. However, unlike organic farming which is growing both in Europe (by 40 to 50% between 2003 and 2010 [12]) and in the US (by 270% between 2000 and 2008 [13]), integrated arable crop production is not expanding because it is perceived by farmers as a complex system which is difficult to implement, labour-consuming, and associated with reduced and unpredictable economic profitability [14,15]. As a consequence, the amount of pesticides sprayed has only decreased slightly in Europe (−3.6% from 2000 to 2007 [16]) and in the US (−7.5% from 2000 to 2007 [17]). Moreover, this decrease can be partly attributed to the substitution of older chemistry, applied at high dosage, by new products that are efficient at lower doses, which actually cannot be considered as a reduction of pesticide reliance. In France, the national action plan, ECOPHYTO 2018, which had set a target of a 50% decrease in pesticide use by the year 2018, is currently far from achieving this goal [18].

So far, assessments of cropping system sustainability have compared few – typically two or three – experimental prototypes that represent conventional, organic or integrated strategies

[19,20]. However, this approach fails to capture the diversity within each of these farming strategies. Given the diversity of crop management options within a conventional, an integrated or an organic strategy, which might lead to contrasted performances, the generic value of experimental results ignoring this variability may be argued. We assessed the sustainability of 48 cropping systems located in regions of intensive arable farming and covering a wide range of pesticide use levels and cultivation techniques such as crop rotations, from monoculture to highly diversified crop rotations, soil tillage (e.g. inversion tillage, shallow tillage or direct drilling), fertilization (mineral or organic fertilizers), or weed management (e.g. only based on herbicide use, including mechanical weeding). More details about the cropping system sample are available in the online SI section (Dataset S1). All the studied cropping systems were followed for between three and 12 years, between 1999 and 2012. Eight cropping systems were organic, 30 were based on integrated farming and 10 were conventional (Figure 1). Using eight sustainability indicators to evaluate the performance of the study systems, our aims were: (i) to identify possible conflicts between the reduction of pesticide reliance and other components of sustainability; and, (ii) to assess the potential of organic and integrated strategies for improving agricultural sustainability.

As the performance of a cropping system depends not only on the combination of management options it implements, but also on the local production situation [21] (including biophysical and socio-economic local aspects), we standardized the indicators of performance and pesticide use, using a ratio of the performances of the cropping systems over those of a local reference system. This enabled us to focus solely on the effects of the management strategies on sustainability indicators. The local references were cropping systems selected as representative of the most widespread crops and practices within each production situation. Pesticide use was measured as the Treatment Frequency Index (TFI), which is a commonly used indicator in Europe to estimate the cropping system dependence on pesticides [22]. In our sample, organic cropping systems did not use any pesticides (synthetic or natural) so their relative TFI, expressed as a ratio of the local reference TFI, was zero. Integrated cropping systems displayed TFI values that were on average half (−47%) of the local references (Table S1).

## Results

Table S1 presents the mean and standard deviation for each performance indicator according to the management strategy (organic, integrated and conventional). The second tab of Dataset S1 provides performance details for each cropping system of the sample.

### Productivity and energy efficiency

Given the primary role of agriculture remains to produce food and other goods, we used an indicator of productivity, expressed as the total yearly amount of energy produced by a cropping system, whatever the crops cultivated (Figure 2a). The productivity of organic cropping systems was below that of their local reference (Figure 3b), ranging from −22% to −76%. For non-organic cropping system, productivity was uncorrelated to relative TFI (Figure 2a and Table 1), with some cropping systems that had a low reliance on pesticides even exceeding the productivity of the local reference. Cropping system productivity may strongly depend on crop type, especially if the whole above-ground biomass is harvested or not. Crops other than grain crops were frequently grown in integrated farming, as they are typically associated with low pesticide requirements and can contribute to weed control in subsequent crops [23]. They typically consist of

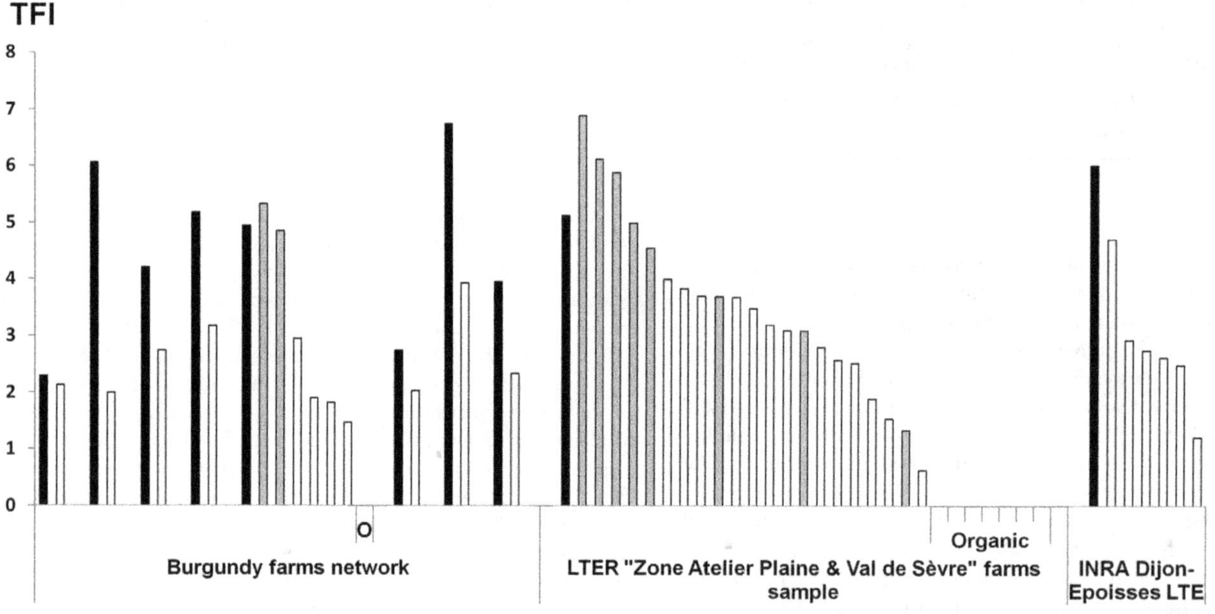


**Figure 1. Distribution of the Treatment Frequency Index (TFI) for the studied arable cropping systems.** Average TFI for each cropping system composing the study sample. At each site, black bars correspond to the local reference, grey bars to conventional cropping systems and white bars to integrated cropping systems. The sample also includes eight organic cropping systems with TFI = 0 and labelled with "O" or "Organic". Details about the cropping systems are available in Dataset S1.

forage crops, dedicated to livestock feeding with limited energy efficiency, or of crops used for non-food applications. However, distinguishing cropping systems based on grain crops or on crops in which all above-ground biomass is harvested did not change the observed pattern. In systems with grain crops only, productivity was not correlated with relative TFI (Table 1), suggesting that a reduction in pesticide use intensity may not be necessarily translated into a decrease in productivity. The second indicator of energy productivity we used was the energy efficiency of cropping systems, resulting from a ratio between energy output and energy input. It evaluated the ability of a cropping system to convert energy inputs into outputs. Organic cropping systems were significantly less energy efficient than other systems (Figures 2b and 3c, Table 2). Despite their energy consumption being lower (Table 2), notably due to their low reliance on nitrogen fertilizers, it was not sufficient to offset their limited productivity. Energy consumption was negatively correlated with relative TFI in integrated and conventional systems which cultivated only grain crops (Table 1). Energy efficiency was also negatively correlated with relative TFI in these systems, although the relationship was weak and only marginally significant ($r_s = -0.35$, $P = 0.07$). The systems with the highest energy efficiency, whether they included crops with all above-ground biomass harvested or not, were mostly integrated systems (Figures 2b and 3c).

## Environmental impact

The environmental impact of cropping systems and their reliance on external inputs were assessed with the indicator I-Pest [24] and with estimates of fuel and nitrogen fertilizer consumption. I-Pest is a predictive indicator that assesses the environmental impacts of pesticide use as the risk of contamination of the air, and surface and ground waters (see Figure S1). As the organic cropping systems composing the sample did not used synthetic or natural pesticide, their cumulated I-Pest was 0. As expected for the rest of the sample, cumulated I-Pest was strongly and positively correlated to relative TFI (Figure 2c and Table 1). Fuel and nitrogen fertilizers together amounted to more than 60% of the total energy inputs for all tested cropping systems. Organic systems consumed more fuel than the rest of the sample (Table 2), with their average consumption exceeding the local references by 17% (Figure 3d). Organic cropping systems had, nonetheless, a lower reliance on N fertilization than the rest of the sample (Figure 3g, Table 2), in line with their lower yield targets and the frequent occurrence of crops with low N requirements used in organic rotations. No relation was detected between fuel consumption and relative TFI in non-organic systems (Figure 2d, Table 1), but a positive correlation was clearly visible between relative TFI and the amount of nitrogen fertilizers applied (Figure 2e).

## Economic sustainability and workload

Economic sustainability was assessed by considering (i) the profitability, i.e. the average semi-net margin over a range of ten real price scenarios for agricultural products, fuel and fertilisers, and (ii) the sensitivity of this profitability in a context of price volatility, i.e. the relative standard deviation of the semi-net margin. The range of price scenarios used for the calculations was set to reflect the variability of the economic context over the last decade. Profitability, when averaged over the ten price scenarios was not correlated with relative TFI for integrated and conventional systems (Figure 2f and Table 1), and no significant difference appeared with organic systems (Mann-Whitney test, $P>0.9$). It suggests that low pesticide use would not necessarily result in lower economic return. The strong variability observed within each class (Figure 3f), most notably for integrated cropping systems, confirmed that strategies to reduce pesticide use could even lead to an increase in profitability. As integrated cropping systems were, in contrast to organic systems, evaluated with a conventional price reference, the most profitable integrated systems were able to efficiently reduce their production costs. No relation was detected between the sensitivity to price volatility and relative TFI in conventional and integrated systems (Figure 2g, Table 1). Sensitivity to price volatility was significantly lower in organic cropping systems than in other systems (Table 2), most probably because: (i) they were based on more diversified crop rotations, which spread risks and buffered semi-net margin at the farming system scale; and, (ii) their crop rotations typically included crops with low N demand, that had reduced reliance on N inputs, whose price is directly related to the volatile price of fossil fuels.

The issue of social sustainability was addressed using the 'workload' indicator, which gives emphasis to the potential for bottlenecks where available workforce is a limiting factor at the farm scale (Figure 2h). Workload was calculated for each technical operation but excluded time devoted to transport and crop monitoring. Workload was found not correlated with relative TFI in non-organic cropping systems (Table 1), and no significant difference was found with the organic group (Mann-Whitney test, $P>0.1$), so that reducing pesticide use does not necessarily imply an increased workload. Indeed, in integrated systems, labour requirements ranged from low to high relative values (Fig 2h). The level of workload was, however, related to the type of fertilization, with cropping systems having organic fertilization requiring an average of 13% greater working time, as compared to mineral fertilizer-based cropping systems (Table 2).

## Crop diversity

Diversification of crop rotations is often presented as an efficient management tool for controlling pests and to improve agricultural sustainability [25,26]. We used a crop sequence indicator, Isc [27], which estimates the consistency of the crop sequence with regard to the potential of input reduction, by addressing effects of crop rotation on pathogens, pests, weeds, soil structure and nitrogen supply of preceding crops. Even if no significant correlation appeared between Isc and relative TFI (Table 3), organic and integrated cropping systems displayed significantly higher Isc values than conventional systems (Table 2). A negative correlation between Isc and productivity suggests that diversifying crop rotation may reduce cropping system productivity (Table 3), but the Spearman correlation test was no longer significant when organic cropping systems were excluded ($P = 0.07$). We did not detect any significant relationship between energy efficiency and crop diversification, whether organic cropping systems were included or not ($P = 0.44$). No correlation was observed between Isc and semi-net margin, but workload appeared to be lower for systems with higher Isc (Table 3). We found the expected negative correlation between Isc and N fertilization rates, and consequently between Isc and energy consumption (Table 3). We focused therefore more particularly on cropping systems including legume crops, which also displayed higher Isc values than the rest of the sample (Table 2). The role of legume in improving energy efficiency at the cropping system scale was clearly demonstrated by the correlation between the frequency of occurrence of legumes in the crop rotation and the energy efficiency ($r_s = 0.37$, $P<0.05$). The sensitivity to price volatility was negatively correlated with the frequency of occurrence of legumes in the crop rotation ($r_s = -0.33$, $P = 0.02$), but positively correlated with the level of N fertilization ($r_s = 0.49$, $P = 5*10^{-4}$). Fostering exogenous N independence therefore appeared as an efficient way to limit income variability.

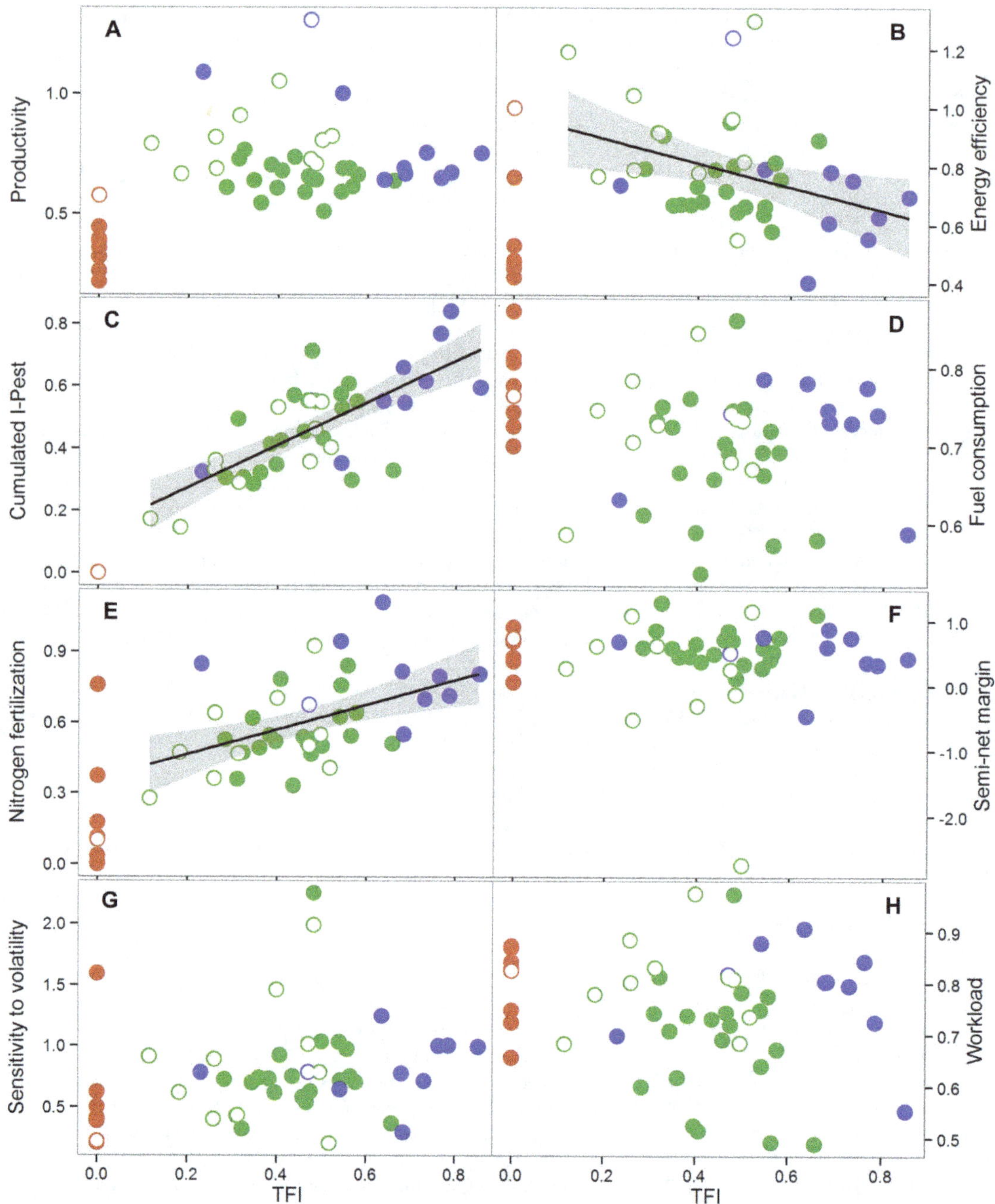


**Figure 2. Relationship between sustainability indicators and relative TFI.** Cropping system performances according to their relative TFI. Conventional, integrated and organic cropping systems are represented by blue, green and red symbols respectively. Filled symbols correspond to the cropping systems with grain crops only and empty symbols refer to the cropping systems including crops for which the whole above-ground biomass is harvested. Each sustainability indicators is expressed as the natural logarithm of the ratio between the cropping system and the local reference indicators. Linear regressions are represented with their standard error for cumulated I-Pest (Pearson correlation test: $r_p = 0.74$, $P = 5*10^{-8}$), nitrogen fertilization (Pearson correlation test: $r_p = 0.48$, $P = 0.002$), and energy efficiency (Pearson correlation test: $r_p = -0.38$, $P = 0.02$). Performance metric included: a) productivity, b) energy efficiency, c) cumulated I-Pest, d) fuel consumption, e) nitrogen fertilization, f) semi-net margin, g) sensitivity to price volatility, h) workload.

## Discussion

This work was aimed at detecting cropping systems able to reconcile low pesticide use and other components of sustainability. Our original multiple dimensions approach, based on a precise description of management practices, was designed to compare and contrast numerous cropping systems from different production situations. This approach, applied at the large-scale, was able to provide generic knowledge about potential trade-offs between the different issues of agricultural sustainability.

### Sustainability of integrated and organic farming

Our results show that achieving a low level of pesticide use is possible without triggering negative side effect on any of the

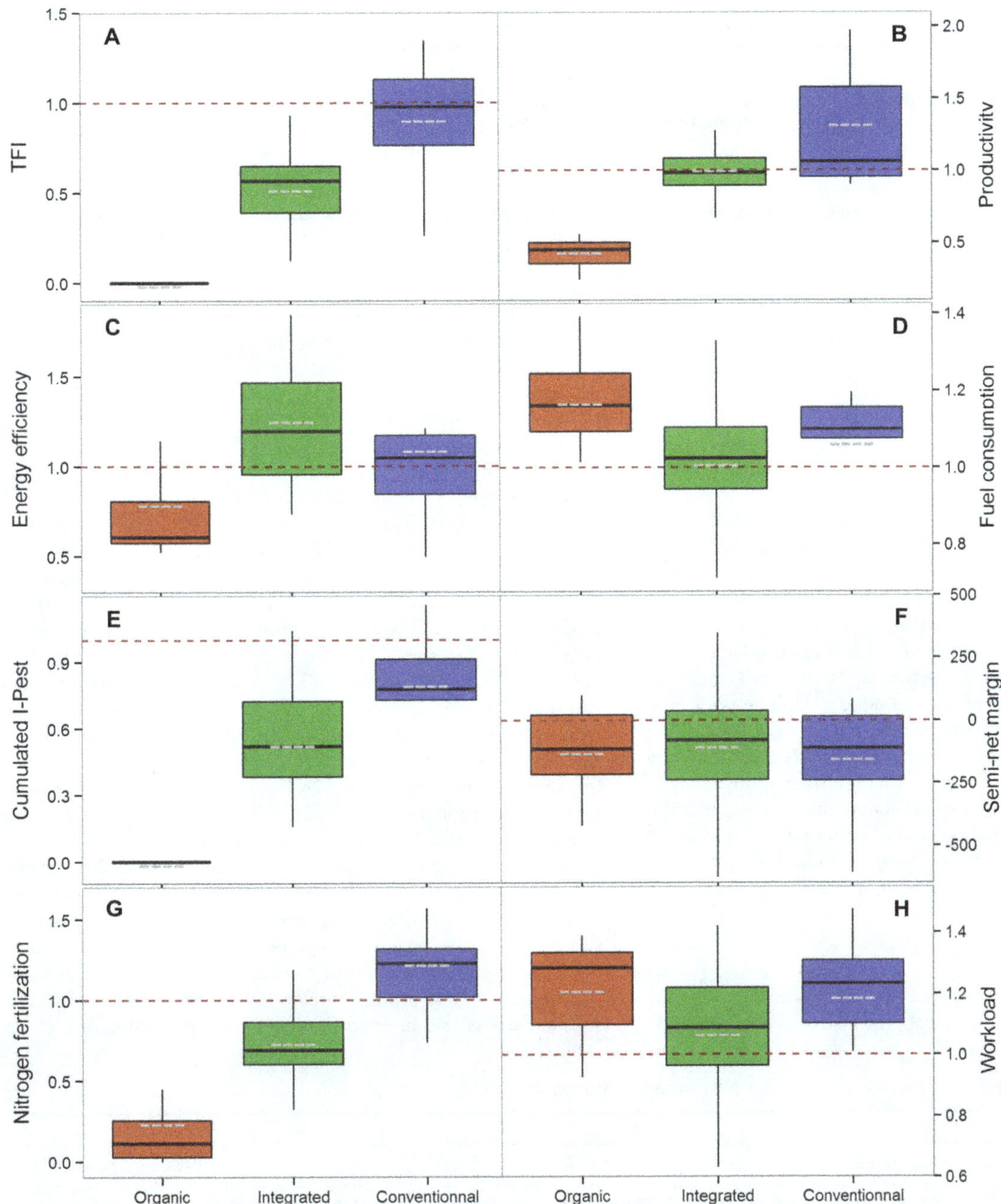


**Figure 3. Cropping systems distribution according to sustainability indicators.** Performance indicators are expressed as a ratio of the local reference indicator, except for semi-net margin, expressed as a difference with the local reference. Conventional, integrated and organic cropping systems are represented by blue, green and red box plots respectively. The horizontal black bars and grey dashed bars correspond to median and mean values respectively. The horizontal red dashed bar recalls the position of the local references. Outliers are not represented. Performance metrics included: a) Treatment Frequency Index, b) productivity (organic farming: one outlier, v = 0.78; integrated farming: two outliers, v1 = 1.48 and v2 = 1.87), c) energy efficiency (organic farming: one outlier, v = 1.72), d) fuel consumption (integrated farming: one outlier, v = 1.37; conventional farming: two outliers v1 = 0.8 and v2 = 0.88), e) cumulated I-Pest (conventional farming: two outliers v1 = 0.38 and v2 = 0.42), f) semi-net margin, g) nitrogen fertilization (organic farming: one outlier, v = 1.13; integrated farming: two outliers, v1 = 1.32 and v2 = 1.52), h) workload (conventional farming: one outlier v = 0.74).

components of cropping system sustainability we assessed in this study. Integrated cropping systems were not only associated with low pesticide use and low risks for contamination of air and water with pesticide residues; they also displayed lower energy consumption than more intensive cropping systems and are likely to improve energy efficiency without impact on productivity and profitability. Lower pesticide usage in arable cropping systems did not imply a heavier workload, another critical point conditioning strongly the adoption of an innovative strategy.

Organic farming prohibits the use of synthetic pesticides and fertilizers, and this approach is often associated with low nitrogen fertilization, as observed in our sample. In addition to the positive

**Table 1.** Rank correlation between TFI and sustainability indicators for integrated and conventional cropping systems.

| **Spearman correlation** | **Productivity** | **Productivity (grain crops only)** | **Energy consumption** | **Energy consumption (grain crops only)** | **N fertilization** | **Pesticides environmental impact** |
|---|---|---|---|---|---|---|
| $r_s$ | *−0.17 (NS)* | *0.06 (NS)* | *0.30 (NS)* | 0.42 | 0.51 | 0.67 |
| *P-value* | *0.3* | *0.8* | *0.06* | 0.03 | $9*10^{-4}$ | $4*10^{-6}$ |
| **Spearman correlation** | **Fuel consumption** | **Energy efficiency** | **Semi-net margin** | **Sensitivity to prices volatility** | **Workload** | **Crop Sequence Indicator (Isc)** |
| $r_s$ | *0.06 (NS)* | −0.40 | *−0.05 (NS)* | *0.22 (NS)* | $9*10^{-3}$ *(NS)* | −0.22 |
| *P-value* | *0.7* | 0.01 | *0.8* | *0.2* | *0.95* | 0.2 |

Spearman rank correlation tests ($\alpha = 0.05$). $r_s$ is the Spearman correlation coefficient. Values of $r_s$ followed by (NS) are not significant.

effects on environmental quality, numerous studies underlined other environmental benefits of organic farming such as effects on pollinator dynamics [28], on landscape floristic composition [29], as well as on soil microbial diversity [5,30]. Here we demonstrated that organic farming does not necessarily affect profitability and workload, and conversely, might strengthen farm financial stability in a variable and unpredictable economic context. Organic cropping systems were however less productive and less energy efficient than integrated systems in our sample. Although highly dependent on crops and production context, productivity in organic farming was already reported as lower than in conventional farming by other comparative studies [31]. The poor land use efficiency associated with organic farming is a key issue in the current land sharing – land sparing debate about the growing competition for land use [32], and notably urban sprawl [33] as well as the necessity to keep natural spaces undisturbed [34,35]. Both aspects – environmental benefits of organic farming and the limited productivity per unit of land – should therefore be considered by decision makers in their incentives for sustainable agriculture.

## Crop diversification

Our results support the hypothesis that crop diversification may be an effective means to enhance cropping system performance. At the cropping system scale, crop diversification provides agronomic advantages, such as the regulation of pests, diseases and weeds [36,37,25]. In our sample, the most diversified cropping systems, which displayed the highest values of the crop sequence indicator, Isc, were indeed less dependent on pesticides. Their low environmental impact on water and air quality makes crop diversification an interesting potential pathway for reducing the damage caused by agriculture on natural resources (e.g., biodiversity [38]), as well as on human health (e.g. neurological degenerative disorders [39]). By mitigating the adverse effects of climate variability, crop diversification may also improve system resilience for productivity [40], with the increasing likelihood of extreme weather events requiring farm adaptation [41]. Economic market volatility is an additional source of variation and risk factor for farm economic stability. We found that crop diversification, particularly through the introduction of legumes in the crop rotation, is likely to limit dependence on inputs that have unstable prices. By allowing a decrease in the use of exogenous N fertilizer across a crop rotation, legume cultivation reduces production cost fluctuations and consequently makes the cropping system less sensitive to market volatility. Legumes come with a supplementary advantage [42] in the face of the considerable amount of fossil energy necessary to produce mineral N fertilizers, and we noted a substantial increase in energy efficiency for crop rotations where

**Table 2.** Significantly different groups for a given performance indicator.

| **Indicator** | **Test designation** | **P-value** | **Statistic W** |
|---|---|---|---|
| Productivity | Difference of productivity between organic cropping systems and the rest of the sample | $2*10^{-8}$ | 318 |
| Productivity | Difference of productivity between cropping systems including crops with the whole above-ground biomass harvested and the rest of the sample | $5*10^{-4}$ | 76 |
| Energy efficiency | Difference of energy efficiency between organic cropping systems and the rest of the sample | 0.005 | 258 |
| Energy consumption | Difference of energy consumption between organic cropping systems and the rest of the sample | 0.001 | 272 |
| Fuel consumption | Difference of fuel consumption between organic cropping systems and the rest of the sample | 0.02 | 72 |
| N fertilization rate | Difference of N fertilization rate between organic cropping systems and the rest of the sample | $2*10^{-4}$ | 285 |
| Sensitivity to price volatility | Difference of sensitivity to price volatility between organic cropping systems and the rest of the sample | 0.01 | 250 |
| Workload | Difference of workload between cropping systems based on organic fertilization and the rest of the sample | 0.045 | 189 |
| Crop sequence indicator | Difference of Isc between organic cropping systems and conventional cropping systems | 0.01 | 69 |
| Crop sequence indicator | Difference of Isc between integrated cropping systems and conventional cropping systems | 0.001 | 255 |
| Crop sequence indicator | Difference of Isc between cropping systems including legumes and the rest of the sample | $6*10^{-6}$ | 63 |

Mann-Whitney tests ($\alpha = 0.05$). All P-values are below 0.05, indicating that the differences between means of the sub-samples are significant for the corresponding indicators.

**Table 3.** Rank correlation between Crop Sequence Indicator Isc and sustainability indicators.

| Spearman correlation | Productivity | Energy consumption | N fertilization | Pesticides environmental impact | Fuel consumption |
|---|---|---|---|---|---|
| $r_s$ | −0.35 | −0.42 | −0.41 | −0.39 | *−0.07 (NS)* |
| *P-value* | 0.01 | 0.003 | 0.004 | 0.006 | *0.7* |
| **Spearman correlation** | **Energy efficiency** | **Semi-net margin** | **Sensitivity to price volatility** | **Workload** | |
| $r_s$ | *−0.05 (NS)* | *0.10 (NS)* | *−0.23 (NS)* | −0.29 | |
| *P-value* | *0.7* | *0.5* | *0.1* | 0.04 | |

Spearman correlation tests ($\alpha = 0.05$). $r_s$ is the Spearman correlation coefficient. Values of $r_s$ followed by (NS) are not significant.

legume crops are more frequent. The most part of legumes introduced as diversification crops are however forage crops, and livestock production is commonly considered more energy consuming than plant production [43]. Conversely, the use of farmyard manure may contribute to reduce mineral fertilizers reliance for grain and forage crop production. The necessity of (i) integrating these situation-dependent parameters into energy balancing calculations, and, (ii) evaluating other environmental indicators [44] will be critical for the assessment of livestock production as a management option for enhancing agricultural sustainability.

A key agronomical advantage of crop diversification is related to the management of weed resistance to herbicides. Crop diversification is an efficient means to alternate herbicide modes of action and to introduce diversified measures of weed control, allowing changing selection pressure on weed communities and thus maintaining a sensitive weed population (i.e. maintaining high herbicide efficiency) [45].

Our results demonstrate a negative correlation between the Isc value and workload. We can nevertheless assume that diversifying crop rotations increases cropping system complexity and time devoted to field observations. Another aspect is that crop diversification may lead to a more evenly distributed workload over the seasons. Crops diversity implies a greater diversity in sowing and harvest periods, which are both times of peak labour that strongly influence task organisation and farmer decision making [15]. By reducing the amplitude of these peaks in labour, crop diversification could contribute to ensuring greater farmer decisional flexibility at the farm scale.

Beyond technical and organizational issues at the farm level, diversifying crop production as a component of an integrated strategy at regional or national scale would inevitably lead to important changes in production volumes, as well as markedly changing agricultural sectors within each production basin. It would definitely require an adaptation in the organisation of the whole agricultural sector and the development of new local markets. These economic and social lock-ins are rightly highlighted as the main limiting constraints hindering crop diversification [46]. However, by creating a particular economic sub-context, niche markets can be attractive and able to support innovation. Promoting such niche markets, for integrated farming development, would be the first step along an accelerating cycle of improvement based on mutually positive feed-backs between production and outlets.

## Materials and Methods

In all cases, the field studies did not involve endangered or protected species.

For future permissions about the private farm network of Burgundy, please contact Marie-Sophie Petit (co-author of the research article, Chambre Régionale d'Agriculture de Bourgogne) and Sandrine Petit (co-author of the research article, INRA).

For future permissions about the private farms survey carried out on the LTER "Zone Atelier Plaine & Val de Sèvre", please contact Nicolas Munier-Jolain (corresponding author, INRA).

### Study areas

The main objective of this study was to highlight potential conflicts between pesticide use and a set of sustainability indicators, so the cropping systems we consider were selected to maximize the contrast across the range of possible pesticide use intensities. The sample of cropping systems we used originates from:

(i) A long term experiment conducted since 2000 at the INRA Dijon-Epoisses farm in Bretenière (Burgundy, eastern France; 47°20′N, 5°2′E) in order to assess Integrated Weed Management-based cropping systems [15,26]. Seven cropping systems were tested between 2000 and 2012 including different combinations of technical levers likely to reduce pesticide reliance.

(ii) An experimental network (bringing together 14 cropping systems) monitored (1) by the local agricultural extension services and coordinated by the Chambre Régionale d'Agriculture de Bourgogne, and (2) by the INRA de Dijon. This network involved contrasting private farms of the Burgundy region, and was developed to test feasibility of innovative cropping systems with reduced pesticide use in a realistic context.

(iii) A survey of private farms carried out in 2010 on the LTER "Zone Atelier Plaine & Val de Sèvre" [47] located in the Poitou-Charentes region (450 $km^2$ study area in western France), and set up to explore a diversity of pesticide reliance, including organic farming, conventional intensive systems, and intermediate, IPM-based systems. Twenty nine varied cropping systems were surveyed in this area.

### Cropping systems classification

Details of cropping systems, including crop sequences, performances, and detailed crop management operations are made available in the Dataset S1 and S2. Cropping systems were considered as conventional, integrated or organic according to the following rules. Cropping systems complying with the organic farming specifications were treated as 'organic'. Other systems were considered as 'integrated', when they were based either on diversified crop rotations including unusual alternative crops for the production situation (i.e. not present in the local reference crop rotation), or when crop management included at least one non-chemical management approach that contributed to the control of pests, diseases or weeds. These included for instance biocontrol, mechanical weeding and false seed bed techniques. Systems that were not classified as 'organic' or 'integrated' were classified as 'conventional'.

## Local reference definition

For each of the 48 systems, a local cropping system reference was selected to reflect the most widespread crop rotation and associated technical management, as well as the typical agricultural performance in the production situation. Using this local cropping system reference made it possible to distinguish the effects of agronomic strategies from the effects of the production situation (soil, climate, economic and social context) when assessing the various components of sustainability of each cropping system. The Dijon-Epoisses experiment included a reference standard system that follows recommendations of local extension services [26], and which was used as the local reference. For each farm of the network across Burgundy, the local reference was defined as the cropping system implemented within the farm before the set-up of the alternative cropping system, even though crop management sequences were slightly updated according to expert appraisal to match with current standards (e.g. active ingredients allowed). For the Zone Atelier "Plaine et Val de Sèvre", local expert knowledge was used to select one system from the survey, with a standard crop rotation for the area, and a crop management representative of local practices. This system was then used as the local reference for all the remaining surveyed systems of the area.

## Assessment of sustainability

The assessment of sustainability at the cropping system scale was based on a range of indicators covering economic, environmental and social issues. The Treatment Frequency Index (TFI) [22] estimates the number of registered doses applied, for each pesticide, per hectare and per crop season. Averaged over the cropping system, this indicator summarizes the level of dependence on pesticides, which should be distinguished from the environmental impact of pesticide use. This indicator is calculated for each pesticide application according to the following formula:

$$TFI = \frac{Application\ rate \times Treated\ surface\ area}{Registered\ dose \times Plot\ surface\ area}$$

The application rate and the registered dose were both expressed for a given commercial product (which possibly contains several active ingredients). The recommended application dose depends obviously on the treated crop and on the targeted pest. Here we defined the registered dose as the lowest application dose which is recommended for a given crop. The TFI for a given crop season was then calculated as the sum of the TFI for each pesticide application performed during this crop season. Productivity was evaluated as the amount of energy harvested yearly. This approach allowed the comparison of different crop rotations that included crops with different yielding potentials and different energy content. For each crop, yields were transformed into the energy metric using their Lower Heating Value (LHV) [48], which corresponds to the amount of energy released per unit of mass by the combustion of the harvested biomass. Energy consumption was estimated from the conversion of inputs into energy according to the Dia'terre reference database [48]. Dia'terre is an assessment tool developed by the Agency for the Environment and Energy Management (ADEME) in the framework of the French Plan for Energy Performance (PPE) to evaluate a carbon-energy balance at the farm scale. The reference database used to design this assessment tool provides energy values for indirect energy consumption associated with the production of farming inputs. For instance, the calculation of the energy cost associated with the production of nitrogen fertilizers integrates the energy necessary from raw material production (e.g. Haber–Bosch process) through materials processing, manufacture and distribution. We used the reference energy cost provided by the carbon calculator Dia'terre to compute the energy balances of the cropping systems. In this way, the inputs necessary for crop production were converted into energy, using the energy cost of fertilizers, pesticides, seeds, water spread for irrigation, fuel consumed by the equipment and the amount of steel necessary to manufacture this equipment, i.e. energy cost of mechanization (see Dataset S3). The energy requirements for preparing farmyard manure are farm-specific and very difficult to quantify precisely. A simplification was consequently required: following previous studies based on energy balancing methods in crop production [49], the energy equivalent of farmyard manure was equated with that of the mineral fertilizers they substituted (using a substitution value related to the fertilizing efficiency of manure). Energy efficiency was computed from the ratio between productivity and energy consumption. For assessing the economic productivity, the gross product derived from the direct conversion of crops yields into economic values. The 'semi-net' margin was calculated as the gross product per hectare from which we subtracted the input costs (fertilizers, pesticides, seeds, fuel, water and mechanisation). This 'semi-net' margin assessed the system profitability without taking into account subsidies or incentives. The sensitivity to price volatility was defined as the relative standard deviation of the semi-net margin calculated over ten contrasting real price scenarios selected between 2000 and 2010, and thus measured the ability of a cropping system to generate a stable income in a variable economic context. The ten scenarios integrated the prices of crops but also the prices of volatile inputs such as fertilizers or fuel. Each price scenario was defined at a given moment between 2000 and 2010, and it therefore reflected the correlations between the prices of crop products and inputs. This approach notably made it possible to integrate better the effects coming from crop diversity (proportion of cereal crops, oil crops or protein crops) on cropping system profitability and economic stability. Fuel consumption and workload were estimated according to in-field cropping operations only, without considering fuel and time consumed for farm-to-field transports, or extra-workload dedicated to equipment maintenance or field observations. The size, fuel requirements and working output of the various equipment types were standardized for all cropping systems, and defined from a national database [50], consistent with the aim of evaluating management strategies, and of ignoring the potential effects of the equipment specifications (See Dataset S4 for the details of the equipment used for the calculations).

Pesticide environmental impact was expressed as cumulated I-Pest [24]. This indicator measures the risk associated with pesticide application for three compartments of the environment, namely the air, the surface water and the groundwater. This risk indicator, ranging from 0 to 1 (maximum risk), and calculated for each active substance application, is based on: (i) field inherent sensitivity to pesticide transfer toward these three compartments; (ii) characteristics of the active substance (e.g. ecotoxicity, mobility, half-life); and, (iii) information about the conditions of the spraying operation (e.g. amount of active substances employed, canopy cover at the date of treatment) in order to calculate three impact factors, one for each compartment. I-Pest index is obtained using fuzzy decision trees that allow the aggregation of these three impact factors into one synthetic indicator. The diagram presented in Figure S1 illustrates how this indicator of pesticide environmental impact was computed for each pesticide active substance that was sprayed within the field.

The crop sequence indicator Isc [27] is used as an additional indicator to quantify the agronomic effects of crop diversification. Isc ranges on a qualitative scale between 0 and 10 (best value) and is calculated as shown in the following equation:

$$Isc = kp \times kr \times kd$$

Isc is based on the assessment of the effects of the previous crop on the current crop (kp), with respect to the development of pathogens, pests and weeds, to soil structure and nitrogen supply. kp, ranging between 1 and 6, was assessed for 470 couples crop/previous crop. kp is corrected by two factors taking into account the crop frequency (kr ranging between 0.3 and 1.2) and the crop rotation whole diversity (kd ranging between 1.0 and 1.4). Isc yields respectively 0.5 for wheat monoculture, 3.3 for a rape/wheat rotation, 5.1 for a rape/wheat/barley rotation, and 7.6 for a maize/wheat/sunflower/spring barley rotation.

### Computation of sustainability indicators at the cropping system level

As a first step each indicator was calculated for each cropping operation composing our database (Dataset S2). These values of indicators were summed over the crop season year, and then averaged across years and across plots, each plot being considered as a replicate of a given cropping system. Each indicator was therefore calculated at the cropping system level, integrating (i) the different crops composing the crop sequence, (ii) the variability of crop production related with the inter-annual climatic variability, and (iii) the possible variation in plot properties.

All sustainability indicators were expressed per hectare and per year. For distinguishing specifically the effects of the management strategy on cropping system sustainability from the effects of the production situation, each indicator computed for a given cropping system was then expressed as a ratio (or as a distance in the case of semi-net margin) between the system indicator and the local reference indicator. To increase the quality of the graphs drawing the relationship between sustainability indicators and pesticide use, values of assessment indicators were translated into natural logarithm (Figure 2), which reduced the visual effect of extreme values.

### Statistical analyses

Spearman and Pearson correlations were estimated using the 'rcorr' correlation matrix function in the *Hmisc* package of R v2.15.0 [51]. The difference between the means of two subsamples for a given indicator was tested with a non-parametric Mann-Whitney test ('wilcox.test' function with two samples) in the *stats* package of R v2.15.0.

## Supporting Information

**Figure S1** Simplified description of the assessment process of pesticide environmental impact in the I-Pest model.

**Dataset S1** Cropping systems details. A.xlsx file describing the cropping systems of the studied sample (e.g. crop rotation, tillage and weed management strategies). This file also provides information about the local reference associated with the evaluation of each cropping system. The second tab provides the respective performances of each cropping system described in the first tab.

**Dataset S2** Cropping operations database. A.xlsx file which provides the details of all cropping operations carried out in each cropping system: type of cropping operation, date (when recorded), application rates (for pesticides, fertilizers, seeds and irrigation) and proportion of the plot surface targeted.

**Dataset S3** Energy balancing database. A.xlsx file with two sheets. The first sheet provides energy cost values for inputs: pesticides active substances, fuel, fertilizers, irrigation water and seeds. The second sheet includes the Lower Heating Values (LHV) for usual crops, that is to say the energy contained in one mass unit of crop harvested.

**Dataset S4** Standard equipment characteristics. A.xlsx file describing the technical characteristics of the standard equipment we associated with each cropping operation. Details include the purchase price, the payback period and the maintenance cost to calculate the mechanization costs, but also the equipment size and weight, the working output, the fuel consumption rate and the energy cost value.

**Table S1** Means and standard deviations for the range of performance indicators according to the management strategy. A.xlsx file summarizing and comparing the performances of organic, integrated and conventional cropping systems which compose the study sample. Significant difference between groups was tested with a Mann-Whitney test.

## Acknowledgments

We thank A. Villard, C. Vivier and M. Geloen for their contribution to the experimental network in Burgundy, and D. Meunier, P. Farcy, P. Chamoy for technical assistance. We particularly want to thank D. Bohan for the precious advice on style and the language corrections.

## Author Contributions

Conceived and designed the experiments: NMJ VB CB SP. Performed the experiments: ML FB. Analyzed the data: ML FB. Contributed reagents/materials/analysis tools: CB MSP. Wrote the paper: NMJ ML. Discussed the results and commented on the manuscript: NMJ ML VB CB SP MSP FB.

## References

1. Foley JA, Ramankutty N, Brauman KA, Cassidy ES, Gerber JS, et al. (2011) Solutions for a cultivated planet. Nature 478: 337–342.
2. Pimentel D (1995) Amounts of pesticides reaching target pests: environmental impacts and ethics. Journal of Agricultural and Environmental Ethics 8: 17–29.

3. Richardson M (1998) Pesticides-friend or foe? Water science and technology 37: 19–25.
4. Tilman D, Cassman KG, Matson PA, Naylor R, Polasky S (2002) Agricultural sustainability and intensive production practices. Nature 418: 671–677.
5. Maeder P, Fliessbach A, Dubois D, Gunst L, Fried P, et al. (2002) Soil fertility and biodiversity in organic farming. Science 296: 1694–1697.
6. Holland JM, Frampton GK, Cilgi T, Wratten SD (1994) Arable acronyms analysed - a review of integrated arable farming systems research in Western Europe. Annals of applied biology 125: 399–438.
7. Ferron P, Deguine JP (2005) Crop protection, biological control, habitat management and integrated farming. A review. Agronomy for Sustainable Development 25: 17–24.
8. Trewavas A (2001) Urban myths of organic farming. Nature 410: 409–410.
9. Pimentel D, Hepperly P, Hanson J, Douds D, Seidel R (2005) Environmental, energetic, and economic comparisons of organic and conventional farming systems. Bioscience 55: 573.
10. Directive 2009/128/EC of the European Parliament and of the Council (2009) Official journal of the European Union. Available: http://eur-lex.europa.eu/LexUriServ/LexUriServ.do?uri=OJ:L:2009:309:0071:0086:en:PDF Accessed 2013 Sep 10.
11. Munier-Jolain N, Dongmo A (2010) Evaluation de la faisabilité technique de systèmes de Protection Intégrée en termes de fonctionnement d'exploitation et d'organisation du travail. Comment adapter les solutions aux conditions locales? Innovations Agronomiques 8: 57–67.
12. European commission (2010) Eurostat Agriculture online database. Available: http://epp.eurostat.ec.europa.eu/portal/page/portal/agriculture/farm_structure/database. Accessed 2013 Jul 19.
13. USDA Economic Research Service (2010) U.S. certified organic farmland acreage, livestock number, and farm operations. Available: http://www.ers.usda.gov/data-products/organic-production.aspx "\l ".UiWVOX8QO89. Accessed 2013 Jul 22.
14. Bastiaans L, Paolini R, Baumann DT (2008) Focus on ecological weed management: what is hindering adoption? Weed Research 48: 481–491.
15. Pardo G, Riravololona M, Munier-Jolain N (2010) Using a farming system model to evaluate cropping system prototypes: Are labour constraints and economic performances hampering the adoption of Integrated Weed Management? European Journal of Agronomy 33: 24–32.
16. Food and Agriculture Organization of the United Nations (FAO) (2013) FAOSTAT Resources: Pesticides Use. Available: http://faostat.fao.org/site/424/DesktopDefault.aspx?PageID=424#ancor. Accessed 2013 Jul 25.
17. U.S. Environmental Protection Agency (EPA) (2011) Pesticides industry sales and usage: 2006 and 2007 market estimates. Washington, D.C.: U.S. Environmental Protection Agency. Available: www.epa.gov/opp00001/pestsales/07pestsales/market_estimates2007.pdf. Accessed 2013 Jul 25.
18. Ministère de l'Agriculture, de l'Agro-alimentaire et de la Forêt (2012) Note de suivi du plan Ecophyto 2018: tendances de 2008 à 2011 du recours aux produits phytopharmaceutiques. Available: http://agriculture.gouv.fr/IMG/pdf/121009_Note_de_suivi_2012_cle0a995a.pdf. Accessed 2013 Jul 26.
19. Reganold JP, Glover JD, Andrews PK, Hinman HR (2001) Sustainability of three apple production systems. Nature 410: 926–930.
20. Deike S, Pallutt B, Christen O (2008) Investigations on the energy efficiency of organic and integrated farming with specific emphasis on pesticide use intensity. European Journal of Agronomy 28: 461–470.
21. Aubertot JN, Robin MH (2013) Injury Profile SIMulator, a qualitative aggregative modelling framework to predict crop injury profile as a function of cropping practices, and the abiotic and biotic environment. I. Conceptual bases. PLoS one 8: e73202.
22. OECD (2001) Environmental Indicators for Agriculture, Volume 3: Methods and Results. Available: www.oecd.org/tad/sustainable-agriculture/40680869.pdf. Accessed 2014 Feb 26.
23. Meiss H, Mediene S, Waldhardt R, Caneill J, Bretagnolle V, et al. (2010) Perennial lucerne affects weed community trajectories in grain crop rotations. Weed Research 50: 331–340.
24. Van der Werf H, Zimmer C (1998) An indicator of pesticide environmental impact based on a fuzzy expert system. Chemosphere 36: 2225–2249.
25. Davis AS, Hill JD, Chase CA, Johanns AM, Liebman M (2012) Increasing cropping system diversity balances productivity, profitability and environmental health. PloS one 7: e47149.
26. Chikowo R, Faloya V, Petit S, Munier-Jolain NM (2009) Integrated Weed Management systems allow reduced reliance on herbicides and long-term weed control. Agriculture, Ecosystems and Environment 132: 237–242.
27. Bockstaller C, Girardin P (2000) Using a crop sequence indicator to evaluate crop rotations. 3rd International Crop Science Congress 2000 ICSC, Hambourg, 17–22 August 2000, p. 195.
28. Andersson GKS, Rundlöf M, Smith HG (2012) Organic Farming Improves Pollination Success in Strawberries. PloS one 7: e31599.
29. Aavik T, Liira J (2010) Quantifying the effect of organic farming, field boundary type and landscape structure on the vegetation of field boundaries. Agriculture, Ecosystems and Environment 135: 178–186.
30. Li R, Khafipour E, Krause DO, Entz MH, de Kievit TR, et al. (2012) Pyrosequencing Reveals the Influence of Organic and Conventional Farming Systems on Bacterial Communities. PloS one 7: e51897.
31. Seufert V, Ramankutty N, Foley JA (2012) Comparing the yields of organic and conventional agriculture. Nature 485: 229–232.
32. Foley JA, Defries R, Asner GP, Barford C, Bonan G, et al. (2005) Global consequences of land use. Science 309: 570–574.
33. Theobald DM (2001) Land-use dynamics beyond the American urban fringe. Geographical Review 91: 544.
34. Phalan B, Onial M, Balmford A, Green R (2011) Reconciling food production and biodiversity conservation: land sharing and land sparing compared. Science 333: 1289–1291.
35. Hulme MF, Vickery JA, Green RE, Phalan B, Chamberlain DE, et al. (2013) Conserving the birds of Uganda's banana-coffee arc: land sparing and land sharing compared. PLoS One 8: e54597.
36. Altieri MA, Nicholls CI, Ponti L (2009) Crop diversification strategies for pest regulation in IPM systems. Integrated pest management Cambridge University Press, Cambridge, UK. pp. 116–130.
37. Krupinsky J, Bailey K, McMullen M, Gossen B, Turkington T (2002) Managing plant disease risk in diversified cropping systems. Agronomy Journal 94: 198–209.
38. Beketov MA, Kefford BJ, Schäfer RB, Liess M (2013) Pesticides reduce regional biodiversity of stream invertebrates. Proceedings of the National Academy of Sciences 110: 11039–11043.
39. Ascherio A, Chen H, Weisskopf MG, O'Reilly E, McCullough ML, et al. (2006) Pesticide exposure and risk for Parkinson's disease. Annals of Neurology 60: 197–203.
40. Di Falco S, Chavas JP (2008) Rainfall shocks, resilience, and the effects of crop biodiversity on agroecosystem productivity. Land Economics 84: 83–96.
41. Reidsma P, Ewert F, Lansink AO, Leemans R (2010) Adaptation to climate change and climate variability in European agriculture: The importance of farm level responses. European Journal of Agronomy 32: 91–102.
42. Nemecek T, von Richthofen JS, Dubois G, Casta P, Charles R, et al. (2008) Environmental impacts of introducing grain legumes into European crop rotations. European Journal of Agronomy 28: 380–393.
43. Pimentel D, Pimentel M (2003) Sustainability of meat-based and plant-based diets and the environment. The American Journal of Clinical Nutrition 78: 660S–663S.
44. Halberg N, van der Werf HMG, Basset-Mens C, Dalgaard R, de Boer IJM (2005) Environmental assessment tools for the evaluation and improvement of European livestock production systems. Livestock Production Science 96: 33–50.
45. Beckie HJ (2009) Herbicide Resistance in Weeds: Influence of Farm Practices. Prairie Soils and Crops 2:3.
46. Meynard JM, Messéan A, Charlier A, Charrier F, Farès M, et al. (2013) Freins et leviers à la diversification des cultures. Etudes au niveau des exploitations agricoles et des filières. Synthèse du rapport d'étude, INRA. Available: http://inra.dam.front.pad.brainsonic.com/ressources/afile/223799-6afe9-resource-etude-diversification-des-cultures-synthese.html. Accessed 2013 Mar 15.
47. Centre d'Etudes Biologiques de Chizé (2009) Zone Atelier «Plaine & Val de Sèvre». Available: http://www.zaplainevaldesevre.fr. Accessed 2013 Jun 5.
48. Agence de l'Environnement et de la Maîtrise de l'Energie (ADEME) (2011) Guide des valeurs Dia'terre. Version du référentiel 1.13. Available: http://www2.ademe.fr/servlet/KBaseShow?sort=-1&cid=96&m=3&catid=24390. Accessed 2013 Jul 12.
49. Hülsbergen KJ, Feil B, Biermann S, Rathke GW, Kalk WD, et al. (2001) A method of energy balancing in crop production and its application in a long-term fertilizer trial. Agriculture, Ecosystems and Environment 86: 303–321.
50. Bureau de Coordination du Machinisme Agricole (BCMA) (2012) Simcoguide online decision tool. Available: http://simcoguide.pardessuslahaie.net/#accueil. Accessed 2013 Apr 12.
51. R Development Core Team (2012). R: A language and environment for statistical computing (R Foundation for Statistical Computing, Vienna, Austria).

# 3

# Additional Nitrogen Fertilization at Heading Time of Rice Down-Regulates Cellulose Synthesis in Seed Endosperm

**Keiko Midorikawa[1], Masaharu Kuroda[2]*, Kaede Terauchi[1], Masako Hoshi[1], Sachiko Ikenaga[3], Yoshiro Ishimaru[1], Keiko Abe[1,4], Tomiko Asakura[1]***

**1** Department of Applied Biological Chemistry, Graduate School of Agricultural and Life Sciences, The University of Tokyo, Bunkyo-ku, Tokyo, Japan, **2** Crop Development Division, NARO Agricultural Research Center, Inada, Joetsu, Niigata, Japan, **3** Field Crop and Horticulture Research Division, NARO Tohoku Agricultural Research Center, Morioka, Iwate, Japan, **4** Food Safety and Reliability Project, Kanagawa Academy of Science and Technology, Takatsu-ku, Kawasaki, Kanagawa, Japan

## Abstract

The balance between carbon and nitrogen is a key determinant of seed storage components, and thus, is of great importance to rice and other seed-based food crops. To clarify the influence of the rhizosphere carbon/nitrogen balance during the maturation stage of several seed components, transcriptome analysis was performed on the seeds from rice plants that were provided additional nitrogen fertilization at heading time. As a result, it was assessed that genes associated with molecular processes such as photosynthesis, trehalose metabolism, carbon fixation, amino acid metabolism, and cell wall metabolism were differentially expressed. Moreover, cellulose and sucrose synthases, which are involved in cellulose synthesis, were down-regulated. Therefore, we compared cellulose content of mature seeds that were treated with additional nitrogen fertilization with those from control plants using calcofluor staining. In these experiments, cellulose content in endosperm from plants receiving additional nitrogen fertilization was less than that in control endosperm. Other starch synthesis-related genes such as starch synthase 1, starch phosphorylase 2, and branching enzyme 3 were also down-regulated, whereas some α-amylase and β-amylase genes were up-regulated. On the other hand, mRNA expression of amino acid biosynthesis-related molecules was up-regulated. Moreover, additional nitrogen fertilization caused accumulation of storage proteins and up-regulated Cys-poor prolamin mRNA expression. These data suggest that additional nitrogen fertilization at heading time changes the expression of some storage substance-related genes and reduces cellulose levels in endosperm.

**Editor:** Sara Amancio, ISA, Portugal

**Funding:** This work was supported by Nissin Food Holdings Co., LTD (http://www.nissinfoods-holdings.co.jp/). It was supported by grant in aid for food science research to T. A. and M. K. from the Tojuro Iijima Foundation for Food Science and Technology (http://www.iijima-kinenzaidan.or.jp/). The funders had no role in study design, data collection and analysis, decision to publish, or preparation of the manuscript.

**Competing Interests:** The authors' received funding from a commercial source (Nissin Food Holdings Co., LTD). Our laboratory was established by the donation from Nissin Food Holdings Co., LTD.

* E-mail: kurodama@affrc.go.jp (MK); asakura@mail.ecc.u-tokyo.ac.jp (TA)

## Introduction

Regulation of carbon (C) and nitrogen (N) metabolism is indispensable for plant growth and development. Carbon and nitrogen species are essential constituents of both macronutrients and signaling metabolites, which influence several cellular processes and gene expression [1–6]. In crop plants, starch and protein content in seed are determinants of yield and quality. These are synthesized using sugars and amino acids from the plant body, and share photosynthetic carbon sources for their synthesis. Thus, control of C/N balance during the reproductive stage is critical for high-yield and high-quality crop production.

The amount and timing of nitrogen fertilization are the most important factors for beneficial control of C/N balance, because the distribution of carbon sources from photosynthesis is generally influenced by plant nitrogen conditions [7,8]. For example, the expression of photosynthetic and carbon fixation-related genes rapidly decreases in rice roots and leaves under low nitrogen conditions [9]. Under these conditions, rice leaves turn to pale green, carbon fixation is reduced, and remobilized nitrogen is used for other metabolic processes. In contrast, additional nitrogen fertilization facilitates maturation and elevates grain yields in commercial rice cultivation [10]. A high-nitrogen condition retards leaf senescence by maintaining nitrogen-containing compounds such as chlorophyll and photosynthetic proteins, thus elevating photosynthetic activity and transport of photosynthetic materials to the seed throughout the seed-maturation period. Such conditions may alter accumulation of seed components. In fact, protein content of rice seeds is elevated under conditions of high nitrogen fertilization [11].

Above results suggest that rhizospheric nitrogen influences global gene expression in the plant body, thereby implying the existence of gene networks that control C/N balance. In addition, gene expression in seeds is strongly influenced by the condition of the plant body. However, no studies have examined the influence of nitrogen on gene expression, metabolic processes, and accumulation of components in rice seeds. In this study, we examined the effects of nitrogen fertilization at heading time of rice, because plant body sizes and numbers of spikelets are fixed before fertilization. Rhizospheric nitrogen may directly affect metabolic processes during seed maturation. Thus, we examined changes in gene expression that correspond with nitrogen

| Experimental plot | | Germination | Active tillering | Heading | Sampling |
|---|---|---|---|---|---|
| | DAY | 0 | 37 | 68 | 78 - 93 |
| | Fertilization (/ 6 rice plants) | N6-P8-K6-Mg2 2.5 g | N6-P8-K6-Mg2 1.5 g | $NH_4Cl$ 400 mg | |
| $+NH_4Cl$ | | + | + | + | |
| Control | | + | + | - | |

**Figure 1. Rice cultivation schedule.** Samples were grown in a growth chamber at 28°C/22°C over a 12-h light/12-h dark cycle. Fertilizer was supplied at planting and 37 days after germination. In addition, 400 mg ammonium chloride ($NH_4Cl$) was supplied at heading time to the "$+NH_4Cl$" group.

fertilization at heading time using DNA microarray analysis. Subsequently, we examined whether these changes in gene expression are correlated with seed components. The present data indicate that nitrogen fertilization at heading time decreases cellulose synthesis.

## Materials and Methods

### Rice Cultivation under Field Conditions

*Oryza sativa* L. cv. Nipponbare was used in all experiments. Rice plants were grown in a paddy field at the NARO Agricultural Research Center, Niigata, Japan. Field trials were conducted for three years.

At heading time, the experimental field was divided into a control plot and a nitrogen-fertilized (N-fertilized) plot using a plastic board. Ammonium chloride was sprayed on the soil surface of the N-fertilized section at a rate of 8 kg/1,000 $m^2$. The sampling area contained 10×10 plants in each section, and all mature grains in the sampling area were harvested and prepared for analysis of nitrogen and amino acid content.

### Rice Cultivation in a Plant Incubator

To improve reproducibility, a plant incubator with fluorescent lamps on inner walls (model FLI2000A, Tokyo Rikakikai, Tokyo, Japan) was used to simulate paddy field conditions.

The schedule of cultivation is shown in Fig. 1. Plastic containers (C-AP fruit 500-1; 173×123×70 mm; Chuo Kagaku, Saitama, Japan) were filled with 500 mL rice nursery soil (Honen Agri, Niigata, Japan) and were supplied with 2.5 g fertilizer containing 0.15 g nitrogen, 0.2 g phosphate, 0.15 g potassium, and 0.05 g magnesium. Six plants were cultivated in each plastic container. Each of the four containers was placed on stainless trays (TRAY SUS No. 9, 367×257×92 mm; AZ ONE, Osaka, Japan); two trays, one for the control plot and the other for the N-fertilized plot, were placed in an incubator. Each plot contained 24 plants and tap water was provided to the depth such that the container was submerged. These plants were cultivated under short-daytime conditions with a 12-h maximum illumination (28°C)/12-h dark (22°C) cycle. Water depth was recovered every 2 days using tap water. The growth of each plant was restricted to the main culm by removing tillers. After 5 weeks, 1.5 g fertilizer was supplied to each container. At heading stage, ammonium chloride was sprayed on the soil surface of the N-fertilized plot tray at a rate of 400 mg/container.

To examine the synthesis of storage materials in more detail, samples were taken at 10, 15, 20, and 25 DAF during incubator cultivation. Spikelets were marked as for field cultivation and were subsequently sampled. Developing grains were collected randomly from 24 plants in each experimental plot. All dehusked grains from the same sampling time were mixed, frozen in liquid nitrogen, and stored at −80°C until use. For microarray analysis, RNAs from 15-DAF grains were used, because storage materials are synthesized actively during this stage [12]. Finally, mature grains were harvested from each plant and their nitrogen content was determined.

### Analysis of Nitrogen Content, Amino Acid Composition, and Protein Composition

Nitrogen content was determined using the Kjeldahl method in field trials. Amino acid analysis was performed at the Japan Food Research Laboratories (http://www.jfrl.or.jp/e/index.html; Tokyo, Japan) using an automated amino acid analyzer and HPLC according to their standard protocol. Total amino acid content was analyzed after hydrolysis of rice samples using hydrogen chloride. Free amino acid content was analyzed using sulfosalicylic

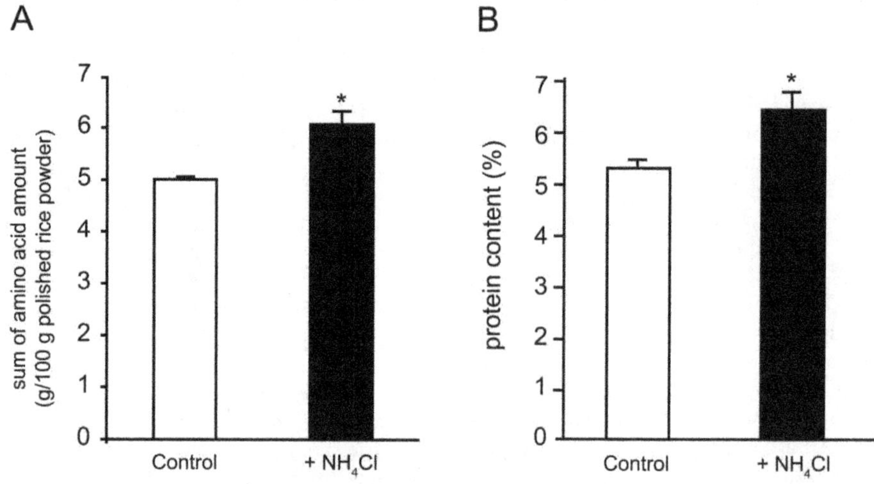


**Figure 2. Amino acid and protein contents of mature polished rice cultivated in a field.** Field trials were conducted for three years and data are expressed as means of the three trials. Filled bars correspond with samples from the N-fertilized plot and unfilled bars correspond with samples from the control plot. Error bars represent standard deviation (SD; *$P<0.01$) (A) Total amino acid content was analyzed after hydrolysis using hydrogen chloride. (B) Protein content was estimated using the Kjeldahl method.

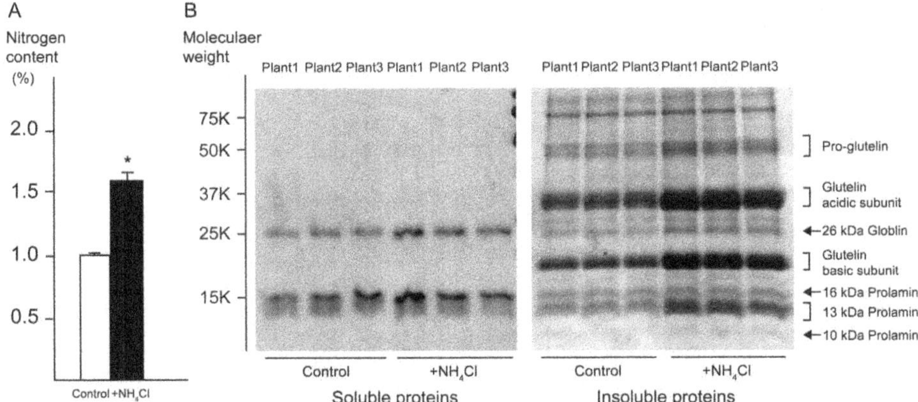


**Figure 3. Analysis of proteins in polished rice from incubator cultivated plants.** Three independent plants were selected from the N-fertilized and control plots. (A) Nitrogen content of polished rice was determined using a Rapid-N analyzer. Error bars represent standard deviation (SD; *$P<0.001$) (B) Protein composition of polished rice was analyzed using SDS-PAGE. Soluble proteins and insoluble proteins were extracted sequentially from the same sample. Major storage proteins in rice seed were extracted as insoluble proteins, as indicated at the right side of the gel.

acid- extractable fraction of rice samples. After harvest of all three trials, polished rice seeds were ground to powder in a cyclone sample mill (Shizuoka Seiki, Shizuoka, Japan), and nitrogen and amino acid content were determined for each trial. The sum of individual amino acid content was calculated, and differences between means were used for statistical analysis.

Nitrogen content was determined using a rapid-N-III automated nitrogen analyzer (Elementar Analysensysteme GmbH, Hanau, Germany) in incubator cultivation. Three independent plants from each fertilization division were selected, and nitrogen content of six polished grains from each plant was determined.

Protein composition of polished rice was analyzed using SDS-PAGE as shown previously with some modifications [13]. Subsequently, six seeds from each plant were ground to a powder, and soluble protein was extracted by adding 2.4 mL globulin extraction buffer containing 10 mM Tris-HCl (pH 7.5), 1 mM EDTA, and 0.5 M NaCl. Insoluble protein was then extracted by adding 2.4 mL of the total protein extraction buffer containing 50 mM Tris-HCl, (pH 6.8), 8 M urea, 4% SDS, and 5% 2-mercaptoethanol. Subsequently, a 10 µL aliquot of extracted protein was used for SDS-PAGE.

## Total RNA Isolation and DNA Microarray Preparation

Seeds at 15 DAF were collected from 24 control plants and 24 N-fertilized plants, immediately frozen in liquid nitrogen and pooled. Three sets of five seeds were selected from each seed pool. Samples were then ground to a fine powder, and RNA was extracted using an RNeasy Plant Mini kit (Qiagen, Hilden, Germany).

DNA microarray analysis was performed using a One-Cycle cDNA Synthesis kit (Affymetrix, Santa Clara, CA), a Sample Cleanup Module (Affymetrix), and a GeneChip IVT Labeling kit (Affymetrix). Experimental procedures were performed according to the manufacturer's protocols. Briefly, after synthesizing cDNA from purified RNA, biotinylated cRNA was transcribed from cDNA using T7 RNA polymerase, fragmented, and was then added to a GeneChip Rice Genome Array containing over 50,000 rice genes (Affymetrix). Fragmented RNA was hybridized with the array at 45°C for 16 h. The array was then washed, labeled with phycoerythrin, and scanned for fluorescence using the Affymetrix GeneChip System. All microarray data were submitted to the National Center for Biotechnology Information (NCBI) Gene Expression Omnibus database (http://www.ncbi.nlm.nih.gov/geo/; GEO Series ID GSE49818). Further, Affymetrix GCOS software was used to convert array images and probe intensities to CEL files. CEL files were then transferred to a personal computer and were analyzed using R statistical language and Bioconductor software. Files were normalized using the distribution-free weighted (DFW) method [14]. The pvclust function was then used to perform clustering analysis of the samples.

Gene expression patterns in rice grown under $+NH_4/-NH_4$ conditions were compared using the "rank products" function [15]. Genes with false discovery rates (FDR) of <0.05 were extracted, analyzed with the "pvclust" function [16], and divided according to their expression patterns. Identified genes were analyzed using the annotation file for the Rice Genome Array, which was downloaded from the NetAffx database on the Affymetrix website.

Gene-annotation enrichment analysis of differentially expressed genes (DEGs) was performed using the Database for Annotation, Visualization and Integrated Discovery (http://david.abcc.ncifcrf.gov/) [17] and Quick GO (http://www.ebi.ac.uk/QuickGO/) [18]. EASE scores from modified Fisher's exact test $P$-values [19] were used to extract statistically over represented GO terms from DEGs. When annotations could not be obtained using the NetAffx

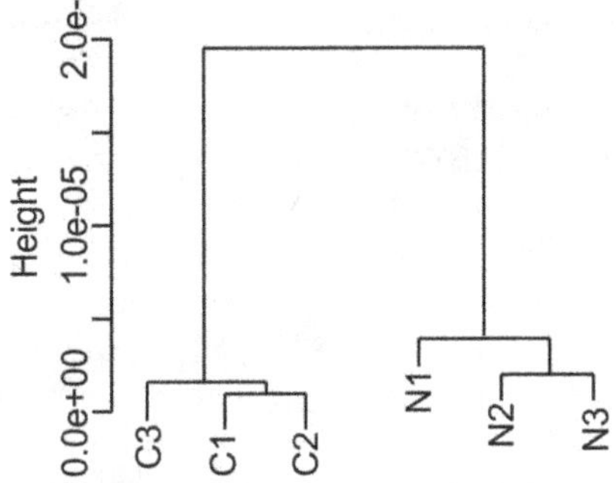


**Figure 4. Cluster dendrogram.** A cluster dendrogram was generated using rice seed gene expression data from six samples and the "pvclust" function. Each sample was prepared at 15-DAF seeds from 24 control or 24 N-fertilized plants and were pooled. Three sets of five seeds were picked from each seed pool. C, control samples; N, N-fertilized samples.

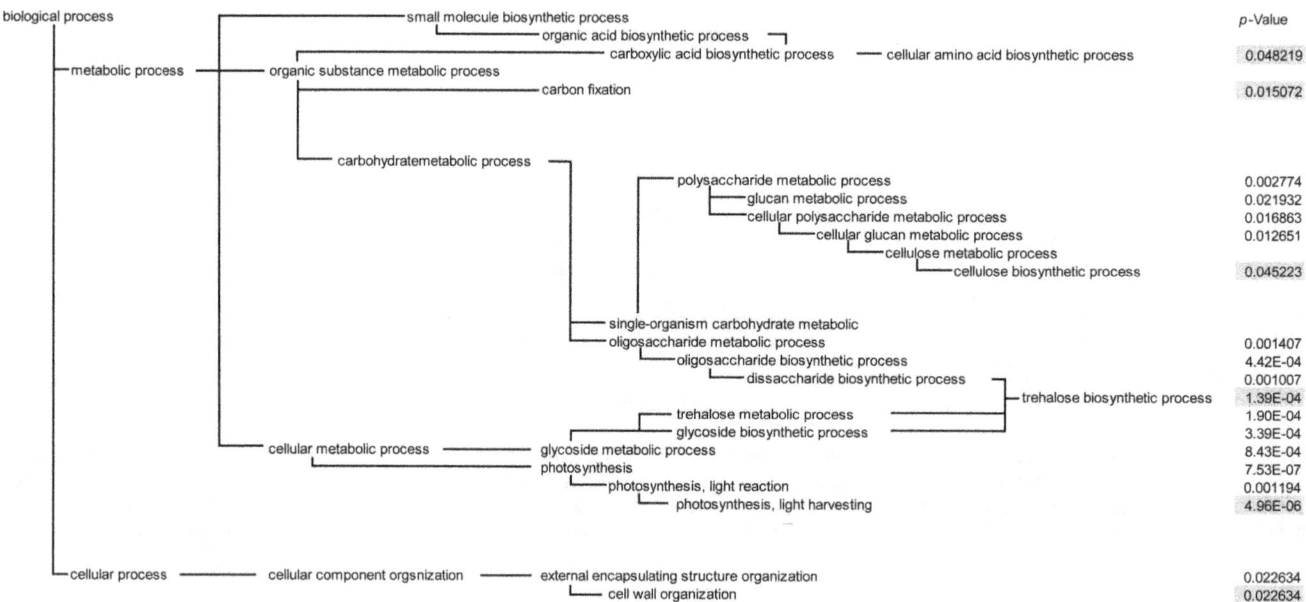


**Figure 5. Significantly enriched categories were identified using QuickGO.** In response to additional fertilization, 1,365 genes were up- or down-regulated; FDR-corrected *P*-values of categories at the deepest hierarchical level are shaded; $^{*}P<0.05$.

**Table 1.** Genes related to synthesis of storage substances in seeds with altered mRNA expression in response to additional fertilizer.

| Probe set ID | RAP-DB ID | RAP-DB Annotation | Gene Expression |
|---|---|---|---|
| Os.26109.1.S1_x_at | Os05g0329100, Os05g0329200, Os05g0329400, Os05g0330600 | Cysteine-poor 13 kDa prolamin | Up |
| Os.25998.1.S1_at | Os05g0329100, Os05g0329200, Os05g0329400, Os05g0330600 | Cysteine-poor 13 kDa prolamin | Up |
| Os.8502.4.S1_at | Os05g0329300 | Cysteine-poor 13 kDa prolamin | Up |
| Os.8502.5.S1_x_at | Os05g0328333, Os05g0329100, Os05g0329200, Os05g0329400, Os05g0330600, Os07g0219300, Os07g0219400, Os07g0220050 | Cysteine-poor 13 kDa prolamin | Up |
| Os.20396.1.A1_at | Os02g0456150 | 11-S plant seed storage protein family protein. | Up |
| Os.20396.1.A1_s_at | Os02g0456150 | 11-S plant seed storage protein family protein. | Up |
| OsAffx.2749.1.S1_at | Os02g0456150 | 11-S plant seed storage protein family protein. | Up |
| Os.17979.1.S1_at | Os02g0244100 | Grain weight 2 (OsGW2). | Up |
| OsAffx.16823.1.S1_at | Os08g0137250 | Fertilization-Independent endosperm (Protein Fertilization-Independent Seed 3) (OsFIE1). | Up |
| Os.12593.1.S1_s_at | Os08g0473600 | Alpha-amylase isozyme 3E precursor (EC 3.2.1.1). | Up |
| Os.10339.1.S1_at | Os03g0141200 | Similar to Beta-amylase PCT-BMYI (EC 3.2.1.2). | Up |
| Os.46618.1.S1_at | Os10g0565200 | Similar to Beta-amylase PCT-BMYI (EC 3.2.1.2). | Up |
| Os.13907.1.S1_at | Os02g0248800 | Similar to Glutelin type-B 2 precursor. | Up |
| Os.29800.1.S1_x_at | Os01g0702900, Os02g0771500 | Sucrose-phosphate synthase (EC 2.4.1.14). | Up |
| Os.12725.1.S1_at | Os06g0160700 | Similar to Starch synthase I, chloroplast precursor (EC 2.4.1.21) (Soluble starch synthase 1) (SSS 1). | Down |
| Os.2623.1.S1_at | Os01g0851700 | Similar to Cytosolic starch phosphorylase (Fragment)/Starch phosphorylase 2. | Down |
| Os.4179.1.S1_at | Os02g0528200 | Branching enzyme-3 precursor (EC 2.4.1.18). | Down |
| OsAffx.13550.1.S1_s_at | Os03g0808200 | UDP-glucuronosyl/UDP-glucosyltransferase family protein. | Down |
| Os.33722.1.S1_at | Os01g0736100 | UDP-glucuronosyl/UDP-glucosyltransferase family protein. | Down |
| Os.9127.1.S1_a_at | Os06g0194900 | Sucrose synthase 2 (EC 2.4.1.13). | Down |
| Os.9860.1.S1_at | Os07g0616800 | Sucrose synthase 3 (EC 2.4.1.13). | Down |

**Table 2.** Differentially expressed genes in significantly enriched GO categories (*$P<0.05$).

| | RAP-DB ID | RAP-DB annotation | Gene Expression |
|---|---|---|---|
| Photosynthesis, light harvesting | Os08g0435900, Os07g0562700, Os07g0558400, Os01g0600900, Os03g0592500, Os04g0457000 | Chlorophyll a/b-binding protein | Up |
| Trehalose biosynthetic process | Os03g0386500 | Trehalose-6-phosphate phosphatase 9 | Up |
| | Os02g0661100 | Trehalose-6-phosphate phosphatase 1 | Down |
| | Os09g0369400 | Trehalose-6-phosphate phosphatase 7 | Down |
| | Os01g0730300 | Trehalose-6-phosphate synthase 3 | Down |
| | Os02g0790500 | Trehalose-6-phosphate synthase 5 | Up |
| | Os05g0517200 | Trehalose-6-phosphate synthase 6 | Up |
| Carbon fixation | Os01g0791033, Os05g0427800, Os12g0207600 | Ribulose bisphosphate carboxylase large chain precursor (EC 4.1.1.39) (RuBisCO large subunit). | Up |
| | Os12g0292400, Os12g0291400 | Ribulose bisphosphate carboxylase small chain | Up |
| Cell wall organization | Os10g0555900, Os10g0548600 | Beta-expansin precursor | Up |
| | Os08g0160500 | Cellulose synthase-like protein F6 (OsCslF6) | Down |
| | Os10g0450900 | Glycine-rich cell wall structural protein 2 precursor | Up |
| | Os07g0208500 | Cellulose synthase A8 (OsCESA8) | Down |
| | Os02g0130200 | Virulence factor, pectin lyase fold family protein | Up |
| | Os07g0252400 | Cellulose synthase A6 (OsCESA6) | Down |
| | Os02g0738600 | Endoglucanase 7 | Down |
| | Os03g0377700 | Cellulose synthase-like A5 (CSLA5) | Down |
| | Os03g0808100 | Cellulose synthase A2 (OsCESA2) | Down |
| | Os03g0837100 | Cellulose synthase A5 (OsCESA5) | Down |
| | Os08g0237000 | Xyloglucan endotransglycosylase/hydrolase protein 8 precursor (End-xyloglucan transferase) (OsXTH8) | Down |
| Cellulose biosynthetic process | Os07g0252400 | Cellulose synthase A6 (OsCESA6) | Down |
| | Os08g0160500 | Cellulose synthase-like protein F6 (OsCslF6) | Down |
| | Os07g0208500 | Cellulose synthase A8 (OsCESA8) | Down |
| | Os03g0808100 | Cellulose synthase A2 (OsCESA2) | Down |
| | Os03g0837100 | Cellulose synthase A5 (OsCESA5) | Down |
| Cellular amino acid biosynthetic process | Os04g0669800 | Methylthioribose kinase | Up |
| | Os01g0720700 | Serine acetyltransferase 1 | Up |
| | Os11g0256000 | Acetolactate synthase, small subunit family protein | Up |
| | Os09g0565700 | Prephenate dehydratase domain containing protein | Up |
| | Os12g0578200 | Chorismate mutase, chloroplast precursor (CM-1) | Up |
| | Os03g0291500 | Asparagine synthase domain containing protein | Up |
| | Os01g0681900 | NADH - Glutamate Synthase 1 | Down |
| | Os03g0279400 | Arginine biosynthesis bifunctional protein ArgJ, chloroplastic | Down |
| | Os02g0510200 | Acetohydroxyacid synthase | Up |
| | Os03g0826500 | Anthranilate synthase alpha 1 subunit | Up |
| | Os03g0389700 | Phospho-2-dehydro-3-deoxyheptonate aldolase 1, chloroplastic | Up |

Only annotated genes are listed. All 1365 differentially expressed genes are listed in Tables S2 and S3.

database, we consulted the following databases: Rice TOGO (http://agri-trait.dna.affrc.go.jp/index.html), RiceXPro (http://agri-trait.dna.affrc.go.jp/index.html), MSU Rice Genome Annotation Project (http://rice.plantbiology.msu.edu/), RAP-DB (http://rapdblegacy.dna.affrc.go.jp/), and UniProt (http://www.uniprot.org/).

## β-glucan Staining

β-glucan staining was performed on 30 mature seeds that were randomly selected from each fertilization plot in the third field trial. Fresh-frozen rice grain sections were prepared for fluorescence microscopy as described in Saito *et al.*, (2008). Frozen sections (5 μm thick) were cut from each seed, soaked in a 10% (w/v) potassium hydroxide solution, and were then stained with 0.05% (w/v) calcofluor white for 1–2 min. Further, sections were observed under a microscope using UV illumination. ImageJ

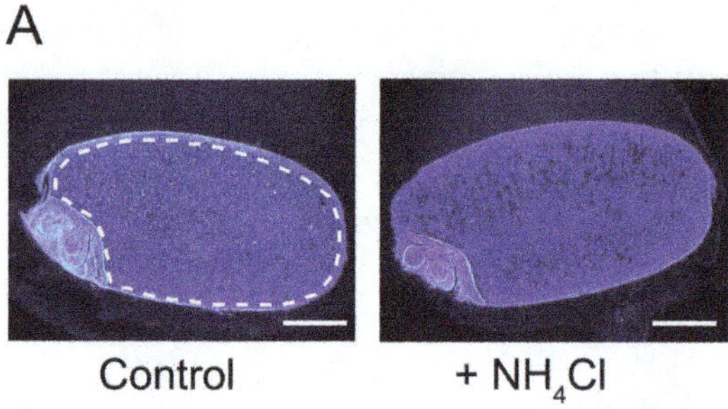


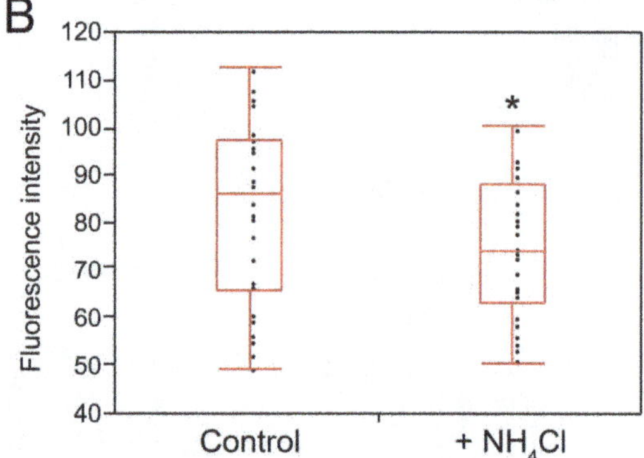


**Figure 6. β-Glucan content of rice grains.** (A) Histochemical staining of mature rice seed. Cross-sections of grains from the control plot (left) and from the N-fertilized plot (+$NH_4Cl$; right) are shown. Sections were stained for β-glucan using calcofluor white. The white-dotted line indicates the area analyzed for fluorescence intensity. The fluorescence intensity of the endosperm was calculated by adjusting the background intensity. Scale bars, 1 mm. (B) Endosperm fluorescence intensity was analyzed using ImageJ. Sections were cut from 30 randomly selected grains from each plot. The horizontal line inside the box plot indicates the median value. The inner box indicates the interquartile range and runs between $25^{th}$ and $75^{th}$ percentiles. The upper line extending from the box indicates the largest value between the $75^{th}$ percentile and the point at 1.5 times the interquartile range. The lower line extending from the box indicates the smallest point between the $25^{th}$ percentile and the point at 1.5 times the interquartile range; $^*P<0.05$.

software (http://rsbweb.nih.gov/ij/) was used to quantify fluorescence intensities. Fluorescence intensity per unit area in the part of endosperm that was surrounded by the dotted lines was measured was adjusted by subtraction of background fluorescence intensity.

## Results and Discussion

In this study, the timing of additional nitrogen fertilization is the key factor. At heading time, plant body growth ceases and number of spikelets are fixed. Therefore, the effects of additional nitrogen (ammonium) are likely to be reflected in seed maturation and storage material synthesis.

In general, enriched nitrogen fertilization elevates protein content, as determined using the Kjeldahl method [10]. In agreement, the sum of amino acid and protein content, were elevated in mature rice treated with additional fertilization in all trials (Fig. 2 and Table S1). As shown in Table S1, amino acid content was also higher in individuals from the N-fertilized plot in all trials. However, free amino acid levels remained low or undetectable in all trials (Table S1). These results indicate that nitrogen fertilization at heading time was mainly utilized for protein synthesis and not for the accumulation of free amino acids.

At the third field trial, we took some preliminary results about developing grains. Grain weight did not differ significantly between plants from the N-fertilized and control plot at every sampling time (data not shown). Although differences were not statistically confirmed, seed nitrogen content was higher in the N-fertilized plot than that in the control plot at every sampling time. It is a reasonable speculation that gene expression in developing grain also changes in response to additional nitrogen fertilization.

To investigate the genomic response to nitrogen fertilization, a plant incubator was used to control growth conditions between replicate experiments. Rice plants from the two experimental plots were morphologically similar at heading stage, although the leaf blades of N-fertilized rice clearly showed a darker green color than those of the control group (Fig. S1). Mature seeds were obtained from all but one control plant. Moreover, the average maturation rate was calculated from the number of mature seeds divided by the number of spikelets, and was found to be 88.0% in the N-fertilization plot and 92.2% in the control plot. Nitrogen content of mature polished grains was much higher in plants from the N-fertilized plot (Fig. 3A), and this difference of seed nitrogen content between two plots was greater in incubator cultivation than in field trial. In incubator cultivation, water was pooled in trays and plant sizes were reduced by removing tillers. Thus, all fertilizer remained in incubator trays, allowing for higher quantities of nitrogen to be utilized for grain development. Accordingly, higher nitrogen content of seeds from the N-fertilized plot was reflected by band density of insoluble proteins in SDS-PAGE analysis (Fig. 3B). Each density corresponding to glutelin and a 13 kDa prolamin was higher in the case of N-fertilized plot.

Dehusked 15-DAF developing seeds from rice plants were subjected to DNA microarray analysis, and data from six samples (three per experimental plot) were normalized using the DFW method and subjected to hierarchical clustering analysis. As shown in Fig. 4, a distinct separation was observed between the N-fertilized and control samples.

Furthermore, DEGs were identified between the control and N-fertilized samples using the rank products method. DEGs with FDR$<$0.05 were extracted, revealing 678 significantly up-regulated genes and 687 significantly down-regulated genes due to additional fertilization (Tables S2 and S3). The extracted DEGs were classified into functional categories according to gene ontology (GO) terms. Significantly enriched categories of DEGs (FDR$<$0.05) are shown in Fig. 5. The hierarchical GO structure indicated a more specific category with a deeper hierarchy. Therefore, most important categories appeared at the lower end of the tree (depth of hierarchy is shown in shadowed areas; Fig. 5). Furthermore, six functional categories were significantly enriched in DEGs. These included "photosynthesis, light harvesting," "trehalose biosynthetic process," "carbon fixation," "cell wall organization," "cellulose biosynthetic process," and "cellular amino acid biosynthetic process." Genes that were predicted to be involved in the synthesis of storage substances were also identified from the list of DEGs (Table 1). The "photosynthesis, light harvesting" category included genes that encode chlorophyll-binding proteins, and Rubisco was identified from the "carbon fixation" category (Table 2). Consistent with these results, the leaf blades of the N-fertilized samples were clearly darker in color than those of the control blades (Fig. S1). Because DNA microarray samples were collected from 15-DAF seeds with green pericarps, the molecules mentioned above are likely localized to the pericarp.

Cell wall-related genes such as those encoding cellulose synthase A catalytic subunit (CESA) and xyloglucan endotransglucosylase/hydrolase, were down-regulated following additional fertilization (Table 2). Genes encoding OsCESA2, OsCESA5, OsCESA6,

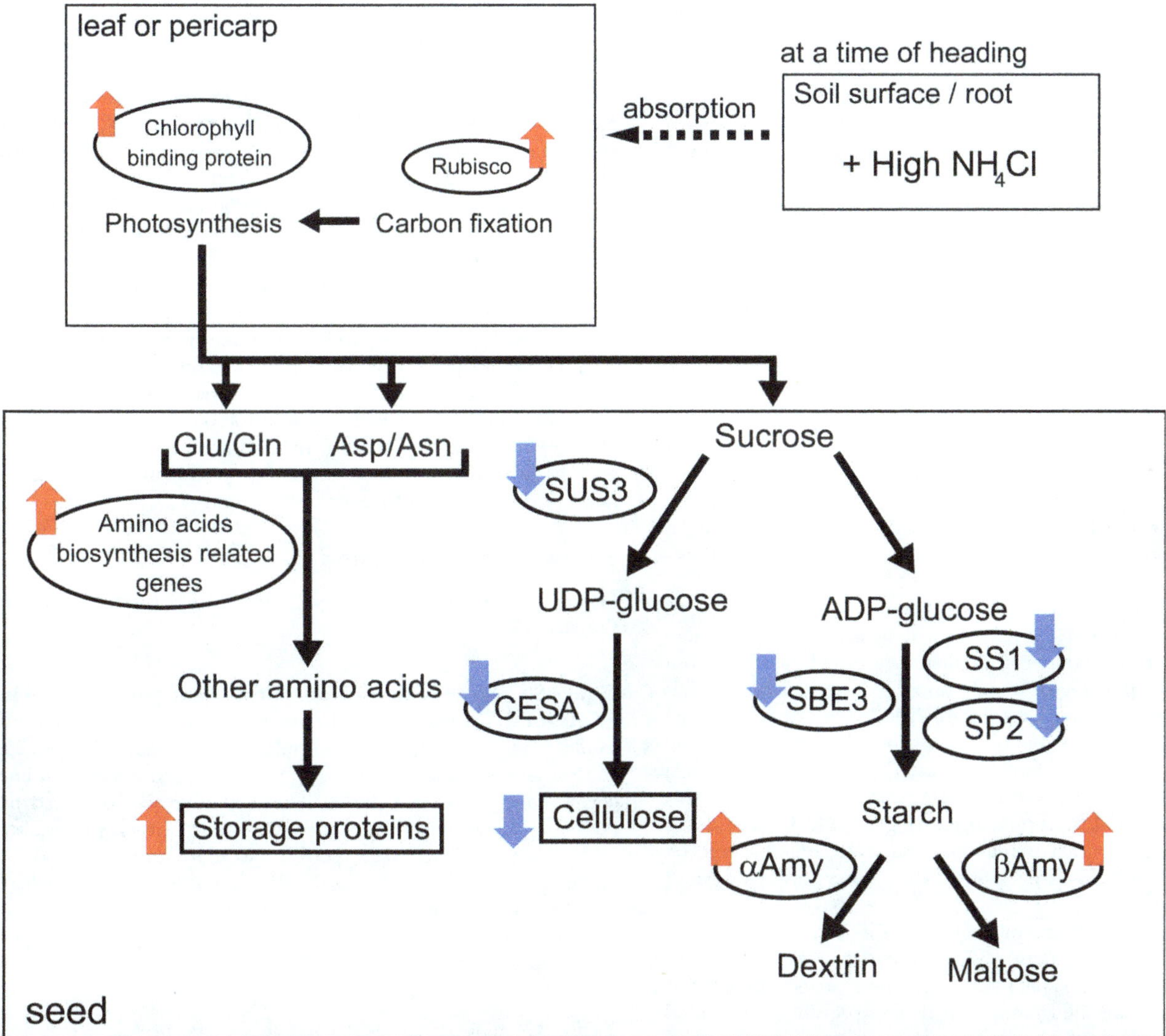


**Figure 7. Changes in gene expression and rice seed compounds with additional fertilization.** Notable change in mRNA expression and seed compounds are summarized. Molecules in the open oval are DEGs from DNA microarray experiments. Rectangles show the compound whose content was measured in this study. Red and blue arrows indicate up- and down-regulation by additional nitrogen fertilization, respectively. SUS3, sucrose synthase 3; CESA, cellulose synthase A catalytic subunit; SBE3, starch-branching enzyme 3; SS1, starch synthase 1; SP2, starch phosphorylase 2; αAmy, α-amylase; βAmy, β-amylase.

OsCESA8, CESA-like protein A5 (*OsCslA5*), and CESA-like protein F6 (*OsCslF6*) are all involved in cell wall synthesis, and were down-regulated in the presence of additional fertilization. Thus, seeds were examined to determine whether β-glucan constituents of cellulose and (1, 3; 1, 4)-β-D-glucan actually decreased. Staining with calcofluor white demonstrated weaker fluorescence intensity in seed endosperms from the N-fertilized plot than from the control plot (Fig. 6). Cell walls of rice endosperm comprise cellulose and hemicellulose [21]. Rice hemicellulose comprises arabinoxylans and (1, 3; 1, 4)-β-D-glucan, also known as mixed-linkage glucan (MLG). Although calcofluor white stains both cellulose and MLG, rice does not accumulate significant amount of MLG in its grains [21,22]. Thus the present data suggest decreased cellulose content of endosperm. In general, cellulose is synthesized from UDP-glucose by CESA, which comprises a cellulose synthase complex of six subunits [23–25]. Although at least 11 genes have been annotated as CESA in the MSU rice genome annotation project, the functions of only three genes, *OsCESA4*, *OsCESA7*, and *OsCESA9*, have been examined [26]. Sucrose synthase (SUS), which is involved in UDP-glucose metabolism, was down-regulated by additional fertilization, and plays a direct role in cell wall biosynthesis by forming a complex with CESA [27–30]. Six SUS homologs are present in the rice genome [31]. Among these, *SUS2* and *SUS3* were extracted from DEGs as genes that were down-regulated by additional fertilization (Table 1). In particular, SUS3 is reportedly localized primarily in the endosperm and in the aleurone layer [31,32], indicating that CESA2, CESA5, CESA6, and CESA8 may form a complex with SUS3 to synthesize cellulose in the endosperm. The relationship between rice quality and cellulose content remains poorly understood. However, it is accepted that rice with a high nitrogen content has less stickiness, greater hardness after cooking, less palatability, and less processing quality [33–35]. Accordingly, cell walls may become harder with increased cellulose content. However, in the present experiments, rice with high nitrogen content contained less cellulose. Thus, the complexities of relationship between cooked-rice properties and rice-seed components require further detailed study.

Other polysaccharide-related genes were identified from DEGs with FDR<0.05 (Table 1). The genes encoding starch synthase 1, starch branching enzyme 3, and starch phosphorylase 2 were down-regulated. These enzymes participate in starch synthesis from UDP-glucose. However, α-amylase and β-amylase, which further process starch into smaller sugars, were up-regulated in plants receiving additional fertilization (Table 1). Although various α-amylases are present in rice, only *RAmy3E* (Os08g0473600) was identified in this study, and is reportedly localized in seeds [36–38]. Because genes associated with starch synthesis were down-regulated concomitantly with increases in mRNA expression of starch-degrading enzymes, the starch content of rice may have decreased. This phenomenon was previously reported in rice kept under high temperature conditions during maturation [39]. Besides, starch synthesis is reportedly suppressed when plant bodies are subjected to carbon starvation, favoring monosaccharide production [36,40].

Some genes of cellular amino acid biosynthetic processes, expect for NADH-glutamate synthase 1 and arginine biosynthesis bifunctional protein, were up-regulated, (Table 2 and Fig. S2). This result indicates that amino acid metabolism was activated to meet the demands of storage protein synthesis. In fact, amino acid and storage protein content were increased in polished rice from the N-fertilized plot (Figs. 2 and 3). Also, DNA microarray analysis showed remarkable increase in mRNA expression of storage proteins, in particular that of Cys-poor prolamins. Glutelin B2 and Os02g0456150 mRNA, annotated as "11 S-plant seed storage protein family protein", were also increased. The latter gene was found to be a homologous pseudogene of glutelin C. Other storage proteins were not extracted in our analysis, suggesting that mRNA response to additional nitrogen fertilization varies between types of storage protein.

Prolamin comprises several molecular species of various sizes, including 10-kDa, Cys-poor 13-kDa, Cys-rich 13-kDa, and 16-kDa prolamins [20,41,42]. Cys-poor prolamins are accumulated at late stages of grain filling [43], and their expression is increased during suppression of glutelins, which are the most abundant storage protein in rice [44]. Both the present data and previous reports suggest that the expression of Cys-poor prolamin is controlled at the mRNA level, and effectively stores excess nitrogen during additional fertilization [43,44]. This may be because Cys-poor prolamin has the simple primary structure with the absence of cysteine residues, and is directly accumulated in the endoplasmic reticulum where protein synthesis occurs. Rice with a high protein content shows poor cooking quality, and prolamins are suggested to associate such phenomena [33–35,45]. Thus, Cys-poor prolamins may be an important gene target in rice breeding, and may reflect cooking properties of rice.

## Conclusion

The present data reveal rapid changes in C/N balance in response to rhizosphere nitrogen fertilization at heading time. Notable changes in mRNA expression and seed compounds are summarized in Figure 7. Leaf color showed a darker green and the mRNA expression of Rubisco and chlorophyll-binding protein in seed pericarp was greater in the N-fertilized plot. In general, such situation contributes to increased total photosynthesis, carbon fixation, and nitrogen absorption in the rice plant body, and can improve final yield of grains per unit area [10]. More amounts of sucrose and glutamine/glutamate and asparagine/aspartic acid may be transported into seeds in N-fertilized plot. However, compounds of seeds were not always increased in the N-fertilized plot. The accumulation of nitrogen compounds such as storage proteins increased, whereas that of polysaccharides such as cellulose decreased, correlating with changes in mRNA expression for synthetic processes of these compounds. The mRNA expression profile for starch biosynthesis and starch degradation suggested that levels of other polysaccharides such as starch may also change. The possible mechanism underlying our result is that the amount of carbon backbones may be a limiting factor in seed and it may be primarily used for nitrogen accumulation in endosperm under high nitrogen fertilization (Fig. 7). This study provides new insights into the relationship between fertilization and grain maturation and contributes to the understanding of nutrient distribution during rice production.

## Supporting Information

**Figure S1 Rice of 10 days after additional nitrogen fertilization.**

**Figure S2 Heat map of differentially expressed genes in additional nitrogen fertilization.** Each column represents results from an independent samples. C1, C2 and C3 indicate control samples, and N1, N2 and N3 N-fertilized ones. Each line corresponds to a single probe. The heat map was prepared by obtaining the Z scores from the signal value of each probe after DFW normalization. More reddish and more greenish stand for higher and lower expression levels than the mean, relatively. Significantly enriched Gene ontology (GO) terms ($P$<0.05) and storage substances related genes are shown on the right side of the heat map.

**Table S1** The amino acid content and the protein content of mature polished rice cultivated in a field.

**Table S2** The gene in which expression significantly down-regulated with the additional fertilizer (FDR 0.05>).

**Table S3** The gene in which expression significantly up-regulated with the additional fertilizer (FDR 0.05>).

## Acknowledgments

We thank Dr. Tamura (Tokyo University of Agriculture) for her helpful discussions and technical assistance.

## Author Contributions

Conceived and designed the experiments: KM MK TA. Performed the experiments: KM MK SI. Analyzed the data: KM KT TA. Contributed reagents/materials/analysis tools: KM MH. Wrote the paper: KM MK YI KA TA.

## References

1. Wang R, Guegler K, LaBrie ST, Crawford NM (2000) Genomic analysis of a nutrient response in Arabidopsis reveals diverse expression patterns and novel metabolic and potential regulatory genes induced by nitrate. Plant Cell 12: 1491–1509.
2. Wang R, Okamoto M, Xing X, Crawford NM (2003) Microarray analysis of the nitrate response in Arabidopsis roots and shoots reveals over 1,000 rapidly responding genes and new linkages to glucose, trehalose-6-phosphate, iron, and sulfate metabolism. Plant Physiol 132: 556–567.

3. Stitt M (2002) Steps towards an integrated view of nitrogen metabolism. J Exp Bot 53: 959–970.
4. Rolland F, Moore B, Sheen J (2002) Sugar Sensing and Signaling in Plants. Plant Cell 14: S185–205.
5. Price J, Laxmi A, St Martin SK, Jang J-C (2004) Global transcription profiling reveals multiple sugar signal transduction mechanisms in Arabidopsis. Plant Cell 16: 2128–2150.
6. Thum KE, Shin MJ, Palenchar PM, Kouranov A, Coruzzi GM (2004) Genome-wide investigation of light and carbon signaling interactions in Arabidopsis. Genome Biol 5: R10.
7. Foyer CH, Parry M, Noctor G (2003) Markers and signals associated with nitrogen assimilation in higher plants. J Exp Bot 54: 585–593.
8. Scheible WR, Gonzalez-Fontes A, Lauerer M, Muller-Rober B, Caboche M, et al. (1997) Nitrate Acts as a Signal to Induce Organic Acid Metabolism and Repress Starch Metabolism in Tobacco. Plant Cell 9: 783–798.
9. Lian X, Wang S, Zhang J, Feng Q, Zhang L, et al. (2006) Expression profiles of 10,422 genes at early stage of low nitrogen stress in rice assayed using a cDNA microarray. Plant Mol Biol 60: 617–631.
10. Matsushima S (1995) 3. Phisiology of high-yielding rice and its cultivation. In: Matsuo T, Kumazawa K, Ishii R, Ishihara K, Hirata H, editors. Science of the rice plant. Tokyo: Food and Agriculture Plicy Resarch Center. 753–766.
11. Taira H (1995) 1. Physical properties. In: Matsuo T, Kumazawa K, Ishii R, Ishihara K, Hirata H, editors. Science of the rice plant. Tokyo: Food and Agriculture Plicy Resarch Center. pp. 1064–1078.
12. Suzuki K, Hattori A, Tanaka S, Masumura T, Abe M, et al. (2005) High-coverage profiling analysis of genes expressed during rice seed development, using an improved amplified fragment length polymorphism technique. Funct Integr Genomics 5: 117–127.
13. Kuroda M, Kimizu M, Mikami C (2010) A simple set of plasmids for the production of transgenic plants. Biosci Biotechnol Biochem 74: 2348–2351.
14. Chen Z, McGee M, Liu Q, Scheuermann RH (2007) A distribution free summarization method for Affymetrix GeneChip arrays. Bioinformatics 23: 321–327.
15. Breitling R, Armengaud P, Amtmann A, Herzyk P (2004) Rank products: a simple, yet powerful, new method to detect differentially regulated genes in replicated microarray experiments. FEBS Lett 573: 83–92.
16. Suzuki R, Shimodaira H (2006) Pvclust: an R package for assessing the uncertainty in hierarchical clustering. Bioinformatics 22: 1540–1542.
17. Huang DW, Sherman BT, Lempicki RA (2009) Systematic and integrative analysis of large gene lists using DAVID bioinformatics resources. Nat Protoc 4: 44–57.
18. Binns D, Dimmer E, Huntley R, Barrell D, O'Donovan C, et al. (2009) QuickGO: a web-based tool for Gene Ontology searching. Bioinformatics 25: 3045–3046.
19. Hosack DA, Dennis G, Sherman BT, Lane HC, Lempicki RA (2003) Identifying biological themes within lists of genes with EASE. Genome Biol 4: R70.
20. Saito Y, Nakatsuka N, Shigemitsu T, Tanaka K, Morita S, et al. (2008) Thin frozen film method for visualization of storage proteins in mature rice grains. Biosci Biotechnol Biochem 72: 2779–2781.
21. Shibuya N (1985) Comparative Studies on Cell Wall Preparations from Rice Bran, Germ, and Endosperm. Cereal Chem 62: 252–258.
22. Burton RA, Fincher GB (2012) Current challenges in cell wall biology in the cereals and grasses. Front Plant Sci 3: 130.
23. Taylor NG, Howells RM, Huttly AK, Vickers K, Turner SR (2003) Interactions among three distinct CesA proteins essential for cellulose synthesis. Proc Natl Acad Sci USA 100: 1450–1455.
24. Brown DM, Zeef LAH, Ellis J, Goodacre R, Turner SR (2005) Identification of novel genes in Arabidopsis involved in secondary cell wall formation using expression profiling and reverse genetics. Plant Cell 17: 2281–2295.
25. Atanassov II, Pittman JK, Turner SR (2009) Elucidating the mechanisms of assembly and subunit interaction of the cellulose synthase complex of Arabidopsis secondary cell walls. J Biol Chem 284: 3833–3841.
26. Tanaka K, Murata K, Yamazaki M, Onosato K, Miyao A, et al. (2003) Three Distinct Rice Cellulose Synthase Catalytic Subunit Genes Required for Cellulose Synthesis in the Secondary Wall. Plant Physiol 133: 73–83.
27. Fujii S, Hayashi T, Mizuno K (2010) Sucrose synthase is an integral component of the cellulose synthesis machinery. Plant Cell Physiol 51: 294–301.
28. King SP, Lunn JE, Furbank RT (1997) Carbohydrate Content and Enzyme Metabolism in Developing Canola Siliques. Plant Physiol 114: 153–160.
29. Usuda H, Demura T (1999) Development of sink capacity of the "storage root" in a radish cultivar with a high ratio of "storage root" to shoot. Plant Cell Physiol 40: 369–377.
30. Amor Y, Haigler CH, Johnson S, Wainscott M, Delmer DP (1995) A membrane-associated form of sucrose synthase and its potential role in synthesis of cellulose and callose in plants. Proc Natl Acad Sci USA 92: 9353–9357.
31. Huang JW, Chen JT, Yu WP, Shyur LF, Wang a Y, et al. (1996) Complete structures of three rice sucrose synthase isogenes and differential regulation of their expressions. Biosci Biotechnol Biochem 60: 233–239.
32. Hirose T, Scofield GN, Terao T (2008) An expression analysis profile for the entire sucrose synthase gene family in rice. Plant Sci 174: 534–543.
33. Ken' ichi O, Akiharu K, Hisashi S (1993) Quality Evaluation of Rice in Japan. Japan Agric Res Q 95–101.
34. Okadome H (2005) Application of instrument-based multiple texture measurement of cooked milled-rice grains to rice quality evaluation. Japan Agric Res Q 39: 261–268.
35. Zhou Z, Robards K, Helliwell S, Blanchard C (2002) Composition and functional properties of rice. Int J Food Sci Technol 37: 849–868.
36. Yu SM, Kuo YH, Sheu G, Sheu YJ, Liu LF (1991) Metabolic derepression of alpha-amylase gene expression in suspension-cultured cells of rice. J Biol Chem 266: 21131–21137.
37. Karrer E, Rodriguez R (1992) Metabolic regulation of rice alpha-amylase and sucrose synthase genes in planta. Plant J 2: 517–523.
38. Umemura T, Perata P, Futsuhara Y, Yamaguchi J (1998) Sugar sensing and α-amylase gene repression in rice embryos. Planta 204: 420–428.
39. Yamakawa H, Hakata M (2010) Atlas of rice grain filling-related metabolism under high temperature: joint analysis of metabolome and transcriptome demonstrated inhibition of starch accumulation and induction of amino acid accumulation. Plant Cell Physiol 51: 795–809.
40. Akihiro T, Mizuno K, Fujimura T (2005) Gene expression of ADP-glucose pyrophosphorylase and starch contents in rice cultured cells are cooperatively regulated by sucrose and ABA. Plant Cell Physiol 46: 937–946.
41. Ogawa M, Kumamaru T, Satoh H, Iwata N, Omura T, et al. (1987) Purification of Protein Body-I of Rice Seed and its Polypeptide Composition. Plant Cell Physiol 28: 1517–1527.
42. Fabian C, Ju Y-H (2011) A review on rice bran protein: its properties and extraction methods. Crit Rev Food Sci Nutr 51: 816–827.
43. Saito Y, Shigemitsu T, Yamasaki R, Sasou A, Goto F, et al. (2012) Formation mechanism of the internal structure of type I protein bodies in rice endosperm: relationship between the localization of prolamin species and the expression of individual genes. Plant J 70: 1043–1055.
44. Shigemitsu T, Saito Y, Morita S (2012) Separation and Identification of Rice Prolamins by Two-Dimensional Gel Electrophoresis and Amino Acid Sequencing. Biosci Biotechnol Biochem 76: 594–597.
45. Tanaka, Y Hayashida, S. Hongo M (1975) The relationship of the feces protein particles to rice protein bodies. Agric Biol Chem 39: 515–518.

# 4

# The Effects of Manure and Nitrogen Fertilizer Applications on Soil Organic Carbon and Nitrogen in a High-Input Cropping System

**Tao Ren[1,2], Jingguo Wang[1], Qing Chen[1*], Fusuo Zhang[1], Shuchang Lu[3]**

1 College of Resources and Environmental Science, China Agricultural University, Beijing, China, 2 College of Resources and Environment, Huazhong Agricultural University, Wuhan, China, 3 Department of Agronomy, Tianjin Agricultural University, Tianjin, China

## Abstract

With the goal of improving N fertilizer management to maximize soil organic carbon (SOC) storage and minimize N losses in high-intensity cropping system, a 6-years greenhouse vegetable experiment was conducted from 2004 to 2010 in Shouguang, northern China. Treatment tested the effects of organic manure and N fertilizer on SOC, total N (TN) pool and annual apparent N losses. The results demonstrated that SOC and TN concentrations in the 0-10cm soil layer decreased significantly without organic manure and mineral N applications, primarily because of the decomposition of stable C. Increasing C inputs through wheat straw and chicken manure incorporation couldn't increase SOC pools over the 4 year duration of the experiment. In contrast to the organic manure treatment, the SOC and TN pools were not increased with the combination of organic manure and N fertilizer. However, the soil labile carbon fractions increased significantly when both chicken manure and N fertilizer were applied together. Additionally, lower optimized N fertilizer inputs did not decrease SOC and TN accumulation compared with conventional N applications. Despite the annual apparent N losses for the optimized N treatment were significantly lower than that for the conventional N treatment, the unchanged SOC over the past 6 years might limit N storage in the soil and more surplus N were lost to the environment. Consequently, optimized N fertilizer inputs according to root-zone N management did not influence the accumulation of SOC and TN in soil; but beneficial in reducing apparent N losses. N fertilizer management in a greenhouse cropping system should not only identify how to reduce N fertilizer input but should also be more attentive to improving soil fertility with better management of organic manure.

**Editor:** Xiujun Wang, University of Maryland, United States of America

**Funding:** The authors are grateful to the National Natural Science Foundation of China (No. 31071858), Innovative Research Team of Beijing Fruit Vegetable Industry, the innovative group grant of NSFC (No. 30821003) and Basic Application and Cutting-edge Technology Research Projects of Tianjin City (09JCYBJC08600). The funders had no role in study design, data collection and analysis, decision to publish, or preparation of the manuscript.

**Competing Interests:** The authors have declared that no competing interests exist.

* E-mail: qchen@cau.edu.cn

## Introduction

Soil organic matter plays a key role in soil biological and chemical processes, and changes in soil organic matter strongly influence soil N turnover because of the importance of available C for microbial immobilization [1-3]. Soils with higher organic matter contents may immobilize more N and reduce N loss to the environment. Otherwise, the depletion of available C will cause more rapid N turnover and losses [4]. In addition, changes in N availability can also alter soil C turnover [5]. There is no doubt that higher crop production in response to mineral N fertilizer application results in greater root exudates and more crop residues, thereby enhancing SOC sequestration in agricultural soils [6]. In addition increasing N fertilizer application can stabilize organic matter [7] and retard the mineralization of older soil organic matter [8]. N fertilization plays a positive role in enhancing the SOC [7], [9-12]. However, the addition of N fertilizer has also been reported to have a negative or no effect on SOC accumulation [13-17]. Changes in the decomposability of fresh plant litter and soil organic matter fractions, the stability of soil aggregates, and/or shifts in the microbial community can be used to explain the decreases in SOC attributed to N fertilizer addition [13], [15]. Therefore, achieving a better understanding of the interaction between N fertilizer and SOC in agricultural soils is essential for maximizing SOC storage and minimizing potential N losses.

Intensive vegetable production systems in northern China differ from other ecosystems in which excessive nutrients and water are applied, which far exceed the resources needed for vegetable growth [18-19]. As shown in previous studies [20-21], these practices have resulted in serious N losses to the environment. Therefore, more work was done to understand how to reduce N fertilizer input with optimal N and irrigation strategies, along with catch crops in the intensive greenhouse vegetable cropping system [22-24]. Nevertheless, a recent survey of the largest greenhouse vegetable production region in northern China showed that the soil C/N ratio in greenhouse soils was lower than that of the adjacent open field soils because of the high accumulation rate of soil N as a result of an excessive N input [25]. The low soil C/N ratio implied that the C levels were insufficient for the cropping system, which would limit N immobilization by soil microorganisms [2], [26-27] and may lead to high N losses [28]. This finding

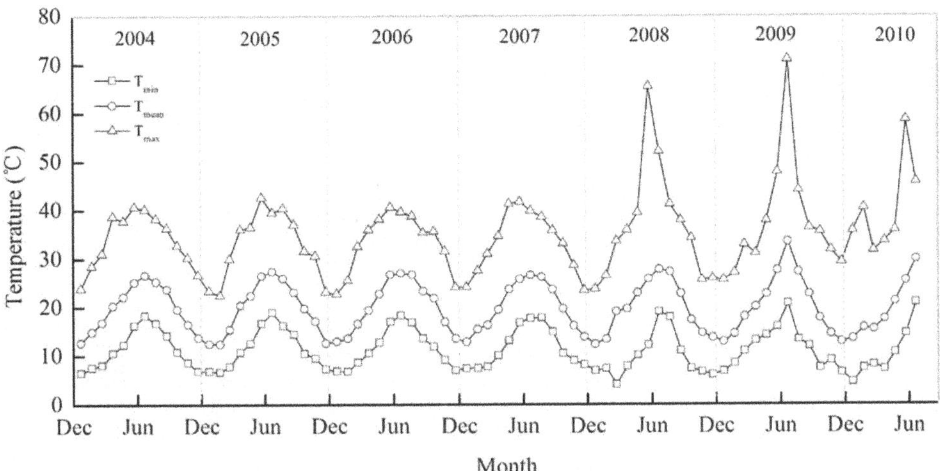


**Figure 1. The monthly mean temperature, minimum and maximum temperature inside greenhouse from 2004WS to 2009AW in the year-round greenhouse tomato planting system in Shouguang, northern China.**

shows that not only optimal N fertilizer management is needed to be studied, but also the soil organic matter content must be improved to enhance soil N retention capacity, which will reduce N losses to the environment and improve N use efficiency.

In a conventional greenhouse vegetable planting system, most of the plant residues are removed at harvest out of fear of infecting the next crop with fungi and other pathogens. Large amounts of organic manure with a low C/N ratio, such as poultry manure and pig manure, are the main soil carbon supplements in the greenhouse vegetable cropping system in northern China [29]. Excessive N fertilizer application is also a significant characteristic of this cropping system. However, it is unclear how the continuous application of poultry manure and excessive mineral N influences the soil organic matter and total N. Whether lower optimized N fertilizer inputs will alter the accumulation of soil organic matter and total N in the greenhouse field? A greenhouse tomato experiment into which different N management strategies were introduced was conducted from 2004 to 2010 in Shouguang county, the largest greenhouse vegetable production region in northern China. In contrast to common farming practice, optimal N management could reduce mineral N fertilizer input by 72% without decreasing the fruit yield [24]. In this experiment, different types of organic manure, including conventional dry chicken manure and wheat straw, were applied. This environment provided an opportunity to gain a better understanding of the influence of organic manure together with mineral N inputs on the SOC and total N pools, and determine if reducing mineral N fertilizer input will alter SOC and TN accumulation. Moreover, it provided an opportunity to analyse the influence of changes in the SOC pool on N losses in the greenhouse vegetable cropping system. Evaluating this system will be of great assistance in improving management practices for maintaining soil fertility and productivity while minimizing potential N losses from the high-input greenhouse vegetable cropping system.

## Materials and Methods

### Ethics statement

No specific permits were required for these field studies. No specific permissions were required for these locations/activities because they were not carried out on privately owned or protected areas. The field studies did not involve endangered or protected species.

### Site description and crop management

The experiment was established in February 2004 in a traditional unheated commercial solar greenhouse (84×8.5 m) in Luojia (36°55′N, 118°45′E), Shouguang, northern China. The greenhouse was constructed from a vertical clay wall and covered with polyethylene film throughout the year. Thus the air temperature inside greenhouse is higher than in outside. The monthly mean temperature, minimum and maximum temperature inside greenhouse from 2004 to 2010 was showed in Figure 1. Groundwater was used for irrigation with an average of 573 mm over 11 applications for every growing season. During construction in 1999, surface soil in the greenhouse was used to build the clay wall. The silt loam that remained in the field was approximately 60 cm lower than the open field. Chicken manure was applied at 30 t DW $ha^{-1}season^{-1}$ for the first 2 years to improve soil fertility. After 2001, chicken manure application was reduced to 15 t DW $ha^{-1}$ $season^{-1}$ until the experiment was established in 2004. According to FAO classification, the soil at the outset of the trial had 637 g sand $kg^{-1}$, 323 g silt $kg^{-1}$ and 40 g clay $kg^{-1}$ from 0 to 0.1m soil depth, 620 g sand $kg^{-1}$, 335 g silt $kg^{-1}$ and 45 g clay $kg^{-1}$ from 0.1 to 0.3 soil depth, 662 g sand $kg^{-1}$, 301 g silt $kg^{-1}$ and 37 g clay $kg^{-1}$ from 0.3 to 0.6 soil depth, respectively.

Tomato (*Lycopersicon esculentum* Mill.) has been the sole crop since the construction of the greenhouse in 1999. There were two cropping seasons per year, namely a winter-spring (WS) and an autumn-winter (AW) tomato crop. For the WS season, 4-week-old tomato seedlings were transplanted by hand into double rows in the middle of February, harvesting was completed in the middle of June. After a 2-month fallow period, the second (AW) crop was transplanted in early August and the final harvest was taken the following January. The tomato vines were removed from the greenhouse after each final harvest to reduce the infecting of disease carry over into the next crop.

### Treatments

From February 2004 the treatments included (1) **CK**, control treatment, where neither organic manure nor mineral fertilizer N was applied. The N from irrigation water was the important source of N input in the CK treatment, ranging from 25 to 177 kg

N $ha^{-1}$ $season^{-1}$ with an average of 102 kg N $ha^{-1}$ $season^{-1}$. The large variation in N input from irrigation water across different growing season was due to different amount of irrigation water and different concentration of $NO_3^-$N in irrigation water. (2) **MN**, organic manure treatment, only organic manure was broadcast as a basal fertilizer with no mineral N fertilizer applied. Organic manure was bought from different poultry farms every growing season and C and N content of chicken manure were different across different growing season. From 2004WS to 2006WS only dry chicken manure was used; and the application rates of dry chicken manure were 8, 11, 8, 11 and 5 t $ha^{-1}$ $season^{-1}$, with an average of 271 kg N $ha^{-1}$ $season^{-1}$. During the autumn-winter season in 2006 (2006AW) and onward, additional chopped wheat straw (1-5 cm long) was added to the soil together with dry chicken manure. From 2006AW to 2009AW the application rates of dry chicken manure were 8, 8, 8, 8, 8, 10 and 8 t $ha^{-1}$ $season^{-1}$; and the rates of wheat straw were 2, 2.5, 4, 2, 4, 4 and 4 t $ha^{-1}$ $season^{-1}$, with an average of 22 kg N $ha^{-1}$ $season^{-1}$. The treatment was also labeled as "MN+S" instead of "MN". (3) **CN**, conventional N treatment, organic manure was applied as in the MN treatment (except in 2006AW). N fertilizer was applied as a side-dressing at a rate of 120 kg N $ha^{-1}$ on 4-6 occasions based on local farmers normal management practice which depended on the weather conditions, tomato cultivar and growth stage. The average mineral N fertilizer input was 635 kg N $ha^{-1}$ $season^{-1}$. From 2006AW, the CN treatment plot was split into CN and CN+S sub-treatments with plot sizes of 21.8 $m^2$ and 32.8 $m^2$, respectively. For the CN sub plot only dry chicken manure was applied; and for the CN+S treatment both wheat straw and dry chicken manure were incorporated. The chicken manure and wheat straw were applied at the same rates and timings as in the MN treatment. The mineral N fertilizer application was the same as for the original CN plot. (4) **RN**, reduced N treatment, chicken manure was applied at the same rate as in the MN and CN treatment, and mineral N fertilizer was applied as a side-dressing based on an N target value and soil mineral N content in the root zone (0-30 cm soil layer) for different growth stages from 2004WS to 2007WS. The equation used was as follows:

$$\textbf{Recommended fertilizer N} = \textbf{N target value} - \textbf{NO}_3^- - \textbf{N in the top 0.3m of the soil profile before the recommendation} - \textbf{NO}_3^- - \textbf{N from irrigation water} \quad (1)$$

N target value is calculated from crop N uptake, the necessary soil $N_{min}$ residue and soil net mineralization [30], which reflects the synchronization of crop N requirement and soil N supply. Here the initial N target values were 300 kg N $ha^{-1}$ for the side-dressing at each stage of fruit cluster development in 2004. From 2005 the target values from transplanting to the third cluster growth stage were changed to 250 and 200 kg N $ha^{-1}$ for the fourth cluster to the end of harvest in the WS seasons. In the AW season the N target value from transplanting to the fourth cluster growth stage was changed to 200 kg N $ha^{-1}$, with 250 kg N $ha^{-1}$ for the fifth and sixth cluster growth stages [24]. Considering crop N uptake in different growth period and soil N supply, N from irrigation water, only 2-4 side-dressing events were applied according to the differences between N target values and soil $N_{min}$ content in the root zone. From 2007AW the N recommendation was simplified based on the experiences of preceding years. Three or four side-dressing events with an interval of 7–10 days at a rate of 50 kg N $ha^{-1}$ were required in April and October. The more detail introduction of optimized N management was reported by Ren et al [24]. Compared with the CN treatment, 71.3% of mineral N fertilizer was reduced without influencing fruit yield, with an average of 182 kg N $ha^{-1}$ $season^{-1}$. From 2006AW, the RN treatment plot was split into RN and RN+S sub-treatments with the same straw amendments and plot sizes as in the CN treatment. During 2006AW and 2007WS growing season, the mineral N fertilizer application rates of RN and RN+S treatment were determined based on the N target values and soil $N_{min}$ content in the root zone before side-dressing. There were litter differences on mineral N fertilizer application rates between RN and RN+S treatment. Since 2007AW, the mineral N fertilizer application was the same for RN and RN+S treatment.

All treatments were set up in a randomized block design with three replicates. Plots were separated for each other by plastic film at a depth of 30 cm. The exogenous C and N inputs are shown in Table 1. The average C input from chicken manure was 2779 kg C $ha^{-1}$ $season^{-1}$, with a range of 1114 to 5445 kg C $ha^{-1}$ $season^{-1}$. Beginning in 2006AW, an average of 1119 kg wheat straw C $ha^{-1}$ $season^{-1}$ was supplied, and the total C input was as high as 3483 kg C $ha^{-1}$ $season^{-1}$. The N sources included mineral N fertilizer, organic manure and irrigation water. In the CK treatment, the N from irrigation water was the only source of N input, with an average of 102 kg N $ha^{-1}$ $season^{-1}$. The average exogenous N inputs for the MN, RN and CN treatments were 354, 527 and 984 kg N $ha^{-1}$ $season^{-1}$, respectively.

Urea was the main mineral N fertilizer. All plots received $P_2O_5$ as calcium monophosphate (12% $P_2O_5$) and $K_2O$ as potassium sulfate (50% $K_2O$) during each growing season. The average input during the past 12 growing seasons were 350 kg $P_2O_5$ $ha^{-1}$ and 563 kg $K_2O$ $ha^{-1}$ per season.

## Soil sampling and analysis

**Soil organic carbon and total N concentrations.** Three soil cores (3.5 cm in diameter) were taken from each plot to a depth of 0.6 m and subdivided into 0-0.1, 0.1-0.3 and 0.3-0.6 m increments on April 18, 2004, and January 20, 2010. Fresh soil cores were taken to the lab immediately, mixed thoroughly to provide a composite sample from each plot, sieved through a 2 mm mesh and then air-dried and stored in plastic bottles. After removing any carbonates with 1 M HCl, the SOC and total N concentrations were determined using a C and N analyzer (vario MACRO CN, Elementar, Germany).

At the same time, soil bulk density in different soil layers was measured by the cutting ring method. In 2004, three samples were collected from the whole greenhouse and the average soil bulk densities were 1420 kg $m^{-3}$, 1450 kg $m^{-3}$ and 1480 kg $m^{-3}$ for the 0-10 cm, 10-30 cm and 30-60 cm soil layers, respectively. The bulk density of each plot monitored in 2010 is shown in Table 2 and no significant differences were observed among the different treatments.

**Soil organic carbon fraction.** Changes in the quantity and quality of the SOM pool are generally difficult to detect in the short term following agricultural management. Labile and recalcitrant SOM separated by different methods provide a more sensitive indicator for evaluating the effect of different management strategies on SOM dynamics [31-33]. Here, soil organic carbon fractionation procedures were carried out as described by Blair et al [31]. Air-dried soil samples containing approximately 15 mg of C were oxidized with 25 mL of 333 mmol $L^{-1}$ $KMnO_4$ for 1 h at 25°C on a shaker at 180 rpm. The samples were then

**Table 1.** Exogenous C and N inputs in the greenhouse tomato production system in Shouguang, northern China (kg $ha^{-1}$ $season^{-1}$).

| Growing season | C[1] | | | | | | N[1] | | | | | |
|---|---|---|---|---|---|---|---|---|---|---|---|---|
| | CK | MN(MN+S)[3] | RN | RN+S | CN | CN+S | CK | MN(MN+S) | RN | RN+S | CN | CN+S |
| 2004WS[2] | 0 | 2860 | 2860 | - | 2860 | - | 56 | 316 | 644 | - | 1186 | - |
| 2004AW[2] | 0 | 5445 | 5445 | - | 5445 | - | 118 | 478 | 638 | - | 1198 | - |
| 2005WS | 0 | 3476 | 3476 | - | 3476 | - | 54 | 370 | 497 | - | 1000 | - |
| 2005AW | 0 | 3902 | 3902 | - | 3902 | - | 177 | 435 | 636 | - | 1155 | - |
| 2006WS | 0 | 1114 | 1114 | - | 1114 | - | 165 | 327 | 465 | - | 927 | - |
| 2006AW | 0 | 4095 | 3423 | 4095 | 5135 | 5807 | 133 | 456 | 651 | 671 | 1200 | 1212 |
| 2007WS | 0 | 2258 | 1698 | 2258 | 1698 | 2258 | 144 | 308 | 509 | 486 | 838 | 848 |
| 2007AW | 0 | 4806 | 3406 | 4806 | 3406 | 4806 | 82 | 417 | 492 | 517 | 872 | 897 |
| 2008WS | 0 | 2110 | 1606 | 2110 | 1606 | 2110 | 25 | 180 | 321 | 330 | 771 | 780 |
| 2008AW | 0 | 3052 | 1606 | 3052 | 1606 | 3052 | 39 | 202 | 382 | 402 | 782 | 802 |
| 2009WS | 0 | 4205 | 2592 | 4205 | 2592 | 4205 | 135 | 441 | 605 | 641 | 1105 | 1141 |
| 2009AW | 0 | 3857 | 2214 | 3857 | 2214 | 3857 | 99 | 322 | 489 | 523 | 769 | 802 |
| Average | 0 | 3432 | 2779 | 3483 | 2921 | 3728 | 102 | 354 | 527 | 510 | 984 | 926 |

[1]C input from chicken manure and wheat straw; N input from mineral fertilizer, organic manure and irrigation water;
[2]WS: winter-spring growing season, AW: autumn-winter growing season;
[3]Since 2006AW chopped wheat straw and dry chicken manure was broadcast as a basal fertilizer and the treatment was labeled as "MN+S" instead of "MN";

**Table 2.** Soil bulk density in the soil profile in Jan. 2010 in a year-round greenhouse tomato planting system in Shouguang, northern China (kg $m^{-3}$).

| Treatment | Soil layer (cm) | | |
|---|---|---|---|
| | 0-10 | 10-30 | 30-60 |
| CK | 1333±41 | 1545±70 | 1589±63 |
| MN(MN+S)[1] | 1296±118 | 1548±79 | 1586±56 |
| RN | 1301±82 | 1502±30 | 1460±42 |
| RN+S | 1285±35 | 1508±126 | 1546±87 |
| CN | 1227±213 | 1584±129 | 1512±127 |
| CN+S | 1283±43 | 1493±114 | 1542±132 |
| Average | 1288 | 1530 | 1539 |

[1]Since 2006AW chopped wheat straw and dry chicken manure was broadcast as a basal fertilizer and the treatment was labeled as "MN+S" instead of "MN";

centrifuged, diluted and spectrophotometrically measured at 565 nm. The oxidized carbon was considered labile C, and the remainder represented the non-labile C.

**Apparent N losses.** The nutrient balance is often used to estimate potential environmental risks [34-35]. The apparent N losses were calculated according to Equation (2), as described by Ren et al [24], as follows:

$$N_{loss} = N_{min\ initial} + N_{manure} + N_{fert} + N_{irri} - N_{crop} - N_{min\ harvest} \quad (2)$$

where $N_{loss}$ = apparent N loss, $N_{min\ initial}$ = soil $N_{min}$ content at 0-0.6 m before transplanting, $N_{manure}$ = total N input from organic manure, $N_{fert}$ = N from mineral fertilizer, $N_{irri}$ = $NO_3^-$-N from irrigation water, $N_{crop}$ = total N uptake by tomato aboveground parts, and $N_{min\ harvest}$ = soil $N_{min}$ content at 0-0.6 m at the end of the harvest.

The parameters were calculated as follows:

Three soil cores (3.5 cm in diameter) were collected from each plot at a depth of 0.6 m and then subdivided into 0-0.3 m and 0.3-0.6 m increments before transplanting and at the end of the harvest for each season. Fresh soil cores were mixed thoroughly to give a composite sample from each plot and then passed through a 2 mm sieve. Next, 12 g subsamples were weighed and extracted by shaking with 100 mL of 1 mol $L^{-1}$ KCl for 1 h. The extract was stored at -18°C until an analysis of the $NO_3^-$-N and $NH_4^+$-N concentrations could be carried out with a continuous flow analyzer (TRAACS Model 2000). The water content of the soil samples was also gravimetrically determined to calculate the soil $N_{min}$ ($NO_3^-$-N+$NH_4^+$-N) content on a dry matter basis. The soil bulk density was used to convert the mineral N in mg per kg of soil to kg per hectare.

The input of N from irrigation water was determined by recording the amounts applied, and water samples were collected during each irrigation event over the entire growing season. The samples were stored frozen until $NO_3^-$-N and $NH_4^+$-N analysis.

Plant samples were collected from each plot at the end of the harvest; divided into leaves, fruit, stems and roots; and weighed before and after drying at 70°C for 48 h. The dried shoots were ground before determining the total N, which was conducted using a modified Kjeldahl method with salicylic acid. N uptake was calculated as the product of dry matter and total N concentration in different parts.

**Data analysis.** Analysis of variance (*ANOVA*) was used to determine the significance of treatment effects based on a randomized complete block design. Multiple comparisons of mean values were performed using either Duncan's multiple range tests or Fisher's protected least significant difference (*LSD*) test at the 0.05 level of probability. Statistical analysis was performed using version 6.12 of the SAS software package (SAS Institute Inc., Cary, NC).

## Results

### Soil organic carbon and total N concentrations

Soil organic carbon (SOC) and total soil N (TN) concentrations in soil profiles to a depth of 60 cm are shown in Figure 2. The highest SOC concentration was seen in the 0-10 cm layer, below which it decreased for all treatments. In April of 2004, the SOC of the soil profile was similar across all N treatments, except there were some variations in the 30-60 cm layer associating with uneven soil fertility. After 6 years, SOC in the CK treatment was significantly lower than that of the other treatments in the 0-10 cm soil layer. However, there were no significant differences among the other treatments. In the 10-30 cm layer, SOC differed significantly between the RN and CN treatments. In comparison to the initial value, SOC decreased significantly in all CK treatment layers after 6 years of cultivation, but the only significant change in SOC for the CN treatment was an increase in the 0-10 cm layer.

Patterns in total N in the soil profile followed a similar pattern to that of the SOC (Figure 1). In January 2010, soil TN in the CK treatment was significantly lower than that of the other treatments in the 0-10 cm layer. In addition, no significant differences were observed among the MN, RN and CN treatments. The TN values did not show significant differences across N treatments in the 10-60 cm layer in 2010.

### Soil organic carbon and total N pool

Total SOC and TN pool in the profile above 60 cm were 46.6-55.1 t C $ha^{-1}$ and 6.8-8.3 t N $ha^{-1}$, respectively (Table 3). After 6 years of cultivation, approximately 13.1 t C $ha^{-1}$ and 2.2 t C $ha^{-1}$ were lost from the CK and MN treatments, respectively. The decreased bulk density was mainly attributed to the decreased SOC in the MN treatment (Table 2). The SOC pool increased to 0.8 t C $ha^{-1}$ and 1.7 t C $ha^{-1}$ in response to the RN and CN treatments, respectively. For the N pool, the only reduction occurred in the CK treatment. Approximately 0.63 t N $ha^{-1}$ was lost over the last 6 years. Average accumulation rates of SOC and TN in the 0-60 cm layer were -2.17, -0.37, 0.14 and 0.28 t C $ha^{-1}$ $a^{-1}$ and -0.10, 0.14, 0.00 and 0.06 t N $ha^{-1}$ $a^{-1}$ for the CK, MN,

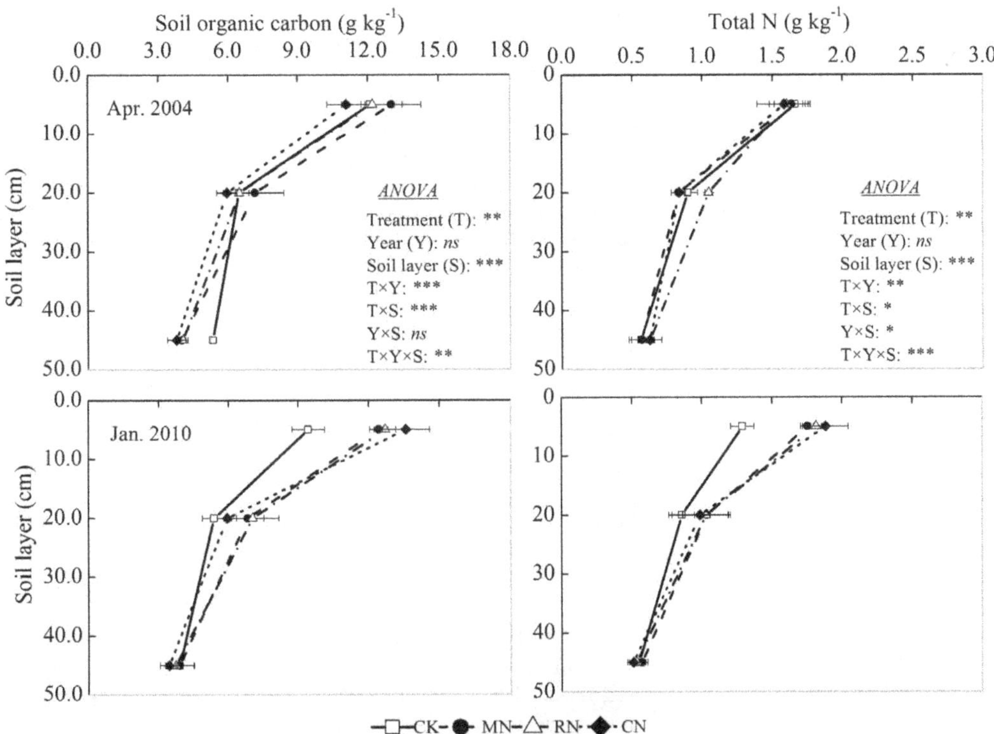


**Figure 2. The distribution of soil organic carbon and total N concentrations in the soil profile with different N treatments in the year-round greenhouse tomato planting system in Shouguang, northern China.** Note: *, ** and *** indicate significant differences at $P<0.05$, $P<0.01$ and $P<0.001$, ns denotes no significant difference.

RN and CN treatments, respectively. The addition of straw over 4 years made little difference to SOC and TN concentration

## Soil organic carbon fractions

Figure 3 shows the distribution of labile and non-labile carbon in different soil layers. In April 2004, the labile C concentrations were similar across all N treatments in the same layer. However, in January 2010, the soil labile C concentration in the CK treatment was significantly lower than that of the other treatments in the 0-30 cm layer. Compared with the initial values in 2004, the soil labile C concentration in the CK treatment did not decrease significantly according to paired-sample T test; however, labile C increased significantly in the RN and CN treatments for the 0-10 cm layer.

Changes in the concentration of the non-labile C fraction in the soil profile were similar to those of the SOC. In January 2010, the soil non-labile C concentration in the 0-10 cm layer increased as the N application increased. When compared with April 2004, it decreased significantly in the CK treatment. Nevertheless, it increased significantly in the CN treatment. For all other treatments, no significant changes were observed in the soil profile.

## Apparent N losses

Figure 4 shows the annual apparent N losses from 2004 to 2009 as obtained estimated from the nitrogen balance. For the CN treatment, the average annual mean apparent N losses were as high as 1529 kg N ha$^{-1}$ a$^{-1}$, accounting for 77% of the annual exogenous N input. A significant decrease in the annual apparent N losses of 633 kg N ha$^{-1}$ a$^{-1}$ occurred in the RN treatment because the rate of N fertilization had been reduced to less than one-third of that in the CN treatment. For the CK treatment, soil N and nitrate in the irrigation water were the major sources of N, which were less input than N removed by plant uptake, resulting in a negative N balance. These changes could indirectly explain the decreased TN in the CK treatment. Although the straw treatments led to the addition of C, the apparent N losses from treatments with and without straw amendments did not differ significantly. The N surpluses were 490 and 1285 kg N ha$^{-1}$ a$^{-1}$ N for RN+S and CN+S, respectively.

## Interactions between C and N in soil

Figure 5 shows the relationship between the changes in SOC and TN concentration after 6 years of cultivation, as well as annual apparent N losses and average N inputs. The SOC and TN concentrations increased in response to N addition but showed a negative response to excessive N application (Figure 5a, b). The total N in the soil did not increase linearly with the increase in applied mineral N, which might be explained by changes in the SOC and its fractions (Figure 5c). The minor changes in SOC in response to the current type and amount of organic manure application, limited N storage in soil and N was close to saturation. With the increase in fertilizer N application, the apparent N losses linearly increased and conventional N fertilizer management caused the highest apparent N losses (Figure 5d).

# Discussion

## Effect of N fertilizer on SOC

Greater root exudates and more crop residues in response to mineral N fertilizer application were the dominant reasons why N fertilizer application improved the SOC [6]. In this experiment, although 71.3% of mineral N fertilizer was cut down in the RN treatment compared with the CN treatment, there were no

**Table 3.** Distribution of the soil organic C and N pools in the soil profile of a greenhouse tomato production system in Shouguang, northern China.

| | | SOC pool (t ha$^{-1}$) | | | | | | N pool (t ha$^{-1}$) | | | | | |
|---|---|---|---|---|---|---|---|---|---|---|---|---|---|
| | | CK | MN (MN+S)[2] | RN | RN+S | CN | CN+S | CK | MN (MN+S)[2] | RN | RN+S | CN | CN+S |
| Apr-04 | 0-10 cm | 17.2±0.2 | 18.5±1.8 | 17.3±1.8 | - | 15.7±1.2 | - | 2.36±0.11 | 2.32±0.17 | 2.27±0.17 | - | 2.25±0.27 | - |
| | 10-30 cm | 18.8±1.3 | 20.8±3.7 | 18.8±2.0 | - | 17.3±1.3 | - | 2.61±0.21 | 2.41±0.15 | 3.04±0.07 | - | 2.43±0.05 | - |
| | 30-60 cm | 23.7±0.6 | 17.8±0.5 | 18.1±0.9 | - | 16.8±1.8 | - | 2.50±0.39 | 2.52±0.33 | 2.79±0.37 | - | 2.78±0.14 | - |
| Jan-10 | 0-10 cm | 12.2±0.9 | 16.0±0.5 | 16.4±0.6 | 17.9±0.9 | 17.5±1.3 | 17.6±1.0 | 1.66±0.11 | 2.26±0.06 | 2.34±0.05 | 2.42±0.09 | 2.4±0.21 | 2.54±0.12 |
| | 10-30 cm | 16.5±1.5 | 20.9±2.2 | 21.6±3.5 | 17.2±1.2 | 18.2±1.3 | 18.6±2.1 | 2.63±0.28 | 3.17±0.52 | 3.15±0.50 | 2.70±0.23 | 3.02±0.61 | 2.95±0.18 |
| | 30-60 cm | 18.0±2.8 | 18.0±2.9 | 17.1±0.1 | 17.2±1.5 | 15.8±1.8 | 18.7±0.7 | 2.55±0.22 | 2.64±0.20 | 2.62±0.14 | 2.68±0.06 | 2.35±0.19 | 2.86±0.41 |
| Δ(0-60 cm)[1] | | -13.1±4.5 | -2.2±1.0 | 0.8±6.1 | - | 1.7±3.6 | - | -0.63±0.50 | 0.82±0.27 | 0.01±0.74 | - | 0.34±0.47 | - |
| Accumulation rate (t ha$^{-1}$ a$^{-1}$) | | -2.17 | -0.37 | 0.14 | - | 0.28 | - | -0.10 | 0.14 | 0.00 | - | 0.06 | - |

[1]Δ(0-60 cm) = SOC (N) pool$_{2004}$- SOC (N) pool$_{2010}$;
[2]Since 2006AW chopped wheat straw and dry chicken manure was broadcast as a basal fertilizer and the treatment was labeled as "MN+S" instead of "MN";

significant differences on fruit yields and plant biomass between the CN and RN treatment [24]; root exudates were presumed to be similar between these two treatments. Besides, in our greenhouse system crop residues were removed from the greenhouse at harvest because of the risks of disease carryover, so organic manure was the major C supplement. The C input from organic manure was the same for the CN and RN treatments in the same growing season (Table 1). Moreover, no significant changes in the root C/N ratio (data not shown), soil microbial community [36] and soil organic matter fractions (Figure 3) between the RN and CN treatments were found. Thus, no significant differences in SOC pool between the CN and the RN treatment were observed in this experiment, which was similar to other work [14], [17]. Apparently, N was not limiting factor in this cropping system and reduced mineral N input would not alter SOC and TN accumulation. All these findings indicate that when organic manure is used, optimizing N fertilizer input over a continuous 6-year period would not affect the SOC or TN contents in the greenhouse vegetable cropping system; but it was helpful in reducing the environmental risks without influencing fruit yields (Figure 4).

## Effect of organic manure on SOC

Organic manure application brought lots of N, with the average of 252 kg N ha$^{-1}$. In contrast to the CK treatment, significant increments on fruit yields were achieved; yet there were no differences on fruit yields and plant biomass between MN+S treatment and RN, CN treatment in most growing seasons [24], demonstrating that N from organic manure was important N source for crop growth and excessive mineral N fertilizer application was wasteful without considering N from organic manure. In addition, Organic manure application is considered to be a consistent method for maintaining soil fertility over the long-term [10], [37]. In this unique production system, the major source of organic carbon is organic manure. If the input of organic manure is excluded, only 44-146 kg C ha$^{-1}$ season$^{-1}$ from residual roots is incorporated [38]. If no organic manure is applied, SOC concentration, especially for stable organic carbon, will decline significantly (Figure 2, 3). These results are similar to those of another long-term greenhouse tomato experiment [39].

Before the start of this experiment, about 210 t ha$^{-1}$ chicken manure had been applied across the whole greenhouse during 1999-2003; and the high organic manure application might build higher soil organic carbon pool in a short time. Therefore, in contrast to the treatments with organic manure application, there was the lack of a big difference on SOC in the CK treatment for the next 6 years. As well, no observable changes in soil organic C was found over 6 years of successive chicken manure applications. Whether it implied the soil organic carbon pool in our study was saturated? Although the coarse-textured soil has lower capacity for C and N stabilization, the saturated soil organic carbon pool in 0-30cm soil layer could be high to 75.6-96.8 t ha$^{-1}$ according to Hassink [40] and Six's C-saturation model [41]. These values were higher than it reported in our study, indicating that soil organic carbon could be improved with optimum management. Quantity and quality of input organic manure significantly influenced soil organic C dynamics. In contrast to farmers' normal manure application, the application rate of chicken manure in the experiment was lower, ranging from 8 t ha$^{-1}$ season$^{-1}$ to 12 t ha$^{-1}$ season$^{-1}$. Whether SOC content would be enhanced with the organic manure application rate increase? Indeed, high rates of organic manure application were conducive to enhanced SOC and SOC fractions [42-43]. Ge et al [39] demonstrated that it would take 10-15y with 75 t ha$^{-1}$ a$^{-1}$ of horse manure to increase

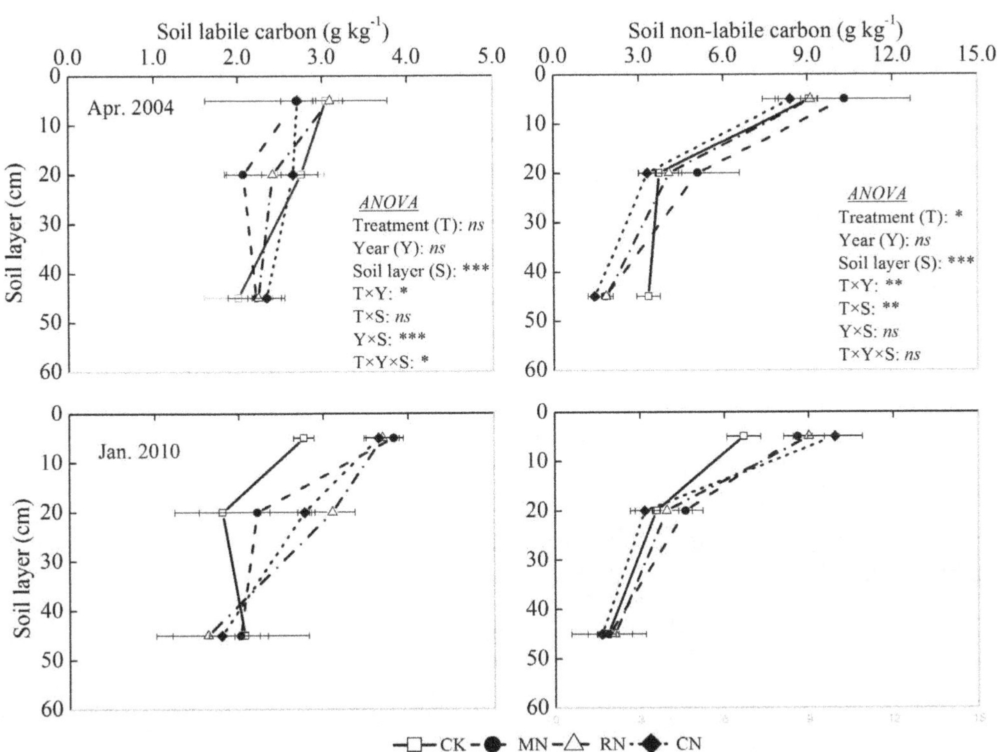


**Figure 3. The distribution of the soil labile and non-labile carbon concentrations in the soil profile with different N treatments in the year-round greenhouse tomato planting system in Shouguang, northern China.** Note: Note: *, ** and *** indicate significant differences at $P<0.05$, $P<0.01$ and $P<0.001$, ns denotes no significant difference.

the soil organic matter content from 24 g $kg^{-1}$ to 30-40 g $kg^{-1}$ in the greenhouse vegetable soil. However, environmental pressures associating with excessive application of organic manure were serious [44]; and in Europe the applications of organic manure are restricted to not exceed 170 kg N $ha^{-1}$ $y^{-1}$ by the legislation. In our experiment N input from organic manure averaged 252 kg N $ha^{-1}$ $season^{-1}$, with 335 kg N $ha^{-1}$ $y^{-1}$ apparent N loss. Obviously it will not be an effective way to enhance SOC relying on the increments of organic manure application rate in greenhouse vegetable cropping system. Changing the type of organic manure input might be an important way to heighten SOC. For chicken and pig manure with high proportions of water-soluble C and easily biodegradable organic compounds, approximately 45-62% of C is evolved as $CO_2$-C within 30 days [45-46]. Plaza et al [47] reported a significant decrease in the total organic C in soils amended with pig manure slurry. However, manure with a greater ratio of C/N, or high content of recalcitrant C could reduce mineralization of bio-labile compounds, thereby enhancing soil organic matter [48-49]. Long term field experiments showed that the benefits in SOC content were higher from application of rice straw and compost than that from pig manure [42], [50].

To improve soil organic carbon content, wheat straw was added from in 2006AW. However, no significant increase in the SOC or the SOC fraction was observed after 4 years of cultivation. This result was similar to that of Antil's study [51], which found that the SOC in bulk soil decreased or was not affected by a slurry + straw treatment in both fallow and cropped plots, even after 28 and 38 yrs. According to the mechanisms of real and apparent priming effects [52], it is assumed that when chicken manure and wheat straw are applied together, microorganisms may first use the C from chicken manure to activate the microbial community; however, when the easily decomposed organic carbon is consumed, activated microorganisms will use the wheat straw carbon. Most of added straw carbon is then utilized by microorganisms, perhaps explaining why there was no effect on SOC when chicken manure and wheat straw were incorporated together. The short duration might be another important reason why no differences were seen. Additional long-term studies should be conducted to determine if the application of a mixture of wheat straw and chicken manure is an effective method of enhancing soil organic carbon in the greenhouse vegetable cropping systems over the long run. Overall, developing an optimum organic manure management system to enhance soil fertility is now one of the important issues in greenhouse vegetable cropping systems in China. In comparison to increasing the application rate, shifting the type of organic manure from pig and chicken manure to manure with a wider ratio of C/N or high content of recalcitrant C may be more practical for enhancing soil organic matter in greenhouse vegetable cropping system.

## Effect of soil organic carbon content on the fate of N

Similar to the results from 2004 to 2007 [24], approximately 77% of the exogenous N input was surplus in the CN treatment. With the exception of immobilized N in soil clays, N leaching [53] and $N_2O$ emissions [21] were the major N loss processes in these vegetable cropping systems. Furthermore, warm and moist conditions with sufficient available nitrate and labile carbon can lead to denitrification loss, which might also be an important N loss process [54]. In any case, an excessive N surplus would lead to high potentially N losses. Therefore, most studies on the prevention of N losses have focused on reducing the N input and on irrigation strategies [24], [55] and catch crops [56]. In the

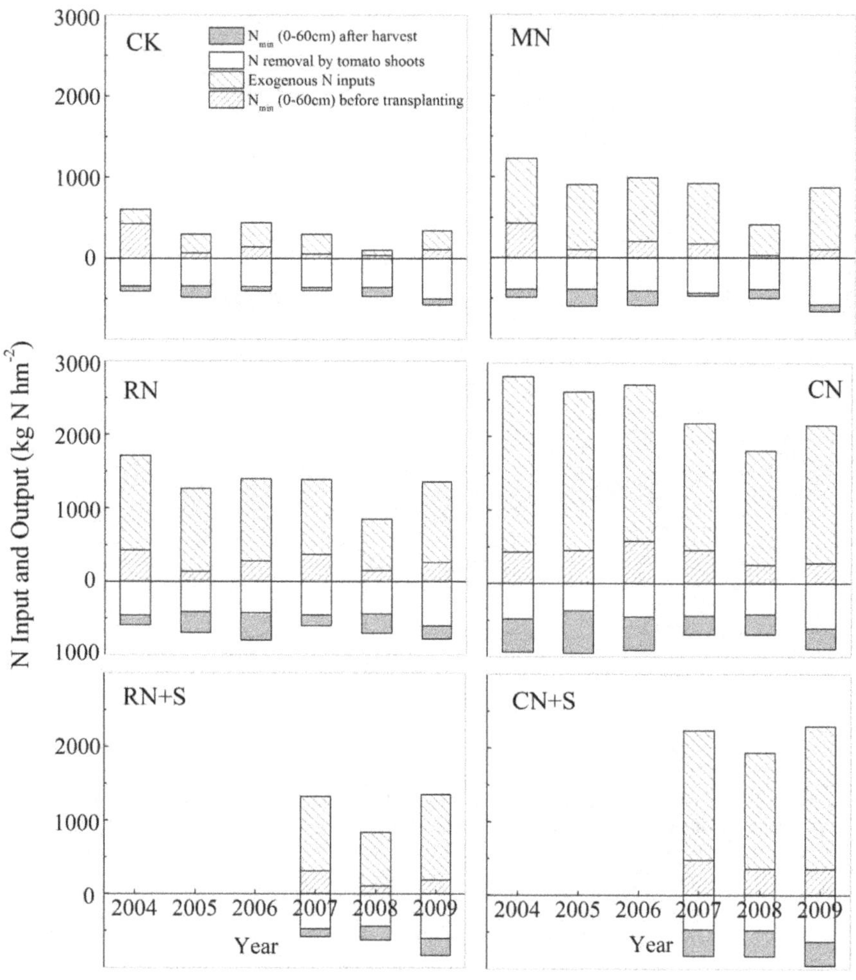


**Figure 4. Apparent N loss with different treatments in the year-round greenhouse tomato planting system from 2004 to 2009 in Shouguang, northern China.**

experiment, the rate of N fertilization in the RN treatment had been reduced to less than one-third of that in the CN treatment and apparent N losses were also decreased. Other than that, SOC plays an important role in regulating soil N turnover and the improvement of SOC is beneficial to increase potential rates of N immobilization and reduce N losses [57]. Yang et al [58] showed that the rate of absolute N change increased linearly with changes in the size of the C pool change and organic N capital was determined by long-term carbon sequestration. A similar tendency was observed in the present study (Figure 3c). In our experiment, SOC concentration was not increased even though high amounts of organic manure were applied during a 6 year period; Changes in organic N were similarly limited. Moreover, the mineralization rate of total organic N might be greater than the retention of exogenous N to soil organic N pool because of the great amount of organic manure applied within several years before the experiment started. Thus more exogenous N was lost to the environment. The low soil C/N ratio indirectly revealed that there was insufficient C in the cropping system would limiting N immobilization by soil microorganisms [2], [26] and leading to high N losses [28]. Therefore, improving soil organic matter content and enhancing potential rates of soil N immobilization according to optimal organic manure management was crucial, as well reducing N fertilizer input, to lower N losses in greenhouse vegetable cropping system.

## Conclusion

Organic manure represents a major organic C source in conventional greenhouse vegetable cropping systems in China. Without additions of organic manure, SOC, particularly stable C is likely to decline. However, no significant increment in SOC was observed with the addition of high amounts of low C/N ratio organic manure or plant residues. Shifting the type of manure from chicken manure to manures with a wider ratio of C/N, or high content of recalcitrant may be more effective in enhancing soil fertility in greenhouse vegetable production. On the basis of organic manure application, optimized N fertilizer inputs according to root-zone N management did not influence the accumulation of SOC and TN in soil; but beneficial in reducing apparent N losses.

The SOC concentration was a dominant limiting factor for soil total N enhancement. Given the current type and quantity of organic manure application, the SOC concentration was unchanged and most applied N was in excess and was lost to the environment. Therefore, integrating nutrient management, including optimized N fertilizer input, as well as enhanced the soil

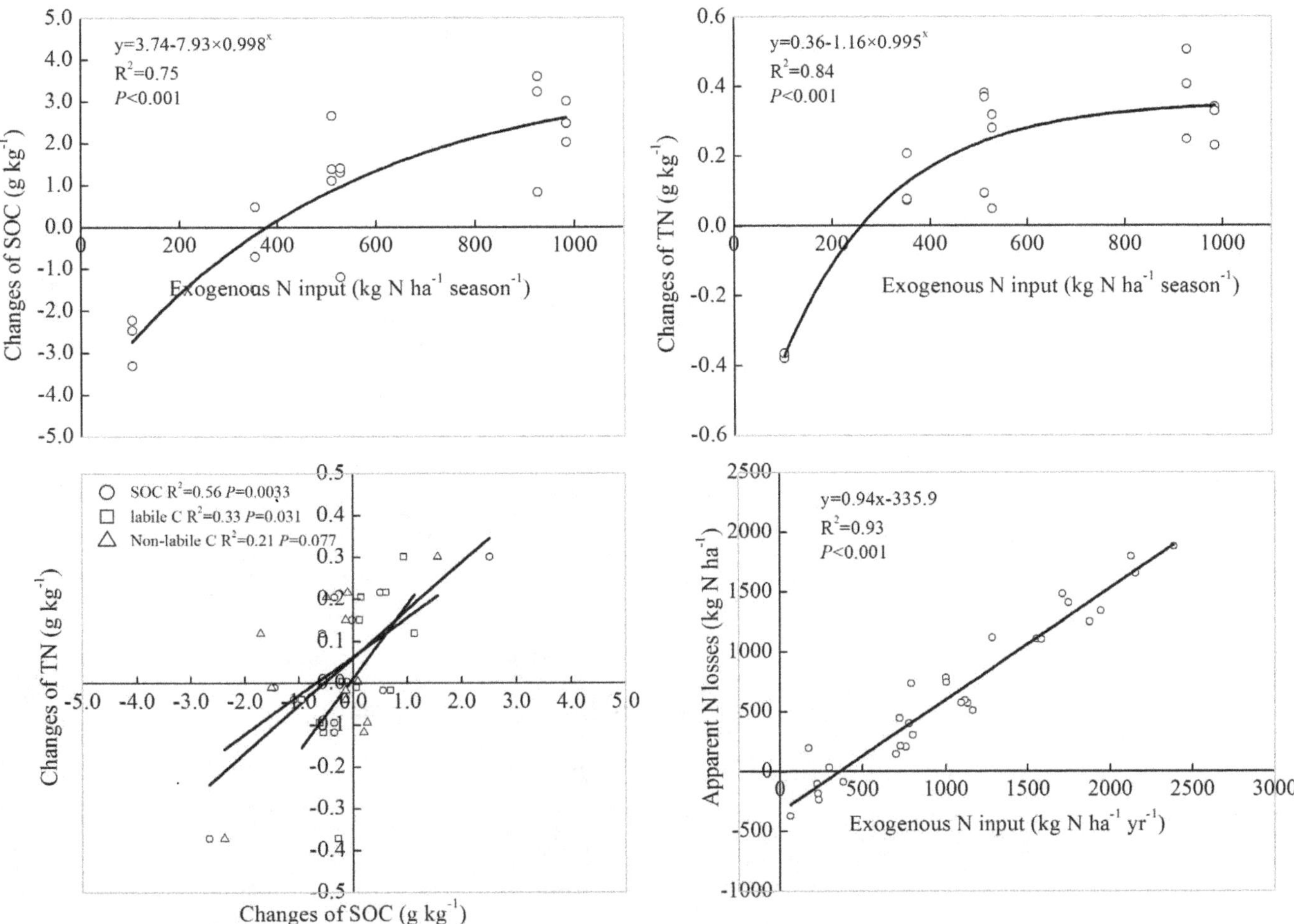


**Figure 5. The relationship between the changes in SOC and TN concentration after 6 years of cultivation, as well as annual apparent N losses and the average N inputs in the year-round greenhouse tomato planting system in Shouguang, northern China.**

organic matter content, should be considered to maintain soil fertility and productivity, minimize potential N losses and achieve sustainable development in greenhouse vegetable cropping systems.

## Author Contributions

Conceived and designed the experiments: QC JW FZ. Performed the experiments: TR. Analyzed the data: TR SL. Contributed reagents/materials/analysis tools: QC JW. Wrote the paper: TR QC.

## References

1. Bird JA, van Kessel C, Horwath WR (2002) Nitrogen dynamics in humic fractions under alternative straw management in temperate rice. Soil Sci Soc Am J 66: 478-488.
2. Accoe F, Boeckx P, Busschaert J, Hofman G, Van Cleemput O (2004) Gross N transformation rates and net N mineralization rates related to the C and N contents of soil organic matter fractions in grassland soils of different age. Soil Biol Biochem 36: 2075-2087.
3. Paré MC, Bedard-Haughn A (2013) Soil organic matter quality influences mineralization and GHG emissions in cryosols: a field-based study of sub-to high Arctic. Global Change Biol 19: 1126-1140.
4. Compton JE, Boone RD (2002) Soil nitrogen transformation and the role of light fraction organic matter in forest soils. Soil Biol Biochem 34: 933-943.
5. Neff JC, Townsend AR, Gleixner G, Lehman SJ, Turnbull JT, et al. (2002) Variable effects of nitrogen additions on the stability and turnover of soil carbon. Nature 419: 915-917.
6. Christopher SF, Lal R (2007) Nitrogen management affects carbon sequestration in North American Cropland soils. Crit Rev Plant Sci 26: 45-64.
7. Swanston C, Homann PS, Caldwell BA, Myrold DD, Ganio L, et al. (2004) Long-term effects of elevated nitrogen on forest soil organic matter stability. Biogeochemistry 70: 227-250.
8. Hagedorn F, Spinnler D, Siegwolf R (2003) Increased N deposition retards mineralization of old soil organic matter. Soil Biol Biochem 35: 1683-1692.
9. Malhi SS, Harapiak JT, Nyborg M, Gill KS, Monreal CM, et al. (2003) Total and light fraction organic C in a thin Black Chernozemic grassland soil as affected by 27 annual application of six rates of fertilizer N. Nutr Cycl Agroecosyst 66: 33-41.
10. Blair N, Faulkner RD, Till AR, Poulton PR (2006) Long-term management impacts on soil C, N and physical fertility part I: Broadbalk experiment. Soil Till Res 91: 30-38.
11. Jagadamma S, Lal R, Hoeft RG, Nafziger ED, Adee EA (2007) Nitrogen fertilization and cropping systems effects on soil organic carbon and total nitrogen pools under chisel-plow tillage in Illinois. Soil Till Res 95: 348-356.
12. Lemke RL, VandenBygaart AJ, Campbell CA, Lafond GP, Grant B (2010) Crop residue removal and fertilizer N: effects on soil organic carbon in a long-term crop rotation experiment on a Udic Boroll. Agr Ecosyst Environ 135: 42-51.
13. Mack MC, Schuur EAG, Bret-Harte MS, Shaver GR, Chapin III FS (2004) Ecosystem carbon storage in arctic tundra reduced by long-term nutrient fertilization. Nature 431: 440-443.
14. Dolan MS, Clapp CE, Allmaras RR, Baker JM, Molina JAE (2006) Soil organic carbon and nitrogen in a Minnesoota soil as related to tillage, residue and nitrogen management. Soil Till Res 89: 221-231.
15. Fonte SJ, Yeboah E, Ofori P, Quansah GW, Vanlauwe B, et al. (2009) Fertilizer and residue quality effects on organic matter stabilization in soil aggregates. Soil Biol Biochem 73: 961-966.
16. Liu LL, Greaver TL (2010) A global perspective on belowground carbon dynamics under nitrogen enrichment. Ecol Lett 13: 819-828.

17. Lu M, Zhou XH, Luo YQ, Yang YH, Fang CM, et al. (2011) Minor stimulation of soil carbon storage by nitrogen addition: A meta-analysis. Agr Ecosys Environ 140: 234-244.
18. Chen Q, Zhang XS, Zhang HY, Christie P, Li XL (2004) Evaluation of current fertilizer practice and soil fertility in vegetable production in the Beijing region. Nutr Cycl Agroecosyst 69: 51-58.
19. He FF, Chen Q, Jiang RF, Chen XP, Zhang FS (2007) Yield and nitrogen balance of greenhouse tomato (*Lycopersicum esculentum* Mill.) with conventional and site-specific nitrogen management in Northern China. Nutr Cycl Agroecosyst 77: 1-14.
20. Song XZ, Zhao CX, Wang XL, Li J (2009) Study of nitrate leaching nitrogen fate under intensive vegetable production pattern in northern China. CR Biol 332: 385-392.
21. He FF, Jiang RF, Chen Q, Zhang FS, Su F (2009) Nitrous oxide emissions from an intensively managed greenhouse vegetable cropping system in Northern China. Environ Pollut 157(5): 1666-1672.
22. Mao XS, Liu MY, Wang XY, Liu CM, Hou ZM, et al. (2003) Effects of deficit irrigation on yield and water use of greenhouse grown cucumber in the North China Plain. Agr Water Manage 61(3): 219-228.
23. Guo RY, Li XL, Christie P, Chen Q, Jiang RF, et al. (2008) Influence of root zone nitrogen management and a summer catch crop on cucumber yield and soil mineral nitrogen dynamics in intensive production systems. Plant Soil 313: 55-70.
24. Ren T, Christie P, Wang JG, Chen Q, Zhang FS (2010) Root zone soil nitrogen management to maintain high tomato yields and minimum nitrogen losses to the environment. Sci Hortic 125: 25-33.
25. Lei BK, Fan MS, Chen Q, Six J, Zhang FS (2010) Conversion of wheat-maize to vegetable cropping systems changes soil organic matter characteristics. Soil Sci Soc Am J 74(4): 1320-1326.
26. Degens BP, Schipper LA, Sparling GP, Vojvodic-Vukovic M (2000) Decreased in organic C reserves in soils can reduce the catabolic diversity of soil microbial communities. Soil Biol Biochem 32: 189-196.
27. Cookson WR, Abaye DA, Marschner P, Murphy DV, Stockdale EA, et al. (2005) The contribution of soil organic matter fractions to carbon and nitrogen mineralization and microbial community size and structure. Soil Biol Biochem 37: 1726-1737.
28. Gundersen P, Callesen I, de Vries W (1998) Nitrate leaching in forest ecosystems is related to forest floor C/N ratios. Environ Pollut 102: 403-407.
29. Zeng XB, Bai LY, Li LF, Su SM (2009) The status and changes of organic matter, nitrogen, phosphorus and potassium under different soil using styles of Shouguang of Shangdong Province. Acta Ecol Sin 29(7): 3737-3746 (in Chinese).
30. Feller C, Fink M (2002) $N_{min}$ target values for field vegetables. Acta Hort 571: 195-201.
31. Blair GJ, Lefroy RDB, Lisle L (1995) Soil carbon fractions based on their degree of oxidation, and the development of a carbon management index for agricultural systems. Aust J Agric Res 46: 1459–1466.
32. Six J, Paustian K, Elliott ET, Combrink C (2000) Soil structure and organic matter: I. Distribution of aggregate-size classes and aggregate-associated carbon. Soil Sci Soc Am J 64: 681-689.
33. McLauchlan KK, Hobbie S (2004) Comparison of labile soil organic matter fractionation techniques. Soil Sci Soc Am J 68: 1616-1625.
34. Öborn I, Edwards AC, Witter E, Oenema O, Ivarsson K, et al. (2003) Element balances as a tool for sustainable nutrient management: a critical appraisal of their merits and limitations within an agronomic and environmental context. Eur J Agron 20: 211-225.
35. Sieling K, Kage H (2006) N balance as an indicator of N leaching in an oilseed rape-winter wheat-winter barley rotation. Agric Ecosyst Environ 15: 261-269.
36. Zhao XC (2011) Effects of fertilization and crop rotation on soil microbial community structure of greenhouse tomato. Master Thesis, China Agricultural University, Beijing China (In Chinese).
37. Edmeades DC (2003) The long-term effects of manures and fertilizers on soil productivity and quality: a review. Nutr Cycl Agroecosyst 66: 165-180.
38. Lei BK, Chen Q, Fan MS, Zhang FS, Gan YD (2008) Changes of soil carbon and nitrogen in Shouguang intensive vegetable production fields and their impacts on soil properties. Plant Nutr Fert Sci 14(5): 914-922 (in Chinese).
39. Ge XG, Zhang EP, Zhang X, Wang XX, Gao H (2004) Studies on changes of filed-vegetable ecosystem under long-term fixed fertilizer experiment (I) changes of soil organic matter. Acta Hortic Sin 31(1): 34-38 (in Chinese).
40. Hassink J (1997) The capacity of soils to preserve organic C and N by their association with clay and silt particles. Plant Soil 191: 77-87.
41. Six J, Conant RT, Paul EA, Paustian K (2002) Stabilization mechanisms of soil organic matter: Implications for C-saturation of soils. Plant Soil 241: 155-176.
42. Liu J, Schulz H, Brandl S, Miehtke H, Huwe B, et al. (2012) Short-term effect of biochar and compost on soil fertility and water status of a Dystric Cambisol in NE Germany under field conditions. J Plant Nutr Soil Sci 175: 698-707.
43. Wang XJ, Jia ZK, Liang LY, Han QF, Ding RX, et al. (2012) Effects of organic manure application on dry land soil organic matter and water stable aggregates. Chin J Appl Ecol 23(1): 159-165 (in Chinese).
44. Ju XT, Kou CL, Zhang FS, Christie P (2006) Nitrogen balance and groundwater nitrate contamination: comparison among three intensive cropping systems on the North China Plain. Environ Pollut 143: 117-125.
45. Ajwa HA, Tabatabai MA (1994) Decomposition of different organic materials in soils. Biol Fert Soils 18: 175-182.
46. Cayuela ML, Velthof GL, Mondini C, Sinicco T, van Groenigen JW (2010) Nitrous oxide and carbon dioxide emissions during initial decomposition of animal by-products applied as fertilizers to soils. Geoderma 157: 235-242.
47. Plaza C, Garcia-Gil JC, Polo A (2005) Effects of pig slurry application on soil chemical properties under semiarid conditions. Agrochimica 49: 87-92.
48. Piccolo A, Spaccini R, Nieder R, Richter J (2004) Sequestration of a biologically labile organic carbon in soils by humified organic matter. Climatic Change 67: 329-343.
49. Adani F, Genevini P, Ricca G, Tambone F, Montoneri E (2007) Modification of soil humic matter after 4 years of compost application. Waste Manage 27: 319-324.
50. Li ZP, Liu M, Wu XC, Han FX, Zhang TL (2010) Effects of long-term chemical fertilization and organic amendments on dynamics of soil organic C and total N in paddy soil derived from barren land in subtropical China. Soil Till Res 106: 268-274.
51. Antil RS, Gerzabek MH, Haberhauer G, Eder G (2005) Long-term effects of cropped vs. fallow and fertilizer amendments on soil organic matter I. organic carbon. J Plant Nutr Soil Sci 168: 108-116.
52. Blagodatskaya E, Kuzyakov Y (2008) Mechanisms of real and apparent priming effects and their dependence on soil microbial biomass and community structure: critical review. Biol Fert Soils 45:115-131.
53. Lin Y (2010) Solute transportation and soil $H^+$ production budgets in a greenhouse vegetable production system. Master Thesis, China Agricultural University, Beijing China (In Chinese).
54. Ryden JC, Lund JL (1980) Nature and extent of directly measured denitrification losses from some irrigated vegetable crop production units. Soil Sci Soc Am J 44:505-511.
55. Zotareli L, Scholberg JM, Dukes MD, Mu~noz-Carpena R, Icerman J (2009) Tomato yield, biomass accumulation, root distribution and irrigation water use efficiency on a sandy soil, as affected by nitrogen rate and irrigation scheduling. Agr Water Manage 96: 23-34.
56. Constantin J, Mary B, Laurent F, Aubrion G, Fontaine A, et al. (2010) Effects of catch crops, no till and reduced nitrogen fertilization on nitrogen leaching and balance in three long-term experiments. Agr Ecosyst Environ 135: 268-278.
57. Schimel DS (1986) Carbon and nitrogen turnover in adjacent grassland and cropland ecosystems. Biogeochemistry 2: 345-357.
58. Yang YH, Luo YQ, Finzi AC (2011) Carbon and nitrogen dynamics during forest stand development: a global synthesis. New Phytol 190(4): 977-989.

5

# Green Manure Addition to Soil Increases Grain Zinc Concentration in Bread Wheat

**Forough Aghili[1]*, Hannes A. Gamper[1], Jost Eikenberg[2], Amir H. Khoshgoftarmanesh[3], Majid Afyuni[3], Rainer Schulin[4], Jan Jansa[5], Emmanuel Frossard[1]**

**1** Institute of Agricultural Sciences, Department of Environmental Systems Science, Swiss Federal Institute of Technology (ETH) Zürich, Switzerland, **2** Paul Scherrer Institute (PSI), Radioanalytics Laboratory, Villigen, Switzerland, **3** College of Agriculture, Department of Soil Sciences, Isfahan University of Technology, Isfahan, Iran, **4** Institute of Terrestrial Ecosystems, Department of Environmental Systems Science, Swiss Federal Institute of Technology (ETH) Zürich, Switzerland, **5** Institute of Microbiology, Academy of Sciences of the Czech Republic, Prague, Czech Republic

## Abstract

Zinc (Zn) deficiency is a major problem for many people living on wheat-based diets. Here, we explored whether addition of green manure of red clover and sunflower to a calcareous soil or inoculating a non-indigenous arbuscular mycorrhizal fungal (AMF) strain may increase grain Zn concentration in bread wheat. For this purpose we performed a multifactorial pot experiment, in which the effects of two green manures (red clover, sunflower), $ZnSO_4$ application, soil $\gamma$-irradiation (elimination of naturally occurring AMF), and AMF inoculation were tested. Both green manures were labeled with $^{65}Zn$ radiotracer to record the Zn recoveries in the aboveground plant biomass. Application of $ZnSO_4$ fertilizer increased grain Zn concentration from 20 to 39 mg Zn $kg^{-1}$ and sole addition of green manure of sunflower to soil raised grain Zn concentration to 31 mg Zn $kg^{-1}$. Adding the two together to soil increased grain Zn concentration even further to 54 mg Zn $kg^{-1}$. Mixing green manure of sunflower to soil mobilized additional 48 µg Zn (kg soil)$^{-1}$ for transfer to the aboveground plant biomass, compared to the total of 132 µg Zn (kg soil)$^{-1}$ taken up from plain soil when neither green manure nor $ZnSO_4$ were applied. Green manure amendments to soil also raised the DTPA-extractable Zn in soil. Inoculating a non-indigenous AMF did not increase plant Zn uptake. The study thus showed that organic matter amendments to soil can contribute to a better utilization of naturally stocked soil micronutrients, and thereby reduce any need for major external inputs.

**Editor:** Raffaella Balestrini, Institute for Sustainable Plant Protection - C.N.R., Italy

**Funding:** This study was supported by the grant 123920 of the Swiss National Science Foundation (SNF) and the Swiss Agency for Development and Cooperation (SDC) in the framework of Research Partnerships with Developing Countries. The funding body and host institutions had no role in study-design, data collection, analysis, and manuscript preparation. JJ was further supported by the J.E. Purkyně Fellowship and the long-term development program RVO61388971.

**Competing Interests:** The authors have declared that no competing interests exist.

* Email: forough.aghili@usys.ethz.ch

## Introduction

Wheat is the third most-grown cereal after maize and rice, draws on the largest cropping area, and ranks first in global cereal trade [1]. It is the crop contributing most calories to large segments of the global human population. Yet, wheat-based diets do not provide sufficient zinc (Zn), which is a leading cause of widespread Zn deficiency [2]. Low grain Zn concentration is cause of low plant-availability of Zn in many soils on which wheat is produced [3,4]. Efforts have been made to breed wheat cultivars efficient in acquiring Zn from soil and effective in (re-) translocating Zn to grains [5,6]. Zinc supply from soil has, however, to be increased for such elite crop cultivars to become fully effective, which can be achieved via appropriate soil fertility management, such as application of mineral Zn fertilizer [7], addition of green manure [8], and/or a strengthening of nutrient acquisition traits, including symbioses of roots with naturally occurring and newly introduced arbuscular mycorrhizal fungi (AMF) [9–12].

Plants and probably also AMF take up Zn from the soil solution in form of free $Zn^{2+}$ and $ZnOH^{+}$ ions, or as Zn-organic ligand complexes [13,14]. The depleted soil solution is replenished with Zn via desorption or dissolution from the soil matrix, followed by dispersion by diffusion [15]. Desorption from the soil matrix is controlled by soil pH, salinity, and clay, carbonate, and organic matter contents [15]. To promote Zn dissolution plants and microorganisms exude soil-acidifying protons and Zn-chelating low-molecular-weight organic compounds [16,17]. These Zn mobilizing activities of roots and microorganisms are promoted under iron (Fe), Zn, and other nutrient deficiencies [18]. The most well known classes of exuded organic Zn ligands are carboxylates, amides, and non-proteinaceous amino acids (= siderophores) [19,20]. A recent study on Zn uptake by wheat in the field [8] suggests that residues from the previous crop in crop rotations and thus possibly also addition of green manure to soil can raise the plant-available Zn in soil and promote Zn uptake in crop plants, mainly by contributing Zn-chelating dissolved organic carbon and amino acids.

Arbuscular mycorrhizal fungi, phylum Glomeromycota), because of their ubiquity and crucial involvement in plant nutrient acquisition and health, are a plant production-relevant group of root symbionts [21,22]. Their hyphae form effective, large-distance conduits for the transport of temporarily and spatially heterogeneously available nutrients of low mobility in soil, such as phosphate (P) and Zn [23]. Crop residue or green manure addition to soil and sporadic rain and irrigation may create nutrient flushes that crop plants with root systems extended by extraradical AMF hyphae may much more effectively utilize [24]. Arbuscular mycorrhizal fungi, whether indigenous or deliberately introduced as inoculants, were repeatedly reported to benefit Zn uptake in crop plants [9,10]. While the phenomenon of increased plant Zn uptake of mycorrhized as opposed to non-mycorrhized plants has already often been demonstrated [10], we still lack information about controlling factors [8] and the possible reinforcing effect of combined enrichment of soils with organic matter and mineral Zn fertilizer.

Here we report findings of a cross-factorial pot experiment on Zn nutrition of bread wheat involving the enrichment of calcareous soil with green manure of red clover and sunflower and inoculating plants with a non-indigenous AMF strain. For comparison and to study its interactions with the other experimental factors also a water-soluble mineral Zn fertilizer ($ZnSO_4$) was applied. To compare the effect of all naturally occurring AMF and possibly associated other soil microorganisms, the soil of half of the experimental units was γ-irradiated and re-inoculated with an AMF-free microbial filtrate of a water suspension of non-sterilized soil. An AMF inoculation treatment was further included in an attempt to assess the potential of this rapidly growing technology and for controlling against non-AMF mediated biotic effects by the soil γ-irradiation and microbial re-inoculation procedure. We predicted that 1) green manure addition to soil improves Zn uptake by wheat, 2) green manure addition improves Zn uptake from mineral Zn fertilizer and 3) inoculation of wheat plants with a foreign AMF strain improves Zn uptake, particularly from soil whose indigenous AMF had been killed by γ-irradiation.

## Materials and methods

### Experimental design

A complete cross-factorial experiment with the factors green manure (3 levels: none, red clover, sunflower), mineral Zn fertilization (2 levels: none, $ZnSO_4$), γ-irradiation (2 levels: non-irradiated; γ-irradiated and re-inoculated with soil microbes, except AMF), and AMF inoculation (2 levels: none, inoculated) was set up with each treatment replicated five times. The 120 experimental pots were arranged randomly on a glasshouse bench.

### Soil origin and physicochemical soil analyses

The soil was collected in September 2009 to a depth of 30 cm from an arable field at the Rudasht research station (32, 29/N, 52, 10/E, 1560 m above sea level), located 65 km south-east of Isfahan in central Iran [permission for soil sampling was given by the Isfahan Agricultural and Natural Resource Research Center and for its import to Switzerland by the Swiss Federal Department of Agriculture]. The soil was air-dried and passed through a 5 mm sieve, before being shipped to Switzerland, where 5 kg were further passed through a 2 mm sieve for physicochemical analyses.

This experimental soil was a Typic Haplocambid, according to the USDA soil taxonomy [25]. The field where the soil was sampled had been under agricultural use with periodical irrigation from a nearby river. The crop rotation prior to soil sampling comprised barley, wheat, sugar beet, and wheat, interrupted by dry fallow periods. The climate in central Iran is semi-arid with a long-term mean annual temperature of 16.8°C, a mean annual rainfall lower than 100 mm, and a potential annual evaporation higher than 2000 mm [26].

The soil texture was a silty clay loam based on the measurements by the hydrometer method [27] and according to the USDA soil classification scheme [6]. The pH was 7.9 and the electrical conductivity (EC) 23 $dSm^{-1}$, as determined by the procedure of Rhoades [28], using an ISE pH meter 720A (Mettler Toledo LLC, Columbus, OH, USA) and an E- 518 conductometer (Metrohm Herisau, Switzerland), respectively. The soil was low in organic carbon [5 g (kg soil) $^{-1}$] and nitrogen [0.6 g (kg soil) $^{-1}$], as estimated by the methods of Walkley and Black [29] and on a Carlo Erba flash combustion CN-analyzer (NA1500, Carlo Erba, Milano, Italy), respectively. The soil contained 300 mg $CaCO_3$ (kg soil)$^{-1}$ as determined after neutralization with hydrochloric acid and back titration with sodium hydroxide [30].

The total Zn concentration of the soil was 80.2 mg Zn (kg soil)$^{-1}$, according to energy dispersive X-ray fluorescence spectrometry (XRF) measurements on a Spectra X-Lab 2000 instrument (SPECTRO Analytical Instruments GmbH, Kleve, Germany). Potentially plant-available Zn was extracted from 25 g of soil, using 50 ml of 0.05 M diethylene triamine-penta-acetic acid (DTPA), adjusted to pH 7.2 [31], and measured on an Agilent 7500 C inductively coupled plasma mass spectrometer (ICP-MS, Agilent Technologies, Santa Clara, California, USA). The amount of DTPA-extractable Zn was 0.45 mg Zn (kg soil)$^{-1}$.

### Partial soil desalination, γ-irradiation and fertilization

Preliminary tests showed that wheat growth on the untreated soil was severely impaired by salinity stress (Forough Aghili et al., unpubl. data). Therefore, we flushed the soil in five rounds, with a volume ratio of 1:1 soil:deionized water each prior to the experiment. This treatment can be considered comparable to repeated flood irrigation, used by farmers at the field site prior to sowing. Soil washing lowered the EC from 23 to 6.7 dS $m^{-1}$, which is close to the threshold value of 5.9 dS $m^{-1}$ for salinity-induced growth depression in wheat [32]. Flushing with water also slightly decreased the soil pH from 7.9 to 7.7. Analyses of the leachate indicated that large amounts of sodium, chlorine and sulfur had been removed, but virtually none of the total soil Zn (data not shown). The DTPA-extractable concentration of Zn in the soil even increased slightly from 0.45 to 0.47 mg Zn (kg soil)$^{-1}$. No spores of AMF could be retrieved from the leachate. The partially desalinated soil was again air-dried and sieved to a particle size of ≤5 mm.

Half of the soil was γ-irradiated with a dose of 25–75 kGy using a $^{60}Co$ source (Studer-Hard, Däniken, Switzerland, http://www.leoni-irradiation-services.com) in order to kill the indigenous soil microbes. To reintroduce soil microbes, but not AMF, 50 ml of a filtrate of a 2.5% (w:v) soil water suspension was added per kilogram soil. This microbial filtrate was prepared by passing the soil suspension twice through Whatman No 1 filter papers, which size-excluded AMF propagules. This procedure was recommended by Thompson [33] for studying the effect of AMF on Zn nutrition and is also utilized in many comparable studies to analyze the relative effect of different soil microorganisms on plant nutrition [34,35]. Indeed, soil γ-irradiation is currently the only way to create a control treatment, free of naturally occurring AMF, since mycorrhiza-defective mutants of wheat are not available [36].

The flushed soils (non-irradiated and γ-irradiated) received 400 mg K, 200 mg P, and 80 mg Fe per kilogram soil in the form of finely ground $K_2SO_4$, $CaHPO_4$ and $FeSO_4$ $7H_2O$ salts.

## Preparation and application of $^{65}Zn$-labeled green manure

We chose red clover (*Trifolium pratense L.*) and sunflower (*Helianthus annuus* L.) for the production of experimental green manures, because they are also often grown before wheat in the region of Isfahan [8]. Furthermore, Soltani et al. [8] found that compared to other possible green manure plants, red clover and sunflower introduced significant amounts of dissolved organic carbon into the soil solution, which was probably important for Zn solubilization.

Seeds of clover and sunflower were surface sterilized in 15% $H_2O_2$ for 15 min, rinsed with distilled water and germinated on moistened filter papers for two days. The most vigorously growing seedlings were transferred to a combined sand bed-hydroponics system, which consisted of compartmented seedling trays filled with silica sand of a particle size of 0.7–1.2 mm, partially immersed in a full-strength Hoagland nutrient solution [37]. After 28 days, the nutrient solution was exchanged for one that contained radioactive $^{65}Zn$ for an additional 26 days of vegetative plant development. While 30 liters of Hoagland nutrient solution were labeled with 5.55 MBq of $^{65}Zn$ for about 400 red clover plantlets, 40 l of Hoagland nutrient solution were labeled with 9.25 MBq of $^{65}Zn$ for about 140 sunflower plantlets. The $^{65}Zn$ radiotracer was added as carrier-free $ZnCl_2$, dissolved in weak HCl (Amersham plc, GE Healthcare, Little Chalfont, UK). The photoperiod in the glasshouse was set to 14 h, the minimal light intensity to 12 kLux, the temperature to 22°C during the day and 16°C at night, and the relative air humidity was allowed to vary between 30 and 40%.

At harvest, the entire shoots of red clover and only the leaf blades of sunflower, without petioles and stems, were collected, air-dried and finely ground to powder. The total tissue C and N concentrations were determined in 4 mg tissue samples using the above-mentioned CN-analyzer. The total P and Zn concentrations were measured in 100 mg subsamples of each of the green manures, using the above-mentioned ICP-MS after incineration at 550°C for 6 h and extracted with 3 ml of hot 14.4 M nitric acid ($HNO_3$). The concentration of radioactive $^{65}Zn$ in the green manures was measured by γ-spectrometry in 300 mg subsamples using IGC2 high purity germanium detectors (ORTEC Advanced Measurements Technologies Inc, Oak Ridge, Tennessee, USA). Selected characteristics of both green manures are presented in Table 1.

The green manures (Table 1) were homogenously mixed into the soils at a dosage of 4 g dry matter per kilogram of dry soil. This dosage corresponds approximately to an addition of 18 tons of dry matter per hectare considering a soil bulk density of 1.5 t $m^{-3}$ and incorporation to a soil depth of 30 cm.

## Mineral Zn fertilization

Half of the soil in each green manure × irradiation treatment combination was fertilized with water soluble Zn in the form of $ZnSO_4$ at a dosage of 4.7 mg Zn (kg soil)$^{-1}$, corresponding approximately to 19 kg Zn ha$^{-1}$.

## AMF inoculation

Finally, half of all pots in each combination of the above treatments were inoculated with approximately 250 spores of a pot culture of the Swiss isolate BEG155 of the AMF species *Claroideoglomus claroideum* [38]. The spores of the inoculum were extracted by wet-sieving and decanting [39] and directly pipetted onto the roots of the transplanted wheat seedlings. This careful spore isolation and inoculation procedure prevented co-inoculation of many other microbes and introduction of organic matter in the form of root pieces, which would have been the case with so-called whole inoculum {= growth substrate, containing fungal spores, hyphae and colonized root fragments [40]}.

Data on AMF root colonization were not systematically collected in this experiment, since microscopic and molecular genetic quantification in a previous experiment with the same wheat cultivar and soil clearly showed a high infection pressure by the naturally occurring AMF (Forough Aghili et al., unpublished data). However, successful colonization of the roots by the AMF isolate BEG155 and by naturally occurring AMF were confirmed in a subset of all experimental treatments by quantitative polymerase chain reaction as in the previous study and additionally by phylogenetic sequence analysis of a 1.6 kb-long ribosomal DNA amplicon of the AMF colonizing the roots of a subset of all experimental units (Anouk Guyer et al., unpublished data).

## Planting, growth conditions, harvest and mineral nutrient analyses

Wheat kernels {*Triticum aestivum* cv. Kavir, a cultivar widely grown in Iran [41]} were surface-sterilized with 15% $H_2O_2$ for 15 min and pre-germinated on moistened filter paper for three days prior to transplantation. Four seedlings were planted into each pot with 550 g soil and later thinned to the most vigorously growing two individuals. Nitrogen in the form of $NH_4NO_3$ was applied weekly, six times, amounting to 600 mg N (kg soil)$^{-1}$ until tillering.

**Table 1.** Mineral nutrient concentrations of the two types of applied green manure prepared from red clover and sunflower.

| | **Red clover** | **Sunflower** |
|---|---|---|
| | **(Mean ± SE)** | **(Mean ± SE)** |
| **N mg (g DM*)$^{-1}$** | 43.3±0.6 | 58.4±0.4 |
| **C/N (mass ratio)** | 9.1±0.1 | 6.3±0.1 |
| **P mg (g DM)$^{-1}$** | 5.0±0.1 | 7.3±0.1 |
| **Zn µg (g DM)$^{-1}$** | 45.6±0.6 | 104.4±2.4 |
| **$^{65}$Zn kBq¶ (g DM)$^{-1}$** | 6.9±0.02 | 15.8±0.1 |

* DM: dry matter.
¶Radioactivity at the time of harvest of the experiment.
The means and associated standard errors (SE) of three replicate subsamples of the nitrogen (N) concentration, carbon (C) to N mass ratio, phosphorus (P) and zinc (Zn) concentrations, and $^{65}Zn$ concentration are listed.

The climatic conditions in the glasshouse were set to a photoperiod of 14 h with a minimum light intensity of 10 kLux, a 22/17°C day/night air temperature regime, and 40–45% relative air humidity. Soil moisture was kept at about 70% soil water holding capacity by daily watering to weight.

Harvest took place at grain maturity, 125 days after planting. The total weight of the grains and straw of the two plants in each pot was determined after drying to constant weight at 60°C for 72 h. Subsamples of 200 mg of finely ground wheat straw and grain were incinerated in porcelain crucibles at 550°C for 6 h and the ashes dissolved in 2 ml of boiling 13 M $HNO_3$. Zinc concentrations were measured using ICP-MS. Measurement accuracy and precision were ensured by regular inclusion of a certified plant reference standard in the analyses (hay powder IAEA-V-10, 24 mg Zn $kg^{-1}$) and by running blank samples with each sample batch. The concentration of radioactive $^{65}Zn$ was measured in the grains and straw by γ-spectrometry as described for the green manure above. Zinc uptake into roots could not be determined because it was not possible to recover the roots quantitatively from the clay-rich soil. Soil pH and the concentrations of total N and DTPA-extractable soil Zn were measured at the end of the experiment, using the above-described analytical procedures.

## Calculations and data analyses

**Plant Zn uptake from green manure, soil and mineral Zn fertilizer.** Plant Zn uptake was calculated as the product of Zn concentration and biomass for both the grain and straw fractions and summed to obtain the amount of Zn contained in the entire aboveground biomass. All radioactivity measurements were related to the harvest time, taking the radioactive decay into account ($^{65}Zn$ half-life time: 243.9 d). The percentage of Zn in the aboveground biomass derived from the green manure [$Zn_{dgm}$ (%)] was calculated as:

$$Zn_{dgm}(\%) = 100 \times SA_p \times (SA_{gm})^{-1} \quad (1)$$

where $SA_p$ [Bq (mg $Zn^{-1}$)] is the specific activity (SA) of Zn in the aboveground biomass and $SA_{gm}$ the specific activity of Zn in the added green manure. The specific activity of Zn in the plant was calculated as:

$$SA = [^{65}Zn] \times [Zn]^{-1} \quad (2)$$

where [$^{65}Zn$] stands for the concentration of radioactivity per kilogram of plant material [Bq (kg DM)$^{-1}$] and [Zn] stands for the concentration of total Zn in the plant [mg Zn (kg DM)$^{-1}$].

Using the equations 1 and 2, the amount of Zn taken up to the aboveground biomass from green manure [$Zn_{dgm}$, μg Zn (kg soil)$^{-1}$] was calculated as:

$$Zn_{dgm} = (Zn_{dgm} \times Zn_{upt}) \times 10^{-2} \quad (3)$$

where $Zn_{upt}$ is the amount of Zn taken up to the aboveground plant biomass per kilogram of soil [μg Zn (kg soil)$^{-1}$].

Knowing $Zn_{dgm}$, it is possible to calculate the percentage of Zn recovered by the aboveground plant biomass from the added green manure [$Zn_{rec\_gm}$ (%)]:

$$Zn_{rec\_gm} = 100 \times Zn_{dgm} \times Zn_{added\_gm}{}^{-1} \quad (4)$$

where $Zn_{added\_gm}$ is the amount of Zn added with the green manure [μg Zn (kg soil)$^{-1}$].

Knowing $Zn_{dgm}$, it is also possible to calculate the amount of Zn in the plant that is derived from non-labeled Zn sources in soil, i.e. in the non-fertilized soil from the native soil Zn pool and in the mineral Zn-fertilized soil from the native soil Zn pool as well as the applied $ZnSO_4$ [$Zn_{dsoil}$, μg Zn (kg soil)$^{-1}$]:

$$Zn_{dsoil} = Zn_{upt} - Zn_{dgm} \quad (5)$$

**Statistical analyses.** The results were analyzed by fixed factor four-way or three-way analyses of variance (ANOVA) with two levels (γ-irradiated and non-irradiated) of 'soil γ-irradiation', three levels (none, red clover and sunflower) of 'green manure addition', two levels [0 and. 4.7 mg Zn (kg soil)$^{-1}$] of 'mineral Zn fertilization', and two levels (with and without inoculation with *C. claroideum*) of 'AMF inoculation'. The interaction terms were sequentially removed from the initially complete ANOVA model when not significant (backward elimination). In addition, a one-way ANOVA was carried out on the saved residuals of a three-way ANOVA with the factors 'soil γ-irradiation', 'mineral Zn fertilization' and 'AMF inoculation' in order to specifically analyze the effect of green manure addition. Compliance of data dispersion with ANOVA model assumptions was verified using Shapiro-Wilcoxon's normality and Levene's equal variance tests. If necessary, data were $log_{10}$ or squareroot transformed. All statistical analyses were carried out using the software package SPSS version 17 (SPSS Inc., Chicago, Ill inois, USA). Means of factor levels were compared by least-significant-difference (LSD) tests at the significance level of $p<0.05$. The figures were prepared in SigmaPlot version 12 [Systat Software Inc. (SSI), San Jose, California, USA].

# Results

## Plant biomass and grain nitrogen concentration

The grain yield of the experimental wheat plants was not significantly affected by any of the treatments (Figure 1A, Table 2), but the large variance may have concealed a similar γ-irradiation effect as on the total aboveground biomass (Figure 1B). The total aboveground biomass (= straw + grains) was significantly larger for the wheat plants grown in γ-irradiated than non-irradiated soil ($F_{1,96} = 10.45$, $p = 0.002$). Inoculation with *C. claroideum* had no significant effect on aboveground biomass production (Table 2).

Both types of green manure significantly increased grain N concentration ($F_{2,70} = 31.41$, $p<0.001$, Table 2), which ranged between 30 and 45 g N (kg DM)$^{-1}$ in the plants grown on γ-irradiated soil and between 25 and 37 g N (kg DM)$^{-1}$ in the plants grown on non-irradiated soil (Figure 2). Soil γ-irradiation also led to a significant increase in grain N concentrations ($F_{1,70} = 10.04$, $p = 0.003$, Table 2). Inoculation with *C. claroideum* and $ZnSO_4$ addition had no significant effects on grain N concentration (Table 2).

## Grain Zn concentration and Zn uptake to grains and aboveground biomass

Grain Zn concentration increased significantly with application of mineral Zn fertilizer ($F_{1,70} = 50.96$, $p<0.001$) and green manure ($F_{2,70} = 13.48$, $p<0.001$), but was not affected by inoculation of *C. claroideum* and soil γ-irradiation (Figure 3A, Table 2). The lowest grain Zn concentration [20±1.64 mg Zn (kg DM)$^{-1}$] was measured in the treatment with neither mineral Zn fertilization, nor addition of green manure, while the highest grain Zn

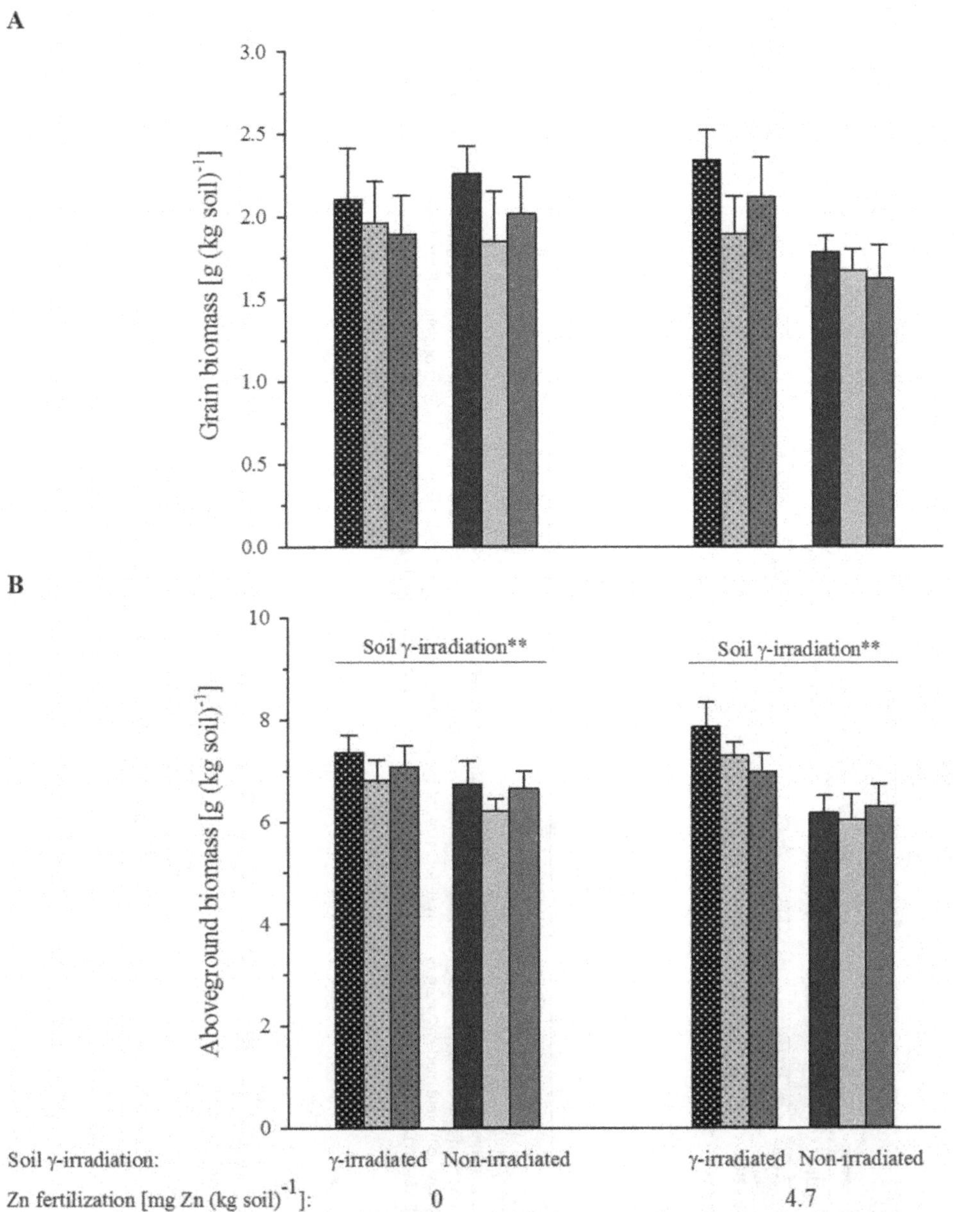


**Figure 1. Grain yield (A) and aboveground biomass (B) of bread wheat grown in a calcareous soil.** Black bar filling denotes the treatments without any added green manure, light grey bar filling the treatments in which the soil was amended with green manure of red clover, and dark grey bar filling those treatments in which green manure of sunflower was added to the soil. Bar hatching highlights the treatment missing the naturally occurring arbuscular mycorrhizal fungi after γ-irradiating the soil. Bars show mean values and associated standard errors of five experimental replicates. The significances (**, $p<0.01$) of soil γ-irradiation from a four-factorial analysis of variance are indicated. For full statistical details see Table 2.

concentration [$54 \pm 2.32$ mg Zn (kg DM)$^{-1}$] was found in the treatment with combined application of $ZnSO_4$ fertilizer and green manure of sunflower (Figure 3A).

The highest Zn concentration in the aboveground biomass [$46.1 \pm 1.43$ mg Zn (kg DM)$^{-1}$] was found in plants grown on non-irradiated soil amended with green manure of sunflower in combination with $ZnSO_4$ (Figure 3B). The lowest aboveground Zn concentration was observed in plants grown on γ-irradiated soil that had neither been fertilized with mineral Zn, nor amended with green manure (Figure 3B). Application of mineral Zn fertilizer reduced the differences between the aboveground biomass Zn concentrations of those plants that had received green manure and those that had not, resulting in a statistically significant interaction between green manure addition and mineral Zn fertilization ($F_{2,96} = 6.24$, $p<0.01$; Table 2). Inoculation with *C. claroideum* had no significant effect on the concentration of Zn in the aboveground biomass (Table 2).

The amount of Zn taken up from the native soil Zn pool and mineral Zn fertilizer that was allocated to grains ranged between 46 and 88 µg Zn (kg soil)$^{-1}$, while the total amount of Zn accumulated in the aboveground biomass ranged between 117 and 367 µg Zn (kg soil)$^{-1}$ (Table 3). The lowest values were again observed in the absence of mineral Zn fertilization and green manure addition, while the highest values were observed after

**Table 2.** Results of fixed factor four-way analyses of variance (ANOVA) on plant and soil parameters.

| ANOVA model | Grain biomass | | Aboveground biomass | | Grain N concentration | | Grain Zn concentration | | Aboveground Zn concentration | | Zn uptake into grains | |
|---|---|---|---|---|---|---|---|---|---|---|---|---|
| **Source of variance** | **F value** | **df** | **F value** | **df** | **F value** | **df** | **F value** | **df** | **F value** | **df** | **F value** | **df** |
| **Full model** | $0.29^{ns}$ | 23 | $2.47^{*}$ | 23 | $4.45^{***}$ | 23 | $5.92^{***}$ | 23 | $8.91^{***}$ | 23 | $2.78^{**}$ | 23 |
| **Green manure (A)** | $0.33^{ns}$ | 2 | $0.26^{ns}$ | 2 | $31.41^{***}$ | 2 | $13.48^{***}$ | 2 | $13.71^{***}$ | 2 | $1.98^{ns}$ | 2 |
| **Zn fertilization (B)** | $0.19^{ns}$ | 1 | $0.12^{ns}$ | 1 | 0.09 | 1 | $50.96^{***}$ | 1 | $125.23^{***}$ | 1 | $40.52^{***}$ | 1 |
| **Soil $\gamma$-irradiation (C)** | $0.20^{ns}$ | 1 | $10.45^{**}$ | 1 | 10.04 | 1 | $2.99^{ns}$ | 1 | $9.15^{**}$ | 1 | $0.17^{ns}$ | 1 |
| **Inoculation (D)** | $0.76^{ns}$ | 1 | $0.13^{ns}$ | 1 | 2.97 | 1 | $1.95^{ns}$ | 1 | $1.04^{ns}$ | 1 | $0.95^{ns}$ | 1 |
| **A×B** | r | 2 | r | 2 | R | 2 | $12.71^{**}$ | 2 | $6.24^{**}$ | 2 | $12.71^{***}$ | 2 |
| **A×C** | r | 2 | $0.61^{ns}$ | 2 | $0.98^{ns}$ | 2 | $3.36^{*}$ | 2 | $4.41^{*}$ | 2 | $1.22^{ns}$ | 2 |

| ANOVA model | Zn uptake into aboveground biomass | | $Zn_{dgm}$ | | $Zn_{rec\_gm}$ | | $Zn_{dsoil}$ | | DTPA-extractable soil Zn | |
|---|---|---|---|---|---|---|---|---|---|---|
| **Source of variance** | **F value** | **df** | **F value** | **df** | **F value** | **df** | **F value** | **df** | **F value** | **df** |
| **Full model** | $4.26^{***}$ | 23 | $5.95^{***}$ | 15 | $8.61^{***}$ | 15 | $6.15^{***}$ | 23 | $16.40^{***}$ | 23 |
| **Green manure (A)** | $4.13^{*}$ | 2 | $39.92^{***}$ | 2 | $72.25^{***}$ | 2 | $3.50^{*}$ | 2 | $85.22^{***}$ | 2 |
| **Zn fertilization (B)** | $53.49^{***}$ | 1 | $21.23^{***}$ | 1 | $22.31^{***}$ | 1 | $77.13^{***}$ | 1 | $168.84^{***}$ | 1 |
| **Soil $\gamma$-irradiation (C)** | $0.01^{ns}$ | 1 | $5.21^{*}$ | 1 | $8.85^{**}$ | 1 | $1.14^{ns}$ | 1 | $0.64^{ns}$ | 1 |
| **Inoculation (D)** | $2.34^{ns}$ | 1 | $8.25^{**}$ | 1 | $4.82^{*}$ | 1 | $1.26^{ns}$ | 1 | $2.10^{ns}$ | 1 |
| **A×B** | $0.49^{ns}$ | 2 | $2.12^{ns}$ | 2 | $0.13^{ns}$ | 2 | $2.70^{ns}$ | 2 | $2.17^{ns}$ | 2 |
| **A×C** | $0.44^{ns}$ | 2 | $0.63^{ns}$ | 2 | $4.64^{*}$ | 2 | $2.95^{ns}$ | 2 | $0.54^{ns}$ | 2 |

DTPA: diethylene triamine-penta-acetic acid.

na: not applied, r: removed from the statistical model, ns: not significant, *, $p<0.05$; **, $p<0.01$; ***, $p<0.001$.

The measurements of the plant parameters {grain biomass, total above ground biomass, grain N concentration, grain Zn concentration, aboveground Zn concentration, Zn uptake to grains, Zn uptake to the aboveground biomass, Zn derived from green manure in the aboveground biomass [$Zn_{dgm}$], proportion of the Zn recovered in the above ground biomass [$Zn_{rec\_gm}$] of the Zn added with the green manure Zn derived from the soil and fertilizer in the aboveground biomass [$Zn_{dsoil}$]}and soil parameter (DTPA-extractable Zn from soil) were made at grain maturity. Effect sizes of the experimental factors and their significant interactions are indicated as F-values alongside with the statistical significance level. The interactions of the factors A×D, B×C, B×D, C×D, A×B×C, A×B×D, A×D×C, B×C×D, A×B×C×D were not significant in any of the 11 ANOVAs and thus these results are not listed.

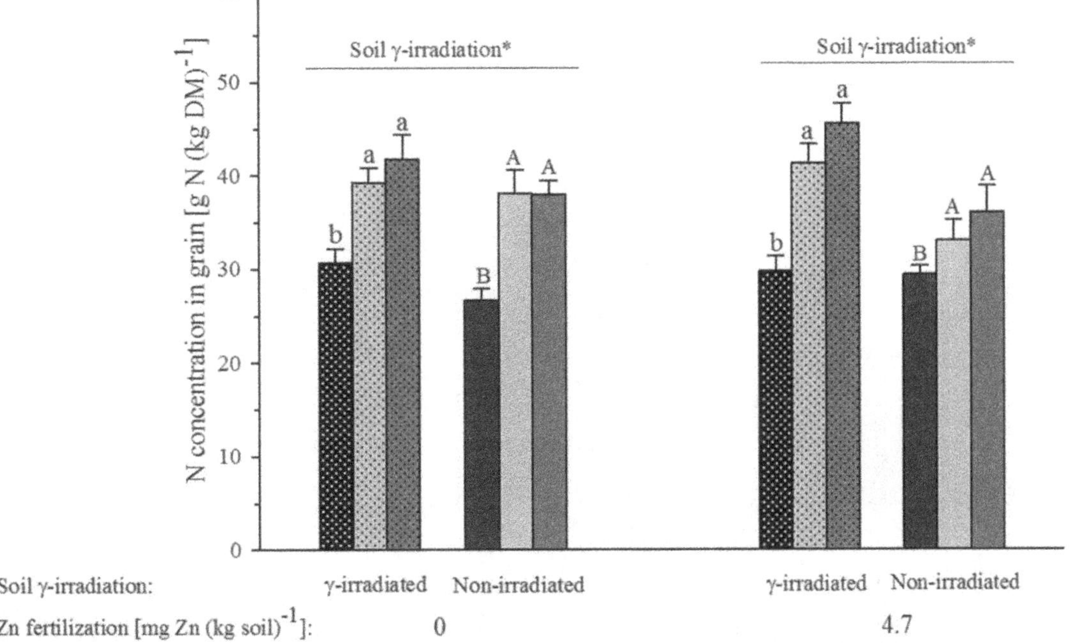


**Figure 2. Grain nitrogen (N) concentration of wheat grown in a calcareous soil.** Bar filling denoting the different green manure addition treatments is the same as in Figure 1. Bars show mean values and associated standard errors of five experimental replicates. The significances (*, $p < 0.05$) of the effect of soil γ-irradiation from a four-factorial analysis of variance are shown. Different letters indicate statistical differences of separate least significant difference tests at $p<0.05$ for the different green manure addition treatments within the combinations of soil γ-irradiation and mineral Zn fertilization. For full statistical details see Table 2.

addition of mineral Zn fertilizer and green manure of sunflower. Adding green manure to the soil tended to increase Zn accumulation in the grains, when no mineral Zn fertilizer had been applied, but this effect was not statistically significant (Table 2). Inoculation with *C. claroideum* had no significant effect on Zn uptake to the grains (Table 2).

## Zn derived from green manure and both, soil and mineral Zn fertilizer, in the aboveground biomass

The Zn from the added green manure recovered in the aboveground biomass ($Zn_{rec\_gm}$) ranged between 2.5% and 4.1% for red clover, and between 1.4% and 2.7% for sunflower (Figure 4). These Zn recoveries from the two types of green manures differed significantly across all treatments ($F_{1,79} = 72.25$, $p<0.001$). Mineral Zn fertilization lowered $Zn_{rec\_gm}$ for both forms of green manure ($F_{1,64} = 22.31$, $p<0.001$, Figure 4 & Table 2), as did soil γ-irradiation ($F_{1,64} = 8.85$, $p<0.01$). The recovered Zn fraction from sunflower residues was also reduced by the inoculation of *C. claroideum* when also mineral Zn fertilizer was applied ($F_{1,18} = 10.84$, $p<0.01$; Figure 4).

The amount of Zn in the aboveground biomass derived from green manure ($Zn_{dgm}$) ranged between 4.5 μg Zn (kg soil)$^{-1}$ in the treatment of combined soil γ-irradiation, mineral Zn fertilization, and addition of green manure of red clover, and 11 μg Zn (kg soil)$^{-1}$ in the treatment with green manure of sunflower but without mineral Zn fertilization, soil γ-irradiation, and AMF inoculation (Table 3). The amount of plant Zn derived from green manure of sunflower ($Zn_{dgm}$) was consistently higher than that derived from red clover ($F_{1,64} = 39.92$, $p<0.001$; Tables 2 & 3).

The amount of Zn derived from the native soil Zn pool ($Zn_{dsoil}$) was significantly increased with green manure addition, when no $ZnSO_4$ was applied ($F_{2,96} = 3.50$, $p = 0.034$, Table 2 & Figure 5). This apparent soil Zn mobilization was more pronounced for green manure produced from sunflower than from red clover (Figure 5). Under conditions of co-application of $ZnSO_4$, the effect of green manure addition was, however, variable and weak and statistically not significant. There was also no significant effect of γ-irradiation, nor any effect by the non-indigenous *C. claroideum* inoculant on the total amount of Zn derived from soil and mineral Zn fertilizer, either (Table 2).

## Changes in DTPA-extractable Zn from soil, total N and pH in soil

The concentration of DTPA-extractable Zn from soil measured after plant harvest was significantly higher in the treatments with green manure than without green manure ($F_{1,96} = 169$, $p<0.001$). Application of mineral Zn fertilizer also significantly increased DTPA-extractable Zn ($F_{2,96} = 85.2$, $p<0.001$; Tables 2 & 3). There was a highly significant correlation between the total Zn taken up into the aboveground biomass (Y) and the DTPA-extractable Zn concentration in the soil (X) according to a double reciprocal regression model [$Y = 1/(0.004 + 2.05/X)$] with $r^2 = 0.56$ and $p<0.0001$.

Green manure addition slightly decreased soil pH, independently of the soil γ-irradiation treatment ($F_{2,96} = 139.4$, $p<0.001$; Table 4). The total soil N content was higher at harvest in the treatments that received green manure than in those that did not ($F_{2,96} = 321.7$, $p<0.001$, Table 2 & 4).

## Discussion

The results confirmed our first prediction, because green manure addition to soil increased Zn uptake by bread wheat. The findings, however, did not unequivocally support our second

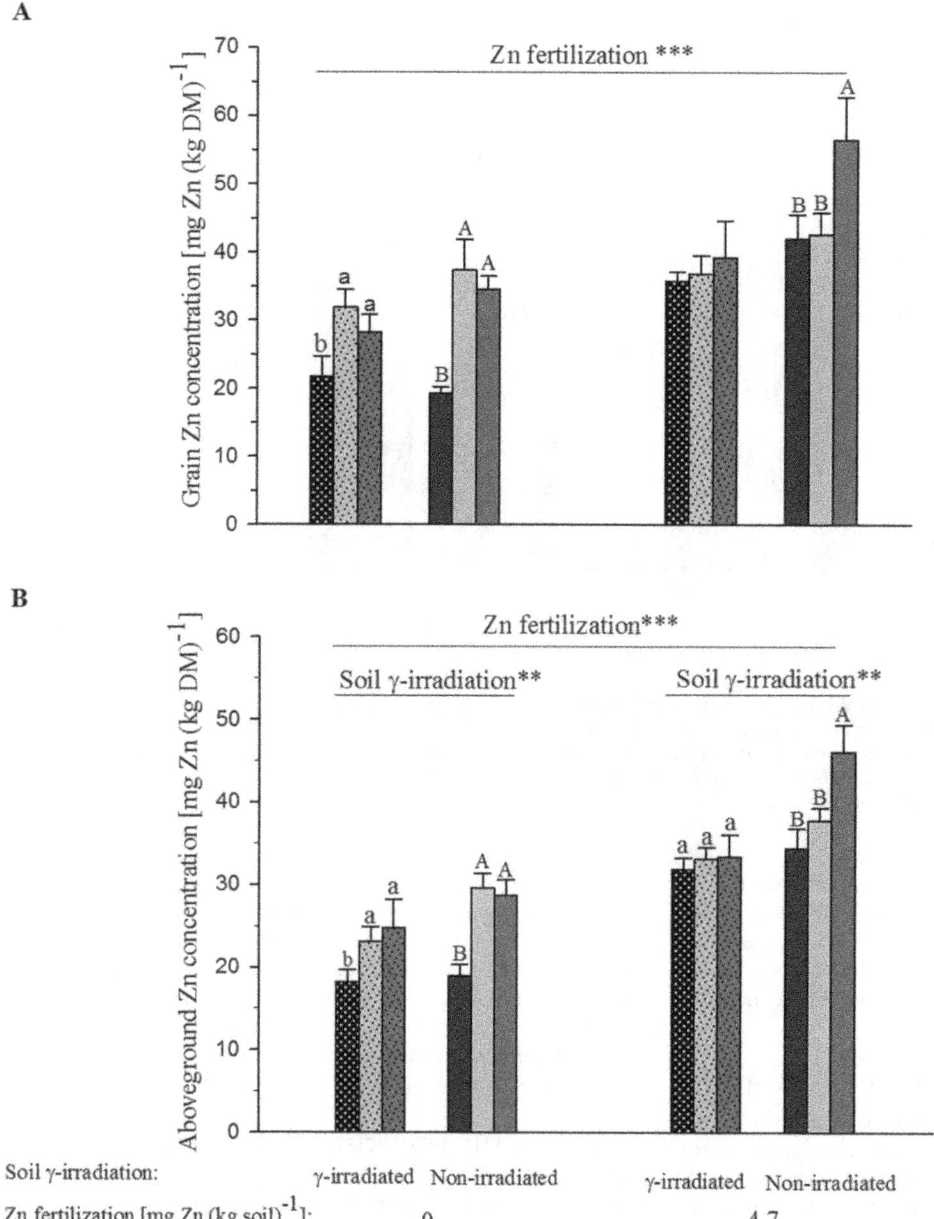


**Figure 3. Grain (A) and aboveground (B) zinc (Zn) concentrations of wheat grown in a calcareous soil.** Bar filling denoting the different green manure addition treatments is the same as in Figure 1. Bars show mean values and associated standard errors of five experimental replicates. The significances (**, $p<0.01$; ***, $p<0.001$) of the effect of mineral Zn fertilization, soil γ-irradiation of fixed factor four-way analyses of variance are shown. Different letters indicate statistical differences of separate least significant difference tests at $p<0.05$ for the different green manure addition treatments within the combinations of soil γ-irradiation and mineral Zn fertilization. For full statistical details see Table 2.

prediction, since utilization of the applied mineral Zn fertilizer by the plants did not improve in response to green manure addition. Also our last prediction that inoculation of a non-indigenous AMF strain would improve Zn transfer from soil to plants was not confirmed, not even in AMF- and thus competitor-free soil after removal of naturally occurring AMF by soil γ-irradiation. These two later findings appear to be related to the fact that Zn availability in the studied soil was neither limiting plant growth, nor yield.

In agreement with previous studies [43,44], we observed a clear increase in grain Zn concentration after application of mineral Zn fertilizer. However, application of mineral Zn fertilizer did neither translate into increased plant growth, nor higher grain yields, showing that Zn availability was not the primary plant growth-limiting factor in this soil. Sadeghzadeh [42] reports DTPA-extractable Zn levels below 0.6 mg Zn (kg soil)$^{-1}$ to limit wheat growth in calcareous soil. Since our soil contained 0.47 mg Zn (kg soil)$^{-1}$ and application of mineral Zn fertilizer did not promote wheat growth, it appears that the threshold for growth-limitation of soil Zn concentration may even be lower, although Zn uptake can obviously be raised by application of mineral Zn fertilizer.

**Table 3.** Effects of green manure addition and $ZnSO_4$ application to soil, soil γ-irradiation and inoculation of a non-indigenous arbuscular mycorrhizal fungus (AMF) on plant and soil parameters.

| Green manure addition | Soil γ-irradiation | Zn added with $ZnSO_4$ | Zn added with green manure | AMF inoculation | Zn uptake into aboveground biomass | $Zn_{dgm}$ | $Zn_{dsoil}$ | DTPA-extractable soil Zn |
|---|---|---|---|---|---|---|---|---|
| | | µg Zn (kg soil)$^{-1}$ | | | µg Zn (kg soil)$^{-1}$ | µg Zn (kg soil)$^{-1}$ | | µg Zn (kg soil)$^{-1}$ |
| **None** | **γ-irradiated** | 0 | 0 | no | 159.3±12.4 | | 159.3±12.4 | 450± 20 |
| | **γ-irradiated** | 0 | 0 | yes | 127.6± 6.9 | | 127.6± 6.9 | 440± 32 |
| | **Non-irradiated** | 0 | 0 | no | 117.4± 13.2 | | 117.4± 13.2 | 460± 35 |
| | **Non-irradiated** | 0 | 0 | yes | 127.2± 10.4 | | 127.2± 10.4 | 480± 42 |
| | **γ-irradiated** | 4700 | 0 | no | 251.6± 16.6 | | 251.6± 16.6 | 2150± 100 |
| | **γ-irradiated** | 4700 | 0 | yes | 227.3± 14.3 | | 227.3± 14.3 | 1960± 93 |
| | **Non-irradiated** | 4700 | 0 | no | 206.6± 13.2 | | 206.6± 13.2 | 2180± 110 |
| | **Non-irradiated** | 4700 | 0 | yes | 228.4± 27.5 | | 228.4± 27.5 | 1990± 160 |
| **Red clover** | **γ-irradiated** | 0 | 183 | no | 152.9± 21.4 | 5.56± 0.59 | 147.4± 21.3 | 1220± 100 |
| | **γ-irradiated** | 0 | 183 | yes | 156.9± 25.3 | 5.86± 1.03 | 151.1± 19.6 | 1500± 120 |
| | **Non-irradiated** | 0 | 183 | yes | 205.0± 18.6 | 8.52± 0.35 | 196.5± 18.4 | 2010± 70 |
| | **Non-irradiated** | 0 | 183 | yes | 176.0± 23.8 | 6.10± 0.42 | 169.9± 20.2 | 2430± 110 |
| | **γ-irradiated** | 4700 | 183 | no | 224.0± 34.1 | 4.51± 0.39 | 219.5± 34.1 | 5050± 120 |
| | **γ-irradiated** | 4700 | 183 | yes | 243.9± 10.6 | 4.93± 0.38 | 239.0± 12.6 | 3550± 74 |
| | **Non-irradiated** | 4700 | 183 | no | 202.9± 25.2 | 5.32± 0.52 | 197.6± 23.8 | 4430± 32 |
| | **Non-irradiated** | 4700 | 183 | yes | 226.7± 22.4 | 6.11± 0.54 | 220.6± 21.5 | 3950± 92 |
| **Sunflower** | **γ-irradiated** | 0 | 418 | no | 227.9± 39.6 | 10.32± 1.39 | 217.6± 31.4 | 3310± 66 |
| | **γ-irradiated** | 0 | 418 | yes | 128.9± 11.4 | 8.38± 0.96 | 120.6± 22.6 | 1750± 120 |
| | **Non-irradiated** | 0 | 418 | no | 189.9± 14.3 | 11.05± 0.87 | 178.9± 16.4 | 3230± 86 |
| | **Non-irradiated** | 0 | 418 | yes | 174.9± 30.2 | 9.45± 1.61 | 165.5± 20.7 | 1950± 33 |
| | **γ-irradiated** | 4700 | 418 | no | 258.9± 25.8 | 8.44± 0.64 | 250.5± 41.3 | 3850± 82 |
| | **γ-irradiated** | 4700 | 418 | yes | 205.8± 20.4 | 5.94± 0.75 | 200.0± 19.5 | 4990± 36 |
| | **Non-irradiated** | 4700 | 418 | no | 367.0± 40.2 | 8.5± 0.64 | 358.5± 37.4 | 4540± 120 |
| | **Non-irradiated** | 4700 | 418 | yes | 227.9± 24.5 | 5.97± 0.80 | 222.0± 26.2 | 4680± 130 |

DTPA: diethylene-triamine-penta-acetic acid

Mean values and associated standard errors of five experimental replicates of Zn uptake to the aboveground biomass, Zn in the aboveground biomass derived from green manure ($Zn_{dgm}$), Zn derived from soil and fertilizer ($Zn_{dsoil}$) and DTPA-extractable Zn from soil at grain maturity are listed.

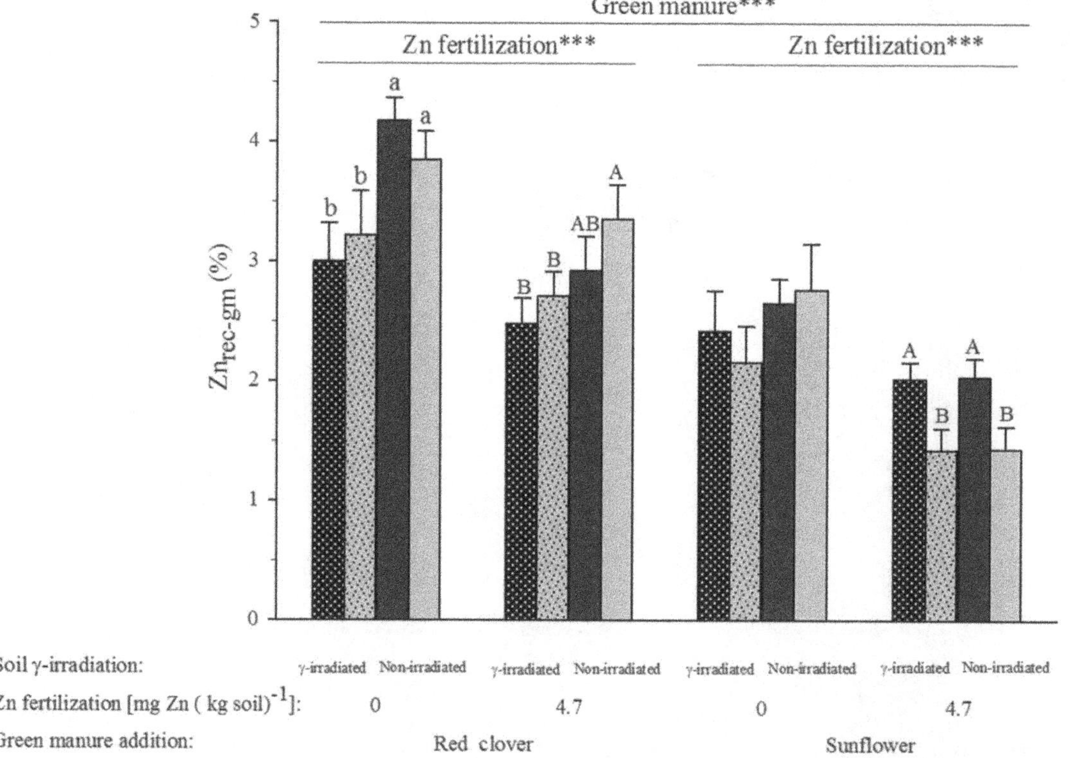


**Figure 4. Percentage of zinc (Zn) recovered from green manure ($Zn_{rec-gm}$) in the aboveground wheat biomass.** Black bar filling denotes treatments with arbuscular mycorrhizal fungus (AMF) inoculation and light grey filling treatments without AMF inoculation. Bar hatching highlights treatments in which the naturally occurring AMF were missing after γ-irradiating the soil. Bars show mean values and associated standard errors of five experimental replicates. The significances (***, $p<0.001$) of the effect of green manure addition and mineral Zn fertilization from a fixed factor four-way analysis of variance are indicated above the bars. For full statistical details see Table 2.

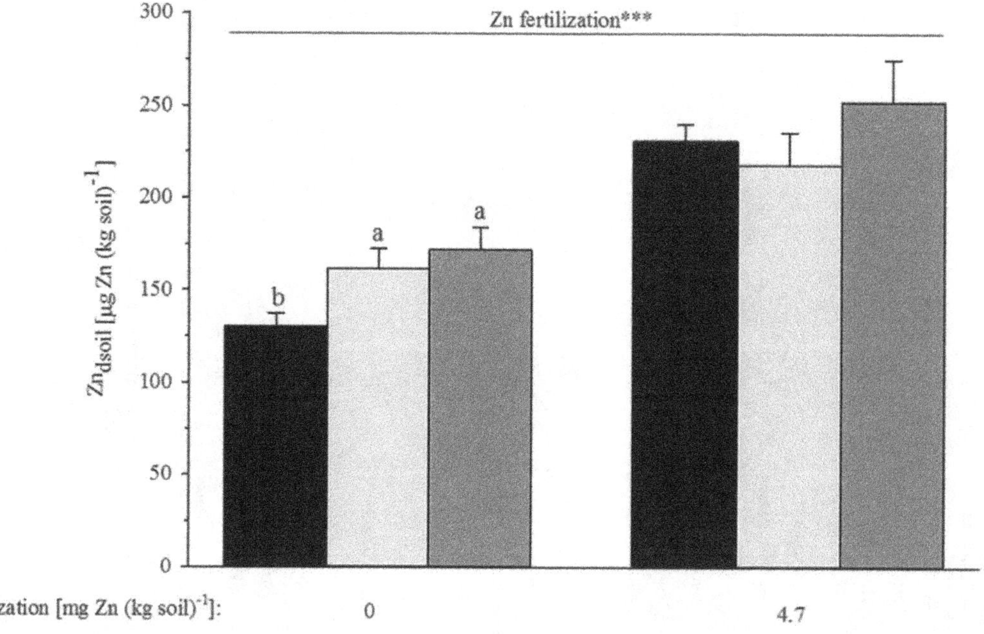


**Figure 5. Zinc (Zn) uptake from soil to the aboveground biomass of bread wheat grown in a calcareous soil after addition of two types of green manure.** The Zn taken up from soil either originated from the native soil Zn pool or from soil as well as applied $ZnSO_4$ fertilizer. Bar filling denoting the different green manure addition treatments is the same as in Figure 1. Bars show mean values and associated standard errors of five experimental replicates. The significance (***, $p<0.001$) of the effect of mineral Zn fertilization from a fixed factorial four-way analysis of variance is indicated above the bars. Different letters indicate statistical differences of separate least significant difference tests at $p<0.05$ for the different green manure addition treatments within the different Zn fertilization levels. For full statistical details see Table 2.

**Table 4.** Total soil nitrogen (N) concentration and pH in water at wheat harvest.

| Soil | Green manure | Total N (g $kg^{-1}$) | pH ($H_2O$) |
|---|---|---|---|
| | | (Mean ± SE) | (Mean ± SE) |
| Non-irradiated | None | $0.55\pm0.02^B$ | $7.75\pm0.03^A$ |
| | + Red clover | $0.87\pm0.02^A$ | $7.50\pm0.02^B$ |
| | + Sunflower | $0.88\pm0.02^A$ | $7.53\pm0.01^B$ |
| γ-irradiated | None | $0.63\pm0.01^b$ | $7.64\pm0.02^a$ |
| | + Red clover | $0.89\pm0.02^a$ | $7.46\pm0.01^b$ |
| | + Sunflower | $0.96\pm0.02^a$ | $7.48\pm0.01^b$ |

DTPA: diethylene-triamine-penta-acetic acid
The values represent the means and associated standard errors (SE) of 20 experimental units. Different superscript letters of the same type following the SE indicate statistical difference among the means of the three green manure treatments within each soil γ-irradiation treatment at $p<0.05$, according to least significance difference tests.

## Mixing green manure to soil increased Zn uptake by wheat via Zn mobilizing effects

The observed increase in Zn uptake from the native soil Zn pool after green manure addition could originate from changes to the soil conditions, root growth, and/or plant physiology. Only addition of green manure to the soil increased the DTPA-extractable Zn from 0.47 to 0.5 or 2 mg Zn (kg soil)$^{-1}$, when green manure of red clover and sunflower were added, respectively. Combining $ZnSO_4$ application with green manure addition to soil increased the DTPA-extractable Zn concentration even further, namely up to 3.8 mg Zn (kg soil)$^{-1}$ when red clover was used and up to 5.0 mg Zn (kg soil)$^{-1}$ when sunflower was used, respectively (Table 3). We thus can conclude that the lack of a plant growth response to application of mineral Zn fertilizer (Figure 1A&B) must be explained by an alternative plant growth-limiting factor and not the >0.6 mg Zn (kg soil)$^{-1}$ of DTPA-extractable Zn [42].

The larger increase in DTPA-extractable soil Zn in response to addition of green manure of sunflower than such of red clover (Table 3) could have been due to the Zn-richer green manure of sunflower than the green manure of red clover. For production of the green manure of sunflower, but not that of red clover, only leaf laminae were used, which could explain the higher mineral nutrient concentrations of it (Table 1). Moreover, the relatively young age (54 day) and bigger sunflower than red clover seeds with own nutrient reserves may have further contributed to nutrient concentration differences of the two types of green manure (Table 1), translating into differences in plant Zn uptake.

There may have been further co-nutritional or root- and microbial growth-mediated Zn uptake stimulation, because the green manure produced of sunflower was also richer in N than that of red clover (Table 1). The green manure of the leaf laminae of the well N-supplied sunflower plants could have released N during its decomposition and stimulated root growth and possibly the synthesis of Zn facilitating N-rich plant and microbial compounds. Elevated root and microbial exudation and necromass in response to green manure of sunflower may also explain the higher DTPA-extractable Zn concentrations of the soil and increased plant Zn uptake. Root- and microbe-derived organic acids, amides and (phyto-) siderophores may have chelated and thereby mobilized Zn for uptake by the wheat plants. More N released from the decomposing green manure of more N-rich sunflower than red clover material (Table1) may have supported the N-demanding biosynthesis of Zn chelating compounds [8,14,45].

Elevated root and microbial activity in response to green manure addition may further have lowered soil pH (Table 4) and thereby mobilized soil Zn. Similarly, stimulation of nitrification upon higher N inputs with the green manure of sunflower than that of red clover and thus release of protons into the soil solution may have raised Zn bioavailability in soil for uptake by the mycorrhized wheat plants.

All in all, the effectiveness of green manure addition to soil to release Zn from the native soil Zn pool for plant uptake was remarkable (Figure 5). This stimulatory effect on plant Zn uptake diminished, however, when additionally $ZnSO_4$ fertilizer was applied. This is probably, because the applied $ZnSO_4$ fertilizer introduced already a high amount of Zn, compared to the amount solubilized by chelating compounds released upon addition of green manure, reducing Zn mobilization form soil.

## Inoculation of a non-indigenous AMF strain did not increase Zn uptake in wheat

A previous experiment (Forough Aghili, unpublished) with the same soil and wheat cultivar and additional analyses on samples of this experiment (Anouk Guyer, unpublished data) showed that the roots of the wheat plants were colonized by naturally occurring AMF of this soil. 1.6 kb-long ribosomal DNA sequences confirmed root colonization by the inoculated Swiss strain of *C. claroideum* of the wheat plants raised in γ-irradiated soil. We suspect that maladaptation of the AMF inoculant to the living conditions in the experimental soil [46] must have compromised symbiotic Zn acquisition by wheat. The inoculated AMF was originally isolated from Swiss arable soil, not calcareous, and with a loamy texture, a lower, near neutral pH, and much richer in organic matter and bioavailable nutrients [47] than the study soil here. The improved Zn recovery from green manure of sunflower after application of mineral Zn fertilizer by the plants inoculated with the foreign AMF strain points at possibly complex soil-root-microbe interactions beneficial to mycorrhized plants. Lack, or just minor positive effects by the non-indigenous AMF inoculant should draw our attention to the fact that AMF inoculants can not be *a priori* considered effective in assisting crop plants to take up Zn from soil when introduced to soils of different physicchemical and biological properties.

γ-Irradiation has been shown to modify the physicochemical soil properties to some degree [48,49]. However, we believe that this did not invalidate our results. While the increased biomass (Figure 1) and N uptake (Figure 2) of the wheat plants grown in γ-irradiated soil can probably be related to released N from decomposing microbial necromass [49], the magnitude of this effect was certainly much smaller than that by the N added with the green manures (Figure 2, Table 2). Likewise, the levels of dissolved organic compounds [48] must have risen in response to γ-irradiation, but again this effect must have been considerably smaller than that caused by mixing 0. 4% (w/w) finely ground green manure to soil. Therefore, following Thompson [33], and until a mycorrhiza symbiosis-defective wheat mutant becomes available, we see no better approach than using γ-irradiation for the study of the influence of entire natural AMF assemblages on plant Zn uptake from real arable soil.

The finding of higher grain and aboveground Zn concentrations in plants raised on native than γ-irradiated soil (Figure 3) points at no Zn uptake stimulatory effect by elevated levels of dissolved organic compounds [48] and N after soil γ-irradiation [49]. The generally increased Zn tissue concentration after addition of green manure to soil (Figure 3) and higher Zn recovery from less N- and Zn-rich green manure of red clover than more N- and Zn-rich green manure of sunflower point at considerable influences by C- rather than N-limited saprotrophic soil microbes on Zn uptake of mycorrhized bread wheat plants. Being obligate biotrophs, AMF rely on saprotrophs for remobilization of mineral nutrients from decomposing green manure that they may as soon as available take up and then possibly deliver to their host plants as observed for the Zn from the green manure of sunflower (Figure 4). Much higher Zn recovery from the Zn- and N-poorer green manure of red clover than the Zn-richer and N-poorer green manure of sunflower (Figure 4) points at faster decomposition of the earlier than latter and thus also different speeds in nutrient recycling from different green manures. However, the two types of green manure appear not to have differed in Zn mobilization from soil (Figure 5).

Native as opposed to γ-irradiated soil after back-addition of an AMF propagule-free microbial filtrate of a water suspension of living soil does certainly not only differ in AMF occurrence. Besides the physicochemical changes discussed above [49], largely stochastic epidemic population growth of saprophytic organisms must occur [35]. Therefore, importance of fungi and microbes conferring similar nutritional benefits to plants as AMF [50] may increase in absence of otherwise naturally occurring AMF and much reduced population sizes of most of other just re-introduced soil microbes. Microbial nutrient competition with plants [34] may be particularly harsh when microbial population growth after inoculation coincides with release of nutrients from new microbial necromass and freshly applied mineral fertilizer, as it must have occurred in the γ-irradiated control soil of this experiment. The re-introduced soil microbes in the filtrate of the soil suspension and the newly inoculated foreign AMF inoculant may thus have mainly sequestered Zn and other nutrients in their newly formed biomass in γ-irradiated soil [34]. This would further explain why the inoculated foreign AMF strain did not assist much in plant Zn uptake.

In summary, this study advanced our knowledge on possible agronomic practices to biofortify cereal grains with Zn. i) It showed that addition of readily decomposable N-rich green manure to calcareous soil can raise grain Zn concentration of bread wheat to levels approaching those of only $ZnSO_4$ fertilizer-fed wheat plants; ii) The study showed that combined addition of N-rich green manure and $ZnSO_4$ fertilizer can considerably raise grain Zn concentration; iii) Mixing green manure to the soil showed that the native soil Zn pool can contribute considerably to total Zn accumulation in the aboveground plant biomass even when no additional mineral Zn fertilizer had been applied to the soil; iv) Lastly, the study emphasizes that care should be taken when trying to utilize AMF inoculants for supporting crop plants in their Zn uptake, when the fungi are not adapted to the prevailing soil conditions.

## Acknowledgments

The authors would like to thank Dr. Laurie P. Mauclaire Schönholzer, Thomas Flura and Max Ruethi for their assistance in running the analytical equipment and Simon Ineichen for practical help in sample processing. We acknowledge critical comments by two anonymous reviewers that helped to clarify the main message of this study.

## Author Contributions

Conceived and designed the experiments: JJ FA EF. Performed the experiments: FA. Analyzed the data: FA EF HAG. Wrote the paper: FA HAG EF RS. Provided the AM fungal inoculant: JJ. Provided the plant seeds and the soil: AHK MA. Enabled gamma-countingL JE. Approved the final manuscript version: JJ FA EF HAG AHK MA JE RS.

## References


1. Curtis BC (2002) Wheat in the world. In B. C. . Curtis, S. . Rajaram, H. Gomez Macpherson (eds.), Bread wheat: Improvement and production. Food and Agriculture Organization of the United Nations, Rome, Italy.
2. Black RE, Victora CG, Walker SP, Bhutta ZA, Christian P, et al. (2013) Maternal and child undernutrition and overweight in low-income and middle-income countries. Lancet 382: 427–451.
3. Welch RM, Graham RD (2004) Breeding for micronutrients in staple food crops from a human nutrition perspective. J Exp Bot 55: 353–364.
4. Cakmak I, Kalayci M, Kaya Y, Torun AA, Aydin N, et al. (2010) Biofortification and localization of zinc in wheat grain. J Agr Food Chem 58: 9092–9102.
5. Cakmak I (2008) Enrichment of cereal grains with zinc: Agronomic or genetic biofortification? Plant Soil 302: 1–17.
6. Khoshgoftar AH, Shariatmadari H, Karimian N, Khajehpour MR (2006) Responses of wheat genotypes to zinc fertilization under saline soil conditions. J Plant Nutr 29: 1543–1556.
7. White PJ, Broadley MR (2011) Physiological limits to zinc biofortification of edible crops. Front Plant Sci 2: 80.
8. Soltani S, Khoshgoftarmanesh AH, Afyuni M, Shrivani M, Schulin R (2013) The effect of preceding crop on wheat grain zinc concentration and its relationship to total amino acids and dissolved organic carbon in rhizosphere soil solution. Biol Fert Soils 50: 239–247.
9. Cavagnaro TR (2008) The role of arbuscular mycorrhizas in improving plant zinc nutrition under low soil zinc concentrations: a review. Plant Soil 304: 315–325.
10. Lehmann A, Veresoglou SD, Leifheit EF, Rillig MC (2014) Arbuscular mycorrhizal influence on zinc nutrition in crop plants – A meta-analysis. Soil Biol Biochem 69: 123–131.
11. Ryan MH, Angus JF (2003) Arbuscular mycorrhizae in wheat and field pea crops on a low P soil: increased Zn-uptake but no increase in P-uptake or yield. Plant Soil 250: 225–239.
12. Ryan MH, Derrick JW, Dann PR (2004) Grain mineral concentrations and yield of wheat grown under organic and conventional management. J Sci Food Agr 84: 207–216.
13. Diesing WE, Sinaj S, Sarret G, Manceau A, Flura T, et al. (2008) Zinc speciation and isotopic exchangeability in soils polluted with heavy metals. Eur J Soil Sci 59: 716–729.
14. Gramlich A, Tandy S, Frossard E, Eikenberg J, Schulin R (2013) Availability of zinc and the ligands citrate and histidine to wheat: does uptake of entire complexes play a role? J Agric Food Chem 61: 10409–10417.
15. Alloway BJ (2009) Soil factors associated with zinc deficiency in crops and humans. Environ Geochem Hlth 31: 537–548.
16. Gao X, Zhang F, Hoffland E (2009) Malate exudation by six aerobic rice genotypes varying in zinc uptake efficiency. J Environ Qual 38: 2315–2321.
17. Daneshbakhsh B, Khoshgoftarmanesh AH, Shariatmadari H, Cakmak I (2013) Phytosiderophore release by wheat genotypes differing in zinc deficiency tolerance grown with Zn-free nutrient solution as affected by salinity. J Plant Physiol 170: 41–46.
18. Dakora FD, Phillips DA (2002) Root exudates as mediators of mineral acquisition in low-nutrient environments. Plant Soil 245: 35–47.
19. Hoffland E, Wei CZ, Wissuwa M (2006) Organic anion exudation by lowland rice (*Oryza sativa L.*) at zinc and phosphorus deficiency. Plant Soil 283: 155–162.
20. Ueno D, Rombola AD, Iwashita T, Nomoto K, Ma JF (2007) Identification of two novel phytosiderophores secreted by perennial grasses. New Phytol 174: 304–310.

21. Parniske M (2008) Arbuscular mycorrhiza: the mother of plant root endosymbioses. Nat Rev Microbiol 6: 763–775.
22. Bonfante P, Genre A (2010) Mechanisms underlying beneficial plant-fungus interactions in mycorrhizal symbiosis. Nat Commun 1: 48.
23. Fitter AH (1991) Costs and benefits of mycorrhizas - implications for functioning under natural conditions. Experientia 47: 350–355.
24. Joner EJ, Jakobsen I (1995) Growth and extracellular phosphatase-activity of arbuscular mycorrhizal hyphae as influenced by soil organic-matter. Soil Biol Biochem 27: 1153–1159.
25. Anonymous (2010). Keys to soil taxonomy. Natural Resources Conservation Service, United States Department of Agriculture, Washington, D.C., 338 p.
26. Mostafazadeh-Fard B, Mansouri H, Mousavi SF, Feizi M (2009) Effects of different levels of irrigation water salinity and leaching on yield and yield components of wheat in an arid region. J Irrig Drain E-Asce 135: 32–38.
27. Gee GW, Bauder JW (1986) Particle size analysis. In: A Klute, editor editors. Methods of soil analysis. Madison, Wisconsin, USA: American Society of Agronomy. pp. 383–411.
28. Rhoades JD (1996) Salinity: Electrical conductivity and total dissolved salts. Methods of soil analysis. Madison, Wisconsin, USA: Soil Science Society of America. pp. 417–435.
29. Walkey A, Black IA (1934) An examination of the Degtjareff method for determinig soil organic matter and proposed modification of the chromic acid titration method. Soil Sci 37: 29–38.
30. Burt R (2004) Soil survey laboratory methods manual: soil survey investigations report Nebraska: United States Department of Agriculture. Natural Resources Conservation Servic.
31. Lindsay WL, Norvell WA (1978) Development of a DTPA soil test for zinc, iron, manganese, and copper. Soil Sci Soc Am J 42: 421–428.
32. Maas EV (1986) Salt tolerance of plants. Appl Agric Res 1: 12–25.
33. Thompson JP (1990) Soil sterilization methods to show VA-mycorrhizae aid P and Zn nutrition of wheat in vertisols. Soil Biol Biochem 22: 229–240.
34. Nazeri N, Lambers H, Tibbett M, Ryan M (2013) Do arbuscular mycorrhizas or heterotrophic soil microbes contribute toward plant acquisition of a pulse of mineral phosphate? Plant Soil 373: 699–710.
35. van de Voorde TFJ, van der Putten WH, Bezemer TM (2012) Soil inoculation method determines the strength of plant-soil interactions. Soil Biol Biochem 55: 1–6.
36. Barker SJ, Stummer B, Gao L, Dispain I, O'Connor PJ, et al. (1998) A mutant in *Lycopersicon esculentum* Mill. with highly reduced VA mycorrhizal colonization, isolation and preliminary characterisation. Plant J 15: 791–797.
37. Hogland DR, Arnon OD (1938) The water-culture method of growing plants without soil. University of California college of agriculture, agricultural experiment station Berkeley, California. USA.
38. Schenck NC, Smith GS (1982) Additional new and unreported species of mycorrhizal fungi (Endogonaceae) from Florida. Mycologia 74: 77–92.
39. Gerdemann JW, Nicolson TH (1963) Spores of mycorrhizal *Endogone* species extracted from soil by wet sieving and decanting. T Brit Mycol Soc 46: 235–244.
40. Cavagnaro TR, Smith FA, Hay G, Carne-Cavagnaro VL, Smith SE (2004) Inoculum type does not affect overall resistance of an arbuscular mycorrhiza-defective tomato mutant to colonisation but inoculation does change competitive interactions with wild-type tomato. *New Phytol* 161: 485–494.
41. Karami M, Afyuni M, Khoshgoftarmanesh AH, Papritz A, Schulin R (2009) Grain zinc, iron, and copper concentrations of wheat grown in central Iran and their relationships with soil and climate variables. J Agr Food Chem 57: 10876–10882.
42. Sadeghzadeh B (2013) A review of zinc nutrition and plant breeding. J Soil Sci Plant Nutr 13: 905–927.
43. Kalayci M, Torun B, Eker S, Aydin M, Ozturk L, et al. (1999) Grain yield, zinc efficiency and zinc concentration of wheat cultivars grown in a zinc-deficient calcareous soil in field and greenhouse. Field Crop Res 63: 87–98.
44. Rengel Z (1999) Zinc deficiency in wheat genotypes grown in conventional and chelator-buffered nutrient solutions. Plant Sci 143: 221–230.
45. Neilands JB (1995) Siderophores: Structure and function of microbial iron transport compounds. J. Biol. Chem 270: 26723–26726.
46. Oliveira RS, Vosatka M, Dodd JC, Castro PML (2005) Studies on the diversity of arbuscular mycorrhizal fungi and the efficacy of two native isolates in a highly alkaline anthropogenic sediment. Mycorrhiza 16: 23–31.
47. Jansa J, Mozafar A, Anken T, Ruh R, Sanders IR, et al. (2002) Diversity and structure of AMF communities as affected by tillage in a temperate soil. Mycorrhiza 12: 225–234.
48. Marschner B, Bredow A (2002) Temperature effects on release and ecological properties of dissolved organic matter (DOM) in sterilized and biologically active soil samples. Soil Biol Biochem 34: 459–466.
49. McNamara NP, Black HIJ, Beresford NA, Parekh NR (2003) Effects of acute gamma irradiation on chemical, physical and biological properties of soils. Appl Soil Ecol 24: 117–132.
50. Kariman K, Barker SJ, Jost R, Finnegan PM, Tibbett M (2014) A novel plant–fungus symbiosis benefits the host without forming mycorrhizal structures. New Phytol 201: 1413–1422.

# 6

# Dryland Soil Hydrological Processes and Their Impacts on the Nitrogen Balance in a Soil-Maize System of a Freeze-Thawing Agricultural Area

**Wei Ouyang[1]*, Siyang Chen[2], Guanqing Cai[1], Fanghua Hao[1]**

**1** School of Environment, State Key Laboratory of Water Environment Simulation, Beijing Normal University, Beijing, China, **2** Marine Monitoring and Forecasting Center of Zhejiang, Hangzhou, China

## Abstract

Understanding the fates of soil hydrological processes and nitrogen (N) is essential for optimizing the water and N in a dryland crop system with the goal of obtaining a maximum yield. Few investigations have addressed the dynamics of dryland N and its association with the soil hydrological process in a freeze-thawing agricultural area. With the daily monitoring of soil water content and acquisition rates at 15, 30, 60 and 90 cm depths, the soil hydrological process with the influence of rainfall was identified. The temporal-vertical soil water storage analysis indicated the local *albic* soil texture provided a stable soil water condition for maize growth with the rainfall as the only water source. Soil storage water averages at 0–20, 20–40 and 40–60 cm were observed to be 490.2, 593.8, and 358 $m^3$ $ha^{-1}$, respectively, during the growing season. The evapo-transpiration (ET), rainfall, and water loss analysis demonstrated that these factors increased in same temporal pattern and provided necessary water conditions for maize growth in a short period. The dry weight and N concentration of maize organs (root, leaf, stem, tassel, and grain) demonstrated the N accumulation increased to a peak in the maturity period and that grain had the most N. The maximum N accumulative rate reached about 500 mg $m^{-2}d^{-1}$ in leaves and grain. Over the entire growing season, the soil nitrate N decreased by amounts ranging from 48.9 kg N $ha^{-1}$ to 65.3 kg N $ha^{-1}$ over the 90 cm profile and the loss of ammonia-N ranged from 9.79 to 12.69 kg N $ha^{-1}$. With soil water loss and N balance calculation, the N usage efficiency (*NUE*) over the 0–90 cm soil profile was 43%. The soil hydrological process due to special *soil* texture and the temporal features of rainfall determined the maize growth in the freeze-thawing agricultural area.

**Editor:** Guoping Zhang, Zhejiang University, China

**Funding:** We are grateful for assistance with the data requirements of the Bawujiu Farm in Heilongjiang Province. The research discussed in this paper benefited from financial support provided by the National Natural Science Foundation of China (Grant No. 41371018, 51121003), the Supporting Program of the "Twelfth Five-year" Plan for Sci & Tech Research of China (2012BAD15B05), and The Fundamental Research Funds for the Central Universities. The funders had no role in study design, data collection and analysis, decision to publish, or preparation of the manuscript.

**Competing Interests:** The authors have declared that no competing interests exist.

* Email: wei@itc.nl

## Introduction

During the agricultural tillage management, the maximum crop production and minimum diffuse nitrogen (N) loading are the priority issues that need to be considered at the same time [1][2]. It is essential to understand the fates of soil water (SW) and N in agricultural systems in order to attain higher crop yields and N usage efficiency [3]. In developing countries, there is much political and commercial pressure for controlling the N application with the sacrifice of the crop harvest [4]. The dryland agriculture in the Sanjian Plain, Northeast China, is a key food base and the most water limited agricultural zone in China [5]. In this freeze-thawing area, the soil N and water efficiency are the keys for dryland tillage sustainability due to the short growing season. However, there are few reports about the dryland soil hydrological process and N use efficiency in freeze-thawing agricultural areas [6].

The soil nutrient N is a major limiting factor for crop growth and production, which is directly related to the N fertilisation and soil N background level [7]. The soil microbial community, which affects the soil N cycle [8], is less active in this freeze-thawing area. The application of chemical N is a basic agricultural practice in worldwide, which causes excessive discharge of N to the aquatic environment [9]. The European Union agricultural landscape contributes about 55% of the diffuse pollution for water eutrophication. This is a major emerging environmental issue in the developing counties [10]. The soil eco-hydrological process is an essential channel for N transport in the soil-crop system.

Plant organs have different allocation rates with N during growth, and the canopy green area has higher N absorption than the other organs during the jointing period [11]. The N accumulations in the upper leaves of the canopy also follow the eco-hydrological pattern in farmland [12]. Some models have been developed to simulate the N absorption in different organs during crop growth with the impacts of the solar radiation, rainfall, temperature, soil hydrology and N concentration [13]. With the temporal pattern of N accumulation in crop organs, the N cycle and efficiency can also be analyzed. The diffuse N discharge from agricultural systems depends on diverse factors, including the climatic, soil properties and agronomic features [14][15]. The tillage practice, soil hydrological process and N usage efficiency are the main potential factors to consider when the goal is to reduce the agricultural diffuse N pollution [16]. The dryland maize farmland in a freeze-thawing agricultural area presents different N discharge patterns due to the special hydrological process [17].

The soil hydrological process is the bridge for diffuse N loss from soil to water bodies [18]. The nitrate-N is the dominant type of diffuse N in the soil profile and it can move swiftly downstream flow with the rainfall through the soil pathways [19]. The soil hydrological process is also closely related to the soil texture, which is the medium for water movement and crop growth. In the freeze-thawing dryland, the soil has short-term, transitional phase and rainfall is the only water source for N movement with crop growth [20]. The soil pattern in the study area is meadow *albic* soil (*Albic Luvisols*) and has a 20–30 cm depth impermeable stratum. The *albic* soil provides a stable soil water (SW) condition in the crop root zone and interfaces with the soil hydrological process. With the consideration of high N efficiency, it is necessary to analyze the special pattern of soil hydrology and its impact on the N efficiency [21]. Evaluating the response of N loss to the combination of rainfall and the soil hydrological process can help to identify the interactions of water with N and reduce diffuse N pollution.

The general objective is to maximize the N efficiency in tillage practices. Typically, of about 20%–70% of N is lost from soil-crop systems and this loss can cause other environmental issues [22]. The quality of soil properties will decline under the combination of intensive application of N fertilizer and low N efficiency; this in turn, will also deliver some risk to water [23]. There are several methods to improve the N efficiency, including accurate fertilizer application and irrigation [24][25]. The dryland soil hydrological process in a freeze-thawing agricultural area presents a special pattern and also affects the N efficiency [26]. This study aims (1) to identify the temporal-vertical dynamics of dryland SW within an *albic* soil layer; (2) to express the dry weight and N accumulation of maize organs in dryland of a freeze-thawing agricultural area; (3) to explore the variation of soil loss, potential N loss, and N use efficiency (*NUE*) in dryland during the growing season.

## Materials and Methods

### 2.1 Ethics statement

The field experiment was conducted in the farmland of Bawujiu Farm, which has been rented by Wang Dongli for 50 years. He is also the contact person for future permission. No specific permissions were required for these locations. The field studies did not involve endangered or protected species. This study did not involve vertebrate. The specific location of our study is 133° 50′–134° 33′ E, 47° 18′47°50′ N.

### 2.2 Study area description

The case study area was conducted on a farm in the north-eastern part of China. The farm has a long history and on the east is adjoined to Russia (Fig. 1). The observation site is located in the southern part of the Farm and the elevation is 48 m. The only water source for dryland tillage is precipitation and maize is the dominant crop. This area the temperate continental monsoon climate has an average annual precipitation of 588 mm and an average yearly temperature of 2.94 °C [27]. From October to the following April, the temperature is below zero and, as a result, the soil freeze-thawing process affects the tillage process. The crop growing season is from May to October.

### 2.3 Field monitoring and soil sample collection

In order to identify soil physical-chemical properties, soil samples at four depths (0–15, 15–30, 30–60 and 60–90 cm) were collected in 2010. The samples from three points chosen at random in each layer were mixed into a composite soil sample and placed in plastic bags for lab measurements. The soil samples were collected before sowing and after harvest. Soil particle size distributions were measured by laser diffraction after the removal of organic residue (MasterSizer S, Malvern Instruments, Malvern, UK). The soil's total nitrogen (TN) and organic carbon (OC) concentrations were measured with a CHN Elemental Analyser (Euro VectorS.P.A EA3000, 136 Milan, Italy) [28]. The ammonia N in rainfall was analyzed with the aid of Nessler's reagent and the nitrate-N concentration was determined by the Ultraviolet Spectro Photometric Method. The soil's pH was determined with a pH meter (METTLER TOLEDO, Switzerland) after mixing the fresh soil with deionised water (1:2.5, w/v) [24]. The soil N stock at four depths was calculated with the following equation,

$$Stock_i = BD_i * H_i * C_i * 10^5 \tag{1}$$

where Stock$_i$ is the soil N storage at layer $i$ (kg/m$^2$), $BD_i$ is the soil bulk density at layer $i$ (g/cm$^3$), $H_i$ is the soil depth at layer $i$ (cm), and $C_i$ is the soil N concentration at layer $i$ (mg/kg).

For the soil hydrological process monitoring, the SW collector head (suction cup: Teflon and quartz; OD 21 mm×L 95 mm; porosity: 2 μm; conductivity: $3.31\times10^{-6}$ mm sec$^{-1}$, PRENART, Danmark) was set at 15, 30, 60 and 90 cm depths. It was covered with a mixed solution of quartz powder and stirred, which can cause a 0.05 MPa vacuum pressure and in direct contact with the soil. The collector tail was connected with PVC pipe and the water samples were collected in bottles by a portable manual vacuum pump. The soil water storage was calculated with soil volumetric water content at different depths. The mean soil water storage at each layer was defined as the average of SW in the upper and lower soil layers. The evapo-transpiration in dryland is calculated with the Bowen ratio and energy balance (BREB). The SW balance was calculated with the following equation [29],

$$SL = R - I_r - ET - \Delta S \tag{2}$$

where $SL$ is the soil water loss (mm), $R$ is the precipitation (mm), $ET$ is the evapo-transpiration (mm), $I_r$ is the canopy interception (mm), and $\Delta S$ is the soil water storage change (mm).

The local meteorological characteristics were monitored with a ZENO station (Coastal, Seattle, WA, USA). Automated soil temperature sensors (Thermistor, Coastal, Seattle, WA, USA) and soil volumetric water content sensors (TDR type, Coastal, Seattle, WA, USA) were placed in each of the four different soil layers. The monitoring sensors were repeated at double at horizontal 0.5 m space.

### 2.4 Crop sample collection and measurement

The local dryland is intensively cultivated with the aid of mechanized equipment and the maize (*Zea mays L.*) planting density is 75,000 plants ha$^{-1}$. The detailed maize planting date, harvest date, relevant tillage and fertilization practices are listed in Table 1. The maize samples in 2011 were collected monthly and the sample in 2010 were only collected in October due to the limited experimental conditions. The detailed process of sample collection and monitoring of crop features (height (CH), leaf area index (LAI) and root depth (RD)) are listed in a previously published paper [24]. The maize samples were taken to the lab

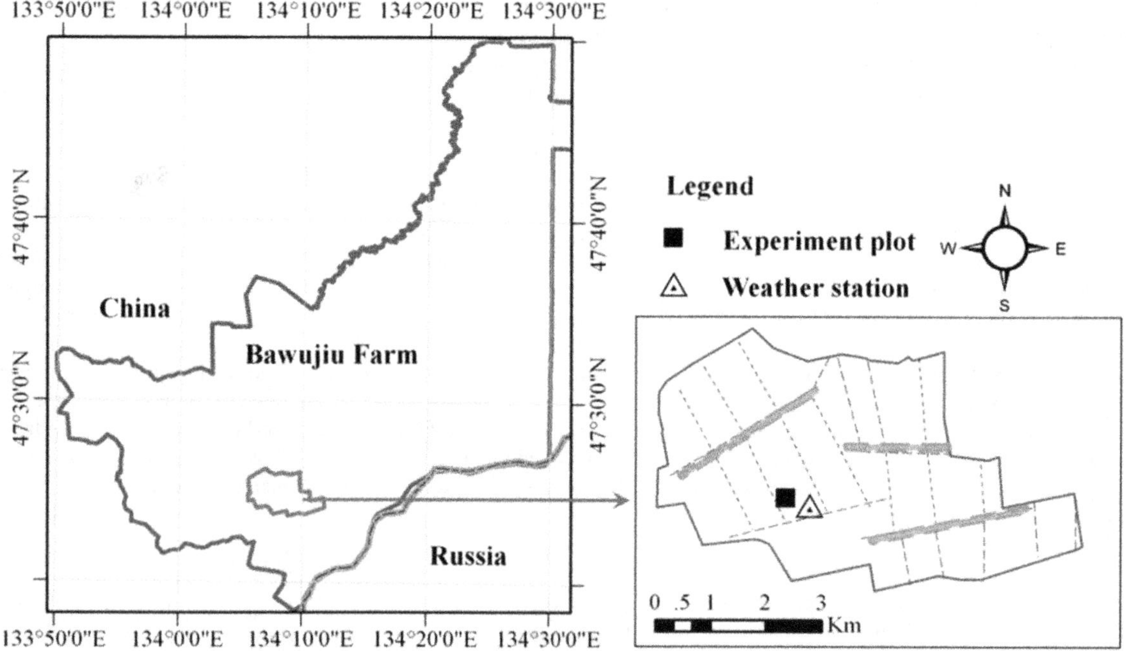


**Figure 1. Location of the experimental dryland site in Northeast of China.**

immediately and the root, leaves, stems, tassels, and grains were separated. The samples were washed clean and then dried for 48 h at 65°C. The dry weight of organ samples was determined after fixing the temperature for 30 min at 105°C. The dried maize samples were crushed and sieved for crop organ N content (%) analysis (Vario EL, Elementar Co. Ltd., Germany). The N uptake amount by crop was calculated based on crop dry weight and plant N content (%).

## 2.5 Nitrogen supply and loss calculation

The N use efficiency (*NUE*) in a crop-soil system is defined as the ratio of N uptake ($N_c$) by a crop to N supply ($N_s$) in a system [30]. The $N_s$ was defined as the total available N to the crop, which included fertilizer N ($N_f$), mineralized N ($N_m$), initial mineralized N in soil ($N_{min\ initial}$), N fixed by soil ($N_x$) and N deposition ($N_d$) from rainfall. The $N_s$ is calculated as:

$$N_s = N_f + N_{\min\ initial} + N_m + N_x + N_d \tag{3}$$

The field $N_m$ is calculated with the crop-soil N equation [31]:

$$\begin{aligned} N_m + (N_{\min\,initial} - N_{\min\,final}) = \\ (N_l + N_c + N_{gl}) - (N_f + N_d + N_x) \end{aligned} \tag{4}$$

where $N_m$ is the mineralized N, $N_{min\ initial}$ is the initial mineralized N concentration in soil, $N_{min\ final}$ is the final mineralized N concentration in soil, $N_l$ is the soil loss N, $N_{gl}$ is the N loss in gas. $N_c$ is the N crop absorption, $N_x$ is the N fixed by soil, and $N_d$ is the N deposition.

**Table 1.** Dryland tillage practices and fertilization information.

| Crop year | Planting date | Harvest date | Tillage prior to planting | Main fertilisation (kg ha$^{-1}$) |
|---|---|---|---|---|
| 2010 | 8/Jun. | 9/Oct. | Tilling | 525 |
| 2011 | 30/May. | 5/Oct. | | 525 |
| Detailed fertilisation data and amounts * (kg ha$^{-1}$) | | | | |
| 2010 | 2011 | N | P | K |
| 8/Jun. | 30/May. | 71.25 | 58.65 | 27.23 |
| 27/Jun. | 20/Jun. | 34.50 | 13.80 | 10.02 |
| 5/Jul. | 1/Jul. | 29.45 | | |

*Based on computation of the N, P, and potassium (K) elements.

The mineralized N is calculated with the following equation,

$$N_m = (N_l + N_c) - (N_f + N_d) - (N_{\min initial} - N_{\min final}) \quad (5)$$

The potential soil water N loss ($N_l$) is defined as the total loss of nitrate and ammonia, which was estimated with the N concentration in SW and the water volume in dryland [32]. Based on the SW loss and the N concentration in SW at different depths, the soil N loss by water was calculated.

### 2.6 Data analyses

Data analysis was performed with Sigma plot 10.0 and SPSS 16.0 software. The differences in dry weight and their percentages on crop organs (including root, leaf, stem, tassel, grain) in different growing stages were tested with ANOVA. The difference in the crop N uptake allocation proportion, the N accumulation in different organs, and the mineralized N concentration in different soil layers were statistically analyzed by ANOVA. The multiple comparisons were analyzed with the Duncan method.

## Results

### 3.1 Temporal-spatial distribution of soil water in dryland

With the soil-maize water monitoring system, the daily precipitation, SW acquisition rate, and SW content in the entire growing period was determined (Fig. 2). The monthly precipitation in the years 2010 and 2011 displayed a similar pattern, except for the lower value in September 2010. The precipitation occurred mainly between July and September, which was also the prime period for maize growth. With the light precipitation in April and May, the maize seed was sowed in moist soil. The precipitation in August and September of 2010 was 64 mm and 120 mm less than in the corresponding months of 2011. The cumulative precipitation from April to September in 2011 was 575 mm, which was 141 mm more than that in 2010. The precipitation was the only water source for the dryland in this freeze-thawing agricultural area, which directly impacted the SW acquisition rate. The SW could be collected only after the rainfall and the vertical difference in the SW acquisition rate was slight (around 0.002 $cm^3\ s^{-1}$). In the entire monitoring period, the biggest acquisition rate occurred at the 90 cm depth after the rainfall in early September. The direct reason for it was that only smaller amount of the water can move deeply in the vertical direction after the absorption by roots.

The SW content increased in a stable manner during the growth period and also coincided with the accumulative precipitation. In all of June, the SW content at the 30 cm depth was the highest, the reason for this was that uptake of water in growing roots resulted in water accumulation in the rhizosphere. Due to the 73.6 mm rainfall on July 3, the SW content at the 30 cm depth increased by 20% to 0.30 $cm^3\ cm^{-3}$. In the next two days, the SW content was transported in the vertical direction and it increased to 0.30 $cm^3\ cm^{-3}$ in the deep layer. This vertical distribution also confirmed the retention role of roots in the early growing season. The surface SW content increased significantly after August due to the precipitation and then was transported to the deep soil layers. Later, the impact of the crop roots decreased and the soil texture was the main factor for SW transport in September.

In order to further analyse SW content fluctuation in the temporal-vertical dimensions, a statistical analysis was performed on the data from the soil layers during the growing season (30 May–30 September) (Table 2). The SW content averages at the 15 cm and 60 cm depths were very close with values of 0.245 and 0.230 $cm^3\ cm^{-3}$, respectively. The mean SW content at the 30 cm and 90 cm depths were higher than at the other two layers with values of 0.272 $cm^3\ cm^{-3}$ and 0.280 $cm^3\ cm^{-3}$, respectively. The close relationship between the SW acquisition rate and the SW content that was shown in Fig. 2, was also confirmed by the statistical results. The SW acquisition rate had the same vertical pattern as the SW content and the higher values appeared at the 30 and 90 cm depths. Before maize seeding (May 30), the SW could be collected when the SW content reached about 0.25 $cm^3\ cm^{-3}$. After maize seeding, the SW could be collected when it reached a content of over 0.30 $cm^3\ cm^{-3}$. The SW acquisition rate was greatly and primarily influenced by rainfall.

### 3.2 Temporal variation of soil water storage

Based on the monitoring of the SW characteristics, the soil volumetric water content (SVW) and soil water storage (SWS) in the top 30 and 90 cm were calculated, respectively (Fig. 3). The SVW and SWS of the 0–90 cm and 0–30 cm depth showed similar temporal trends. In the vertical dimension, the SVW of the same two depths (0–30 cm and 0–90 cm) had similar trends in most periods, except in the first month which revealed a difference due to thawing. The thawing process from the surface to the deep layer and the tillage in preparation for sowing in the top soil decreased the SVW at 0–30 cm depth. There was almost no difference in the SVW between the two depths from mid-May to August. Before the crop harvest period in September, the deeper soil had a higher SVW content than the surface layer due to less root activity and strong precipitation. The SWS of the 0–30 cm depth had a seasonal variation and had a more direct relationship with precipitation events. Before the rainfall on 1 July, the SWS remained at a stable level from mid-May to the end of June. The SWS had a periodic decrease when no rainfall occurred in late July. The SWS reached its summit with the large rainfall in the later of August.

The SWS was calculated with the vertical variation in the SVW. The statistical characteristics of SWS and their proportions at each depth are listed in Table 3. The soil water storage averages at 0–20, 20–40 and 40–60 cm were 490.2, 593.8, and 358 $m^3\ ha^{-1}$. The vertical differences indicated the middle level has a higher soil water content as a consequence of soil texture and crop root zone influences. Because precipitation was the only water source for dryland, the SWS at the three layers in the dry period was smaller than the average for the entire growing season, which also revealed the impact of maize ET on SWS. The SWS in the vertical dimension had the same trend in the entire period and the middle layer had the largest value (548.9 $m^3\ ha^{-1}$). The differences in SWS between the two periods become greater when going from the surface to the deeper layer, an observation derived from the rainfall contribution to the SWS in deeper layer.

Based on daily monitoring data, the accumulative patterns of evapo-transpiration (ET), rainfall, and water loss in the growing season were compared (Fig. 4). The ET was at a low level in the first month of the growing period and then assumed a stable increasing rate beyond the 50th day after sowing. The accumulated SW loss was consistent with the fluctuation in the rainfall. The ET and rainfall stayed with the SW loss over the first 22 days. The accumulated rainfall amount increased by 79.8 mm; a large increase occurred on the 35th day when the ET and water loss increased slightly. The rainfall increased intensively after 80 days, which also caused an increase in the SW loss. The accumulated amounts of rainfall increased from 269.3 mm after 80 days to 440.9 mm after 120 days, and the SW loss increased from 47.9 mm after 75 days to 126.3 mm after 120 days. The temporal pattern demonstrated that the SW loss and ET of dryland were strongly correlated with the rainfall. The ET was also a maize

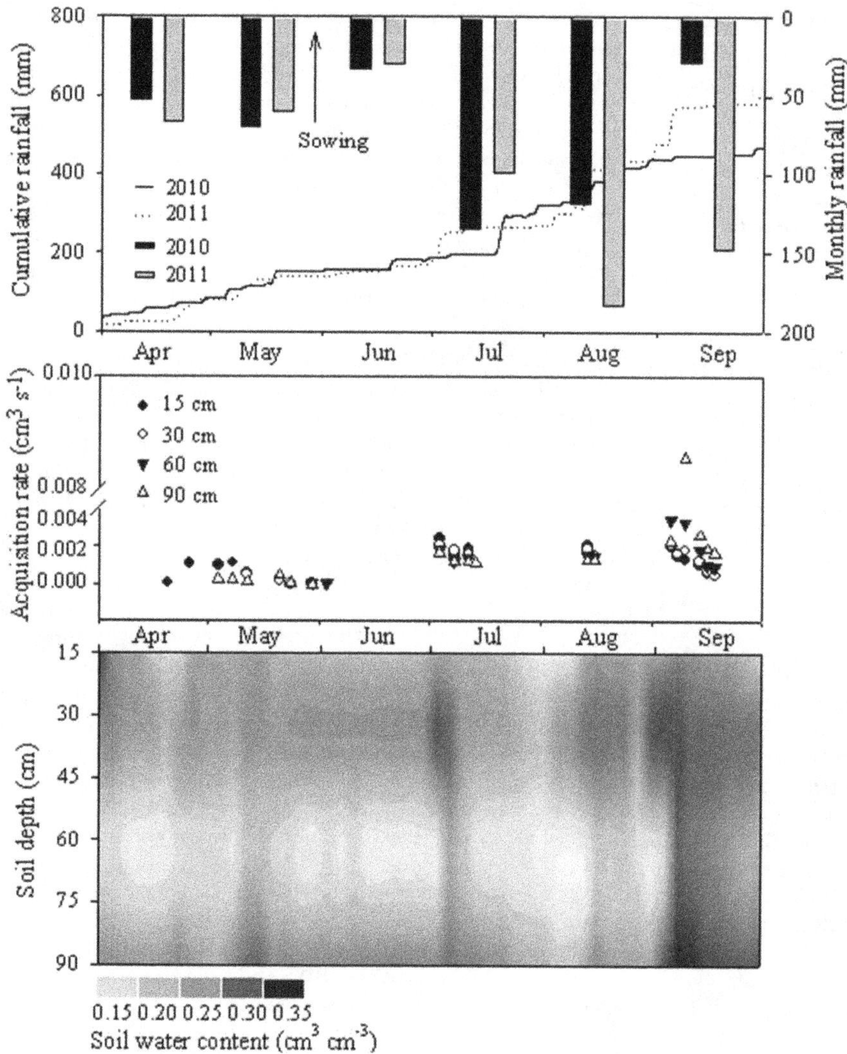


**Figure 2. Daily precipitation in 2010 and 2011, and acquisition rate, and soil hydrological pattern of dryland in 2011.**

growth feature and its similar pattern with rainfall also proved it was the dominant factor for crop growth.

## 3.3 Variation of crop biomass in the growing period

The dry weight of the different maize organs (root, leaf, stem, tassel, and grain) in growing stages were analyzed (Fig. 5). Also shown in the Fig. 5 are the percentages of the total dry weight comprised by each organ in each of four growing seasons. The dry organ weight varied greatly from the seedling stage to maturity, a development period which is relatively short in this freeze-thawing area. The temporal pattern of dry weight in 2011 revealed the changes that occurred over four growing sections. The dry weight of grains increased significantly in the grain filling and maturity sections. The dry weight percentage did not change as intensively as the weight, and it was also found that the grains comprised nearly half of the maize weight in the maturity period. With the N concentration in dried organs, the N absorption amount was quantified along with the organ dry weight. Due to logistical limitations, the crop organ sample in 2010 was only collected in the maturity period. The grain dry weights were larger in 2010 than in 2011. Comparing the dry weights of the organs in the maturity stage of the two years, the dry weights of the roots and tassel were significantly but their percentages differed little.

## 3.4 Nitrogen uptake by crop and accumulation

The temporal N dynamics of crop uptake was then analyzed with the samples obtained from the maize organs (Fig. 6). As the maize grew, the N uptake in leaf first increased and then decreased with the maximum occurring in the jointing period. From grain filling to maturity, N uptake by leaf diminished insignificantly with a relative flat trend compared with that from the jointing to grain filling stages. As maize grew N uptake in stem decreased significantly decrease from 1 to 2 with a probable transfer of much N to tassel. The difference of N uptake by stem in the grain filling and maturity stages was insignificant, but the N uptake by grain increased significantly ($P<0.001$). The N uptake by leaf decreased insignificantly, which indicated that a large amount of N was transferred to tassel in the early time period and then transferred from tassel to grain (N uptakes by tassel in grain filling and maturity were 8.12 g $kg^{-1}$ and 3.57 g $kg^{-1}$, respectively). There were insignificant differences in N uptakes of stem and leaf during the maturity stage in 2010 and 2011. The N uptake by grain in 2010 was 9.1% more than that in 2011.

**Table 2.** Statistical analysis of soil water content and acquisition rate at four depths during maize growing period.

| Index / Depth (cm) | Soil water content ($cm^3\ cm^{-3}$) | | | | Soil water acquisition rate ($cm^3\ s^{-1}$) | | | |
|---|---|---|---|---|---|---|---|---|
| | 15 | 30 | 60 | 90 | 15 | 30 | 60 | 90 |
| Minimum | 0.210 | 0.235 | 0.205 | 0.240 | 0.00008 | 0.00005 | 0.00009 | 0.00003 |
| Maximum | 0.308 | 0.333 | 0.306 | 0.342 | 0.00241 | 0.00208 | 0.00324 | 0.00856 |
| Mean | 0.245 | 0.272 | 0.230 | 0.280 | 0.00125 | 0.00114 | 0.00140 | 0.00167 |
| Std. Error | 0.018 | 0.020 | 0.024 | 0.024 | 0.00077 | 0.00078 | 0.00091 | 0.00220 |
| Coefficient of variance | 0.074 | 0.072 | 0.106 | 0.088 | 0.615 | 0.685 | 0.648 | 1.320 |

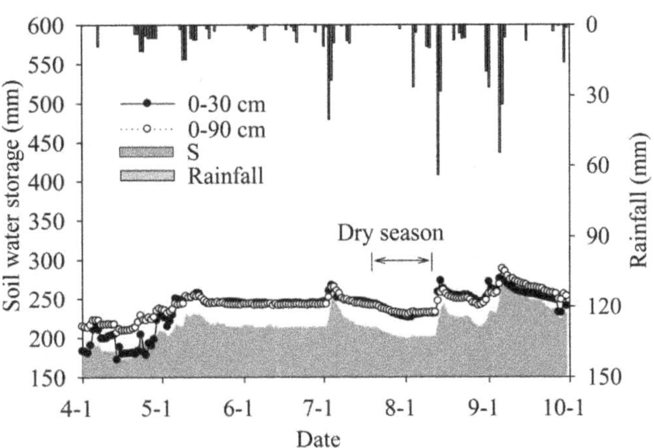


**Figure 3. Temporal patterns of daily rainfall, soil water storage (SWS) and volumetric water content at 30 and 90 cm depths.**

The accumulated N in different organs gradually increased in the early stages and reached a summit in the maturity period. From the jointing to grain filling period, the N in stem increased about 4.23 times and the amount in leaf increased less than five times. The N amounts in crop organs were not only related to the N uptake efficiency, but they also depended on the organ dry weight. According to the analysis in the maturity period, grain accumulated most of N, which occurred in a short time section. The total N amount of maize organs in 2010 was 309.9 kg $ha^{-1}$, which was larger than in 2011; the difference in grain was the dominant factor. The differences of N accumulation in leaf and stem during the maturity stages in the years 2010 and 2011 were insignificant.

Seasonal dynamics of N accumulative rates in organs during the growth period were analyzed (Fig. 7). From June to August, the N accumulative rates of stem and leaf increased rapidly and then declined. When the stem and leaf accumulated less N, the absorption efficiency in grain increased to 383 mg $m^{-2}d^{-1}$. The leaf was the most effective organ in uptaking N and the accumulation rate was about 500 mg $m^{-2}d^{-1}$.

## 3.5 Variations of soil N stocks in the growing season

The N concentrations at four depths before sowing and after harvest were analyzed, which demonstrated the temporal-vertical soil N dynamics. The N (nitrate and ammonia) concentrations at different depths were first analyzed after the growing season (Table 5). Before the sowing, the ammonia-N stock at four depths ranged from 18.8 to 21.3 kg N $ha^{-1}$ with no occurrence of a significant difference. After the harvest, the deeper layer experienced more loss than the surface layer. The middle two layers had close values and the 12.5 kg N $ha^{-1}$ in top layer was the largest value. The nitrate-N concentration in soil was much larger than the ammonia-N. At the beginning of the growing season, the nitrate-N concentration was more than 121 kg $ha^{-1}$ at all depths and had slight differences at 60–90 cm depth. The nitrate-N concentration decreased to 90 kg $ha^{-1}$ after the harvest. The upper two layers had similar patterns and the concentrations ranged between 88–90 kg $ha^{-1}$. There was no significant difference in the concentration in the deeper two layers and their concentrations ranged from 63.2 to 68.0 kg $ha^{-1}$. The nitrate-N was more variable than ammonia-N and had a larger stock gap.

**Table 3.** Dryland soil water storage and its proportion at three depths.

| Depth/cm | Item \ Duration | Entire growth season (30 May.-30 Sep.) | Dry season (20 Jul.–12 Aug.) |
|---|---|---|---|
| 0–20 | SWS/m$^3$ ha$^{-1}$ | 490.2±34.6 | 441.1±19.5 |
| | Proportion/% | 23.4±1.04 | 23.3±0.4 |
| 0–40 | SWS/m$^3$ ha$^{-1}$ | 1084±62.9 | 990.0±55.9 |
| | Proportion/% | 51.8±2.6 | 52.0±0.76 |
| 0–60 | SWS/m$^3$ ha$^{-1}$ | 1442±111 | 1303.0±70.6 |
| | Proportion/% | 68.8±1.6 | 68.5±0.55 |

### 3.6 Nutrient loss and supply in soil-crop system

Based on field water monitoring and SW content in the 0–90 cm profile, the soil water loss (SL) and the leached N were calculated (Table 6). The rainfall and evapo-transpiration (ET) were calculated in the four stages of the maize growing season. The SL had a closer correlation with rainfall than with ET. The rainfall was nearly 100 mm in the jointing stage, but the SL was 11.5 mm due to the intensive ET. Moving from the grain filling stage to the maturity stage, the ET increased and more SL occurred under the higher rainfall event. The leached N was also calculated with the N concentration in soil water. The nitrate-N was the dominant N lost from the 90 cm soil profile, especially in the grain filling and maturity stages. The leached ammonia-N had similar temporal patterns and the largest amount was 0.18 kg ha$^{-1}$.

Based on the N monitoring of rainfall and the survey of N fertilization, the N in rainfall and fertilizer was calculated. With the SWS and N concentration in SW in 0–90 cm profile, the *NUE* in the maize-soil system was calculated (Table 7). The N input from rainfall and fertilizer in the entire maize growing season was 16.5 and 120 kg ha$^{-1}$, respectively. The soil N supply in the entire growth period was about 641 kg ha$^{-1}$. With the nitrate and ammonia concentrations in SW and water volume, the N loss was estimated at 15.6 kg ha$^{-1}$. Based on the N balance, it was calculated that the *NUE* was 43%. On the other hand, the excessive N from dryland was an important potential source for N discharge to the environment.

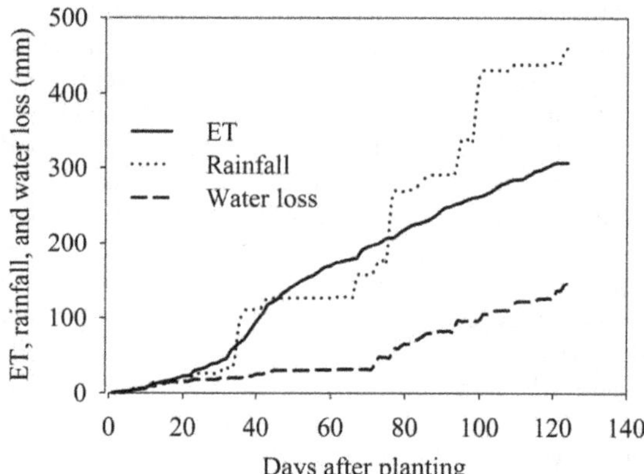


**Figure 4. Cumulative evapo-transpiration (ET), rainfall, and water loss throughout maize growth period.**

## Discussion

### 4.1 The dryland soil hydrological process in a freeze-thawing area

Rainfall is the only water source in dryland and it has a direct impact on SW content and storage [33]. The field monitoring done in four stages as part of our study indicated that the temporal patterns of SW content and acquisition rate closely followed the rainfall amount (Fig. 2). The soil roughness in a cultivated area is increased by tillage, which enhances the vertical transport of SW [34]. The SL was 146.7 mm (32% of rainfall in the growing season). However, the vertical distribution of SW content and soil water storage demonstrate that the water has a short vertical movement range. The SW can be sampled only after the rainfall events, but the acquisition rates showed slight vertical differences. These two findings both demonstrated that the vertical movement of water is slow even with large volumes and the speed can affect the range.

Compared with the long growing periods in other dryland agricultural areas, the local maize growing season is short and has a higher efficiency in the use of water and N [35]. A proper SW condition in the root zone is essential for crop growth in the maize seeding and grain filling stages. The SWS can remain at a stable level (about 1000 m$^3$ ha$^{-1}$ of 40 cm profile) even in the dry season without precipitation (Fig. 3). All of these factors indicate that the *albic* soil texture can provide a positive water condition for crop growth [24]. The SW also controls surficial N movements in the dryland and provides a strong signal for crop absorption [36]. Crop transpiration will increase when the maize is near its maturity period, and the precipitation normally increases and meets this need during the same period (Fig. 4). The combined conditions of special soil texture and precipitation patterns directly determine the dryland soil hydrological pattern, which is the critical factor for maize growth in this freeze-thawing area.

### 4.2 Crop growth and N accumulation differences in plant organs

The quantification and assessment of the organ biomass in crops can help in understanding the impacts of SW and N on crop growth. This information is also necessary for assessing N leaching [37]. The crop growth is measured by the organ dry weight (Fig. 5), which indicates the crop growth rate is higher than in other drylands under warming temperatures. The yearly precipitation in 2011 was more than in 2000 (Fig. 2), but did not lead to the higher grain dry weight in 2011 (Fig. 5). The difference proves that a higher maize yield is based on a combination of climate, soil and tillage. The grain filling period in this freeze-thawing area occurs in August, at a time when temperature and precipitation are at

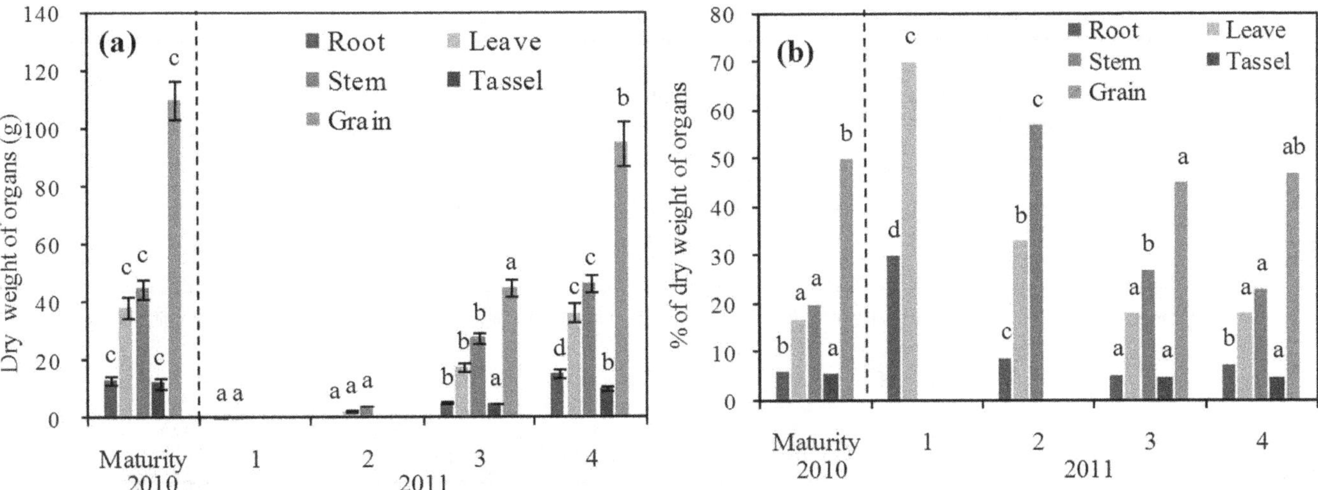


**Figure 5. Dry weights (a) and percentage of maize organ weights (b) at different growth stages (Same letters in same organ between different growth stages indicate no significant difference.** 1-Seedling, 2-Jointing, 3-Grain filling, 4-Maturity).

their highest levels. The active ET facilitates the transport of water and N from soil to crop, which finally causes a higher grain filling performance [38].

The absorbed N by maize organs in the entire growth period had different allocation rates (Fig. 7), which is determined by the photosynthesis of the organs and the eco-hydrological process. The crop accumulative rate in its canopy, stem and grain varied significantly, which means the N absorption capability for tissue production of dedicated organs shifts due to the N flux from soil to above ground biomass [39]. The soil N concentration in the vertical direction proved that the subsurface has a significant variance due to the root uptake. The accumulation rates of different organs over time also follow the typical sigmoid curve [40]. The temporal patterns of N in maize organs also indicate the N absorption is not only related to the soil N concentration, but that there is a complicated systemic feedback signal within the soil-crop system [41]. The SW availability, ET and N concentration are the key signaling factors. As the soil N is transported in crops with water, there are plenty of field observations and models to identify the growth stress reaction to the SW and N conditions [42][43]. This information provides detailed guidelines for N and water adjustment in the soil-crop system.

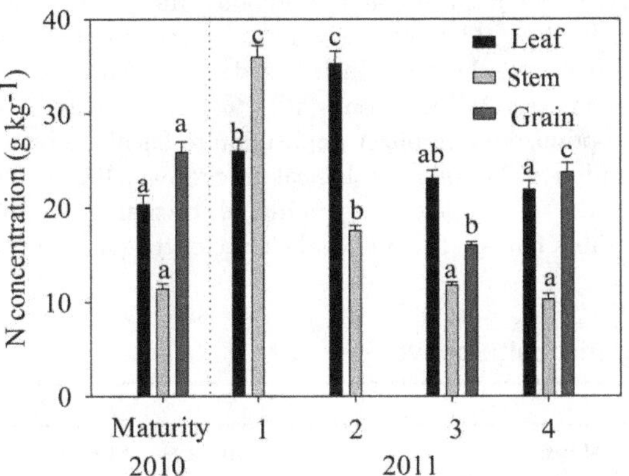


**Figure 6. Seasonal dynamics of N concentration in maize organs (Same letters in same crop organs indicates no significant change between different growth stages.** 1-Seedling, 2-Jointing, 3-Grain filling, 4-Maturity).

## 4.3 Implication for N leach control and soil N optimization

The N uptake efficiency of a crop is the main component for *NUE* and the temporal patterns in Table 4 indicated that the highest amount of grain uptake of N occurs in the graining filling and maturity periods. The maize uptake of N will increase until the optimum levels of water and N in soils are reached [44]. Overlaying this information with the leached N loading, it was found that the intensive N uptake and loss occurred at same time. The strong ET and rainfall in that time period were the reason for the N transport in the crop-soil system [45]. The soil N gap analysis revealed that the observed N gap was similar in the vertical direction, which also demonstrates that the soil N can move among the soil depths with hydrological processes and maintain the N concentration balances [3]. The temporal N analysis in crop organs, soil and leached water explain the maize growth in this freeze-thawing area, which also provides a guideline for N fertilization while, also considering SW conditions.

The N pollution to groundwater due to excessive N fertilizer application is an important environmental concern [46]. The

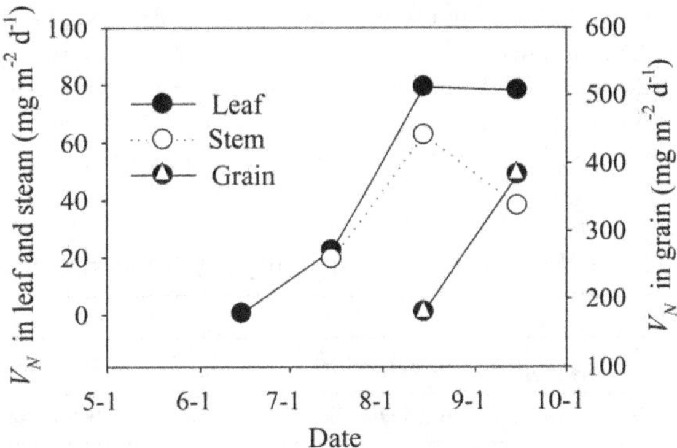


**Figure 7. Seasonal dynamics of N accumulative rates (VN) in maize organs.**

**Table 4.** Temporal pattern of N amount (kg $ha^{-1}$) of maize organs during growing season*.

| Year | Growth stage | Leaf | Stem | Grain |
|---|---|---|---|---|
| 2010 | Mature | 58.6f | 38.3ef | 213de |
| 2011 | Seedling | 0.14a | | |
| | Jointing | 6.85c | 5.81a | |
| | Grain filling | 30.6d | 24.6d | 54.1b |
| | Maturity | 60.0f | 36.0e | 169c |

*Same lowercase letter in one column indicats no significant difference ($P<0.001$); Same uppercase letter in one row indicates no significant difference ($P<0.001$).

**Table 5.** Mineral N stocks (kg $ha^{-1}$) and variations at each depth of growing season.

| Depth (cm) | Before sowing | | After harvest | | Stock gap | |
|---|---|---|---|---|---|---|
| | $NO_3^-$-N | $NH_4^+$-N | $NO_3^-$-N | $NH_4^+$-N | $NO_3^-$-N | $NH_4^+$-N |
| 0–15 | 147d | 21.3b | 90.1d | 12.5c | −48.9c | −8.64bc |
| 15–30 | 144d | 21.2b | 88.1d | 8.12b | −53.8cd | −11.2c |
| 30–60 | 130cd | 19.3b | 68.0c | 9.02b | −65.3e | −9.79c |
| 60–90 | 121c | 18.8b | 63.2c | 5.74a | −56.8d | −12.69c |

Same letters in one column indicates no significant difference ($P<0.001$).

**Table 6.** Amounts of soil water loss (SL) and N leached below maize root zone (90 cm).

| Growth stage | Period | R | ET | SL | leached N (kg $ha^{-1}$) | |
|---|---|---|---|---|---|---|
| | | (mm) | | | $NO_3^-$-N | $NH_4^+$-N |
| Seedling | 30 May to 29 Jun. | 29.8 | 44.6 | 19.6 | 0.03 | 0.01 |
| Jointing | 30 Jun. to 25 Jul. | 96.5 | 119.0 | 11.5 | 3.36 | 0.12 |
| Grain filling | 26 Jul. to 23 Aug. | 165 | 67.3 | 48.6 | 5.53 | 0.18 |
| Maturity | 24 Aug. to 30 Sep. | 168 | 76.6 | 67.5 | 6.26 | 0.10 |

yearly *NUE* is about 43% in this area, which is a relatively high value in comparison with similar dryland agriculture areas [47]. The N leached from farmland mainly depends on the difference between the N obtained by the soil-water-crop system and the N uptake by the crop [48]. As Table 6 shows, the leached N is mainly related to the loss of SW, which is the gap between rainfall and ET. The rainfall in August 2011 was above the long term monthly value, which caused the highest SW and N loss during the growing season. Under the normal climatic conditions, the *NUE* in this area can reach 50%. The nitrate-N is the dominant form of leached N to the groundwater, which is a direct consequence of fertilization and soil nitrification [49]. Models have been developed to optimize the fertilizer application to farmland based on a consideration of the soil hydrological process, which can help to maximize the crop yield with minimal discharge to the environment [50]. The model with local characteristic parameters,

**Table 7.** The N balance and usage efficiency (*NUE*) in crop-soil system during the maize growth period.

| Rainfall | Fertilizer | Initial soil N | | Final soil N | | N loss | | Soil supply N | *NUE* |
|---|---|---|---|---|---|---|---|---|---|
| (kg N $ha^{-1}$) | | (kg N $ha^{-1}$) | | (kg N $ha^{-1}$) | | (kg N $ha^{-1}$) | | (kg N $ha^{-1}$) | (kg $kg^{-1}$) |
| | | $NO_3^-$-N | $NH_4^+$-N | $NO_3^-$-N | $NH_4^+$-N | $NO_3^-$-N | $NH_4^+$-N | | |
| 16.5 | 120 | 542.9±29.76 | 80.6±4.29 | 309.4±18.4 | 35.4±4.37 | 15.2 | 0.41 | 641±14.4 | 0.43±0.03 |

should be applied in the tillage practice to obtain an improved *NUE*.

## Conclusion

The results of this study have shown that combined conditions of soil texture, soil hydrology and precipitation were basic factors for maize growth in the freeze-thawing area in Northeast China. The rainfall significantly affected the pattern of SW content and the allocation of N in the maize organs. Precipitation was the only water source for the dryland, but the variation in SW acquisition rates and SW content remained at a stable level due to the special *albic soil* texture. The dry weight and N accumulation of maize organs in the entire growing period proved that the soil hydrological condition was also enhanced.

The N loss was dependent on both the hydrological process and the N in the water, which was then decided by the rainfall water depth and the applied N. The $NO_3$-N was the dominant leachate, which was closely correlated with the amount of applied N fertilizer. The *NUE* was about 43%, which was also related to the initial soil N and maize uptake. In order to improve the *NUE* and decrease the amounts of leached nitrogen, it would be better to improve the fertilizer application amounts while giving consideration to rainfall patterns and N accumulation rates.

## Author Contributions

Conceived and designed the experiments: WOY. Performed the experiments: SYC GQC. Analyzed the data: WOY SYC. Contributed reagents/materials/analysis tools: FHH. Wrote the paper: WOY.

## References


1. Buckley C, Carney P (2013) The potential to reduce the risk of diffuse pollution from agriculture while improving economic performance at farm level. Environ Sci Policy 25(1): 118–126.
2. Quemada M, Baranski M, Nobel-de Lange MNJ, Vallejo A, Cooper JM (2013) Meta-analysis of strategies to control nitrate leaching in irrigated agricultural systems and their effects on crop yield. Agric Ecosyst Environ, 174(15): 1–10.
3. Ferrant S, Oehler F, Durand P, Ruiz L, Salmon-Monviola J, et al. (2011) Understanding nitrogen transfer dynamics in a small agricultural catchment: Comparison of a distributed (TNT2) and a semi distributed (SWAT) modeling approaches. J Hydrol 406(1–2): 1–15.
4. Panda RK, Behera S (2003) Non-point source pollution of water resources: Problems and perspectives. J Food Agric Environ 1(3–4): 308–311.
5. Wang XB, Dai KA, Wang Y, Zhang XM, Zhao QS, et al. (2010) Nutrient management adaptation for dryland maize yields and water use efficiency to long-term rainfall variability in China. Agric Water Manage 97(9): 1344–1350.
6. Yang YG, Xiao HL, Wei YP, Zhao LJ, Zou SB, et al. (2011) Hydrologic processes in the different landscape zones of Mafengou River basin in the alpine cold region during the melting period. J Hydrol 409(1–2): 149–156.
7. Grzebisz W (2013) Crop response to magnesium fertilization as affected by nitrogen supply. Plant Soil 368(1–2): 23–39.
8. Matejek B, Huber C, Dannenmann M, Kohlpaintner M, Gasche R, et al. (2010) Microbial N turnover processes in three forest soil layers following clear cutting of an N saturated mature spruce stand. Plant Soil 337(1–2): 93–110.
9. Moreau P, Ruiz L, Vertes F, Baratte C, Delaby L, et al. (2013) CASIMOD'N: An agro-hydrological distributed model of catchment-scale nitrogen dynamics integrating farming system decisions. Agric Syst 118: 41–51.
10. Kersebaum KC, Steidl J, Bauer O, Piorr HP (2003) Modelling scenarios to assess the effects of different agricultural management and land use options to reduce diffuse nitrogen pollution into the river Elbe. Phys Chem Earth 28: 537–545.
11. Hirel B, Le Gouis J, Ney B, Gallais A (2007) The challenge of improving nitrogen use efficiency in crop plants: towards a more central role for genetic variability and quantitative genetics within integrated approaches. J Exp Bot 58: 2369–2387.
12. Pask AJD, Sylvester-Bradley R, Jamieson PD, Foulkes MJ (2012) Quantifying how winter wheat crops accumulate and use nitrogen reserves during growth. Field Crop Res 126(14): 104–118.
13. Bonato O, Schulthess F, Baumgärtner J (1999) Simulation model for maize crop growth based on acquisition and allocation processes for carbohydrate and nitrogen. Ecol Modell 124(1): 11–28.
14. Binder DL, Dobermann A, Sander DH, Cassman KG (2002) Biosolids as nitrogen source for irrigated maize and rainfed sorghum. Soil Sci Soc Am J 66(2): 531–543.
15. Favaretto N, Norton LD, Johnston CT, Bigham J, Sperrin M (2012) Nitrogen and Phosphorus Leaching as Affected by Gypsum Amendment and Exchangeable Calcium and Magnesium. Soil Sci Soc Am J 76(2): 575–585.
16. Lam QD, Schmalz B, Fohrer N (2012) Assessing the spatial and temporal variations of water quality in lowland areas, Northern Germany. J Hydrol 438–439(17): 137–147.
17. Friend AD, Stevens AK, Knox RG, Cannell MGR (1997) A process-based, terrestrial biosphere model of ecosystem dynamics (Hybrid v3.0). Ecol Modell 95(2–3): 249–287.
18. Dorioz JM, Ferhi A (1994) Non-point pollution and management of agricultural areas: Phosphorus and nitrogen transfer in an agricultural watershed. Water Res 28(2): 395–410.
19. Keeney DR, Follett RF (1991) Managing Nitrogen for Groundwater Quality and Farm Profitability, SSSA, Madison, WI. 1–7.
20. Kim DG, Vargas R, Bond-Lamberty B, Turetsky MR (2012) Effects of soil rewetting and thawing on soil gas fluxes: a review of current literature and suggestions for future research. Biogeoscience 9(7): 2459–2483.
21. Lewis DR, McGechan MB, McTaggart IP (2003) Simulating field-scale nitrogen management scenarios involving fertiliser and slurry applications. Agr Syst 76(1): 159–180.
22. Dawson JC, Huggins DR, Jones SS (2008) Characterizing nitrogen use efficiency in natural and agricultural ecosystems to improve the performance of cereal crops in low-input and organic agricultural systems. Field Crop Res 107: 89–101.
23. Schroder JL, Zhang HL, Girma K, Raun WR, Penn CJ, et al. (2011) Soil Acidification from Long-Term Use of Nitrogen Fertilizers on Winter Wheat. Soil Sci Soc Am J 75(3): 957–964.
24. Caliskan S, Ozkaya I., Caliskan ME, Arslan M (2008) The effects of nitrogen and iron fertilization on growth, yield and fertilizer use efficiency of soybean in a Mediterranean-type soil. Field Crop Res 108(2): 126–132.
25. Razzaghi F, Plauborg F, Jacobsen SE, Jensen CR, Andersen MN (2012) Effect of nitrogen and water availability of three soil types on yield, radiation use efficiency and evapotranspiration in field-grown quinoa. Agric Water Manage 109: 20–29.
26. Hao FH, Chen SY, Ouyang W (2013) Temporal rainfall patterns with water partitioning impacts on maize yield in a freeze-thaw zone. J Hydrol 486(12): 412–419.
27. Ouyang W, Huang HB, Hao FH, Shan YS, Guo BB (2012) Evaluating spatial interaction of soil property with non point source pollution at watershed scale: the phosphorus indicator in Northeast China. Sci Total Environ 432, 412–421.
28. Jackson JE (1991) A user's guide to principal components. Wiley-Interscience, New York.
29. Moroizumi T, Hamada H, Sukchan S, Ikemoto M (2009) Soil water content and water balance in rainfed fields in Northeast Thailand. Agric Water Manage 96(1): 160–166.
30. Huggins DR, Pan L (1993) Nitrogen efficiency component analysis: an evaluation of cropping system differences in productivity. Agronomy J 85: 898–905.
31. Meisinger JJ, Randall GW (1991) Estimating nitrogen budgets for soil-crop systems. In: Follet, R.F., et al. (Eds.), Managing Nitrogen for Groundwater Quality and Farm Profitability. SSSA, Madison,WI, USA, 85–124.
32. Vázquez N, Pardo A, Suso ML, Quemada M (2006) Drainage and nitrate leaching under processing tomato growth with drip irrigation and plastic mulching. Agric Ecosyst Environ 112(4): 313–323.
33. Tanveer SK, Wen XX, Lu XL, Zhang JL, Liao YC (2013) Tillage, Mulch and N Fertilizer Affect Emissions of CO2 under the Rain Fed Condition. PLOS ONE: 8(9): e72140.
34. Dörner J, Horn R (2009) Direction-dependent behaviour of hydraulic and mechanical properties in structured soils under conventional and conservation tillage. Soil Till Res 102(2): 225–232.
35. Murungu FS, Chiduza C, Muchaonyerwa P, Mnkeni PNS (2011) Mulch effects on soil moisture and nitrogen, weed growth and irrigated maize productivity in a warm-temperate climate of South Africa. Soil Till Res 112(1): 58–65.
36. Gheysari M, Mirlatifi SM, Homaee M, Asadi ME, Hoogenboom G (2009) Nitrate leaching in a silage maize field under different irrigation and nitrogen fertilizer rates. Agric Water Manage 96(6): 946–954.
37. Wegehenkel M, Mirschel W (2006) Crop growth, soil water and nitrogen balance simulation on three experimental field plots using the Opus model—A case study. Ecol Modell 190(1-2): 116–132.
38. Katerji N, Campi P, Mastrorilli M (2013) Productivity, evapotranspiration, and water use efficiency of corn and tomato crops simulated by AquaCrop under contrasting water stress conditions in the Mediterranean region. Agric Water Manage 130: 14–26.
39. Strullu L, Cadoux S, Preudhomme M, Jeuffroy MH, Beaudoin N (2011) Biomass production and nitrogen accumulation and remobilisation by Miscanthus × giganteus as influenced by nitrogen stocks in belowground organs. Field Crop Res 121(3): 381–391.

40. Shoji S, Gandeza AT, Kimura K (1991) Simulation of crop response to polyolefin-coated urea: II. Nitrogen uptake by corn. Soil Sci Soc Am J 55(5): 1468–1473.
41. Nacry P, Bouguyon E, Gojon A (2013) Nitrogen acquisition by roots: physiological and developmental mechanisms ensuring plant adaptation to a fluctuating resource. Plant Soil, 370(1–2): 1–29
42. Kovács GJ (2005) Modelling of adaptation processes of crops to water and nitrogen stress. Phys Chem Earth, Parts A/B/C 30(1–3): 209–216.
43. Overman AR, Scholtz RV (2013) Accumulation of Biomass and Mineral Elements with Calendar Time by Cotton: Application of the Expanded Growth Model. PLOS ONE 8(9): e72810.
44. Kerbiriou PJ, Stomph TJ, Van Der Putten PEL, Van Bueren ETL, Struik PC (2013) Shoot growth, root growth and resource capture under limiting water and N supply for two cultivars of lettuce (Lactuca sativa L.). Plant Soil 371(1–2): 281–297.
45. Asseng S, Turner NC, Keating BA (2001) Analysis of water- and nitrogen-use efficiency of wheat in a Mediterranean climate. Plant Soil 233(1): 127–143.
46. Fang QX, Yu Q, Wang EL, Chen YH, Zhang GL, et al. (2006) Soil nitrate accumulation, leaching and crop nitrogen use as influenced by fertilization and irrigation in an intensive wheat-maize double cropping system in the North China Plain. Plant Soil 284(1–2): 335–350.
47. Ouyang W, Wei XF, Hao FH (2013) Long-term soil nutrient dynamics comparison under smallholding land and farmland policy in northeast of China. Sci Total Environ 450–451(15): 129–139.
48. Hall JA, Bobe G, Hunter JK, Vorachek WR, Stewart WC, et al. (2013) Effect of Feeding Selenium-Fertilized Alfalfa Hay on Performance of Weaned Beef Calves. PLOS ONE 8(3): e58188.
49. Lteif A, Whalen JK, Bradley RL, Camire C (2010) Nitrogen transformations revealed by isotope dilution in an organically fertilized hybrid poplar plantation. Plant Soil 333(1–2): 105–116.
50. Sierra J, Brisson N, Ripoche D, Noel C (2003) Application of the STICS crop model to predict nitrogen availability and nitrate transport in a tropical acid soil cropped with maize. Plant Soil 256(2): 333–345.

# Effects of Fertilization and Clipping on Carbon, Nitrogen Storage, and Soil Microbial Activity in a Natural Grassland in Southern China

**Zhimin Du[1,2], Yan Xie[1,2], Liqun Hu[1], Longxing Hu[1,2], Shendong Xu[3], Daoxin Li[3], Gongfang Wang[3], Jinmin Fu[1,2]***

1 Key Laboratory of Plant Germplasm Enhancement and Specialty Agriculture, Wuhan Botanical Garden, Chinese Academy of Sciences, Wuhan, Hubei, China, 2 Graduate University of Chinese Academy of Sciences, Beijing, Hebei, China, 3 National Dalaoling Forest Park, Yichang, Hubei, China

## Abstract

Grassland managements can affect carbon (C) and nitrogen (N) storage in grassland ecosystems with consequent feedbacks to climate change. We investigated the impacts of compound fertilization and clipping on grass biomass, plant and soil (0–20 cm depth) C, N storage, plant and soil C: N ratios, soil microbial activity and diversity, and C, N sequestration rates in grassland *in situ* in the National Dalaoling Forest Park of China beginning July, 2011. In July, 2012, the fertilization increased total biomass by 30.1%, plant C by 34.5%, plant N by 79.8%, soil C by 18.8% and soil N by 23.8% compared with the control, respectively. Whereas the clipping decreased total biomass, plant C and N, soil C and N by 24.9%, 30.3%, 39.3%, 18.5%, and 19.4%, respectively, when compared to the control. The plant C: N ratio was lower for the fertilization than for the control and the clipping treatments. The soil microbial activity and diversity indices were higher for the fertilization than for the control. The clipping generally exhibited a lower level of soil microbial activity and diversity compared to the control. The principal component analysis indicated that the soil microbial communities of the control, fertilization and clipping treatments formed three distinct groups. The plant C and N sequestration rates of the fertilization were significantly higher than the clipping treatment. Our results suggest that fertilization is an efficient management practice in improving the C and N storage of the grassland ecosystem via increasing the grass biomass and soil microbial activity and diversity.

**Editor:** Ting Wang, Wuhan Botanical Garden, Chinese Academy of Sciences, Wuhan, China, China

**Funding:** Funding came from the "Strategic Priority Research Program - Climate Change: Carbon Budget and Relevant Issues" of the Chinese Academy of Sciences (No. XDA0505040704), and the National Natural Science Foundation of China (No. 31272194). The funders had no role in study design, data collection and analysis, decision to publish, or preparation of the manuscript.

**Competing Interests:** The authors have declared that no competing interests exist.

* E-mail: jfu@wbgcas.cn

## Introduction

The grasslands in China cover an area of 3.92 million km$^2$ and provide 9% to 16% of the total C in the world grasslands [1,2,3]. Concerns about global warming has increased an attention to understand the role of potential C and nitrogen (N) sink in grasslands in mitigating the emission of greenhouse gases (i.e. $CO_2$ and $N_2O$) [4–6]. The C and N sequestration in terrestrial ecosystems constitutes a major mitigation strategy against the global warming [7]. China's grasslands make an important contribution to the world C and N storage and may have significant effects on C and N cycles worldwide [2]. Natural grasslands of southern China cover an area of 79.58 million km$^2$, and probably have a high yield owning to good hydrothermal conditions [8], which can be an important C and N pool.

The processes of C and N sequestration can be greatly affected by grassland managements [9], and good managements are critical for grasslands to enhance C and N sequestration [10–12]. Compound fertilizers or organic amendments affected grasslands C and N storage via increasing plant biomass [10,13,14]. Dersch and Böhm [15] reported that N, phosphorus (P), and potassium (K) fertilizers combined with farmyard manure application enhanced C storage to about 5.6 Mg ha$^{-1}$ after 21 years in Australia. The N fertilization and cover cropping can increase soil organic C and total N by increasing the amount of plant residues returned to the soil [11,16]. Similarly, the application of manure can increase soil organic C and total N levels [17,18]. Clipping was found to affect the grassland C and N storage via reducing plant biomass [9] and changing grass species [19]. Particularly, the potentially dominant plants (i.e. usually larger than their neighbors) often lose a higher proportion of their biomass than their neighbors after clipping [9].

Soil microorganisms exert a dominant influence on the net C and N balance of terrestrial ecosystems by controlling soil organic matter (SOM) decomposition and plant nutrient availability [20,21]. The grassland SOM mainly derived from roots, senescent leaves and stems of the vegetations [22]. The processes and functions of breakdown of the plants residues in soil are greatly impacted by soil microorganisms [23]. Agricultural managements can affect soil microorganisms' condition and ultimately affect the C and N cycling in ecosystems [24,25]. Microbial populations were significantly increased in the soils amended with green manure throughout two-year experiment [26]. Soil microbial diversity and/or activity may be a sensitive indicator of ecosystem change, as it can be quickly affected by disturbances [27,28].

Zhong and Cai [29] demonstrated that soil microbial diversity and average well color development (AWCD) which reflects total microbial activity [30] in the NPK treatment were increased in response to fertilization. Soil microbial biomass, populations and diversity were increased by optimum and balanced fertilization [31,32]. On the other hand, the clipping significantly reduced soil microbial and respiration rate in both warmed and un-warmed plots of tallgrass prairie [33]. Above-ground biomass removal could significantly reduce C inputs from vegetation to soil and lead to significant N loss, resulting in substrate limitation to soil microorganisms [34,35].

Understanding the fate of stored C and N and their potential for anthropogenic manipulation is critically important to evaluate the future state of the atmosphere or terrestrial ecosystems and manage the foreseen global change [36,37]. However, the effects of management in relation to soil microorganisms on the redistribution and cycling of C and N within the plant-soil system were unclear. The objective of this study was to investigate the effects of compound fertilizer and clipping on C and N storage and distribution within a natural grassland ecosystem.

## Materials and Methods

### Site description and experimental design

The study was conducted in the National Dalaoling Forest Park near the dam of the Three Gorges Reservoir in China from July 2011 to September 2012. The experimental site located at approximately 110°56′E, 31°4′N and 1696 m asl. The climate in this region is of a northern subtropical type, with a warm, humid summer, and an obvious altitudinal change. Maximum, minimum, and mean annual temperature was 19.2°C in July, −2.7°C in January, and 8.5°C, respectively. The mean annual precipitation was 1446.8 mm. Although the majority of precipitation occurs in summer, there was still 179.6 mm in winter occurring as snow and sleet [38].

The experimental site had a vegetation coverage of more than 60%, composed of over 20 grass species, but was dominated by *Festuca arundinacea* Schreb (approximately 30% of total above-ground biomass), *Potentilla freyniana* (50% of total above-ground biomass) and *Lysimachia clethroides* Duby (10% of total above-ground biomass). In this humid ungrazed montane meadow, all grasses were shallow rooted in the depth of 0–20 cm, with the maximum density occurred in the 0–10 cm soil layer. The soil in the 0–20 cm depth zone had a pH of 5.8, 12.9 g organic matter per kg soil, 1.1 g total N per kg soil, and 0.4 g total P per kg soil.

The grassland was exposed to three treatments: (i) control; (ii) fertilization; and (iii) clipping. Five replicated plots were conducted for each of the three treatments and arranged in a randomized complete block design. Each plot was measured 10 m by 5 m and fenced on June 20, 2011 to prevent the rabbits or other animals from grazing. The grassland was untreated in the control. In the fertilization treatment, compound fertilizers (15-15-15, N-$P_2O_5$-$K_2O$) were applied on July 15, 2011 and May 15, 2012 (600 kg per ha for each time). In the clipping treatment, the grassland vegetations were clipped to 3 to 5 cm with sickles on July 15, 2011 and May 15, 2012, respectively. The clippings were left *in situ*.

### Plant and soil sampling and analysis

Plant biomass was assessed five times from 2011 to 2012, in May (late spring 2012), July (middle summer 2011, 2012) and September (early autumn 2011, 2012). In each plot, five random 1 m×1 m quadrats were assigned. One quadrat was selected each time for plant sampling. Shoot including living and standing dead within the quadrat was collected. After the shoot was removed, the litter was picked up. Then, five soil sub-samples (7 cm in diam, and 0-5-10-20 cm depths) were collected from each quadrat using a soil auger and pooled together to be a composite sample. The roots (including roots and rhizomes) in the pooled soil cores were picked up and washed with deionized water three times to get rid of residual soil. Shoots, litters, and roots were killed at 105°C for 30 minutes and dried to constant weight at 80°C. The C and N concentrations of the plant samples of July 1, 2011 and July 18, 2012 were measured based on the methods described by Lu [39]. Each of the fresh composite soil samples of July 1, 2011 and July 18, 2012 was sieved (2 mm wire mesh) and divided into two sub-samples: one was kept in the refrigerator at 4°C until microbial analysis and the other was air-dried for the analysis of soil organic C and total N concentrations [39]. Plant and soil C, N storage was calculated as Post's and Tian's methods [40,41].

### Soil microbial populations' analysis

Traditional culture techniques were used to determine the distribution of the main physiological groups [42]. Microbial populations of bacteria, fungi and actinomyces were determined by soil dilution plating on beef extract peptone medium, Martin's medium and Gause's No. 1 synthetic medium, respectively. All media were made up as the methods described by Dong et al. [43] and all microbes were cultivated in a 28°C incubator for 2, 3 and 7 d, respectively.

### BIOLOG analysis and calculation of microbial activity and diversity indices

The soil microbial activity and functional diversity was evaluated using BIOLOG ECO microplate (BIOLOG Inc., Hayward, CA, USA). Soil sample kept in the refrigerator (equal to 10.0 g dried soil) was suspended in 90 mL of sterile 0.85% NaCl solution, shaken at 220 rpm for 30 min and held for 5 min. Then the suspensions were diluted to a final dilution of $10^{-3}$ with sterile 0.85% NaCl solution. Each well of BIOLOG ECO plates was inoculated with 125 μL of the diluted soil extracts and incubated at 25°C. Optical density of the wells was read with BIOLOG Micro Station reader (MicroLog release 4.20) at 590 nm every 12 h for 7 days. The optical density readings were corrected for the water controls in subsequent analysis. Negative readings after the correcting were adjusted to zero. Soil microbial activity measured as AWCD was calculated by the method described by Garland and Mills [30]. The substrate richness, Shannon's diversity index, Shannon's evenness index, McIntosh's diversity index, and McIntosh's evenness index were calculated using the data at 72 h, since the highest rate of microbial growth was observed at this incubation time [44,45]. Formulas used for the above indices calculations were described by Magurran [46] and Staddon et al. [47]. Principal component analysis (PCA) was performed on BIOLOG data divided by the AWCD [30].

### C and N sequestration rates

Changes of C (or N) sequestration were estimated by calculating the difference of C (or N) storage between July, 2012 and July, 2011. C and N sequestration rates (*CSR*, g C $m^{-2}$ $yr^{-1}$ and *NSR*, g N $m^{-2}$ $yr^{-1}$) were calculated using the following equations, respectively:

$$CSR = \frac{C_{rn} - C_{r0}}{n-0} \quad (1)$$

$$NSR = \frac{N_{rn} - N_{r0}}{n - 0} \quad (2)$$

where $C_{r0 \text{ and } n}$ is C storage (g C $m^{-2}$) under the certain management (the fertilization, clipping or the control) during the first and second years in which C storage was measured, respectively; $N_{r0 \text{ and } n}$ is N storage (g N $m^{-2}$) under the certain treatment (the fertilization, clipping or the control) during the first and second years in which N storage was measured, respectively; n is the number of years of duration of the experiment.

### Data analysis

Management effects and interactions between the variables were determined by the analysis of variance using SPSS 13.0 (SPSS, Inc.). Significantly different means were separated using Fisher's protected least significant difference (LSD) test ($p<0.05$). PCA analysis was performed using the Canoco 4.5 software package [48]. The data in July, 2011 were not shown since no obvious difference was found in effect of clipping and fertilization on soil microbial activity and diversity.

## Results

### Biomass

The shoot, root, litter and total biomass of the three treatments varied considerably through different growing season in 2011 and 2012 (Fig. 1). There was no significant difference in biomass among the treatments in July, 2011 (before the implementation of fertilization and clipping). As time went by, the fertilization treatment increased the shoot, root, and total biomass than the control. The clipping treatment, on the contrary, had the less shoot, root, litter and total biomass compared to the control. The total biomass was much higher in middle summer (July) relative to late spring (May) and early autumn (September) for each year. The total biomass in the fertilization treatment in July, 2012 was 30.1% and 73.4% greater than that in the control and clipping treatments, respectively.

### Plant and soil C, N storage and C: N ratios

Plant C storage ranged from 258.6 to 295.2 g C $m^{-2}$, and soil C storage ranged from 2086.9 to 2752.5 g C $m^{-2}$ for the three treatments as measured in July, 2011. Plant N storage ranged from 5.8 to 7.0 g N $m^{-2}$, and soil N storage ranged from 199.8 to 223.0 g N $m^{-2}$ for the three treatments as measured in July, 2011. No difference in C and N storage was found among the three treatments in July, 2011 (data were not presented).

The shoot and root C storage was the highest in the fertilization treatment, followed by the control and clipping in July, 2012 (Table 1). Shoot and root C storage in the fertilization treatment was 1.4 and 1.4 times more than that in the control, 2.2 and 1.9 times more than that in the clipping, respectively. Root C storage was similar between the control and clipping treatment. The shoot and root N storage of the fertilization treatment was 1.8 and 2.0 times more than the control, 3.0 and 2.8 times more than the clipping in July, 2012, respectively (Table 1). No difference in litter C and N storage was observed among the three treatments. The C: N ratios of all plant parts were decreased due to the fertilization treatment, but no difference was found between the clipping and the control (Table 1). Both plant C and N storage increased linearly as the plant biomass increased (both $p<0.001$; Fig. 2a, b).

Fertilized soil stored more C and N compared to the control and clipping treatments regardless soil depths in July, 2012 (Table 2). The soil had less C and N storage in the 0–5 and 10–20 cm zone in the clipping treatment. The fertilization increased soil C by 18.8% and soil N by 23.8%, respectively, when compared to the control. Whereas the clipping decreased soil C and N by 18.5% and 19.4% compared to the control, respectively. No significant difference in soil C: N ratio among the three treatments was found (Table 2). Both soil C and N storage increased linearly as the plant biomass increased (both $p<0.001$; Fig. 2e, i). Furthermore, the plant C storage increased linearly as the soil C storage increased

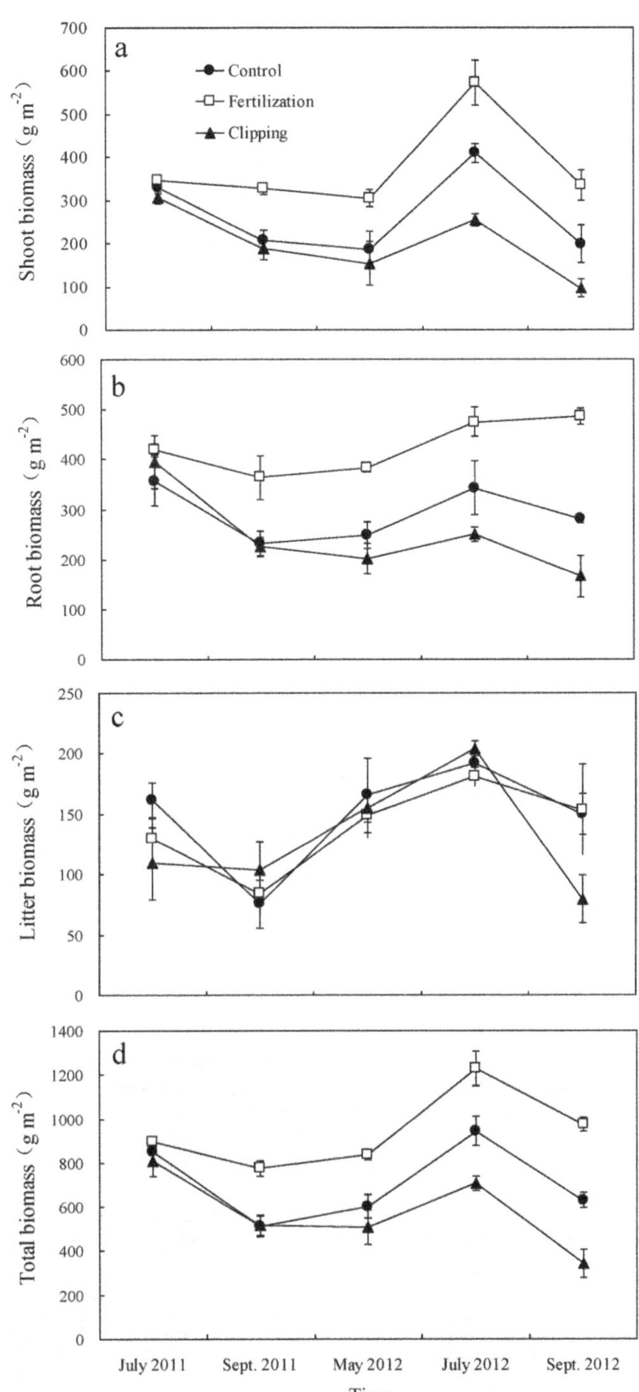


**Figure 1. The shoot (a), root (b), litter (c) and total biomass (d) variation under the different treatments over time. Vertical bars represent standard error (SE).** $n=5$.

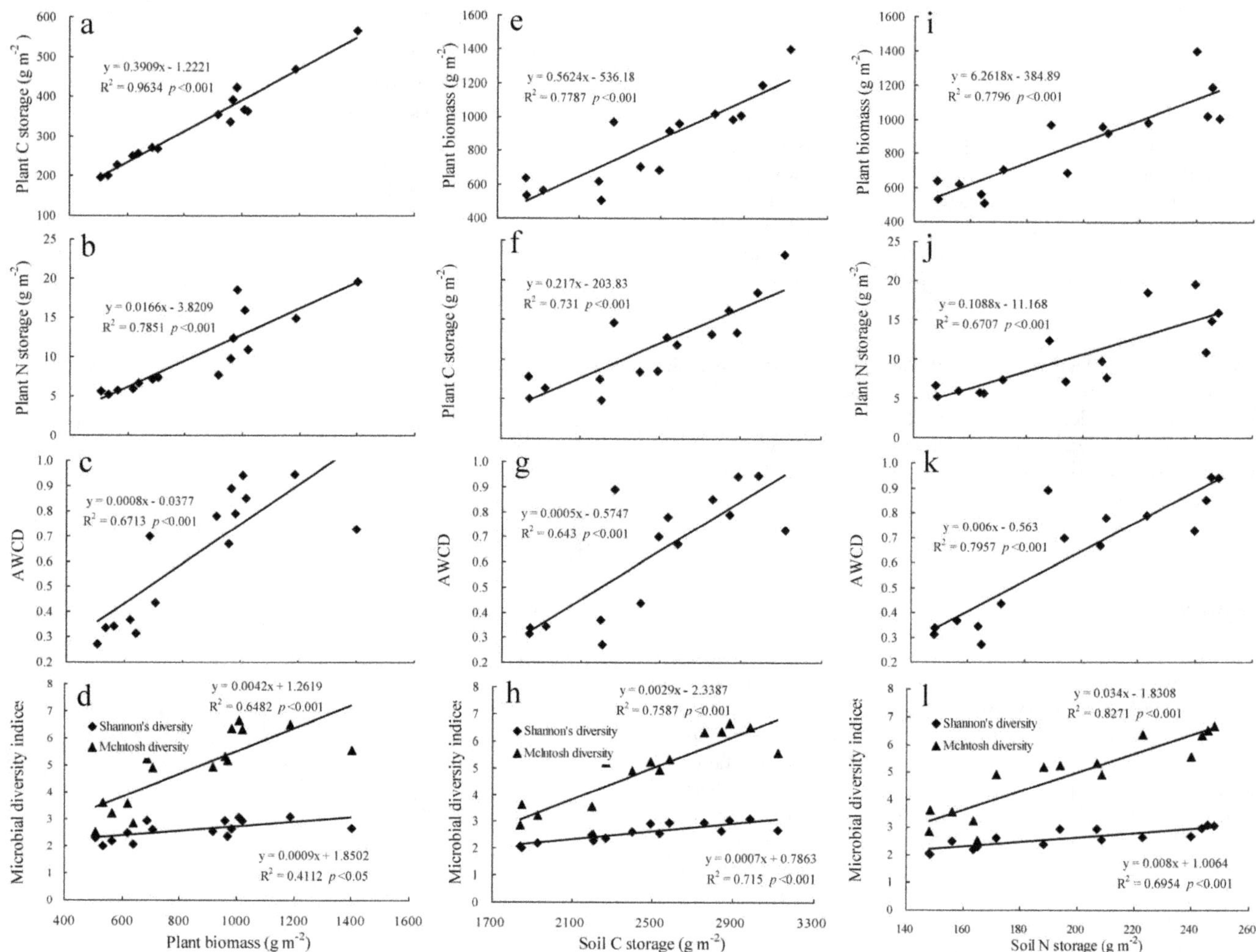


**Figure 2. The relationships of plant biomass with plant C/N storage (a/b), AWCD (c) and soil microbial diversity indices (d), and the relationships of soil C/N storage with plant biomass (e/i), plant C/N storage (f/j), AWCD (g/k) and soil microbial diversity indices (h/l) across the three treatments in July, 2012.**

**Table 1.** Carbon and nitrogen storage and C: N ratio among plant parts under different grassland treatments in July, 2012.

| Treatments | Shoot | Root | Litter | Total |
|---|---|---|---|---|
| | --------C storage (g C m$^{-2}$)-------- | | | |
| Control | 170.8±9.3b | 139.3±21.8b | 15.2±1.0a | 325.2±24.1b |
| Fertilization | 233.0±22.9a | 195.8±19.3a | 8.7±1.9a | 437.5±37.0a |
| Clipping | 108.0±7.1c | 102.7±4.9b | 16.0±2.6a | 226.7±12.1c |
| | --------N storage (g N m$^{-2}$) -------- | | | |
| Control | 5.5±0.6b | 2.8±0.5b | 0.6±0.0a | 8.9±1.0b |
| Fertilization | 10.0±1.1a | 5.6±0.9a | 0.4±0.1a | 16.0±1.5a |
| Clipping | 3.3±0.1b | 2.0±0.2b | 0.5±0.1a | 5.8±0.2b |
| | ----------C:N ratio---------- | | | |
| Control | 32.1±2.2a | 50.1±2.8a | 26.7±1.5a | 37.3±2.5a |
| Fertilization | 23.6±1.9b | 37.3±4.2b | 20.3±1.8b | 27.9±2.2b |
| Clipping | 32.8±2.0a | 51.5±2.0a | 30.3±1.8a | 38.8±1.1a |

The data represent means ± SE ($n = 5$). Letters a, b, c in a column indicate statistical significance base on Fisher's protected LSD test ($p<0.05$) among the three different treatments.

**Table 2.** Carbon and nitrogen storage and C: N ratio of soils at the 0–5, 5–10, and 10–20 cm depths under different grassland treatments in July, 2012.

| Treatments | 0–5 cm | 5–10 cm | 10–20 cm | 0–20 cm |
|---|---|---|---|---|
| | --------C storage (g C m$^{-2}$) -------- | | | |
| Control | 869.6±21.2b | 573.1±26.7b | 1014.1±47.5b | 2456.9±56.5b |
| Fertilization | 992.0±29.3a | 732.9±39.1a | 1193.8±29.5a | 2918.8±62.5a |
| Clipping | 690.3±29.1c | 527.5±21.6b | 784.4±52.4c | 2002.2±82.7c |
| | --------N storage (g N m$^{-2}$) -------- | | | |
| Control | 64.8±5.3b | 44.3±3.2b | 84.2±2.0b | 193.9±6.7b |
| Fertilization | 79.6±2.5a | 62.2±3.4a | 98.3±1.5a | 240.1±4.5a |
| Clipping | 50.0±2.3c | 40.6±2.1b | 66.1±3.5c | 156.2±3.6c |
| | ----------C: N ratio---------- | | | |
| Control | 13.8±1.2a | 13.1±0.9a | 12.0±0.6a | 12.7±0.4a |
| Fertilization | 12.5±0.5a | 11.9±0.9a | 12.2±0.3a | 12.2±0.3a |
| Clipping | 13.9±0.9a | 13.4±1.3a | 11.9±0.4a | 12.8±0.4a |

The data represent means ± SE ($n = 5$). Letters a, b, c in a column indicate statistical significance base on Fisher's protected LSD test ($p<0.05$) among the three different treatments.

($p<0.001$; Fig. 2f), and the plant N storage increased linearly as the soil N storage increased ($p<0.001$; Fig. 2j).

The high, medium and low values of total C and N storage (including plant and 0–20 cm soil zone) were observed in the fertilization, control and clipping grassland ecosystems, respectively (Tables 1, 2). The fertilization increased total C and N storage by 20.6% and 26.3%, respectively. The clipping reduced the total C and N storage by 19.9% and 20.1%, respectively. Furthermore, the plants had lower C and N storage and most C and N was stored in the soils. The 0–20 cm zone soils of the control, fertilization, and clipping treatments held 88%, 87% and 90% of the total C storage, and 96%, 94% and 96% of the total N storage, respectively.

## Soil microbial populations, activity and diversity

There was no difference in soil microbial number, activity and diversity among the three treatments in July, 2011 (data were not presented). On July 18, 2012, the number of bacteria, fungi and actinomyces in the fertilization treatment was 1.5, 1.3 and 1.6 times more than the control, and 2.3, 2.0 and 2.4 times more than the clipping treatment, respectively (Table 3). There was no difference in bacteria and actinomyces numbers between the control and the clipping. However, the clipping had fewer fungi than the control. The soil microbial activity (measured as AWCD)

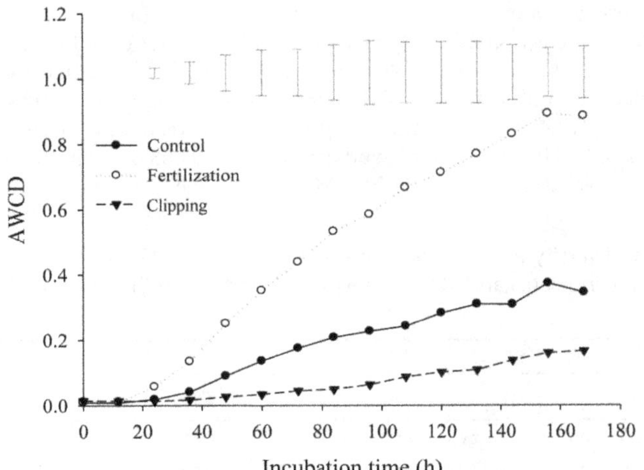


**Figure 3. Change of average well color development (AWCD) of soil microbial community during the incubation time in July, 2012.** Vertical bars represent Fisher's protected LSD ($p<0.05$).

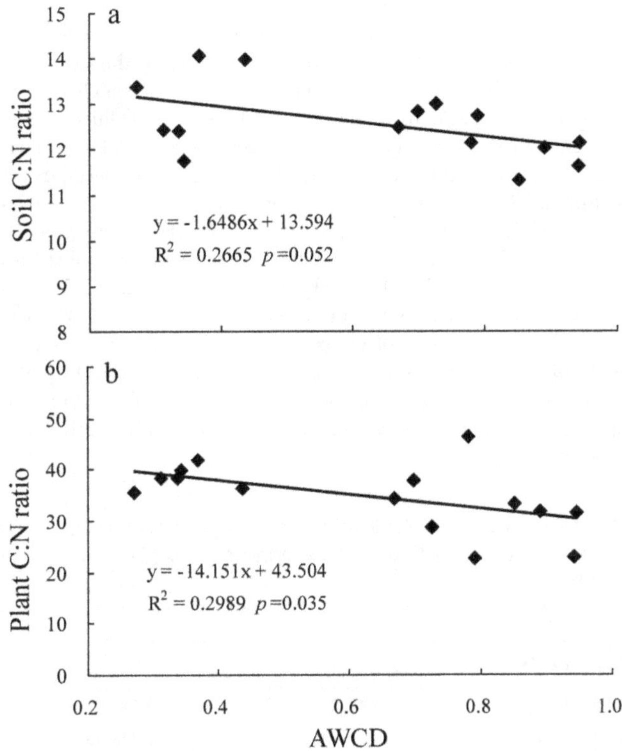


**Figure 4. The relationships of AWCD of soil microbial community with soil C: N ratio (a) and plant C: N ratio (b) in July, 2012.**

**Table 3.** The microbial community structure in grassland soils under different treatments in July, 2012. (CFU $g^{-1}$ dry weight soil).

| Treatments | bacteria number ($\times 10^6$) | fungi number ($\times 10^2$) | actinomyces number ($\times 10^4$) |
|---|---|---|---|
| Control | 2.8±0.1b | 2.7±0.2b | 1.2±0.1b |
| Fertilization | 4.3±0.2a | 3.6±0.1a | 1.9±0.1a |
| Clipping | 1.9±0.1b | 1.8±0.1c | 0.8±0.0b |

The data represent means ± SE ($n = 5$). Letters a, b, c in a column indicate statistical significance base on Fisher's protected LSD test ($p<0.05$) among the three different treatments.

in July, 2012 was increased with incubation time for all the three treatments, and ranked in the order of fertilization > control > clipping (Fig. 3). The AWCD was positively associated with plant biomass, soil C and N storage ($p<0.001$, Fig. 2c, g, k). Both soil and plant C: N ratios decreased linearly with increasing AWCD ($p = 0.052$, $p = 0.035$ Fig. 4a, b). Soil microbial diversity, evenness indices and substrate richness in July, 2012 calculated from the BIOLOG data (72 h) were affected by the fertilization and clipping treatments (Table 4). The higher levels of substrate richness, Shannon's evenness index, and McIntosh diversity and evenness indices were detected in the fertilization treatment, but lower level in the clipping treatment, compared to that in the control. Both the Shannon's and McIntosh diversity indices were positively associated with plant biomass, soil C and N storage ($p< 0.001$, Fig. 2d, h, l).

The PCA analysis indicated that the first two principal components accounted for 51.4% of the total variance (Fig. 5). The soil microbial community of the control, fertilization and the clipping treatments formed three distinct groups. The control was distinctly separated from the fertilization and clipping treatments by Factor 2. The fertilization and clipping treatments were distinctly separated by Factor 1. Furthermore, the factor loading plot also showed that the affinity of soil microbes for the substrates depended on the grassland treatments. The substrates including *L*-Arginine (A4), *L*-Phenylalanine (C4), N-Acetyl-D-Glucosamine (E2), Glucose-1-Phosphate (G2), Phenylethyl-amine (G4) and D-Malic Acid (H3) were favored by soil microbes of the fertilization treatment. The substrates including *L*-Threonine (E4), α-D-Lactose (H1), and D,L-α-Glycerol phosphate (H2) were favored by soil microbes of the clipping treatment, while the substrates including i-Erythritol (C2), 2-Hydroxy Benzoic Acid (C3), γ-Hydroxybutyric Acid (E3) and α-Ketobutyric Acid (G3) were favored by soil microbes of the control. Compared to the control, the fertilization treatment increased the utilization level of the substrates of amines (G4) and amino acids (A4 and C4), while the clipping treatment decreased the utilization level of the substrates of carboxylic acids.

### Plant and soil C, N sequestration rates

The plant, soil (0–20 cm) and total C, N sequestration rates during July, 2011 to July, 2012 were ranked in descending order of fertilization > control > clipping (Table 5). There was no statistical difference in the grassland soil (0–20 cm) and the total C and N sequestration rates among the three treatments. The plant C and N sequestration rates of the fertilization treatment were significantly higher than the clipping treatment ($p<0.05$).

## Discussion

This investigation was conducted on the natural grassland with vegetation coverage of more than 60%, composed of over 20 species of grass. In grassland ecosystems, the immobilization of C and N in the soil is the basic solution for C and N sequestration. Schleinger [49] reported that the below-ground C pool generally had much slower turnover rate than above-ground C. The data collected on July 1, 2011 exhibited that soil C and N (i.e. 15 days prior to imposing experiment) was similar among the three treatments. However, soil C and N on July 18, 2012 (i.e. at the end of the two-year experiment) increased by 18.8% and 18.5%, respectively, in the fertilization relative to the control. Previous studies investigated the effect of fertilizer on the grassland C and N storage and indicated that the accumulation of soil C and N were attributed to the increase of plant biomass [50–52].

As the main source of soil organic matter, the increase of grass biomass (including the shoots, roots, senescent litter) may be the first step to enhance the C and N sequestration in the soil [10,53]. The results of this study indicated that shoot and root biomass was greater in the fertilization treatment vs. the control, but less in the clipping treatment vs. the control. The increase in plant and soil C, N storage was significantly associated with a greater grass biomass (Fig. 2a, b, e, i). Data collected in July, 2012 indicated that the plant and soil C was significantly related to grass biomass ($r^2 = 0.9634$, 0.7787, both $p<0.001$). The correlation coefficient was 0.886 between plant N and biomass, and 0.883 between soil N and biomass (both $p<0.001$). Meanwhile, the plant C and N

**Table 4.** Effects of fertilization and clipping on microbial functional diversity as evaluated by substrate richness (*S*), Shannon's diversity index (*H'*), Shannon's evenness index (*E (S)*), McIntosh diversity index (*U*) and McIntosh evenness index (*E (M)*) in July, 2012 (72 h).

| Treatments | *S* | *H* | *E (S)* | *U* | *E (M)* |
|---|---|---|---|---|---|
| Control | 24.6±0.5b | 2.7±0.1a | 0.8±0.0b | 5.1±0.1b | 0.8±0.0b |
| Fertilization | 27.0±0.6a | 2.9±0.1a | 0.9±0.0a | 6.3±0.2a | 0.9±0.0a |
| Clipping | 21.8±0.4c | 2.2±0.1b | 0.7±0.0c | 3.2±0.2c | 0.5±0.0c |

The data represent means ± SE ($n = 5$). Letters a, b, c in a column indicate statistical significance base on Fisher's protected LSD test ($p<0.05$) among the three different treatments.

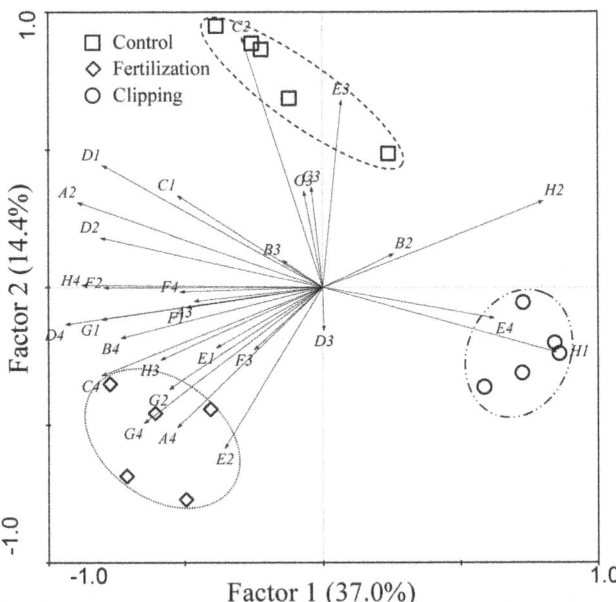


**Figure 5. Principle components analysis of biological data in July, 2012 (72 h).**

storage were positively associated with soil C and N storage (both $p<0.001$). Because the C and N stored in the plant was ultimately transferred into the soil in the forms of plant residues [54]. The beneficial effect of the fertilization on plant biomass could be contributed to the input nutrient [14,55]. The compound fertilizer provided the essential elements of N, P, and K for plant growth, which improved the shoot, root, and total biomass. On the contrary, the clipping limited the plant growth by damaging photosynthesis organs, causing a slow-growing period and a decrease in grass biomass consequently [9]. In addition, the variation of grass biomass indicated that the biomass could be affected by seasonal variation, management types and management time. From the perspective of increasing biomass, spring fertilization could give better results compared with summer fertilization.

The nutrient for plant growth is mainly derived from decomposition of SOM and plant residues input to the soil [56]. Similar to previous studies [9,57], we found that the plant of the fertilization treatment grew rapidly and sequestrated more C (carbohydrates) through photosynthesis and more N through passively and/or actively uptake, while the clipping treatment reduced the plant C and N uptake. Cheng et al [35] also found that the clipping decreased the plant N uptake in the tallgrass prairie. But Ruess et al [58] have reported that the clipping stimulated uptake rates of both ammonium ($NH_4^+$) and nitrate ($NO_3^-$), and ultimately accumulated more total plant N. This inconsistence was perhaps caused by the differences of soil nutrient conditions and plant species. The plant C: N ratio of the fertilization treatment was much lower than the control and the clipping treatments (Table 2). Chen [54] indicated that the SOM and plant residues with lower C: N ratios could be decomposed by microorganisms more easily due to improved soil microbial activity. We also found that the soil and plant C: N ratios decreased linearly with increasing AWCD ($p=0.035$, $p=0.052$). In addition, more humus was predicted for plant substrates with lower C: N ratios [59]. Thus the fertilization grassland could have a higher humification degree and sequestrated C and N in soil for a longer time than the control and the clipping treatments.

Soil microbial activity was involved in the mineralization of soil organic matter and plant residues [60,61]. Previous studies [56,62] indicated that the plant residues provided soil microorganisms with the major resource of nutrients and energy and controlled the soil microbial activity and composition. The results of this study exhibited that the fertilization and clipping treatments affected the grass biomass and ultimately changed the soil microbial activity, diversity and the C substrates utilized by soil microorganisms. The increase in soil microbial activity and diversity in the fertilization treatment was similar to Zhong and Cai's [29] and Marschner et al's study [14].

Our study suggested that the C and N cycles in the grassland ecosystem are determined not only by plant biomass, but also by soil microbial activity. The fertilization treatment, especially spring fertilization, improved the plant growth, increased the soil microbial activity, and ultimately increased the plant and soil C, N storage and sequestration rates. While the clipping treatment had the opposite effect and reduced the C, N storage and sequestration rates in grassland. Grassland management including fertilization was crucial to the grassland recovery from barren and overgrazing [63,64]. Other studies [65,66] indicated that the soil C may reach a new equilibrium in approximately decades after management changes, so continuous fertilization treatment was needed.

**Table 5.** Carbon and nitrogen sequestration rates of plant and soil (0–20 cm) under different grassland treatments from July, 2011 to July, 2012.

| Treatments | Plant | Soil (0–20 cm) | Total |
|---|---|---|---|
| | -----C sequestration rate (g C $m^{-2}$ $yr^{-1}$) ------ | | |
| Control | 66.6±33.6ab | 9.0±328.3 | 75.5±229.3 |
| Fertilization | 142.3±43.3a | 166.3±380.3 | 308.6±379.8 |
| Clipping | −31.9±39.2b | −84.7±194.6 | −116.6±510.3 |
| | -----N sequestration rate (g N $m^{-2}$ $yr^{-1}$) ------ | | |
| Control | 2.9±1.0ab | −5.7±43.6 | −2.9±43.8 |
| Fertilization | 9.0±2.1a | 28.2±15.7 | 37.2±16.2 |
| Clipping | −0.4±0.7b | −11.1±9.4 | −11.5±10.0 |

The data represent means ± SE ($n=5$). Letters a, b, c in a column indicate statistical significance base on Fisher's protected LSD test ($p<0.05$) among the three different treatments.

## Conclusions

The fertilization and the clipping treatments exhibited remarkable effect on the grass biomass, C and N storage, and soil microbial activity and diversity of grassland ecosystems in National Dalaoling Forest Park. After the two-year experiment, the compound fertilizer increased grass biomass, improved soil microbial activity and diversity and increased C and N sequestration in grassland ecosystem. The clipped plots had a lower level of C, N storage, which were mainly attributed to less grass biomass. The soil C, N storage was increased linearly with increasing grass biomass, plant C and N storage and AWCD. The principal component analysis indicated that the soil microbial communities of the control, fertilization and the clipping treatments formed three distinct groups, respectively. Previous and our results suggested that the fertilization might improve soil C and N slowly. So, to improve soil C, continuous fertilization management of grassland is needed. Long-term *in situ* studies to extrapolate the effect of the plant biomass on C and N dynamics combined to soil microbial activity and diversity and soil microclimate might contribute to a better understanding of C and N cycling and mitigation of global warming.

## Acknowledgments

The authors gratefully acknowledge the help of Dr Cheng Xiaoli in providing valuable revisions of the manuscript.

## Author Contributions

Conceived and designed the experiments: ZD SX JF. Performed the experiments: ZD YX LiqunH. LongxingH. Analyzed the data: ZD. Contributed reagents/materials/analysis tools: DL GW JF. Wrote the paper: ZD.

## References

1. Scurlock JMO, Hall DO (1998) The global carbon sink: a grassland perspective. Global Change Biol 4: 229–233.
2. Ni J (2002) Carbon storage in grasslands of China. J Arid Environ 50: 205–218.
3. Deng YL, Wang YK (2010) Assessment on water retention function of grassland ecosystems in the Upper Yangtze River Basin. Asian J Water Environ Pollut 7: 1–6.
4. Powlson DS, Whitmore AP, Goulding KWT (2011) Soil carbon sequestration to mitigate climate change: a critical re-examination to identify the true and the false. Eur J Soil Sci 62: 42–55.
5. Lal R (2003) Global potential of soil carbon sequestration to mitigate the greenhouse effect. Crit Rev Plant Sci 22: 151–184.
6. César Izaurralde R, Rosenberg NJ, Lal R (2001) Mitigation of climatic change by soil carbon sequestration: issues of science, monitoring, and degraded lands. Adv Agron 70: 1–75.
7. Dhillon RS, von Wuehlisch G (2013) Mitigation of global warming through renewable biomass. Biomass Bioenerg 48: 75–89.
8. Lv SH (2005) Status and development prospects of meadow resource in southern of China. J Sichuan Grassland 6: 37–41 (In Chinese with English abstract).
9. Klimeš L, Klimešová J (2001) The effects of mowing and fertilization on carbohydrate reserves and regrowth of grasses: do they promote plant coexistence in species-rich meadows? Evol Ecol 15: 363–382.
10. Smith P, Martino D, Cai Z, Gwary D, Janzen H, et al. (2008) Greenhouse gas mitigation in agriculture. Phil Trans R Soc B: Biol Sci 363: 789–813.
11. Sainju UM, Senwo ZN, Nyakatawa EZ, Tazisong IA, Reddy KC (2008) Soil carbon and nitrogen sequestration as affected by long-term tillage, cropping systems, and nitrogen fertilizer sources. Agr Ecosyst Environ 127: 234–240.
12. Qiu LP, Wei XR, Zhang XC, Cheng JM (2013) Ecosystem carbon and nitrogen accumulation after grazing exclusion in semiarid grassland. PLoS One 8: e55433.
13. Anindo DO, Potter HL (1994) Seasonal variation in productivity and nutritive value of Napier grass at Muguga, Kenya. E Afr Agric For J 59: 177–185.
14. Marschner P, Kandeler E, Marschner B (2003) Structure and function of the soil microbial community in a long-term fertilizer experiment. Soil Biol Biochem 35: 453–461.
15. Dersch G, Böhm K (2001) Effects of agronomic practices on the soil carbon storage potential in arable farming in Austria. Nutr Cycl Agroecosys 60: 49–55.
16. Sainju UM, Singh BP, Whitehead WF (2002) Long-term effects of tillage, cover crops, and nitrogen fertilization on organic carbon and nitrogen concentrations in sandy loam soils in Georgia, USA. Soil Till Res 63: 167–179.
17. Rochette P, Gregorich E (1998) Dynamics of soil microbial biomass C, soluble organic C and $CO_2$ evolution after three years of manure application. Can J Soil Sci 78: 283–290.
18. Collins HP, Rasmussen PE, Douglas CL (1992) Crop rotation and residue management effects on soil carbon and microbial dynamics. Soil Sci Soc Am J 56: 783–788.
19. Fynn RWS, Morris CD, Edwards TJ (2009) Effect of burning and mowing on grass and forb diversity in a long-term grassland experiment. Appl Veg Sci 7: 1–10.
20. Paul EA (1996) Soil microbiology and biochemistry. San Diego, California: Academic Press. 512 p.
21. Liski J, Nissinen A, Erhard M, Taskinen O (2003) Climatic effects on litter decomposition from arctic tundra to tropical rainforest. Global change biol 9: 575–584.
22. Joffre R, Ågren GI (2001) From plant to soil: litter production and decomposition. In: Roy J, Saugier B, Mooney HA, editors. Terrestrial global productivity. San Diego, California, , USA: Academic Press. pp. 83–99.
23. Larkin RP (2003) Characterization of soil microbial communities under different potato cropping systems by microbial population dynamics, substrate utilization, and fatty acid profiles. Soil Biol Biochem 35: 1451–1466.
24. Smith JL, Papendick RI, Bezdicek DF, Lynch JM (1992) Soil organic matter dynamics and crop residue management. In: Metting FB, editor. Soil microbial ecology: applications in agricultural and environmental management. New York: Marcel Dekker Inc. pp. 65–94.
25. Filser J, Fromm H, Nagel RF, Winter K (1995) Effects of previous intensive agricultural management on microorganisms and the biodiversity of soil fauna. Plant Soil 170: 123–129.
26. Sekiguchi H, Kushida A, Takenaka S (2007) Effects of cattle manure and green manure on the microbial community structure in upland soil determined by denaturing gradient gel electrophoresis. Microbes Environ 22: 327–335.
27. Kennedy AC, Smith KL (1995) Soil microbial diversity and the sustainability of agricultural soils. Plant Soil 170: 75–86.
28. Fox CA, MacDonald KB (2003) Challenges related to soil biodiversity research in agroecosystems-issues within the context of scale of observation. Can J Soil Sci 83: 231–244.
29. Zhong WH, Cai ZC (2007) Long-term effects of inorganic fertilizers on microbial biomass and community functional diversity in a paddy soil derived from quaternary red clay. Appl Soil Ecol 36: 84–91.
30. Garland JL, Mills AL (1991) Classification and characterization of heterotrophic microbial communities on the basis of patterns of community-level sole-carbon-source utilization. Appl Environ Microb 57: 2351–2359.
31. Ebhin Masto R, Chhonkar PK, Singh D, Patra AK (2006) Changes in soil biological and biochemical characteristics in a long-term field trial on a sub-tropical inceptisol. Soil Biol Biochem 38: 1577–1582.
32. Chang EH, Chung RS, Tsai YH (2007) Effect of different application rates of organic fertilizer on soil enzyme activity and microbial population. Soil Sci Plant Nutr 53: 132–140.
33. Zhang W, Parker KM, Luo Y, Wan S, Wallace LL, et al. (2005) Soil microbial responses to experimental warming and clipping in a tallgrass prairie. Glob Change Biol 11: 266–277.
34. Wan SQ, Luo YQ (2003) Substrate regulation of soil respiration in a tallgrass prairie: results of a clipping and shading experiment. Global Biogeochem Cy 17: 1054.
35. Cheng XL, Luo YQ, Su B, Wan SQ, Hui DF, et al. (2011) Plant carbon substrate supply regulated soil nitrogen dynamics in a tallgrass prairie in the Great Plains, USA: results of a clipping and shading experiment. J Plant Ecol 4: 228–235.
36. Melillo JM, Prentice IC, Farquhar GD, Schulze ED, Sala OE (1996) Terrestrial biotic responses to environmental change and feedbacks to climate. In: Climate change 1995: the science of climate change. . New York: Cambridge University Press. pp. 445–481.
37. Tans PP, Bakwin PS (1995) Climate change and carbon dioxide. Ambio 24: 376–378.
38. Zhang QF, Zheng Z, Jin YX (1990) Studies on the forest succession in Dalao ridge, Hubei province. Acta Phytoecologica et geobotanica sinica 14: 110–117 (In Chinese with English abstract).
39. Lu RK (2000) Analytical methods of soil agrochemistry. Beijing: China Agricultural Science and Technology Press. 638p (In Chinese).
40. Tian G, Granato TC, Cox AE, Pietz RI, Carlson CR, Jr., et al. (2009) Soil carbon sequestration resulting from long-term application of biosolids for land reclamation. J Environ Qual 38: 61–74.
41. Post WM, Emanuel WR, Zinke PJ, Stangenberger AG (1982) Soil carbon pools and world life zones. Nature 298: 156–159.
42. Zhou J, Guo WH, Wang RQ, Han XM, Wang Q (2008) Microbial community diversity in the profile of an agricultural soil in northern China. J Environ Sci 20: 981–988.
43. Dong M, Wang YF, Kong FZ, Jiang GM, Zhang ZB, et al. (1996) Survey, observation and analysis of terrestrial biocommunities. Beijing: Standards Press. 290 p (In Chinese).

44. Garland JL (1996) Analytical approaches to the characterization of samples of microbial communities using patterns of potential C source utilization. Soil Biol Biochem 28: 213–221.
45. Haack SK, Garchow H, Klug MJ, Forney LJ (1995) Analysis of factors affecting the accuracy, reproducibility, and interpretation of microbial community carbon source utilization patterns. Appl Environ Microb 61: 1458–1468.
46. Magurran AE (1988) Ecological diversity and its measurement. Princeton, NJ: Princeton university press. 179 p.
47. Staddon WJ, Duchesne LC, Trevors JT (1997) Microbial diversity and community structure of postdisturbance forest soils as determined by sole-carbon-source utilization patterns. Microb Ecol 34: 125–130.
48. Ter Braak CJF, Šmilauer P (2002) CANOCO reference manual and CanoDraw for Windows user's guide: software for canonical community ordination (version 4.5). Ithaca, New York: Microcomputer Power. 500 p.
49. Schlesinger WH (1995) Soil respiration and changes in soil carbon stocks. In: Woodwell GM, Mackenzie FT, editors. Biotic feedbacks in the global climatic system: will the warming feed the warming. New York: Oxford University Press. pp. 159–168.
50. Conant RT, Paustian K, Elliott ET (2001) Grassland management and conversion into grassland: Effects on soil carbon. Ecol Appl 11: 343–355.
51. Silveira ML, Liu K, Sollenberger LE, Follett RF, Vendramini J (2012) Short-term effects of grazing intensity and nitrogen fertilization on soil organic carbon pools under perennial grass pastures in the southeastern USA. Soil biol Biochem 58: 42–49.
52. Thornley JHM, Fowler D, Cannell MGR (1991) Terrestrial carbon storage resulting from $CO_2$ and nitrogen fertilization in temperate grasslands. Plant Cell Environ 14: 1007–1011.
53. Kuzyakov Y, Domanski G (2000) Carbon input by plants into the soil. Review. J Plant Nutr Soil Sci 163: 421–431.
54. Chen HM. (2005) Environmental soil science. Beijing: Science Press. 516 p (In Chinese).
55. Pyšek P, Lepš J (2009) Response of a weed community to nitrogen fertilization: a multivariate analysis. J Veg Sci 2: 237–244.
56. Thibodeau L, Raymond P, Camiré C, Munson AD (2000) Impact of precommercial thinning in balsam fir stands on soil nitrogen dynamics, microbial biomass, decomposition, and foliar nutrition. Can J Forest Res 30: 229–238.
57. Porporato A, D'odorico P, Laio F, Rodriguez-Iturbe I (2003) Hydrologic controls on soil carbon and nitrogen cycles. I. Modeling scheme. Adv Water Resour 26: 45–58.
58. Ruess RW, McNaughton SJ, Coughenour MB (1983) The effects of clipping, nitrogen source and nitrogen concentration on the growth responses and nitrogen uptake of an East African sedge. Oecologia 59: 253–261.
59. Nicolardot B, Recous S, Mary B (2001) Simulation of C and N mineralisation during crop residue decomposition: a simple dynamic model based on the C: N ratio of the residues. Plant Soil 228: 83–103.
60. Magill AH, Aber JD (2000) Variation in soil net mineralization rates with dissolved organic carbon additions. Soil biol Biochem 32: 597–601.
61. Zak DR, Tilman D, Parmenter RR, Rice CW, Fisher FM, et al. (1994) Plant production and soil microorganisms in late-successional ecosystems: a continental-scale study. Ecology 75: 2333–2347.
62. Palm CA, Myers RJK, Nandwa SM (1997) Combined use of organic and inorganic nutrient sources for soil fertility maintenance and replenishment. In: Buresh RJ, Sanchez PA, Calhoun F, editors. Replenishing soil fertility in Africa. Madison, Wisconsin: SSSA special publication. pp. 193–217.
63. Chen Q, Hooper DU, Lin S (2011) Shifts in species composition constrain restoration of overgrazed grassland using nitrogen fertilization in Inner Mongolian steppe, China. PloS One 6: e16909.
64. Goetz H (1969) Composition and yields of native grassland sites fertilized at different rates of nitrogen. J Range Manage 22: 384–390.
65. Dumanski J, Desjardins RL, Tarnocai C, Monreal C, Gregorich EG, et al. (1998) Possibilities for future carbon sequestration in Canadian agriculture in relation to land use changes. Climatic Change 40: 81–103.
66. Smith P, Powlson DS, Glendining MJ, Smith JU (1998) Preliminary estimates of the potential for carbon mitigation in European soils through no-till farming. Global change biol 4: 679–685.

8

# Water Consumption Characteristics and Water Use Efficiency of Winter Wheat under Long-Term Nitrogen Fertilization Regimes in Northwest China

**Yangquanwei Zhong, Zhouping Shangguan***

State Key Laboratory of Soil Erosion and Dryland Farming on the Loess Plateau, Northwest A & F University, Yangling, Shaanxi, P.R. China

## Abstract

Water shortage and nitrogen (N) deficiency are the key factors limiting agricultural production in arid and semi-arid regions, and increasing agricultural productivity under rain-fed conditions often requires N management strategies. A field experiment on winter wheat (*Triticum aestivum* L.) was begun in 2004 to investigate effects of long-term N fertilization in the traditional pattern used for wheat in China. Using data collected over three consecutive years, commencing five years after the experiment began, the effects of N fertilization on wheat yield, evapotranspiration (ET) and water use efficiency (WUE, i.e. the ratio of grain yield to total ET in the crop growing season) were examined. In 2010, 2011 and 2012, N increased the yield of wheat cultivar Zhengmai No. 9023 by up to 61.1, 117.9 and 34.7%, respectively, and correspondingly in cultivar Changhan No. 58 by 58.4, 100.8 and 51.7%. N-applied treatments increased water consumption in different layers of 0–200 cm of soil and thus ET was significantly higher in N-applied than in non-N treatments. WUE was in the range of 1.0–2.09 kg/$m^3$ for 2010, 2011 and 2012. N fertilization significantly increased WUE in 2010 and 2011, but not in 2012. The results indicated the following: (1) in this dryland farming system, increased N fertilization could raise wheat yield, and the drought-tolerant Changhan No. 58 showed a yield advantage in drought environments with high N fertilizer rates; (2) N application affected water consumption in different soil layers, and promoted wheat absorbing deeper soil water and so increased utilization of soil water; and (3) comprehensive consideration of yield and WUE of wheat indicated that the N rate of 270 kg/ha for Changhan No. 58 was better to avoid the risk of reduced production reduction due to lack of precipitation; however, under conditions of better soil moisture, the N rate of 180 kg/ha was more economic.

**Editor:** Raffaella Balestrini, Institute for Plant Protection (IPP), CNR, Italy

**Funding:** The study was sponsored by the National Natural Science Foundation of China (41390463, 61273329) and the Important Direction Project of Innovation of CAS (KZCX2-YW-JC408). The funders had no role in study design, data collection and analysis, decision to publish, or preparation of the manuscript.

**Competing Interests:** The authors have declared that no competing interests exist.

* E-mail: shangguan@ms.iswc.ac.cn

## Introduction

Northwest China is a vast semi-arid area with average annual precipitation in the range of 300–600 mm and more than 90% of the land is cropland [1]. This means that water is the primary factor limiting crop yields. In addition, world food demand is expected to double during 2005–2050 [2], thus it is important to increase food production with lower water use [3], particularly in water shortage regions. Currently, water stress and nutrient deficits are the main factors limiting primary production in arid and semi-arid environments [4–8]. Therefore, many rain-fed farming experts have focused on how to increase crop water use efficiency (WUE, i.e. the ratio of grain yield to total ET in the crop growing season) by irrigation and fertilization.

In the 1990 s, many studies on effects of limited irrigation on crop yields and WUE showed that by reducing irrigation volume, crop yield could be generally maintained and product quality improved [9–13],and appropriate irrigation management can increase crop yield and WUE [14–16]. There are several sources of soil water in irrigated or high water-table areas, however, precipitation is the only source of soil water for crop growth in many rain-fed farming systems of arid and semi-arid regions. Therefore, new methods need to be devised to improve WUE in this non-irrigated farming system.

N fertilization is a common practice to increase grain production, but its performance depends on soil water status [17–19]. The importance of increasing crop yield and improving soil quality through fertilization has been confirmed. The increasing use of N fertilizer could significantly increase maize production [8,20], and already affects a large proportion of the world's food production [21,22]. Fan et al. [1] reported that inorganic N and phosphorus (P) fertilization increased grain yields by 50–60% in China, and reports from Europe showed that N fertilizers can increase crop yield significantly [23]. N fertilization is well known to improve soil fertility [24,25]; however, using excessive N fertilizer can decrease the N utilization rate, which not only causes a huge waste of resources and economic losses, but can also adversely impact the environment [26–28]. Balancing the N rate, WUE and yield is an important problem in dryland farming systems. Better understanding of interactions among precipitation, fertilization and crops production is essential for efficient utilizations of water resources and N fertilizers, and sustainable food productions in rain-fed cropping systems experiencing climate change [1]. Long-term fertilization experiments are

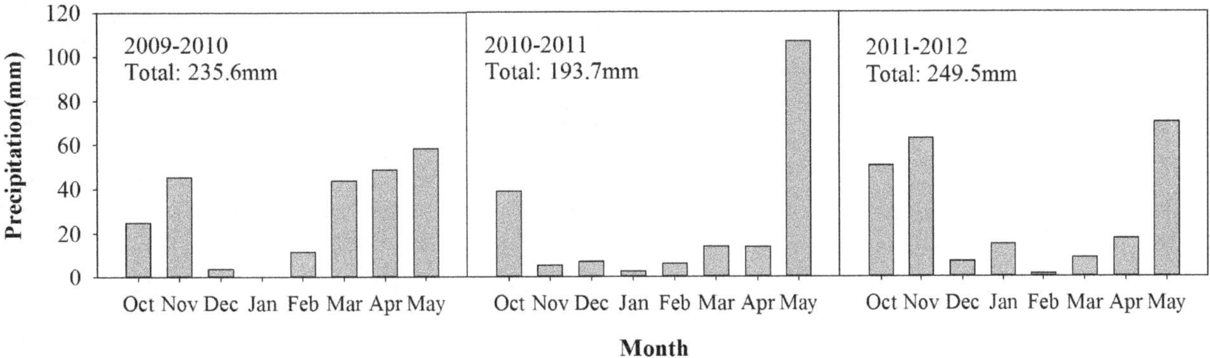


**Figure 1. Monthly and total precipitation during the 2009–2012 wheat growing seasons.** Monthly precipitation of wheat growing seasons in 2009–2010, 2010–2011, 2011–2012 in Shaanxi, Yangling.

valuable to follow crop yield, soil fertility, WUE and risk management over time [29,30]. Various long-term experiments have examined how to increase yield and WUE of wheat, using irrigation, organic or inorganic fertilizer, soil tillage and crop management [31–33]. However, few experiments have been done on evapotranspiration (ET) and WUE under circumstances with only N fertilizer and without irrigation in northwest China. With China's urbanization, increasing numbers of farmers have abandoned farms to urban construction and this has led to a loss of labor. Thus, most farmland in northwest China region still uses traditional cropping practices that all fertilizer applied once prior to planting [1], lack of careful management of irrigation and other tasks. Kang et al. [14] reported that difference in yield and WUE are also related to regional variability in environment and crop varieties, so information specific to a region is needed for developing and refining the agricultural performance in this region. In these circumstances, it is very important to determine the advantages and disadvantages of long-term N fertilization on yield of different varieties.

This study examined two different water-sensitive cultivars of winter wheat (*Triticum aestivum* L.) to investigate effects of N fertilizers on crop yield, ET and WUE, using the most common management of farmers in northwest China. The objectives were (1) to investigate impacts of traditional long-term N fertilization on yields of two different water-sensitive wheat cultivars; (2) to examine the effect of N fertilizers on total ET, and soil water consumption from different soil layers of the two cultivars; and (3) to establish relationships among crop yield, WUE and ET and determine optimum N fertilizer rates in northwest China. This study may compensate for some of the lack of long-term influence only N fertilizer on crop production, and the results should provide guidelines to farmers in the region on choosing appropriate cultivars and obtaining high yields with appropriate N application.

## Materials and Methods

### Experiment site and climatic conditions

The study commenced in October 2004 in an experiment field of the Institute of Soil and Water Conservation of the Northwest A & F University, Yangling, Shaanxi (34°17′56″N, 108°04′7″E). Located on the southern boundary of the Loess Plateau, the experiment site has a temperate and semi-humid climate with a mean annual temperature of 13°C and a mean annual precipitation of 632 mm, of which about 60% occurs during July–September.

### Experiment design

The study adopted a randomized block design with three replications. Two winter wheat (*Triticum. aestivum* L.) cultivars were used: Zhengmai No. 9023 (ZM) is water sensitive and poorly drought-tolerant and Changhan No. 58 (CH) is drought-tolerant and suitable for drought prone environments. The thousand-kernel weights of ZM and CH were 43.58 and 43.61 g, respectively. N treatments were applied at five rates: 0, 90, 180, 270 and 360 kg/ha (N0, N90, N180, N270 and N360, respectively). Plot size was 2 m×3 m with 20 rows (15-cm spaces) of wheat sown at 90 seeds/row. Wheat was sown in early October and harvested in early June the following year. The seeding rate was 130 kg/ha. Immediately before sowing, the fertilizer was evenly spread on the soil surface and then incorporated into the upper 15 cm soil by chiseling. N was applied as urea and P (75 kg $P_2O_5$/ha) as super phosphate. No potassium fertilizer was applied, and the site was ploughed to bury weeds before sowing.

### Measurements

In all treatments, the volumetric soil water content was measured every 10 cm for 0–100 cm of soil and every 20 cm for 100–300 cm with a neutron moisture meter (CNC100, Super Energy, Nuclear Technology Ltd., Beijing, China). The 3-m-long neutron gauge access tube was buried vertically in the center of each plot at the beginning of the study. Soil water was measured during the first week of every wheat growing month except January and February. If any precipitation occurred just before or during the measurement period, then measurements were postponed for several days until the soil moisture attained a normal degree. The yields and the thousand-kernel weights of wheat in all plots were measured at harvest time in early June. Since this paper aimed to test the cumulative effects of N fertilization, we chose data from three wheat growing years with different precipitation characteristics: 2009–2010, 2010–2011 and 2011–2012.

### Calculation and statistics

ET of winter wheat was calculated using the following equation [33]:

$$ET = \Delta S + P + I - R - D$$

**Table 1.** Nitrogen effects on wheat yield and thousand-kernel weights of two cultivars in three years.

| Varieties | Treatments | 2010 | | | 2011 | | | 2012 | | |
|---|---|---|---|---|---|---|---|---|---|---|
| | | Yield(kg/ha) | | Thousand-kernel (g) | Yield(kg/ha) | | Thousand-kernel (g) | Yield(kg/ha) | | Thousand-kernel (g) |
| ZM | N0 | 4716 | c | 49.5 a | 2974 | c | 47.0 a | 5906 | bc | 49.1 a |
| | N90 | 6355 | ab | 42.5 b | 5499 | ab | 42.2 ab | 7272 | a | 42.3 bcd |
| | N180 | 7597 | a | 42.3 b | 6355 | ab | 42.1 ab | 7953 | a | 41.6 cd |
| | N270 | 7527 | a | 42.5 b | 6482 | ab | 40.5 b | 7923 | a | 41.1 d |
| | N360 | 7519 | a | 43.9 b | 6390 | ab | 41.2 ab | 7512 | a | 40.4 d |
| CH | N0 | 4162 | c | 46.1 ab | 3391 | c | 42.2 ab | 5407 | c | 46.5 ab |
| | N90 | 5862 | ab | 42.3 b | 4886 | b | 41.1 ab | 6926 | ab | 45.5 abc |
| | N180 | 6594 | ab | 42.3 b | 5879 | ab | 41.9 ab | 7777 | a | 40.4 d |
| | N270 | 6334 | ab | 37.9 c | 6748 | a | 40.5 b | 8199 | a | 40.1 d |
| | N360 | 6126 | b | 37.1 c | 6808 | a | 40.3 b | 8018 | a | 38.6 d |

Values are means of three replicates for each treatment. Different letters indicate statistical significance at $P<0.05$ within the same column.

Where ΔS is soil water storage change, P is precipitation, I is irrigation rate, R is surface runoff and D is deep water percolation (all in mm).

No irrigation was used and so $I=0$. Precipitation in the three growing seasons is shown in Fig. 1, and measured surface runoff was negligible during these years. Deep percolation was calculated as the difference between soil moisture content and field moisture capacity when the soil water content at this depth was more than the field water holding capacity. In the study site, deep water percolation did not occur.

WUE was defined as follows:

$$WUE = Y/ET$$

Where Y is grain yield (kg/ha).

All data concerned were analyzed by SPSS 16.0 Statistical software. ANOVA was adopted to determine whether treatments were significantly different at $P<0.05$. Duncan's multiple range test was used to differentiate treatment means at $P<0.05$.

## Results and Discussion

### Wheat yield

The yields and thousand-kernel weights of the two wheat cultivars at the different N rates in the different years are presented in Table 1. Grain yield was in the range of 2.94–7.92 t/ha for ZM and 3.39–8.19 t/ha for CH. Grain yields of the two cultivars both differed significantly between N0 and the other N rates, and increased as N rates increased, but did not differ significantly among the N-applied treatments. The lack of significant differences may due to yield in three replications being affected by other factors in field experiment. Usually the significance of increase yield is hard to attain in agricultural research, Morell et al. [36] reported that grain yield mostly 1000 kg/ha higher than control, but still have no statistical difference. The yields of wheat slightly decreased at N360, except for yield of CH in 2011. In 2010, 2011 and 2012, the wheat yields of ZM increased by up to 61.1, 118.0 and 34.7%, respectively, in the N-applied treatments compared to treatment without N fertilization; and corresponding yields of CH increased by up to 58.4, 100.8 and 51.7%. The highest yields of ZM and CH were both for treatments of N180, N270 and N270 in 2010, 2011 and 2012, respectively. Thus, N application significantly increased yields of wheat, but an excessive N rate had no positive effect on grain yield. Previous research has shown similar results with N fertilizer application significantly increasing maize and wheat yield compared to unfertilized treatments [14,15]. Bassoa et al. [23] examined the long-term wheat response to N in rain-fed Mediterranean environments, and showed that yield response was stronger for 120 than 60 and 90 kg N/ha. Many other studies have demonstrated a parabolic relationship between N and grain yield, i.e. when N rate surpassed a certain threshold, the grain yield greatly decreased. In China, there have been many experiments on different wheat cultivars and fertilizer regimes that have shown the maximum N rate is 150–225 kg/ha. At excessive N rates, the leaf protein and chlorophyll contents decrease, and then photosynthesis also decreases [34]. Tinsina et al. [35] also showed that wheat yield was higher at 120 than 180 kg/ha in Bangladesh. Morell et al. [36] showed no additional wheat yield responses to N fertilizers at N rates >100 N kg/ha. All these studies demonstrated that N application could increase wheat yield, but excessive N had no yield benefit.

At the same N rates, the yields of the two cultivars did not differ significantly. In 2010, the rainfall, which was evenly distributed

**Table 2.** Water consumption from soil (ΔS) or precipitation and its ratio to total evapotranspiration (ET) and WUE.

| Year | Varieties | Treatments | Evapotranspiration (ET)(mm) | | Soil water consumption (ΔS) | | Precipitation | | WUE ($kg/m^3$) | |
|---|---|---|---|---|---|---|---|---|---|---|
| | | | | | Amount (mm) | Ratio (%) | Amount (mm) | Ratio (%) | | |
| 2009–2010 | ZM | N0 | 325.72 | e | 90.12 | 27. 7 | 235.6 | 72.3 | 1.45 | de |
| | | N90 | 348.58 | de | 112.98 | 32.4 | | 67.6 | 1.83 | abcd |
| | | N180 | 361.57 | bcd | 125.97 | 34.8 | | 65.2 | 2.09 | a |
| | | N270 | 385.25 | abc | 149.65 | 38.9 | | 61.2 | 1.96 | ab |
| | | N360 | 385.59 | abc | 149.99 | 38.9 | | 61.1 | 1.97 | ab |
| | CH | N0 | 321.39 | e | 85.79 | 26.7 | | 73.3 | 1.29 | e |
| | | N90 | 352.69 | cde | 117.09 | 33.2 | | 66.8 | 1.66 | bcde |
| | | N180 | 373.65 | bcd | 138.05 | 36.9 | | 63.1 | 1.76 | abc |
| | | N270 | 411.98 | a | 176.38 | 42.8 | | 57.2 | 1.48 | cde |
| | | N360 | 395.49 | ab | 159.89 | 40.4 | | 59.6 | 1.52 | cde |
| 2010–2011 | ZM | N0 | 298.40 | d | 104.70 | 35.1 | 193.7 | 64.9 | 1.00 | d |
| | | N90 | 323.28 | bc | 129.58 | 40.1 | | 59.9 | 1.70 | ab |
| | | N180 | 344.98 | ab | 151.28 | 43.8 | | 56.2 | 1.84 | ab |
| | | N270 | 345.60 | ab | 151.90 | 44.0 | | 56.1 | 1.88 | a |
| | | N360 | 337.41 | ab | 143.71 | 42.6 | | 57.4 | 1.89 | a |
| | CH | N0 | 308.48 | cd | 114.78 | 37.2 | | 62.8 | 1.10 | cd |
| | | N90 | 344.01 | ab | 150.31 | 43.7 | | 56.3 | 1.42 | bc |
| | | N180 | 351.92 | a | 158.22 | 45.0 | | 55.0 | 1.67 | ab |
| | | N270 | 351.75 | a | 158.05 | 44.9 | | 55.0 | 1.92 | a |
| | | N360 | 344.63 | ab | 150.93 | 43.8 | | 56.2 | 1.84 | ab |
| 2011–2012 | ZM | N0 | 373.92 | c | 124.42 | 33.3 | 249.5 | 66.7 | 1.58 | a |
| | | N90 | 412.28 | bc | 162.78 | 39.5 | | 60.5 | 1.77 | a |
| | | N180 | 456.45 | a | 206.95 | 45.3 | | 54.7 | 1.75 | a |
| | | N270 | 442.06 | ab | 192.56 | 43.6 | | 56.4 | 1.80 | a |
| | | N360 | 440.44 | ab | 190.94 | 43.3 | | 56.6 | 1.71 | a |
| | CH | N0 | 381.35 | c | 131.85 | 34.6 | | 65.4 | 1.43 | a |
| | | N90 | 464.53 | a | 215.03 | 46.3 | | 53.7 | 1.50 | a |
| | | N180 | 469.35 | a | 219.85 | 46.8 | | 53.5 | 1.66 | a |
| | | N270 | 461.05 | a | 211.55 | 45.9 | | 54.1 | 1.79 | a |
| | | N360 | 441.13 | ab | 191.63 | 43.4 | | 56.6 | 1.82 | a |

Values are means of three replicates for each treatment. Different letters indicate statistical significance at $P<0.05$ within the same column. ΔS has the same significance as total ET.

through the growing season, provided a more favorable environment for ZM, the water-sensitive cultivar, and so its yields were higher than those of CH for all treatments. In 2011, precipitation was the lowest in the whole growing season of all years, despite 106 mm of rainfall in May when wheat filled its seeds. However, such high rainfall at seed filling was unfavorable for wheat yield. Sheng and Wang [37] found that high soil-water contents at the seed-filling stage of wheat can result in lower thousand-kernel weights and grain yields. So drought during the growing season and too much rainfall from the seed-filling to the ripening stages led to the lower yield of wheat in 2011 compared to 2010 and 2012. Before the 2011–2012 growing season, the summer of 2011 received a lots of rain, 672.7 mm during June–September, larger than 421.6 mm in 2009 and 436.7 mm in 2010 summer. This caused total water consumption in 2011–2012 to be higher than previously and so the yields of wheat were the highest among the three years, although precipitation did not differ greatly from the growing season of 2009–2010. Shangguan et al. [38] reported that the fallow efficiencies, expressed as the ratio of soil water accumulation to precipitation received during the period of fallow, were important for yield in the next growing season. The importance of soil-water storage during the fallow period for increasing grain yields of post-fallow crops are supported by many studies on dryland including the Southern Great Plains in the USA [39–41] and the Loess Plateau [38]. In 2011 and 2012, the yield of ZM was higher than that of CH at N0, N90 and N180, and

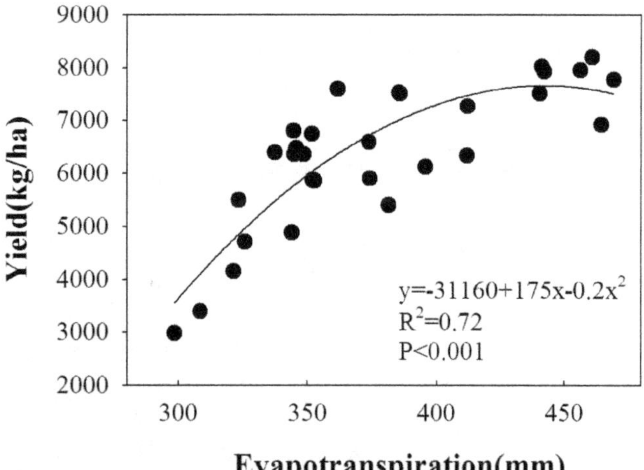


**Figure 2. Relationship between wheat evapotranspiration (ET) (mm) and grain yield (Y) (kg/ha) for winter wheat in northwest China.** The relationship between ET and yield is shown by the equation.

drought-tolerant CH showed higher yield only at the higher N rate in dry years. This is because CH was developed in recent years and prefers high fertilizer levels – consequently its cultivation has greatly expanded in northwest areas. CH is sensitive to N, and high rates of N result in higher yields; however, in contrast ZM is a poorly drought-tolerant cultivar but can produce higher yield at lower N rates. These cultivar characteristics have been demonstrated by many physiological indices in our previous studies [42]. Overall, the water consumption characteristics differed between the wheat varieties, leading to the different production performance. N180 resulted in higher yields of ZM in 2010 when rainfall was evenly distributed over the growing season. However, when rainfall was unevenly distributed or there was a lack of rainfall in the growing season, appropriate increases in the amount of N for CH could result in higher wheat yield to avoid the risk of reduced production.

The thousand-kernel weights of wheat decreased with increased N rates, consistent with many other research results [34]. There were two reasons for this: first, N can increase numbers of wheat tillers and panicles as well as flag leaf photosynthetic rates, but large and thick leaves would shade one another, affecting starch assimilation and transportation to kernels and resulting in lower thousand-kernel weights; secondly, N application could delay the flowering stage of wheat, thereby shortening the grain-filling stage and leading to lower thousand-kernel weights.

## Water consumption characteristics

**Total ET.** In dryland farming, ET is supplied partly from precipitation in the growing season and partly from soil-water storage before planting. However, the relative contribution between precipitation and crop-consumed soil water to ET differs significantly among crops.

The total ET, ΔS or rainfall and its ratio to total ET at the different N rates are shown in Table 2. Total ET behaved differently between years and cultivars. Total ET was in the range of 298.40–442.46 mm for ZM and 361.57–469.35 mm for CH. Total ET was significantly higher in the N-applied than non-N treatments, except for treatment N90. Total ET were highest in 2011–2012 of CH and ZM. In the N-applied treatments, the ET of ZM increased by up to 18.4, 15.8 and 22.1% in 2009–2010, 2010–2011 and 2011–2012, respectively, and correspondingly for CH by 28.0, 14.1 and 23.1%. The ET slightly decreased for N360, indicating that ET could not increase further if too much N was applied. Zhou et al. [15] showed that N fertilizer application decreased water storage in 0–200 cm of soil and particularly so after wheat harvesting. Hunsaker et al. [43] showed that wheat ET was significantly higher than in low N treatments. One explanation for N increasing ET of wheat is that N promotes wheat to grow more and produce longer roots, enabling more soil water to be absorbed; another explanation is that N fertilization increases the leaf area index and transpiration rates of wheat [44]. However, too much N makes soil environments stressful by increasing N concentration in soil solution, thus preventing roots from absorbing water. The total ET in this study was considerably lower than that reported for the southern high plains of the USA [45,46] and the North China Plain [47], but was close to that for the Loess Plateau [48]. These differences are likely due to different climatic conditions, like temperature and precipitation and also attributed to different field management.

In all the experiment years, CH had higher ET than ZM but not significantly at the same N rates – probably due to different characteristics of the varieties. As a drought-tolerant cultivar with long roots, CH is capable of absorbing deep soil water, thus presenting higher ET than ZM. A deep-growing root system will favor taking up deep soil water under water-limited conditions. Research on dryland crops has shown that deep soil-water utilization is probably limited by root density [11,48,49]. The rainfall was less during 2010–2011 than 2009–2010 and 2011–2012 growing seasons so that the ratio of soil water consumption was higher in the former than the other two years (Table 2). As the N application rates increased, the ratios also increased, indicating that N application helped plants utilize deeper soil water.

The relationship between grain yields and seasonal ET was best described by a quadratic function obtained by regression analysis ($Y = -31160+175x-0.2x^2$; Fig. 2). Grain yield did not increase when ET exceeded a certain critical value, e.g. about 430 mm in the present study. Grain yield required a minimum ET of 244 mm for winter wheat (Fig. 2). This minimum ET value is higher than the 84 mm for wheat in the North China Plain [47] and 156 mm in the Mediterranean region [12], as well as higher than the 206 mm of dryland and irrigated wheat reported by Musick et al. [41] in US southern plains. These differences are likely due to such different climates and crop management. This result may indicate that the crop yield in this area will more relies on precipitation and soil water storage.

**Water consumptions in the different soil layers.** Water consumption in the different layers of the soil profile in 40-cm increments is plotted with depth in Fig 3. N applications had a significant effect on ΔS, as well as on ET (Table 2), since N application increased water consumption in the different layers above 200 cm, except during the 2011–2012 growing season for both cultivars. In 2011–2012, at 200 cm soil depth, the N treatments still had higher soil water consumption than N0 treatment, likely due to the large amount of rainfall in this year. A similar result was found by Zhou et al. [15], with N fertilizer application decreasing water storage at soil depths of <200 cm after wheat harvesting. N application increased water consumption in the different soil layers; however, in the same soil layers, water consumption did not differ significantly among the different N rates except for some layers of CH.

Soil water consumption clearly changed with the different N rates (Fig. 3). The trends of water consumption were similar in all treatments as soil layers became deeper. N application increased water consumption in all soil layers. Generally, water consumption

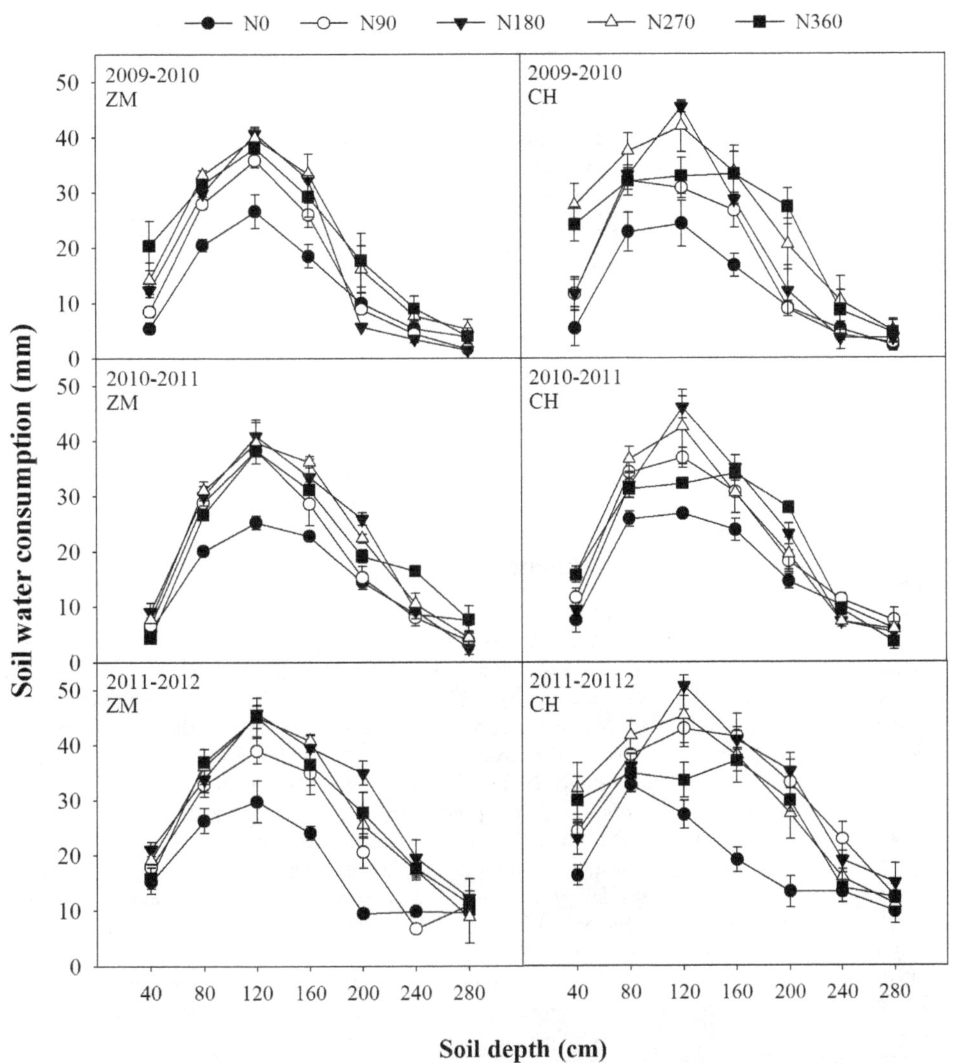


**Figure 3. Wheat evapotranspiration (ET) (mm) of two cultivars in different soil layers and different nitrogen (N) treatments in three years.** Water ET trends as soil depth increased with influence of N fertilizer for two cultivars in three years. Standard error bars are also shown.

of CH was higher than that of ZM, which was true of the total soil water consumption in all layers (Table 2). Water was mainly consumed in layers of 40–160 cm deep, with the highest water consumption for 100–140 cm. Water was stably absorbed for soil layers <120 cm in the N0 treatment, and <160 cm in the N-applied treatments, shown by ΔS of N treatments at 160 cm being higher than for the non-N treatment at 120 cm. This was probably because N application could promote roots to grow longer and stronger, and which were able to absorb deeper soil water. In addition, the N-applied treatments had greater effects on CH than ZM (Fig. 3), showing that CH was sensitive to low N.

## WUE

WUE was in the range of 1–2.09 kg/m$^3$ for ZM and 1.1–1.92 kg/m$^3$ in the three years (Table 2). As the N rates increased the WUE increased, and at N360 the WUE increased slightly but not significantly. Zhou et al. [15] reported that grain yields and WUE did not significantly differ between N rates of 120 and 240 kg/ha. The above indicated that excessive N application had no favorable effect on WUE. In 2009–2010 and 2010–2011, the N-applied treatments significantly improved WUE. However, in

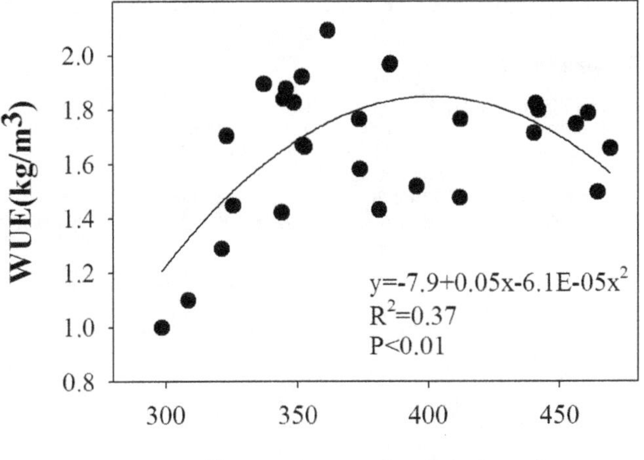


**Figure 4. Relationship between evapotranspiration (ET) (mm) and WUE (Y) for winter wheat in northwest China.** The relation between ET and WUE is shown by the equation.

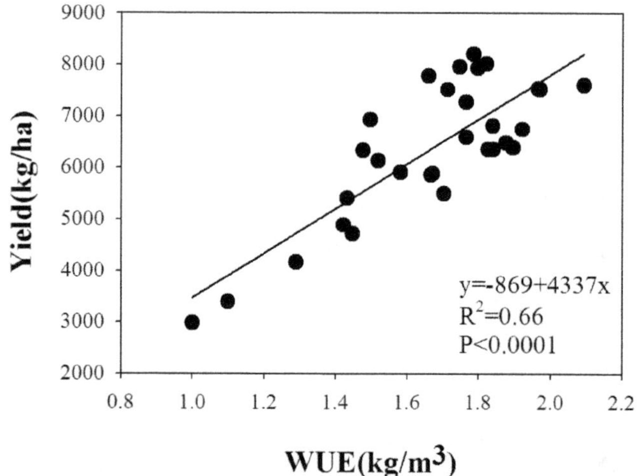


**Figure 5. Relationship between WUE and grain yield (Y) for winter wheat in northwest China.** The relation between WUE and yield could be deduced from the liner equation.

2011–2012, WUE did not significantly differ between the N-applied and non-N treatments. Compared to 2009–2010 and 2010–2011, the WUE of the non-N treatment was higher in 2011–2012. This may be due to the higher soil water content before sowing and the higher rainfall in the growing season in 2012 that increased the grain yield in the non-N treatment. Consequently higher WUE in non-N treatment reduced the difference between non-N and N-applied treatments.

The WUEs obtained in the present study were higher than those of irrigated winter wheat (0.40–0.88 kg/$m^3$) [45,46] and of irrigated wheat in the US southern plains (0.82 kg/$m^3$) [41], as well as higher than these of irrigated wheat in the Loess Plateau (0.73–0.93 kg/$m^3$) [14], demonstrating that in dryland farming systems N fertilizer can be a useful way to increase WUE. However, the results of the present study were similar to those of winter wheat in the North China Plain (0.84–1.39 kg/$m^3$) [12] and of N-fertilized winter wheat in Yangling, Shaanxi (0.8–1.5 kg/$m^3$) [15]. These differences are caused by different climate or water, fertilizer and crop management. N fertilizer application significantly increases the yield and WUE of both wheat and maize, indicating that N fertilizer application is an effective way to increase grain yield in the study region. Deng et al. [50] reviewed four published research reports and found that N fertilizers increased WUE of wheat and potato by an average of 20% in north central and northwest China.

Regression analysis produced a quadratic relationship between ET and WUE (Fig. 4) and correlations were calculated between WUE and wheat yields (Fig. 5). The yields increased linearly with WUE: WUE reached a maximum value at the ET of 401 mm and then decreased (Fig. 4). However, the maximum WUE did not correspond to the maximum grain yield in the study – the higher WUE means that the crop can gain high yield using less water. This is an important method to obtain a balance between higher yields and lower water supplies of wheat in arid and semi-arid regions by increasing its WUE. In the present study, although there were significant differences between N treatments, both cultivars had relatively higher WUEs at the N rate of 270 kg/ha in a dry year.

## Conclusions

N fertilization affected the grain yields, thousand-kernel weights, ET and WUE of the two different water-sensitive wheat cultivars, ZM and CH. The most common pattern of farming in northwest China was used in the present study, with long-term different rates of N fertilization and no irrigation during the wheat growing season. We concluded that (1) in this dryland farming system, increased N fertilization resulted in higher wheat yields in a situation of low precipitation; the drought-tolerant CH showed a yield advantage in a drought environment with high N fertilizer rates; (2) N application affected water consumption in the different soil layers, and promoted absorption and utilization of water from deeper soil layers; and (3) comprehensive consideration of yield and WUE of wheat indicated that the N rate of 270 kg/ha for CH was better to avoid the risk of reduced production due to lack of precipitation; however, under conditions of better soil moisture, the N rate of 180 kg/ha was more economic.

## Author Contributions

Conceived and designed the experiments: YZ ZS. Performed the experiments: YZ. Analyzed the data: YZ. Contributed reagents/materials/analysis tools: YZ. Wrote the paper: YZ ZS.

## References

1. Fan TL, Stewart BA, Wang YG, Luo JJ, Zhou GY (2005) Long-term fertilization effects on grain yield, water-use efficiency and soil fertility in the dry land of Loess Plateau in China. Agr Ecosyst Environ 106: 313–329.
2. Borlaug NE (2009) Foreword. Food Sec. 1, 1–11
3. Perry C, Steduto P, Allen RG, Burt CM (2009) Increasing productivity in irrigated agriculture: agronomic constraints and hydrological realities. Agr Water Manage 96: 1517–1524
4. Li SX, Wang ZH, Malhi SS, Li SQ, Gao YJ, et al. (2009) Nutrient and water management effects on crop production, and nutrient and water use efficiency in dry land areas of China. Adv Agron 102: 223–265.
5. Hooper DU, Johnson L (1999) Nitrogen limitation in dryland ecosystems: responses to geographical and temporal variation in precipitation. Biogeochemistry 46: 247–293.
6. Rockström J, De Rouw A (1997) Water, nutrients and slope position in on-farm pearl millet cultivation in the Sahel. Plant Soil 195: 311–327.
7. Austin AT (2011) Has water limited our imagination for arid land biogeochemistry? Trends Ecol Evol 26: 229–235.
8. Zand-Parsa S, Sepaskhah A, Ronaghi A (2006) Development and evaluation of integrated water and nitrogen model for maize. Agr Water Manage 81: 227–256.
9. Li YS (1982) Evaluation of field soil moisture condition and the ways to improve crop water use efficiency in Weibei region. Journal of Agronomy Shananxi 2: 1–8. (in Chinese)
10. Shan L (1983) Plant water use efficiency and dryland farming production in North West of China. Newslett Plant Physiology 5: 7–10. (in Chinese)
11. Hamblin AP, Tennant D (1987) Root length density and water uptake in cereals and grain legumes: how well are they correlated. Crop Pasture Sci 38(3): 513–527.
12. Zhang H, Oweis T (1999) Water-yield relations and optimal irrigation scheduling of wheat in the Mediterranean region. Agr Water Manage 38: 195–211.
13. Zhang J, Sui X, Li J, Zhou D (1998) An improved water use efficiency for winter wheat grown under reduced irrigation. Field Crops Res 59: 91–98.
14. Kang SZ, Zhang L, Liang YL, Hu XT, Cai HJ, et al. (2002) Effects of limited irrigation on yield and water use efficiency of winter wheat in the loess plateau of China. Agr Water Manage 55: 203–216.
15. Zhou JB, Wang CY, Zhang H, Dong F, Zheng XF, et al. (2011) Effect of water saving management practices and nitrogen fertilizer rate on crop yield and water use efficiency in a winter wheat–summer maize cropping system. Field Crops Res 122. 157–163.
16. Guo YQ, Wang LM, He XH, Zhang Y, Chen SY, et al. (2008) Water use efficiency and evapotranspiration of winter wheat and its response to irrigation regime in the north China plain. Agr Forest Meteorol 148: 1848–1859.
17. Halvorson AD, Nielsen DC, Reule CA (2004) Nitrogen management nitrogen fertilization and rotation effects on no-till dry land wheat production. Agron J 96: 1196–1201.

18. Turner NC (2004) Agronomic options for improving rain fall-use efficiency of crops in dryland farming systems. J Exp Bot 55: 2413–2415.
19. Turner NC, Asseng S (2005) Productivity, sustainability, and rainfall-use efficiency in Australian rainfed Mediterranean agricultural systems. Aust J Agric Res 56, 1123–1136 56: 1123–1136.
20. Kirda C, Topcu S, Kaman H, Ulger AC, Yazici A, et al. (2005) Grain yield response and N-fertilizer recovery of maize under deficit irrigation. Field Crops Res 93: 132–141.
21. Pimentel D, Hurd L, Bellotti A, Forster M, Oka I, et al. (1973) Food production and the energy crisis. Science 182: 443–449.
22. Erisman JW, Sutton MA, Galloway J, Klimont Z, Winiwarter W (2008) How a century of ammonia synthesis changed the world. Nat Geosci 1: 636–639.
23. Bassoa B, Cammarano D, Troccoli A, Chen DL, Joe T (2010) Long-term wheat response to nitrogen in a rainfed Mediterranean environment: Field data and simulation analysis. Eur J Agron 33: 132–138.
24. Hai L, Li XG, Li FM, Suo DR, Guggenberger G (2010) Long-term fertilization and manuring effects on physically-separated soil organic matter pools under a wheat-wheat-maize cropping system in an arid region of China. Soil Biol Biochem 42: 253–259.
25. Malhi S, Nyborg M, Goddard T, Puurveen D (2011) Long-term tillage, straw management and N fertilization effects on quantity and quality of organic C and N in a Black Chernozem soil. Nutr Cycl Agroecosys 90(2): 227–241.
26. Godfray HCJ, Beddington JR, Crute IR, Haddad L, Lawrence D, et al. (2010) Food security: the challenge of feeding 9 billion people. Science 327: 812–818.
27. Schindler D, Hecky R (2009) Eutrophication: more nitrogen data needed. Science 324: 721–722.
28. Hvistendahl M (2010) China's push to add by subtracting fertilizer. Science 327: 801–801.
29. Dawe D, Dobermann A, Moya P, Abdulrachman S, Bijay S, et al. (2000) How widespread are yield declines in long-term rice experiments in Asia. Field Crops Res 66: 175–193.
30. Regmi AP, Ladha JK, Pathak H, Pasuquin E, Bueno C, et al. (2002) Yield and soil fertility trends in a 20-year rice–rice–wheat experiment in Nepal. Soil Sci Soc Am J 66: 857–867.
31. Huifang H, Jiayin S, Dandan Z, Xuanqi L (2012) Effect of irrigation frequency during the growing season of winter wheat on the water use efficiency of summer maize in a double cropping system. Maydica 56(2).
32. Shen JY, Zhao DD, Han HF, Zhou XB, Li QQ (2012) Effects of straw mulching on water consumption characteristics and yield of different types of summer maize plants. Plant Soil Environ 58(4): 161–166.
33. Zhou XB, Chen YH, Ouyang Z (2011) Effects of row spacing on soil water and water consumption of winter wheat under irrigated and rain-fed conditions. Plant Soil Environ 57(3): 115–121.
34. Shangguan Z, Shao M, Dyckmans J (2000) Effects of nitrogen nutrition and water deficit on net photosynthetic rate and chlorophyll fluorescence in winter wheat. Aust J Plant Physiol 156(1): 46–51.
35. Tinsina J, Singh U, Badaruddin M, Meisiner C, Amin MR (2001) Cultivar, nitrogen, and water effects on productivity and nitrogen-use efficiency and balance for rice-wheat sequences of Bangladesh. Field Corps Res 72: 143–161.
36. Morell FJ, Lampurlane J, Alvaro FJ, Martıne C (2011) Yield and water use efficiency of barley in a semiarid Mediterranean agro-ecosystem: Long-term effects of tillage and N fertilization. Soil Till Res 117: 76–84
37. Sheng HD, Wang PH (1985) The relationship between the weight of 1000-seeds and soil water content in winter wheat season. Acta University of Agricultural Boreali occidentalis 13: 73–79. (in Chinese)
38. Shangguan ZP, Shao MA, Lei TW, Fan TL (2002) Runoff water management technologies for dryland agriculture on the Loess Plateau of China. Int J Sust Dev World 9: 341–350.
39. Johnson WC (1964) Some observations on the contribution of an inch of seeding-time soil moisture to wheat yields in the Great Plains. Agron J 56: 29–35.
40. Unger PW (1972) Dryland winter wheat and grain sorghum cropping systems, northern High Plains of Texas. Texas Agricultural Experiment Station 11–26.
41. Musick JT, Jones OR, Stemart BA, Dusek DA (1994) Water-yield relationships for irrigated and dry land wheat in the US southern plains. Agron J 86: 980–986.
42. Zhang XC, Shangguan ZP (2007) Effects of application nitrogen on photosynthesis and growth of different drought resistance winter wheat cultivars. Chinese Journal of Eco-Agriculture. 15(6). (in Chinese)
43. Hunsaker DJ, Kimball BA, Pinter Jr P, Wall G, LaMorte RL, et al. (2000) $CO_2$ enrichment and soil nitrogen effects on wheat evapotranspiration and water use efficiency. Agr Forest Meteorol 104: 85–105
44. Rahman MA, Chikushi J, Saifizzaman M, Lauren JG (2005) Rice straw mulching and nitrogen response of no-till wheat following rice in Bangladesh. Field Crops Res. 91: 71–81.
45. Howell TA, Steiner JL, Schneider AD, Evett SR (1995) Evapotranspiration of irrigated winter wheat-southern high plains. Trans ASAE 38: 745–759
46. Schneider AD, Howell TA (1997) Methods, amount, and timing of sprinkler irrigation for winter wheat. Trans ASAE 40: 137–142
47. Zhang H, Wang X, You M, Liu C (1999) Water-yield relations and water use efficiency of winter wheat in the north China plain. Irrigation Sci 19: 37–45
48. Jupp AP, Newman EI (1987) Morphological and anatomical effects of severe drought on the roots of Loium-perene L. New Phytol 105: 393–402
49. McIntyre BD, Riha SJ, Flower DJ (1995) Water uptake by pearl millet in a semiarid environment. Field Crop Res 43: 67–76
50. Deng XP, Shan L, Zhang HP, Turner NC (2006) Improving agricultural water use efficiency in arid and semiarid areas of China. Agr Water Manage 80: 23–40.

9

# Development of Composite Indices to Measure the Adoption of Pro-Environmental Behaviours across Canadian Provinces

**Magalie Canuel[1]*, Belkacem Abdous[2,3], Diane Bélanger[2,4], Pierre Gosselin[1,2,4]**

**1** Institut national de santé publique du Québec (INSPQ), Québec City, Canada, **2** Centre de recherche du Centre hospitalier universitaire de Québec, Québec City, Canada, **3** Département de médecine sociale et préventive de l'Université Laval, Québec City, Canada, **4** Institut national de la recherche scientifique, Centre Eau Terre Environnement, Québec City, Canada

## Abstract

***Objective:*** The adoption of pro-environmental behaviours reduces anthropogenic environmental impacts and subsequent human health effects. This study developed composite indices measuring adoption of pro-environmental behaviours at the household level in Canada.

***Methods:*** The 2007 Households and the Environment Survey conducted by Statistics Canada collected data on Canadian environmental behaviours at households' level. A subset of 55 retained questions from this survey was analyzed by Multiple Correspondence Analysis (MCA) to develop the index. Weights attributed by MCA were used to compute scores for each Canadian province as well as for socio-demographic strata. Scores were classified into four categories reflecting different levels of adoption of pro-environmental behaviours.

***Results:*** Two indices were finally created: one based on 23 questions related to behaviours done inside the dwelling and a second based on 16 questions measuring behaviours done outside of the dwelling. British Columbia, Quebec, Prince-Edward-Island and Nova-Scotia appeared in one of the two top categories of adoption of pro-environmental behaviours for both indices. Alberta, Saskatchewan, Manitoba and Newfoundland-and-Labrador were classified in one of the two last categories of pro-environmental behaviours adoption for both indices. Households with a higher income, educational attainment, or greater number of persons adopted more indoor pro-environmental behaviours, while on the outdoor index, they adopted fewer such behaviours. Households with low-income fared better on the adoption of outdoors pro-environmental behaviours.

***Conclusion:*** MCA was successfully applied in creating Indoor and Outdoor composite Indices of pro-environmental behaviours. The Indices cover a good range of environmental themes and the analysis could be applied to similar surveys worldwide (as baseline weights) enabling temporal trend comparison for recurring themes. Much more than voluntary measures, the study shows that existing regulations, dwelling type, households composition and income as well as climate are the major factors determining pro-environmental behaviours.

**Editor:** Judi Hewitt, University of Waikato (National Institute of Water and Atmospheric Research), New Zealand

**Funding:** This study was funded by the Green Fund for Action 21 of the 2006-2012 Climate Change Action Plan of the Quebec government. The funders had no role in study design, data collection and analysis, decision to publish, or preparation of the manuscript.

**Competing Interests:** The authors have declared that no competing interests exist.

* Email: magalie.canuel@inspq.qc.ca

## Introduction

A significant source of pollution to our natural environment comes from domestic activities and behaviours. For example household-generated waste in Canada accounts for around a third of total waste and household energy use and municipal water consumption for 17% and 57%, respectively [1-3]. Also, 46% of greenhouse gas emissions (GHG), which contribute to climate change, come from direct and indirect household emissions [4]. The impacts of such household pollution can be important.

Municipal waste can impact the environment in various ways including soil and water contamination from leachate in landfills disposal and the production of greenhouse gas emissions (GHG) and air pollution, either from landfills or the incineration process. When solid waste are recycled or composted instead of being landfilled or incinerated, the demand for energy and new-resources can be reduced significantly [3].

The production of energy can impact the environment in various ways, depending on the technology. In Canada, energy production and consumption accounts for around 80% of all GHG emission [5]. A household can reduce its emission of GHG by reducing electric power use. For instance high energy efficiency electronic devices or cleaner energy sources will generate less pollution and GHG.

Water shortages are happening worldwide and one way to limit their occurrence is through water conservation behaviours. In most homes, more than 60% of water use comes from toilet flushing, showers and baths, making water-saving devices like low-flow shower head an efficient way of reducing water consumption. In summer, water use can increase by 50% for yard activities such as watering the lawn. There are behaviours that households can implement to decrease their water consumption in summer time like using sprinklers with a timer or adopting the use of a rain barrel [6].

It thus becomes clear that addressing sustainability concerns has to take into account not only industry or agriculture, but also household behaviours, their impacts on ecosystems and ultimately on human health. Monitoring trends of household behaviours can inform policy and research agendas on the development of incentives or other mechanisms such as information campaigns to reduce domestic pollution and facilitate adaptative measures to minimize related health risks. The adoption of several pro-environmental behaviours, i.e. actions that contribute to the preservation of the environment, should be encouraged to significantly reduce the anthropic impact on the environment.

In Canada, the Households and the Environment Survey (HES) was designed to measure household behaviours with respect to the environment. The HES is a periodic survey conducted by Statistics Canada, the federal government statistical agency, and administered across Canadian provinces. The survey covers 12 broad themes including energy use and heating, water use, transportation decisions, motor vehicle use, recycling and composting (Figure 1) [7]. While this survey provides various estimations of up to 83 Canadian practices (Figure 1) as well as some information on their socio-demographic characteristics, survey reports are limited to analyses of simple cross-tabulation frequencies for some of the 83 separate behaviours [7-13].

It is difficult to follow up on such a wide array of relevant behaviours and their trends over time, unless they are summarized in some way. A composite index is a tool which can be useful to that purpose as it incorporates several aspects of an issue and allow for monitoring across several themes simultaneously, thus facilitating the measurement of trends [14]. While other environmental indices exist, such as the environmental sustainability index [15], to our knowledge no index currently exists to reflect trends of pro-environmental behaviours at the household level in Canada.

This study thus sets out to develop a composite index that summarizes pro-environmental behaviours at the household level across Canadian provinces based on the HES (2007) given the periodicity and geographical coverage of the survey. Pro-environmental behaviours are defined as actions that contribute to the preservation of the environment and can have a positive impact on the health of the population. This study will serve as baseline of the trend of the composite index over time, given the periodicity and geographical coverage of the survey.

## Materials and Methods

### Ethics statement

This research did not require the approval of an ethics review board as we used an existing and anonymized database made available to universities by Statistics Canada. Statistics Canada obtained consent previous to survey administration. No new data was collected for this study.

### Survey

The Households and the Environment Survey (HES) is conducted by Statistics Canada. It was designed to address the needs of the Canadian Environmental Sustainability Indicators project. The project reports on air quality, water quality and greenhouse gas emissions in Canada using indicators to identify areas of importance to Canadians and monitor progress [16].

The survey aimed Canadian households with at least one person aged 18 year or older. The HES covers all 10 of the provinces and excludes the 3 northern territories, Indian reserves and members of the Canadian Armed Forces. The survey was first conducted in 1991 and since 2005 has been carried out biennially. In the present study, the 2007 HES database was used in its Public Use Microdata Files format (PUMF) [16]. As a sub-sample of the dwellings that were part of the Canadian community health survey (CCHS), the sampling allocation for the HES followed that of the CCHS closely. The CCHS used a multistage stratified cluster design in which the dwelling is the final sampling unit. Three sampling frames were used to select the sample of households: 50% of the sample came from an area frame, 49% came from a list frame of telephone numbers and 1% came from a Random Digit Dialing sampling frame [16].

From the 40 584 households selected in the 2007 CCHS, a sub-sample of 29 957 households were selected for the HES. Of those, 21 690 households responded to the survey resulting in an overall response rate of 72%. The survey is representative of 12 932 350 households, corresponding to 97% of all Canadian households [16]. The questionnaire was administered to the 21 690 households by telephone interview spread over a 6-month period, from October 2007 to February 2008.

### Questionnaire

The person with the best knowledge of environmental household practices was asked to respond on behalf of the household. The main questionnaire covered 12 themes and included 121 questions (figure 1) [16]. Among the questions, 83 measured behaviours and 7 measured socio-demographic characteristics. The other 31 questions covered knowledge, reasons for not adopting the behaviour, or served to specify some characteristics (e.g. of a good) or to filter for the next question.

### Database

The PUMF was used for the analysis and unlike the master file, applies privacy measures to protect personal information [16]. In the PUMF, data were mostly coded as categorical variables. Three different labels (don't know, not stated, and refusal) were used to classify households who did not participate despite eligibility or to protect the anonymity of the household. A 'valid skip' label was used when the provision of a response was not appropriate. For example, a household who answered 'no' to the question for car ownership was allocated a 'valid skip' label for subsequent questions on the characteristics of the car.

### Sampling weights

Sampling weights were applied to ensure that any derived composite index is representative of the study population. They were used when proportions and averages were estimated and to weight the relative frequencies of the Burt matrix in the MCA (see Statistical analysis below).

### Variables selection

This study focuses on everyday pro-environmental behaviours, defined as actions that contribute to the preservation of the

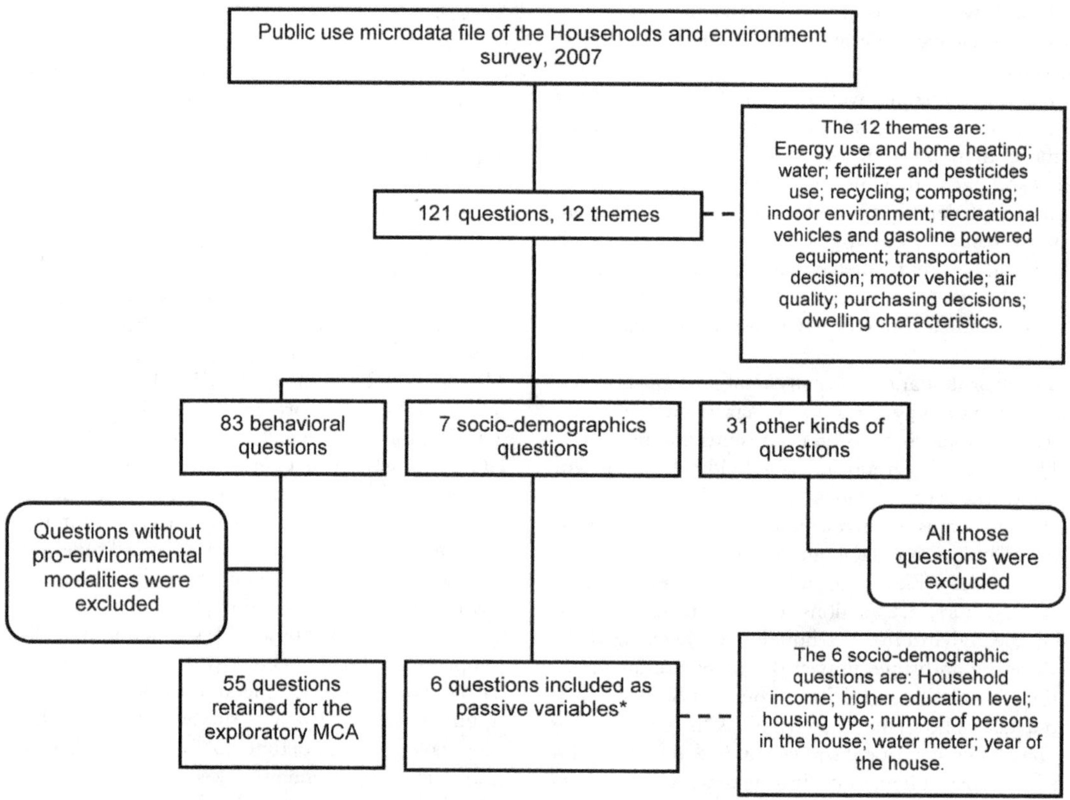


**Figure 1. Number and type of questions selected to develop the composite index. Legend:** *The composite index was also created for the 7th socio-demographic variable, the census metropolitan area (n = 33), but is not presented in this article.

environment and can have a positive impact on the health of the population. For example, air pollutants can be reduced when households adopt behaviours that decrease their energy consumption such as the use of energy-efficient appliances or when they use more sustainable transport options such as public or active transport.

Based on the above definition, a panel of four environmental health experts applied progressive development consensus after iterations, based on a nominal group technique [17] to evaluate HES variables for exclusion. These were either: variables not measuring a behaviour or questions with no clearly pro-environmental response option. Socio-demographic variables were kept as passive variables with zero mass and no influence on the analysis. They support and complement the interpretation of the map representation of the active variables [18].

## Statistical analysis

Given that the data was mostly categorical, the indices in this paper were developed by multiple correspondence analysis (MCA) [18]. Several authors have used Multiple Correspondence Analysis (MCA) as a weighting method for the construction of a composite index [19–23]. MCA is a data reduction procedure for categorical variables (nominal or ordinal) as much as Principal Components Analysis is for quantitative variables [18]. It enables the exploration of associations within a set of variables by transforming the whole data set into dummy variables to form an indicator matrix or upon construction of a matrix from all two-way cross-tabulations among the variables (Burt matrix). This transformed data is treated as a cloud in a space equipped with the classical Chi-square distance. This distance is used in the assessment of homogeneity and variance (inertia) of rows or columns of the indicator or Burt matrix. The most crucial step of MCA is its use of singular value decomposition and weighted least squares techniques to find low-dimensional best fitting subspaces with minimal inertia and information loss [18].

MCA was conducted using the 'ca' package of the R statistical software [24]. First, the HES database was converted to a Burt matrix taking into consideration the sampling weights. A Burt matrix is a square symmetric categories-by-categories matrix formed from all two-way contingency tables of pairs of variables [18].

Then, an exploratory MCA was performed to project data onto maps where potential outliers were identified and excluded from subsequent analyses. MCA was then applied again to determine the most relevant factorial axes that would serve to build the composite index. There are no universal rules for the determination of the number of dimensions to retain in MCA. However, since the first factorial axis captures the most important part of the total inertia, it plays a central role in the computation of a composite index.

As recommended by Asselin [23], we sought questions having the property of First Axis Ordering Consistency (FAOC). To this end, we projected all the questions on the first axis and tried to identify those having an ordinal structure consistent with respect to this axis, i.e. all questions with pro-environmental responses improving from left to right (or conversely).

The computation of the index score was performed as follows: first, the score of any household was obtained by taking the average of its category-weights generated by the MCA. Then for each province we took the average over all household scores as the value of its composite index. The sampling weight was used in this final step. Coordinates were missing for excluded responses.

The 10 average provincial scores were grouped into categories reflecting different levels of adoption of pro-environmental behaviours. First we applied a cluster analysis and then we used a dendrogram plot using SAS version 9.2 (SAS Institute, Cary, NC) to determine such groups. The categories limits generated for the provincial index were used as reference categories for indices on other socio-demographic variables.

Finally, others indices based on various socio-demographic variables were constructed (Figure 1). Household scores were calculated by taking the average of its category-weights generated by the MCA. Then the index score of the socio-demographic category (e.g. household with annual income less than $40,000) is set as the average of the corresponding household scores.

## Results

### Multiple Correspondence Analysis

Of the 121 questions in the survey, 55 were kept by the Expert Panel for use in the MCA. These represented 285 response possibilities. On the MCA map projection there was a clear opposition between the missing data (don't know, refusal, not stated) located far from the map center and the other responses which gathered close to the center (Figure S1). Excluding the missing data rebalanced the model (179 remaining responses) (Figure 2). However, since pro-environmental behaviours were spread over both sides of the first axis, we failed to find any meaning to this first dimension.

We then screened the projected responses to identify questions following an ordinal structure, (i.e all pro-environmental responses of a question have negative coordinates on the first factorial axis (or conversely)). Twenty-three such questions with pro-environmental responses deteriorating from left to right on the first axis (group A), and 16 questions with opposite ordinal structure (group B) were identified. The remaining 16 (of 55) questions were excluded from the analysis because their responses were not sufficiently discriminating (i.e. the pro- and anti-environmental responses were on the same side of the axis or they were grouped close together on the map). As well the majority of these questions (10/16) had at least two responses with a contribution of zero to the first axis (Table S1).

These exploration steps led us to consider two separate composite indices. Group A included 96 responses but after excluding missing data, 52 responses were used in the MCA. The majority of excluded responses had frequencies lower than 2.0% and two responses had frequencies of 4.6% and 4.7%. After exclusion of missing data, some responses still looked like extreme values on the map (Figure 3). They were kept in the analysis as they are 2 of the 3 responses for all questions concerning recycling. Excluding these responses would have resulted in the exclusion of all recycling questions. Responses used in this analysis had a frequency of 7.5% or higher, except for two responses with frequencies of 2.5% and 3.5% (responses on recycling).

For group A, the first dimension explained 32.6% of the inertia while the second explained 16.1% (Table 1). Given that the first factorial axis plays a central role in the construction of this composite index, only the first dimension was selected to construct the index. This group respects the FAOC as pro-environmental responses are located on the left of the first axis as opposed to others responses deteriorating to the right (Figure 3). Also, we noted that the retained questions were associated with five themes of the survey: energy use and home heating, water, recycling, composting and, purchasing decisions. All 23 questions assessed behaviours practiced inside the dwelling and thus the first axis measures these behaviours. Twelve of 15 responses contributing the most to the first factorial axis concerned recycling (Table S2).

The second group of 16 questions (group B) consisted of 86 responses, 41 of which were missing values. The 45 remaining responses used for the MCA had frequencies of 7.0% or higher while excluded responses had frequencies lower than 3.5%. For group B, the first dimension explained 62.1% of the inertia while the second explained only 12.4% (Table 2). Again, the first dimension was selected for the construction of the index and pro-environmental responses were located on the right of the first axis with other responses deteriorating from right to left (Figure 4). The 16 questions cover five themes of the survey: water, fertilizer and pesticide use, recreational vehicles and gasoline powered equipment, transport decisions and air quality, all behaviours being practiced outdoors. Of note, 9 of the 15 responses contributing the most to the first factorial axis concern households with no lawn or garden (i.e. the application of fertilizers or pesticides, yard waste and watering of the lawn or the garden) (Table S3).

Because two distinct behavioural categories resulted from the MCA, two composite indices were created instead of one. The first index (group A) is named the 'Indoor Index' and the second one (group B) the 'Outdoor Index'. Questions included for each index are presented in supporting information, Table S4 and Table S5.

### Composite indices by province

The map representations of the final coordinates generated by the MCA are shown in Figure 3 and Figure 4. Coordinates and other results of the MCA are available in supporting information, Table S2 and Table S3. The coordinates of the first dimension were used to construct each of the two composite indices. Coordinates are missing for responses that have been excluded. Only 0.9% and 0.4% of coordinates are missing for the indoor and outdoor indices, respectively.

For the Indoor Index, the households belonging to a province with negative coordinates tend to adopt more pro-environmental behaviours than those of a province with positive coordinates. In contrast, for the Outdoor Index provinces with positive coordinates adopt more outdoor pro-environmental behaviours than those with negative coordinates.

The cluster analysis and dendogram plot resulted in the classification of each province into one of four categories reflecting different levels of adoption of pro-environmental behaviours: 1) adopting the most; 2) adopting slightly fewer; 3) adopting much fewer and; 4) adopting the fewest. The provincial coordinates and the categories generated from the cluster analysis are shown in Table 3 and Table 4. Maps of the Canadian provinces with their categories of pro-environmental behaviours are shown in Figure 5 and Figure 6.

None of the 10 provinces were classified in both indices as adopting the most pro-environmental behaviours. For the Indoor Index, Ontario (ON), Prince Edward Island (PEI) and Nova Scotia (NS) rated in the top category, British Columbia (BC) and Québec (QC) in the next, the three Prairie provinces and New Brunswick (NB) in the third and Newfoundland and Labrador (NL) in "adopting the fewest" category (Figure 5). For the Outdoor Index, QC scored in the top category with BC, NS, NB and PEI following in second, and Manitoba (MN), ON and NL in third, followed by Alberta (AB) and Saskatchewan (SK) in the bottom category (Figure 6).

Four provinces (BC, QC, NS and PEI) were classified in the top two categories for both indices while four provinces (AB, SK, MN and NL) were classified for both indices, in the two lower categories.

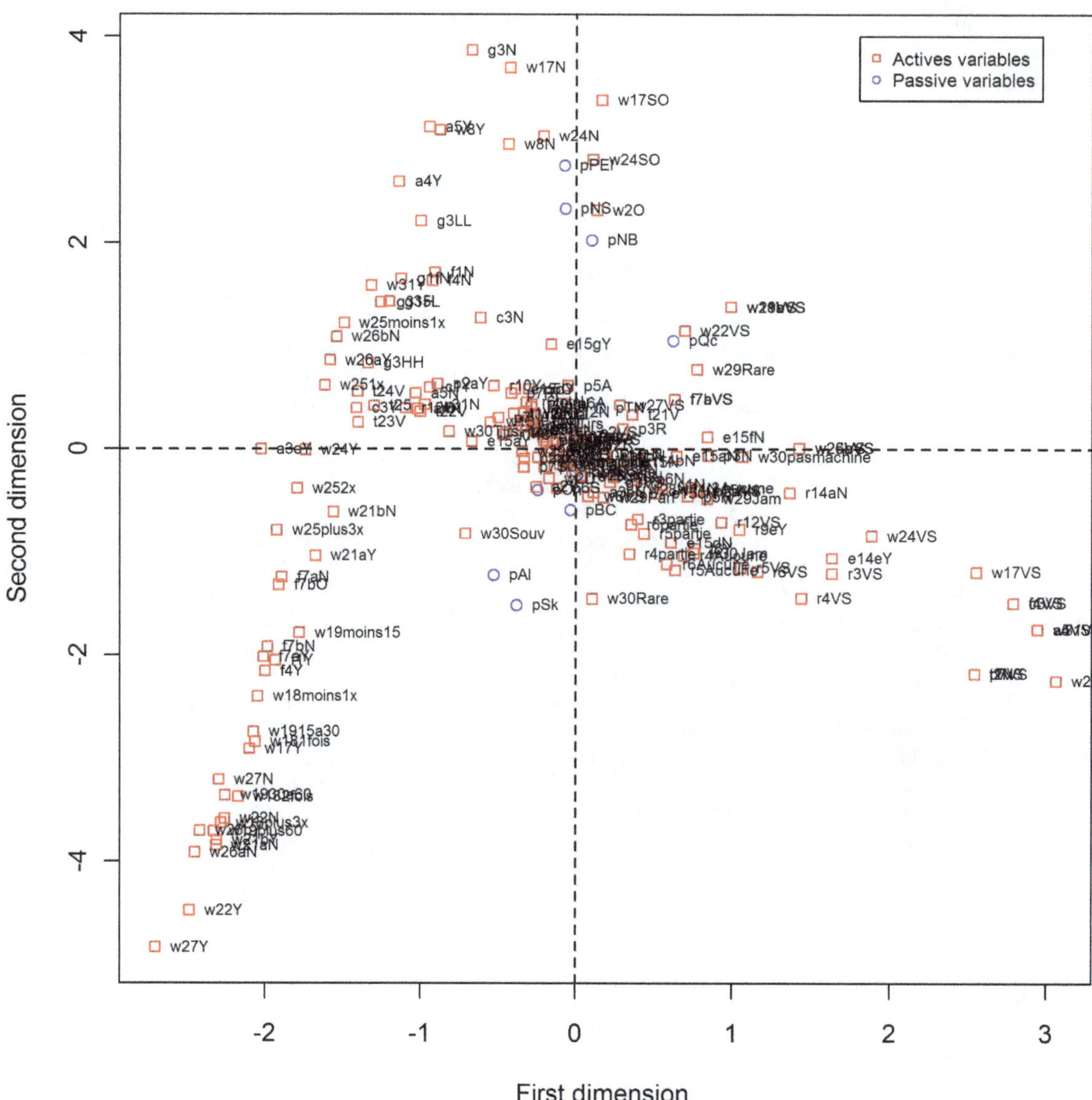


**Figure 2. Map representation of the MCA results on the 55 questions without extreme responses.**

## Composite indices by socio-demographic variables

The coordinates and the classification for the six comparison variables are shown in Table 5. For household income, educational attainment and number of persons in the household, there were oppositions in the classification of the responses. Households with a higher income, or higher educational attainment, or greater number of persons adopted more indoor pro-environmental behaviours, while those with a lower household income, educational attainment, or number of people, adopted more outdoor such behaviours. As well, households with water meters tended to adopt more indoor pro-environmental behaviours than those without, but for outdoors behaviours, the opposite applied – not having a water meter was associated with better adoption of pro-environmental behaviours. And finally, the dwelling's year of construction did not influence the adoption of pro-environmental behaviours as there was no trend on either index (Table 5).

## Discussion

This study sought to develop a composite index which measures the overall adoption of pro-environmental behaviours among Canadian households. MCA, our main analytical technique, was used to aggregate survey data and to provide weights to the responses in the construction of the index. Our approach is similar to other studies in different fields [19–23]. This was followed by a cluster analysis to classify the provinces, as well as an exploration of relationships with socio-demographic factors.

The MCA generated two indices based on 39 of the 55 behavioural questions, an Indoor Index and an Outdoor Index, each reflecting environmental behaviours for 5 of the 12 survey themes. Retaining both indices allowed for better representation of the survey; together they cover 9 themes out of 12 (water use is in both) whereas one single index would have covered only 5, excluding important environmental themes such as fertilizer and pesticide use. As well, because the provincial classifications were different for each index and varied as well in the classification by

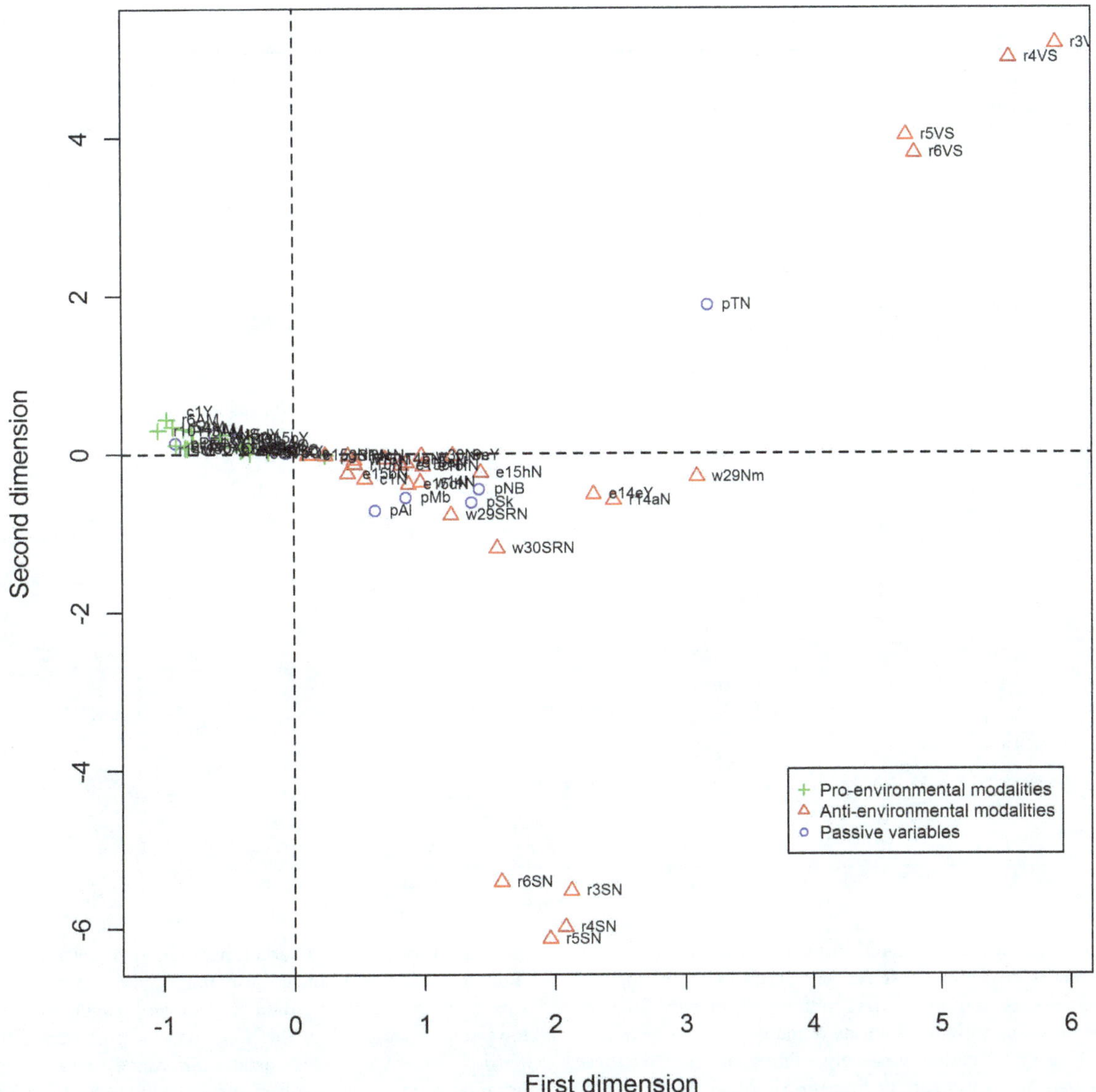


**Figure 3. Map representation of the MCA results on the 23 questions of the group A (Indoor).**

socio-demographics factors for each index (e.g., household income) it was deemed justifiable to keep both indices.

Most (19/23) questions included in the Indoor Index were asked to all households with the exception of questions on recycling where only those households with access to a program were asked to respond. For the Outdoor Index, most questions (11/16) concerned watering of the lawn or the garden, and the use of fertilizers or pesticides. These (11) questions were answered only by households having a yard. However, even if households living in an apartment did not have to answer these questions, they were still recorded in the Index as households adopting pro-environmental behaviours (i.e., most valid skips were classified as pro-environmental responses and some as anti-environmental ones).

## The Indoor Index

One likely explanation for PEI and NS being classified in the top category for the Indoor Index is that nearly 100% of their households recycle and the proportion that compost is substantially above the Canadian average, as reported by Statistics Canada. In these two provinces, households are obligated by law to recycle and compost [25]. Moreover, questions regarding recycling contributed the most to the Indoor Index.

Recycling and composting are also common in ON but its good ranking is also related to the proportion of households that adopt water conservation behaviours (i.e. use water-efficient shower heads and toilets, run dishwasher and washing machine only when full) [7]. Provinces have been slowly adopting a provincial plumbing code requiring that new buildings use water-saving fixtures, with the exception of NL [26;27]. ON however was the first to adopt such a code in 1996 [26;28], and saw an increase in new residential construction from 1996 to 2002 [29], likely contributing to the higher proportion of households practicing water conservation behaviours [28]. This is an example where building codes may be effective in beneficially influencing the passive uptake of pro-environmental practices.

QC's good classification in the Indoor Index is in part due to its proportion of households adopting recycling behaviours being higher than the Canadian average. There were four questions on

**Table 1.** Explained inertia by each dimension for group A: Indoor Index, 2007.

| Dimension | Inertia | Inertia (%) | cumulative Inertia (%) | *scree plot* |
|---|---|---|---|---|
| 1 | 0,0270 | 32,6 | 32,6 | ************************* |
| 2 | 0,0133 | 16,1 | 48,8 | ************ |
| 3 | 0,0078 | 9,4 | 58,2 | ******* |
| 4 | 0,0032 | 3,8 | 62,0 | *** |
| 5 | 0,0030 | 3,6 | 65,6 | *** |
| 6 | 0,0026 | 3,2 | 68,8 | ** |
| 7 | 0,0024 | 2,9 | 71,7 | ** |
| 8 | 0,0021 | 2,5 | 74,2 | ** |
| 9 | 0,0019 | 2,3 | 76,5 | ** |
| 10 | 0,0018 | 2,1 | 78,7 | ** |
| 11 | 0,0017 | 2,0 | 80,7 | ** |
| 12 | 0,0016 | 1,9 | 82,6 | * |
| 13 | 0,0015 | 1,8 | 84,4 | * |
| 14 | 0,0014 | 1,7 | 86,1 | * |
| 15 | 0,0014 | 1,7 | 87,7 | * |
| 16 | 0,0013 | 1,6 | 89,3 | * |
| 17 | 0,0012 | 1,5 | 90,8 | * |
| 18 | 0,0012 | 1,4 | 92,2 | * |
| 19 | 0,0011 | 1,3 | 93,5 | * |
| 20 | 0,0011 | 1,3 | 94,8 | * |
| 21 | 0,0010 | 1,2 | 96,1 | * |
| 52 | 0 | 0,0 | 100,0 | |

recycling which contributed significantly to the first dimension, thus contributing to QC's classification. Despite QC having the lowest proportion of households that compost [9] or participate in alternative recycling activities such as donations of furniture and clothing, QC's classification was only slightly affected as these behaviours had only moderate or low contributions to the Index.

In AB, MN, SK and NB, the proportion of households that adopted indoor pro-environmental behaviours is below the Canadian average (data not shown), explaining their lower classification in the Indoor Index. NL had only a few variables above the Canadian average and had most often the lowest proportion of all provinces. For example, the proportion is below the average for all four questions on water conservation and for all questions on recycling. In this province, there is no provincial plumbing code requiring the use of water-saving fixtures in new buildings [26;27]. Also, the proportion of households with access to a recycling program is only 71% [25].

## The Outdoor Index

Results for the Outdoor Index show a pattern with respect to Coastal proximity, with coastal provinces, with the exception of NL, rating in the two higher categories, and the continental provinces in the two lower categories with the two lowest rated provinces situated in the Prairies. The climate of the Prairies grasslands is characterized by hot summers combined with low precipitation and periodic drought. The climatic region of the Maritimes however is the one with the greatest annual precipitation [30-32], a pattern which is likely reflected in the frequency of watering lawn and or garden. Although watering of the lawn or garden is around the Canadian average in NL, its inhabitants own more recreational vehicles, use more gas and burn more yard waste on the property (data not shown) which may explain its lower classification than the other coastal provinces.

Also, there was an important difference in the proportion of households that used fertilizers and pesticides and QC had, by far, the lowest proportion. QC was the first province to adopt a provincial law in 2006 prohibiting the sale of pesticides for cosmetic purposes [7;33]. The Prairies on the other hand had the highest proportions of households that used pesticides or fertilizers in 2007 [7;10]. Subsequently, other jurisdictions have adopted similar laws begging the question of whether their classifications in the Outdoor Index will change over time.

It should also be noted that QC and BC have the highest proportion of households living in an apartment [10]. Given that most households living in an apartment do not have a backyard, they do not water neither lawn nor garden, nor do they use pesticides outdoors. Hence, they passively adopt pro-environmental behaviours and are considered as such by the MCA. In fact, these responses, recorded as 'valid skip', had the highest contribution to the Index, likely contributing to the higher classification for BC and QC on the Outdoor Index. Such passive behaviours or external factors were not excluded from the Index as they significantly contribute to the preservation of environmental resources.

## Indices for socio-demographic variables

For most socio-demographic variables, there were oppositions in the classification of the modalities, which means that it is not the

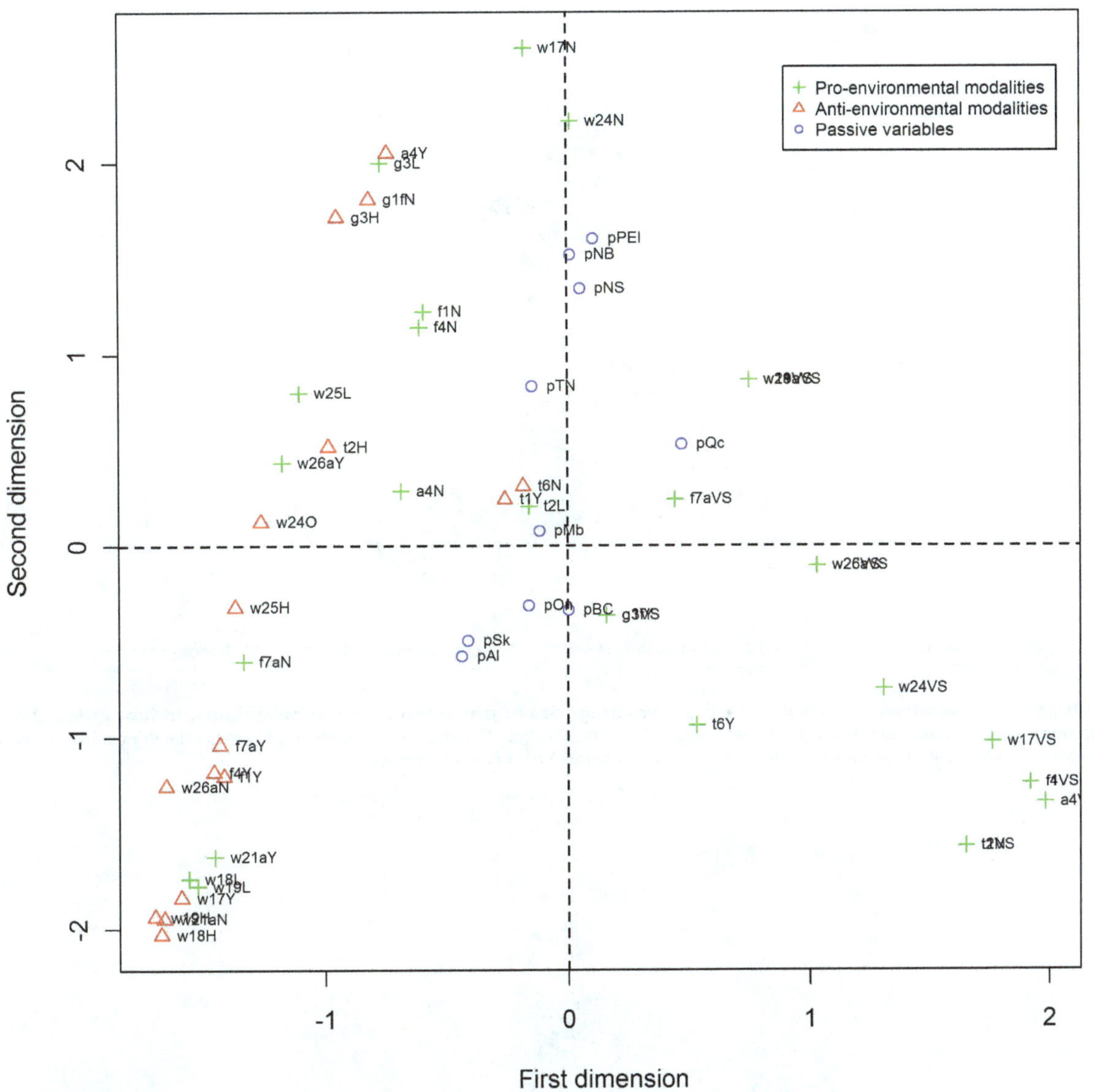


**Figure 4. Map representation of the MCA results on the 16 questions of the group B (Outdoor).**

**Table 2.** Explained inertia by each dimension for group B: Outdoor Index, 2007.

| Dimension | Inertia | Inertia (%) | Cumulative inertia (%) | *scree plot* |
|---|---|---|---|---|
| 1 | 0,2173 | 62,1 | 62,1 | ************************* |
| 2 | 0,0434 | 12,4 | 74,5 | ***** |
| 3 | 0,0162 | 4,6 | 79,1 | ** |
| 4 | 0,0129 | 3,7 | 82,8 | * |
| 5 | 0,0111 | 3,2 | 86,0 | * |
| 6 | 0,0094 | 2,7 | 88,7 | * |
| 7 | 0,0066 | 1,9 | 90,6 | * |
| 8 | 0,0061 | 1,7 | 92,3 | * |
| 9 | 0,0041 | 1,2 | 93,5 | |
| | | | | |
| 45 | 0 | 0,0 | 100,0 | |

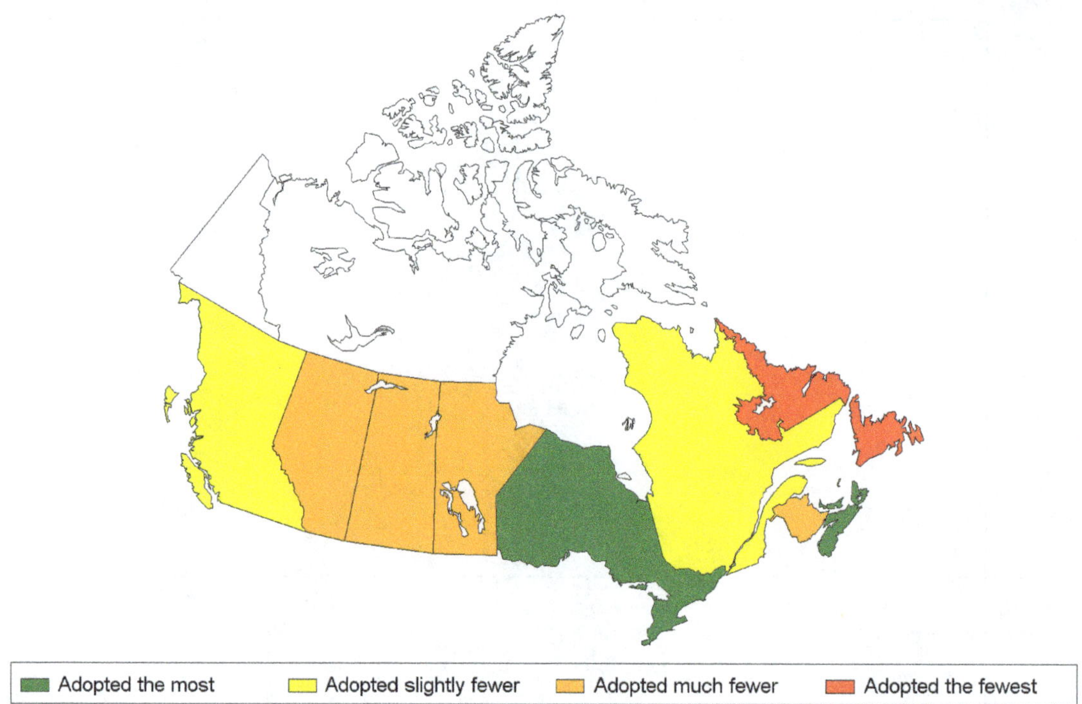


**Figure 5. Provinces' classification according to the four categories of pro-environmental behaviours, Indoor Index, 2007. Legend**: from left to right – British-Columbia, Alberta, Saskatchewan, Manitoba, Ontario, Quebec, New-Brunswick, Nova-Scotia. Prince-Edward-Island is North of the two latter provinces and Newfoundland-and-Labrador is located North-East of Quebec.

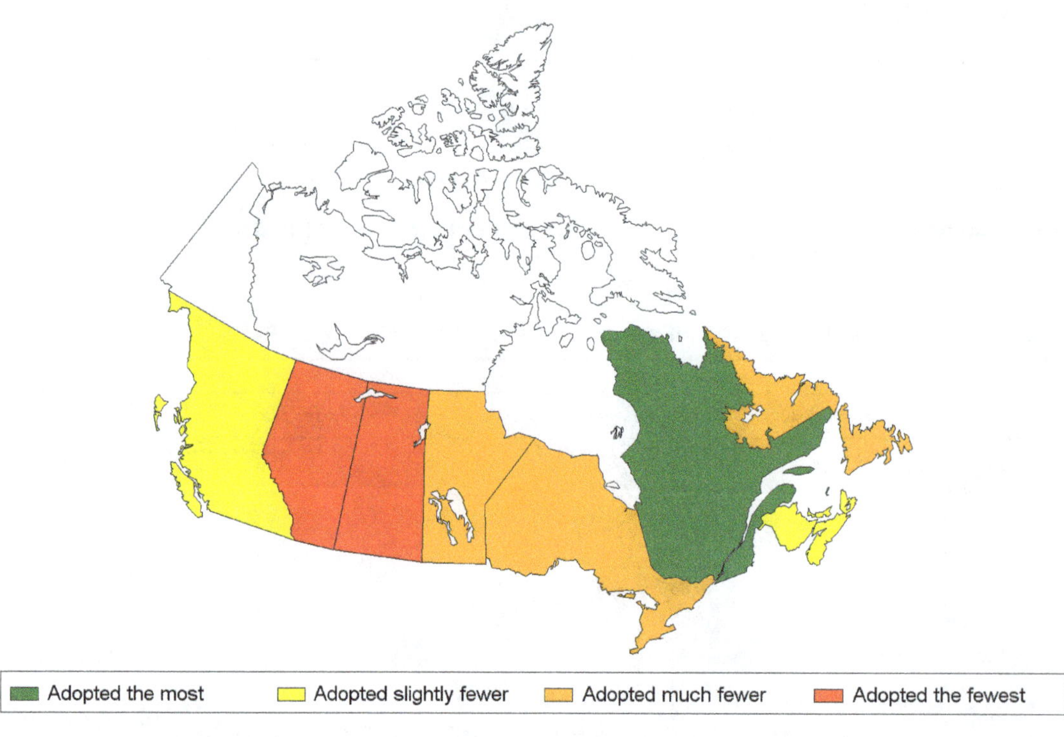


**Figure 6. Provinces' classification according to the four categories of pro-environmental behaviours, Outdoor Index, 2007. Legend**: from left to right – British-Columbia, Alberta, Saskatchewan, Manitoba, Ontario, Quebec, New-Brunswick, Nova-Scotia. Prince-Edward-Island is North of the two latter provinces and Newfoundland-and-Labrador is located North-East of Quebec.

**Table 3.** Provinces' coordinates on the Indoor Index, 2007.

| Provinces | Coordinates | Categories[a] |
|---|---|---|
| Prince-Edward-Island | −0,0262 | ++ |
| Nova-Scotia | −0,0179 | ++ |
| Ontario | −0,0130 | ++ |
| British-Columbia | −0,0055 | + |
| Quebec | −0,0029 | + |
| Alberta | 0,0159 | − |
| Manitoba | 0,0225 | − |
| Saskatchewan | 0,0363 | − |
| New-Brunswick | 0,0381 | − |
| Newfoundland-and-Labrador | 0,0853 | − |

[a]Categories are: adopted the most pro-environmental behaviours (++), adopted slightly fewer (+), adopted much fewer (−) and adopted the fewest (−).

same households that adopt pro-environmental behaviours on both indices. Higher income households may be more able to maintain and repair their housing and also invest in environmentally friendly products such as water and energy efficient appliances or fixtures, which can be more expensive than their regular counterparts [34;35]. Access to such products may contribute to the better classification on the Indoor Index for higher income households. On the other hand, lower income households may be less willing to pay water taxes linked to consumption levels, or to buy chemical products for their lawn or garden. Furthermore, those lower income households live more frequently in apartments where they do not have a yard, and they also own fewer recreational vehicles (data not shown). All these factors likely weigh in on the higher classification attributed to lower versus higher income households on the Outdoor Index.

In Canada, income is usually positively associated to educational attainment [36]. Also, the number of persons in a household will influence the household income. In the HES database, there was a significant correlation between households' income and education level as well as one with the households' income and the number of persons in the households (data not shown). This may explain why the indices by educational level and by number of persons in a household are similar to the one by household income. Any one of these three socio-demographic variables could potentially be used as a surrogate for the other two for future data collection for following Index trends over time.

Studies have shown that water meters with appropriate pricing are an incentive to reduce water consumption [2]. The US EPA estimated a 20% reduction in water consumption with universal metering [37] and a Canadian study also estimated a similar reduction according to structured water pricing [38]. While our results showed that households with water meters tended to score higher on the Indoor Index, households without a water meter scored higher on the Outdoor Index, which is in contrast to the other studies. We estimated that only 9% of households living in an apartment have water meters as opposed to 58% for all other types of dwellings in Canada (data not shown). As stated earlier, a household living in an apartment passively adopts more outdoor pro-environmental behaviours for lack of a lawn or garden to maintain with only a few having a water meter, possibly explaining the discrepancy between our results and those of other studies.

## Factors that can lead to pro-environmental behaviours

There is a wide variety of measures or instruments than can be introduced by governments to influence households behaviours, from economic instruments to direct regulation, labeling, information campaigns and provision of environment-friendly public goods such as public transportation or bicycle paths [39].

This study has identified factors which seem to influence the uptake of beneficial environmental behaviours at the household level. Investment in infrastructure is one of them. The physical or material possibility to act pro-environmentally must indeed be available [40], such as what might be needed for Newfoundlanders to improve their recycling profile.

Regulation is frequently used to efficiently influence the environmental impacts of household decision-making [39] and in our study it also seems to be an important incentive for the adoption of pro-environmental behaviours. This was seen both in the case of building codes requiring the installation of water efficient shower heads and toilets, and in the case of the ban on pesticides for lawn care. In Ontario, the ban of cosmetic pesticides decreased significantly the concentration of some pesticides, mainly herbicides, in the majority of streams under surveillance near urban areas with limited agriculture activities [41].

To encourage a reduction in water consumption, both price and non-price policies should be used. Volumetric water charges are associated with both water-saving behaviours and adoption of water-efficient devices [39]. However, in a study in several OECD countries, Canada had the highest proportion of households that did not know how they were charged for residential water consumption, thus reducing the price effect on water-saving behaviours [39]. In our study, presence of water meters was an incentive to water-saving behaviours only for indoor behaviours. Climate was also another factor that could be influential. Hence, public information on the environmental impact of water consumption and on measures households can adopt to save water should be combined to economic measures according to the OECD [39] and this study.

Other than governmental measures, household characteristics may play a role in the adoption of environment-related behaviours such as income, household composition and dwelling characteristics [39]. According to the OECD survey, low income households and tenants households make fewer financial investments in water efficiency, as can be expected. Grants targeted at those households to correct the economic imbalance are thus recommended by the agency. Moreover, our study showed

**Table 4.** Provinces' coordinates on the Outdoor Index, 2007.

| Provinces | Coordinates | Categories[a] |
|---|---|---|
| Quebec | 0,1038 | ++ |
| Prince-Edward-Island | 0,0261 | + |
| Nova-Scotia | 0,0123 | + |
| New-Brunswick | 0,0052 | + |
| British-Columbia | 0,0012 | + |
| Manitoba | −0,0243 | − |
| Newfoundland-and-Labrador | −0,0273 | − |
| Ontario | −0,0350 | − |
| Saskatchewan | −0,0887 | − |
| Alberta | −0,0962 | − |

[a]Categories are: adopted the most pro-environmental behaviours (++), adopted slightly fewer (+), adopted much fewer (−) and adopted the fewest (−).

**Table 5.** Coordinates and categories of pro-environmental behaviours for other socio-demographic variables, Indoor and Outdoor Indices, 2007.

| | Indoor Index | | Outdoor Index | |
|---|---|---|---|---|
| | Coordinates | Categories[a] | Coordinates | Categories[a] |
| **Household income** | | | | |
| Less than $40,000 | 0,0253 | − | 0,1533 | ++ |
| $40,000 to less than $80,000 | −0,0080 | + | −0,0133 | − |
| $80,000 and over | −0,0273 | ++ | −0,1567 | − |
| **Highest education level** | | | | |
| Secondary diploma or less | 0,0228 | − | 0,0927 | ++ |
| Postsecondary certificate or diploma | −0,0030 | + | −0,0134 | − |
| University | −0,0159 | ++ | −0,0451 | − |
| **Dwelling type** | | | | |
| Apartment | 0,0370 | − | 0,4335 | ++ |
| Others | −0,0144 | ++ | −0,1501 | − |
| **Number of persons in the dwelling** | | | | |
| One | 0,0260 | − | 0,2058 | ++ |
| Two | −0,0073 | + | −0,0255 | − |
| Three | −0,0104 | ++ | −0,0723 | − |
| Four or more | −0,0163 | ++ | −0,1383 | − |
| **Water meter** | | | | |
| Yes | −0,0201 | ++ | −0,1682 | − |
| No | 0,0123 | − | 0,1497 | ++ |
| **Year the dwelling was built** | | | | |
| Before 1946 | −0,0089 | + | 0,0034 | + |
| Between 1946 and 1960 | −0,0040 | + | −0,0130 | − |
| Between 1961 and 1977 | −0,0009 | + | 0,0025 | + |
| Between 1978 and 1983 | −0,0054 | + | −0,0218 | − |
| Between 1984 and 1995 | −0,0084 | + | −0,0364 | − |
| Between 1996 and 2000 | −0,0088 | + | −0,0434 | − |
| Between 2001 and 2005 | −0,0099 | ++ | −0,0948 | − |
| 2006 or latter | 0,0042 | + | 0,0384 | + |

[a]Categories are: adopted the most pro-environmental behaviours (++), adopted slightly fewer (+), adopted much fewer (−) and adopted the fewest (−).

households from both income groups (high or low) or dwelling type (owned or rented) have to improve their act in different domains and that programs should target them accordingly. In short, Canadians remain very dependent for many such actions on where they live and what the climate brings to their yards, or not.

## Limits of the study

We used data from a survey that has been created to address the needs of Statistics Canada and the federal government. Thus, we were limited to its content. The questionnaire does not cover all behaviours that can impact the environment and public health. Also, the indices developed here measure the behaviours available in the survey and retained after the analysis by an expert group for their potential positive impacts on health, and not all existing pro-environmental behaviours. The classification could have been different if other behaviours had been included.

Three themes of the survey were not covered by the indices, namely dwelling characteristics, motor vehicle and indoor environment. However, we believe they would not have much impact in the indices. First, there were no behaviours measured in the dwelling characteristics theme and some of the characteristics were included as passive variables in the indices (e.g. year the building was built). The same happened for the motor vehicle theme (focused on the characteristic of the car), yet we used another theme to include the number of vehicles owned by the households in the outdoor index. For the indoor environment theme, only 2 of the 5 questions measured behaviours and they both concerned the type of chemical products used to clean windows and the dwelling. Although every small action is important for the environment, those questions were excluded as some other practices, such as agriculture, use similar products in much larger quantities [42].

A good standing in the classification does not mean that there is no place for improvement. Indeed, a high proportion of households that adopt pro-environmental behaviours on one question can compensate for a lower proportion on another question of the same index. Also, the provinces were compared to each other and not classified in relation to a gold standard.

Furthermore, it was the MCA that attributed the weight for each modality. Thus, a modality with a higher weight has more impact in the index. For instance, all four recycling questions had the highest contribution to the indoor index. Further studies should investigate if the inclusion of only one of those recycling question or a composite index of those four questions would be more appropriate. The same reasoning should also be applied to questions related to the watering of the lawn or of the garden. Households without a garden or a lawn are rewarded for every question on that subject which at the end can impact greatly the province classification. For example, they were not only rewarded for not watering their lawn, but they were also rewarded for the question concerning the watering duration and the number of time they water. Because the MCA attributed the weights, those household without a garden or lawn had a higher 'reward' that households with a garden or lawn that did not water them.

One general limit of MCA is that the first dimension usually explains a low proportion of the total inertia in the data set and the other dimensions explain less than the first [18]. In this study, the first dimension explained 33% and 62% of the total inertia for the indoor and outdoor index respectively. By using only the first dimension, these indices might not properly reflect all of the behaviours, especially for the indoor one. However, using more than one dimension to build the index would not considerably increase the total inertia explained but would in return increase its complexity. Composite indices are indeed built to simplify the analysis.

Because of the study design, based on households, it was also not possible to evaluate the impact of personal attributes, like age and gender, on the adoption of environmental behaviours. The association between environmental behaviours and age is not clear. Studies have observed all possible trends, from older people adopting more pro-environmental behaviours to the opposite trends or no trend at all [43–45]. Also, women would be more likely to take pro-environmental actions than men, although some studies have found the opposite depending on behaviour and region [43–46]. In our study we found that socio-demographic characteristics like household income and a higher level of education did not have the same influence on outdoor behaviours compared to indoor behaviours. Hence, some differences could also be expected between indoor and outdoor behaviours for age and gender.

Because the survey was not meant to measure attitudes or values, we cannot associate the classification of the province to any difference in values or perception. However, others studies have showed cultural differences across Canadian provinces [47–51]. In Canada, French speaking people are at majority in the province of Quebec but a minority in the rest of Canada as opposed to English speaking Canadian that are a majority in the rest of Canada [52;53]. Several studies have observed differences of values and attitudes in terms of personality, political perspective, priorities and social issues between English-Canadians and French-Canadians [47–51]. Differences in those values could also explain some differences in the adoption of pro-environmental behaviours but further studies are required to confirm it.

Attitudes and values can also be different in immigrants compared to the native born. The former usually have a smaller ecological footprint [54–57]. For example, several studies, mostly from United States, have observed that immigrants have lifestyles that are less demanding on the environment: they consume less, possess fewer luxury items like SUVs, they carpool or use public transportation more often and live in smaller houses [54–57]. In 2006, around 55% of all Canadian immigrants were in Ontario, followed by 18% in British-Columbia and 14% in Quebec [58]. British-Columbia and Quebec had a good classification on both indices. However, we could not estimate the impact of immigration on these classifications, as immigration rules and influx have changed significantly over the last decades [58].

Despite those limits, the indices still give a good idea of the global adoption of pro-environmental behaviours with potential positive impacts on health in Canada and remain easy to explain and understand. The main sectors in which households can have an impact are covered by the indices, like air and soil quality as well as water conservation. The weighting methods used (i.e. MCA) are also more appropriate to assign weights as opposed to an equal weights or expert opinion approach that are often criticized for being arbitrary or simplistic [22]. Others similar indices could be created as the survey is performed every two years. The results obtained with the 2007 indices could serve as the baseline for surveillance purposes, as the survey has been more comprehensive since that date.

## Conclusion

MCA was successfully applied in creating Indoor and Outdoor composite Indices of environmental health relevance based on a readily available periodic Statistics Canada dataset. The Indices cover a good range of environmental themes at the household level and the analysis, particularly the indices weights obtained in the MCA, could be applied to similar surveys worldwide (as baseline weights) enabling temporal trend comparisons for recurring themes. Results uncovered provincial patterns of pro-environmental behaviours adoption with certain provinces scoring consistently higher and others consistently lower, as well as the associations between socio-demographic factors and the indices. Much more than voluntary measures, this study shows that existing regulations, dwelling type, household composition and income as well as climate are the major factors determining pro-environmental behaviours.

## Supporting Information

**Figure S1 Map representation of the MCA results on the 55 questions with extreme responses.**

**Table S1** Results of the MCA on the 55 questions without extreme responses (exploratory analysis). **Legend:** N/A: Results are not available for supplementary variables. Qlt: Quality (i.e. the sum of the squared correlations for the first two dimensions in this case). Inr: Inertias. K: Principal coordinates for the first dimension. Cor: Squared correlation with the first dimension. Ctr: Contributions of the modality to the explained inertia of the first dimension. All cells are multiplied by 1000. Results are the same on rows and on column when a Burt table is used.

**Table S2** Results of the MCA for the Indoor Index, 2007. **Legend:** N/A: Results are not available for supplementary variables. Qlt: Quality (i.e. the sum of the squared correlations for the first two dimensions in this case). Inr: Inertias. K: Principal coordinates for the first dimension. Cor: Squared correlation with the first dimension. Ctr: Contributions of the modality to the explained inertia of the first dimension. All cells are multiplied by 1000. Results are the same on rows and on column when a Burt table is used.

**Table S3** Results of the MCA for the Outdoor Index, 2007. **Legend:** N/A: Results are not available for supplementary variables. Qlt: Quality (i.e. the sum of the squared correlations for the first two dimensions in this case). Inr: Inertias. K: Principal coordinates for the first dimension. Cor: Squared correlation with the first dimension. Ctr: Contributions of the modality to the explained inertia of the first dimension. All cells are multiplied by 1000. Results are the same on rows and on column when a Burt table is used.

**Table S4** Questions and responses selected for the Indoor Index, 2007.

**Table S5** Questions and responses selected for the Outdoor Index, 2007.

## Acknowledgments

The authors thank Mr. Yves Lafortune of Statistics Canada for relevant comments on a preliminary version of this study and Ms Sandra Owens for her contribution to the redaction of this article. Also, thanks to Mr. Gaston Quirion of Laval University Library for facilitating access to the Statistics Canada survey database.

## Author Contributions

Conceived and designed the experiments: MC BA DB PG. Analyzed the data: MC BA. Wrote the paper: MC BA DB PG.

## References

1. Natural Resources Canada (2011) Energy Efficiency trends in Canada, 1990 to 2009Ottawa (On)54 p.
2. Environment Canada (2011) Ottawa (On)Municipal Water Use Report24 p.
3. Mustapha I, Tait M, Trant D (2012) Human Activity and the Environment. Waste management in Canada. Statistics CanadaOttawa (On)Report No: 16-201-x, 46 p.
4. Milito AC, Gagnon G (2008) Greenhouse gas emissions-a focus on Canadian households. EnviroStats 2(4):3–6.
5. Statistics Canada (2008) Human Activity and the Environment: Annual Statistics 2007 and 2008Ottawa (On)159 p.
6. Environment Canada (2013) Wise Water Use. Available: http://www.ec.gc.ca/eau-water/default.asp?lang = En&n = F25C70EC-1. Accessed 6 may 2014.
7. Statistics Canada (2009). Households and the environment, 2007.Ottawa (On)report no: 11-526-x,102 p.
8. Hardie D, Alasia A (2009) Domestic Water Use: The relevance of Rurality in Quantity Used and Perceived Quality. Rural and Small Town Canada Analysis Bulletin 7(5):1–31.
9. Mustapha I (2013) Composting by households in Canada. EnviroStats 7(11):1–6.
10. Lynch MF, Hofmann N (2007) Canadian lawns and gardens: Where are they the greenest? EnviroStats 1(2): 9–14.
11. Birrell C (2008) Energy-efficient holiday lights. EnviroStats 2(4):19–20.
12. Nelligan T (2008) Household's use of water and wastewater services. EnviroStats 2(4):17–8.
13. Babooram A (2008) Canadian participation in an environmentally active lifestyle. EnviroStats 2(4):7–12.
14. Nardo M, Saisana M, Saltelli A, Tarantola S, Hoffman A, et al. (2005) Handbook on Constructing Composite Indicators: Methodology and User GuideParis (Fr)Organisation for Economic Co-operation and Development Publishing162 p.
15. Esty DC, Levy M, Srebotnajk T, de Sherbinin A (2005) 2005 Environmental Sustainability Index: Benchmarking National Environmental Stewardship. New Haven: Yale Center for Environmental Law & Policy, 403 p.
16. Statistics Canada (2010) Microdata User Guide – Households and the environment survey, 2007Ottawa (On)53 p.
17. Stewart DW, Shamdasani PN, Rook DW (2007) Focus Groups: Theory and Practise. 2nd ed.Thousand OaksSAGE Publications200 p.
18. Greenacre M (2007) Correspondence analysis in practice. 2nd EditionNew YorkChapman & Hall/CRC284 p.
19. Dossa LH, Buerkert A, Schlecht E (2011) Cross-Location Analysis of the Impact of Household Socioeconomic Status on Participation in Urban and Peri-Urban Agriculture in West Africa. Hum Ecol Interdiscip J 39(5): 569–581.
20. Charreire H, Casey R, Salze P, Kesse-Guyot E, Simon C, et al. (2010) Leisure-time physical activity and sedentary behaviour clusters and their associations with overweight in middle-aged French adults. Int J Obes (Lond) 34(8):1293–1301.
21. Cortinovis I, Vella V, Ndiku J (1993) Construction of a socio-economic index to facilitate analysis of health data in developing countries. Soc Sci Med 36(8): 1087–1097.
22. Howe LD, Hargreaves JR, Huttly SR (2008) Issues in the construction of wealth for the measurement of socio-economic position in low-income countriesEmerg Themes Epidemiol 5(3): 14 p.
23. Asselin LM (2002) Composite indicator of Multidimensional Poverty - TheoryQuébec (Qc)Institut de Mathématique Gauss33 p.
24. Greenacre M, Nenadic O (2010) Package 'ca' - Simple, Multiple and Joint Correspondence AnalysisR project, 20 p.
25. Munro A (2010) Recycling by Canadian Households, 2007.Statistics Canada, Ottawa (On)34 p.
26. Oaks(2012) Province and Territory Water Efficiency and Conservation Policy Information. Available: http://www.allianceforwaterefficiency.org/2012-Province-Information.aspx. Accessed 13 November 2013.
27. Kinkead J, Boardley A, Kinkead M (2006) An analysis of Canadian and other water conservation practices and initiativesMississauga (On)Canadian Council of Ministers of the Environment274 p.
28. Gibbons WD (2008) Who uses water-saving fixture in the home? EnviroStats 2(3): 8–12.
29. Statistics Canada (2011) CANSIM Table 027-0017: Canada Mortgage and Housing Corporation, mortgage loan approvals, new residential construction and existing residential properties, monthly. Available: http://www5.statcan.gc.ca/cansim/a26?lang = eng&retrLang = eng&id = 0270017&paSer = &pattern = &stByVal = 1&p1 = 1&p2 = 37&tabMode = dataTable&csid = . Accessed 13 November 2013.
30. Bonsal B, Koshida G, O'Brien EG, Wheaton E (2013). Droughts. Available: http://www.ec.gc.ca/inre-nwri/default.asp?lang = En&n = 0CD66675-1&offset = 8&toc = hide . Accessed 13 july 2012.
31. Environment Canada (2010) Water and climate change. Available: http://www.ec.gc.ca/eau-water/default.asp?lang = En&n = 3E75BC40-1. Accessed 13 November 2013.
32. Mekis É, Vincent LA (2011) An overview of the second generation adjusted daily precipitation dataset for trend analysis in Canada. Atmosphere-Ocean 49(2): 163–77.
33. Ministère du Développement durable, de l'Environnement, de la Faune et des Parcs (2011) The pesticides Management Code - Highlights. Available: http://www.mddep.gouv.qc.ca/pesticides/permis-en/code-gestion-en/index.htm. Accessed 13 July 2012.
34. Canada mortgage and Housing Corporation (2013) Reducing energy cost. Available: https://www.cmhc-schl.gc.ca/en/inpr/afhoce/afhoce/afhostcast/afhoid/opma/reenco/index.cfm. Accessed 22 November 2013.
35. BChydro (2013) Buy, build, or rent an efficient home. Available: http://www.bchydro.com/powersmart/residential/guides_tips/green-your-home/whole_home_efficiency/energy_efficient_home.html. Accessed 22 November 2013.
36. Human Resources and Skills Development Canada (2007) What difference does learning make to financial security. Indicators of Well-Being – Special ReportGovernment of Canada14 p.
37. U.S. Environmental Protection Agency (1998) Washington (DC)Water Conservation Plan Guidelines208 p.
38. Reynaud A, Renzetti S, Villeneuve M (2005) Residential water demand with endogenous pricing: The Canadian CaseWater Resour Res 41(w11409)11 p.
39. OECD (2013) Greening Household Behaviour: Overview from the 2011 Survey, OECD Studies on Environmental Policy and Household Behaviours.OECD Publishing306 p.
40. Kollmuss A, Agyeman J (2002) Mind the Gap: Why Do People Act Environmentally and What Are the Barriers to Pro-Environmental Behaviour? Environmental Education Research Aug;8(3):239.
41. Todd A, Struger J (2014) Changes in acid herbicide concentrations in urban streams after a cosmetic pesticides ban. Challenges, 5:138–151.
42. Environment Canada (2013) Ammonia Emissions. Available: https://www.ec.gc.ca/indicateurs-indicators/default.asp?lang = en&n = FE578F55-1. Accessed 30 January 2014.
43. Mainieri T, Barnett EG, Valdero TR, Unipan JB, Oskamp S (1997) Green buying: The influence of environmental concern on consumer behavior. The Journal of Social Psychology Apr;137(2):189–204.
44. Melgar N, Mussio I, Rossi M (2013) Environmental Concern and Behavior: Do Personal Attributes Matter?Facultad de Ciencias Sociales, Universidad de la Republica; 21 p.
45. Xiao C, Hong D (2010) Gender differences in environmental behaviours in China. Population and Environment Sep;32(1):88–104.
46. Lopez A, Torres CC, Boyd B, Silvy NJ, Lopez RR (2007) Texas Latino College Student Attitudes Toward Natural Resources and the Environment. Journal of Wildlife Management Jun;71(4):1275–80.
47. Baer DE, Curtis JE (1984) French Canadian-English Canadian Differences in Values: National Survey Findings. Canadian Journal of Sociology/Cahiers canadiens de sociologie 9(4):405–27.
48. Baillargeon JP (1994) The Cultural Practices of Anglophones in Quebec. Recherches Sociographiques May;35(2):255–71.
49. Gibson KL, McKelvie SJ, Man AF (2008) Personality and Culture: A Comparison of Francophones and Anglophones in Québec. The Journal of Social Psychology Apr;148(2):133–65.

50. Wu Z, Baer DE (1996) Attitudes toward family and gender roles: A comparison of English and French Canadian women. Journal of Comparative Family Studies 27(3):437–52.
51. Young N, Dugas E (2012) Comparing climate change coverage in Canadian English and French-language print media: environmental values, media cultures, and the narration of global warming. Canadian journal of sociology 37(1):25–54.
52. Corbeil JP (2012) Ottawa (On)French and the francophonie in Canada. Census in brief no. 1, Statistics Canada12 p.
53. Corbeil JP (2012) Linguistic Characteristics of Canadians. Language, 2011 Census of Population, Statistics CanadaOttawa (On)22 p.
54. Atiles JH, Bohon SA (2003) Camas Calientes: Housing Adjustments and Barriers to Social and Economic Adaptation among Georgia's Rural Latinos. Southern Rural Sociology 19(1):97–122.
55. Blumenberg E, Smart M (2010) Getting by with a little help from my friends and family: immigrants and carpooling. Transportation May;37(3):429–46.
56. Bohon SA, Stamps K, Atiles JH (2008) Transportation and Migrant Adjustment in Georgia. Population Research and Policy Review Jun;27(3):273–91.
57. Price CE, Feldmeyer B (2012) The Environmental Impact of Immigration: An Analysis of the Effects of Immigrant Concentration on Air Pollution Levels. Population Research and Policy Review Feb;31(1):119–40.
58. Statistics Canada (2011) Immigration in Canada: A portrait of the Foreign-born Population, 2006 Census: Data tables, figures and maps. Available: http://www12.statcan.ca/census-recensement/2006/as-sa/97-557/tables-tableaux-notes-eng.cfm. Accessed 30 January 2014.

10

# Effect of Feeding Selenium-Fertilized Alfalfa Hay on Performance of Weaned Beef Calves

**Jean A. Hall[1]*, Gerd Bobe[2,3], Janice K. Hunter[2¤a], William R. Vorachek[1], Whitney C. Stewart[2¤b], Jorge A. Vanegas[4], Charles T. Estill[2,4], Wayne D. Mosher[2], Gene J. Pirelli[2]**

**1** Department of Biomedical Sciences, College of Veterinary Medicine, Oregon State University, Corvallis, Oregon, United States of America, **2** Department of Animal and Rangeland Sciences, College of Agricultural Sciences, Oregon State University, Corvallis, Oregon, United States of America, **3** Linus Pauling Institute, Oregon State University, Corvallis, Oregon, United States of America, **4** Department of Clinical Sciences, College of Veterinary Medicine, Oregon State University, Corvallis, Oregon, United States of America

## Abstract

Selenium (Se) is an essential micronutrient in cattle, and Se-deficiency can affect morbidity and mortality. Calves may have greater Se requirements during periods of stress, such as during the transitional period between weaning and movement to a feedlot. Previously, we showed that feeding Se-fertilized forage increases whole-blood (WB) Se concentrations in mature beef cows. Our current objective was to test whether feeding Se-fertilized forage increases WB-Se concentrations and performance in weaned beef calves. Recently weaned beef calves (n = 60) were blocked by body weight, randomly assigned to 4 groups, and fed an alfalfa hay based diet for 7 wk, which was harvested from fields fertilized with sodium-selenate at a rate of 0, 22.5, 45.0, or 89.9 g Se/ha. Blood samples were collected weekly and analyzed for WB-Se concentrations. Body weight and health status of calves were monitored during the 7-wk feeding trial. Increasing application rates of Se fertilizer resulted in increased alfalfa hay Se content for that cutting of alfalfa (0.07, 0.95, 1.55, 3.26 mg Se/kg dry matter for Se application rates of 0, 22.5, 45.0, or 89.9 g Se/ha, respectively). Feeding Se-fertilized alfalfa hay during the 7-wk preconditioning period increased WB-Se concentrations ($P_{Linear}<0.001$) and body weights ($P_{Linear}=0.002$) depending upon the Se-application rate. Based upon our results we suggest that soil-Se fertilization is a potential management tool to improve Se-status and performance in weaned calves in areas with low soil-Se concentrations.

**Editor:** Pascale Chavatte-Palmer, INRA, France

**Funding:** Funded in part by Animal Health and Disease Project Formula Funds, Oregon State University, Corvallis, Oregon 97331-4802, United States of America (JAH, Principal Investigator). The funders had no role in study design, data collection and analysis, decision to publish, or preparation of the manuscript. No additional external funding received for this study.

**Competing Interests:** The authors have declared that no competing interests exist.

* E-mail: Jean.Hall@oregonstate.edu

¤a Current address: Deep Springs College, Dyer, Nevada, United States of America
¤b Current address: Texas AgriLife Research, San Angelo, Texas, United States of America

## Introduction

Selenium (Se) is an essential micronutrient of cattle. Provision of adequate Se is important to prevent Se-responsive diseases in growing cattle such as nutritional myodegeneration and Se-responsive unthriftiness [1]. Many parts of the world, including Oregon, USA, are known to have soil conditions conducive to deficient forage-Se content, potentially leading to clinical signs of Se deficiency in livestock grazing or fed crops raised on them [2]. Soils are the major source of Se for plants and soil-Se exits in various forms including selenides, elemental Se, selenites, selenates and organic Se compounds [3]. Soil-Se content varies considerably depending upon geographic location. Low soil pH and high concentrations of sulfur and phosphorus from fertilization decrease Se availability for plants. Leaching from the topsoil in areas of high rainfall or irrigation also lowers forage-Se content. Plant species also differ in their ability to incorporate Se from soil. Most forage plants are categorized as non-Se accumulator plants.

The bioavailability of Se is not straightforward because of wide variation in Se content of foods (determined by a combination of geographical and environmental factors) and chemical forms in which Se may be absorbed and metabolized [4]. In general, organic forms are absorbed and retained more efficiently than inorganic forms [5,6]. Selenium is normally present in the diet in organic forms, e.g., as selenomethionine (SeMet) or selenocysteine (SeCys) [7]. Inorganic Na-selenite and Na-selenate are present in the diet in very small amounts.

Although the essentiality of Se has been known for five decades, the most effective method of Se delivery to cattle for optimum performance is still being investigated. Several means of administering Se to deficient ruminants are available [8]. For example, there are a number of injectable preparations, which often include vitamin E. Selenium can also be added to feed, mineral, and protein supplements. Sustained-release boluses with a life of several months may be used. Because of their weight, these boluses stay in the rumen whereby they gradually release Se. Selenium supplemented by these methods is usually inorganic Na-selenite or Na-selenate. One limitation of supplementing with inorganic Se in salt or feed is the apparent short duration of Se storage in the animal [8,9]. Other limitations to these methods of Se delivery include individual variation or sporadic intake, extra labor requirements, added expenses, and seasonal grazing practices that result in limited access to Se for extended periods of time. Therefore, animals may be Se deficient by the end of the grazing season.

Agronomic biofortification is defined as increasing the bioavailable concentrations of essential elements in edible portions of crop plants through the use of fertilizers. The potential for using Se-containing fertilizers to increase forage Se concentrations and, thus, dietary Se intake has been demonstrated in Finland, New Zealand, and Australia where it has proven to be both effective and safe [10–14]. The predominant chemical form of Se in Se-fortified grains and hays is SeMet [7]. We previously reported that Se-replete beef cows fed Se-fertilized forage for 6 wk had elevated WB-Se concentrations for 20 wk, which ensured adequate WB-Se while grazing forage on Se-deficient soils [15]. We have also shown that the FDA-approved supplementation rate for sheep (0.3 mg of Se/kg of diet as fed, which is equivalent to 0.7 mg of Se/d or 4.9 mg of Se/wk per sheep) for organic Se supplementation was equally effective as supranutritional rates of Na-selenite supplementation (14.7 and 24.5 mg of Se/wk) in increasing whole-blood (WB) Se concentrations, demonstrating the greater oral bioavailability of organic Se in sheep. In addition, short-term exposure to Se-fertilized forage results in whole-body Se status sufficient to maintain adequate WB-Se concentrations throughout grazing periods when there is limited access to Se supplements [16]. In another study Stewart et al. [17] showed that growth and survival was better in lambs from ewes receiving Se-yeast at 5 times the FDA-allowed supplementation rate compared to lambs from ewes receiving the FDA-allowed supplementation rate or no Se.

The transition period between weaning and movement to a feedlot is one of the most stressful times for beef calves. Because Se plays an important role in the immune response in cattle [18], calves may have greater Se requirements during the transitional period. Performance in weaned beef calves is enhanced if a preconditioning program is utilized before calves enter the feedlot. Several weeks in a preconditioning program are recommended to reduce the stress associated with weaning, dehorning, castration, and vaccination with the goal of reducing morbidity and mortality after arrival at the feedlot. Preconditioning also reduces the number of calves pulled to sick pens, and improves weight gain and feed efficiency of calves after arrival at a feedlot [19–21].

The objectives of this study were to evaluate WB-Se status and performance in weaned beef calves fed alfalfa hay fertilized with Se at increasing rates for 7 wk in a preconditioning program prior to entering the feedlot. We hypothesized that feeding weaned beef calves forage fertilized with increasing amounts of Na-selenate would improve both WB-Se status and growth rate.

## Materials and Methods

### Animal Ethics Statement and Study Design

The experimental protocol was reviewed and approved by the Oregon State University Animal Care and Use Committee (ACUP Number: 4051). This was a prospective clinical trial of 7-wk duration (August 29 through October 14, 2010) involving 60 weaned beef calves, primarily of Angus breeding. The calves ranged in age from 4.5 to 6 mo (166±2 d; mean ± SEM) and originated from the Oregon State University Beef Ranch, Corvallis, OR, USA. Body weights at weaning ranged from 181 to 310 kg (239±3.6 kg, mean ± SEM), and body condition scores ranged from 6 to 7 (1 to 9 scale). There were 27 heifers and 33 steer calves in the study.

Corvallis is located at an elevation of 72 m, midway in the Willamette Valley, 74 km east of the Oregon Coast, and 137 km south of Portland. Like the rest of the Willamette Valley, Corvallis falls within the Marine West Coast climate zone with some Mediterranean characteristics. Temperatures are mild year round, with warm, dry, sunny summers and mild, wet winters with persistent overcast skies. Spring and fall are also moist seasons with persistent cloudiness, and light rain falling for extended periods. Winter snow is rare, but occasionally does fall, usually in the form of heavy wet snow, ranging between a dusting to several cm that does not persist on the ground for more than a day. During the mid-winter months after extended periods of rain, thick persistent fogs can form, sometimes lasting the entire day. Rainfall total is surprisingly variable, ranging from an average of 168.7 cm per year in the far northwest hills to 110.9 cm per year at Oregon State University, which is located in the center of Corvallis. Typical distribution of precipitation includes about 50 percent of the annual total from December through February, lesser amounts in the spring and fall, and very little during summer. Rainfall tends to vary inversely with temperatures, with the cooler months being the wettest, and the warmer summer months being the driest. Because of its close proximity to the coast range, Corvallis can experience slightly cooler temperatures, particularly in the hills, compared with the rest of the Willamette Valley. Despite this, temperatures dropping below freezing are a rare event. Average monthly temperatures for September are 25.1°C (high) and 9.0°C (low).

Using a randomized complete block design, calves were blocked at the time of weaning by body weight (BW) and then assigned to one of 4 treatment groups of 15 calves each. Ear tags were used to identify the calves. All calves were put together in a large dry field and fed non Se-fortified grass hay for 4 d. Calves were then placed by treatment group into dry barn lots (11–15 $m^2$/calf; concrete flooring in open lots that were strip cleaned once weekly; dirt flooring in loafing sheds with 5–6 $m^2$/calf; concrete bunks with 64–97 cm of feeder space/calf; all measurements exceeded requirements [22]) with continuous access to water, feed bunks, and shelter. Calves were fed alfalfa hay once daily. The alfalfa hay was grown in fields fertilized with sodium-selenate at an application rate of 0, 22.5, 45.0, or 89.9 g Se/ha. Calves were transitioned to their respective alfalfa hay sources over a 10-d period. Alfalfa hay was fed as follows: 0.64 kg/head/d 1; 0.79 kg/head/d 2; 1.59 kg/head/d 3; 3.2 kg/head/d 4 to 7; 4.77 kg/head/d 8 to 11; and 6.4 kg/head/d 12. In addition, grass hay was offered for the first 10 d, and then discontinued. By the beginning of the third wk, calves were consuming on average 6.4 kg (as fed) of alfalfa hay per head, which was approximately 2.6% of their BW. A specified quantity of hay was offered each day and intake was uniform among groups. Hay was available all day, but by the next morning bunkers were empty and less than 5% of hay (visual estimate) was wasted on the ground.

In addition, calves were fed grain-based concentrate (0.23 kg as fed/head/d for 5 wk and then 0.46 kg as fed/head/d for 2 wk; **Table 1**) containing a coccidiostat (Rumensin® 80; Elanco Animal Health Co, Indianapolis, IN; 0.0195%) and added Se (0.200 mg/kg). Grain concentrate was offered once a day beginning the day calves were placed into the barn lots; grain was placed into the bunkers and consumed before hay was fed. The grain concentrate consisted of 34% steam flaked corn, USDA grade 2; 28% rolled barley; 19% wheat middlings; 10% dried distiller's grains with solubles from an ethanol plant; and 4% dried distiller's grains containing mainly wheat. The ration was formulated for growing beef calves in the 200 to 300 kg weight range to achieve a target average daily gain of 0.5 kg/d.

**Table 1.** Alfalfa hay and grain concentrate nutrient composition (dry matter basis).

| Nutrient | Alfalfa Hay | Grain Concentrate |
|---|---|---|
| Dry matter, g/kg | 906 | 944 |
| Crude protein, g/kg | 183 | 158 |
| Acid detergent fiber, g/kg | 351 | 84 |
| Neutral detergent fiber, g/kg | 406 | 148 |
| Nonfiber carbohydrates, g/kg[1] | 323 | 608 |
| Fat, g/kg | 10 | 34 |
| Ash, g/kg | 78 | 52 |
| Calcium, g/kg | 15.3 | 7.4 |
| Phosphorus, g/kg | 2.7 | 5.4 |
| Magnesium, g/kg | 4.5 | 3.1 |
| Potassium, g/kg | 14.9 | 7.5 |
| Sodium, g/kg | 1.5 | 3.1 |
| Copper, mg/kg | 12 | 13 |
| Iron, mg/kg | 357 | 115 |
| Manganese, mg/kg | 43 | 61 |
| Zinc, mg/kg | 22 | 67 |

[1]Nonfiber carbohydrates calculated by difference.

Prior to this study, dams and calves had free-choice access to a mineral supplement containing 120 mg/kg Se from sodium-selenite. After weaning and during this study, all calves had free-choice access to the same type of mineral supplement, however Se was not added to the mixture. The mineral supplement (dry matter basis) was in loose granular format and contained 57.0 to 64.0 g/kg calcium; 30.0 g/kg phosphorus; 503 to 553 g/kg salt (NaCl); 50.0 g/kg magnesium; 50 mg/kg cobalt; 2,500 mg/kg copper; 200 mg/kg manganese; 200 mg/kg iodine; 6,500 mg/kg zinc (Wilbur-Ellis Company, Clackamas, OR). During the first 10 d, one bloat block containing 13 mg/kg Se (Bloat Guard® POL6.6 Pressed, SWEETLIX® Livestock Supplement System; Mankato, MN) was also offered to each group of calves. Routine farm management practices, including vaccinations and deworming, were the same for all treatment groups with the exception that one calf in the 45.0 g Se/ha group was castrated at the beginning of the trial.

## Selenium Fortified-Alfalfa Hay and Other Selenium Analyses

The soil was enriched with Se by mixing sodium-selenate (RETORTE Ulrich Scharrer GmbH, Röthenbach, Germany) with water and spraying it onto the soil surface of an alfalfa field at an application rate of 0, 22.5, 45.0, or 89.9 g Se/ha immediately after the first cutting of hay in June 2010. Fields were approximately 1.2 ha each. The application rates were chosen based on work with Selcote Ultra® (10 g Se/kg as 1:3 $Na_2SeO_4$:$BaSeO_4$; Terralink, Vancouver, British Columbia, Canada) in previous studies [15,16]. Second-cutting alfalfa hay was harvested 40 d after Se application and then analyzed for nutrient and Se content. Alfalfa yield was approximately 4.9 ton/ha. To determine whether Se remained in the soil after harvesting second-cutting alfalfa hay, third-cutting alfalfa hay was also harvested from the respective field plots after another 50 d and analyzed for Se content. A Penn State forage sampler was used to take 25 cores from random bales in each alfalfa hay source (0, 22.5, 45.0, or 89.9 g Se/ha). This sampling regime was repeated 3 times (all samples collected mid way through the feeding trial) for each alfalfa hay source. Core samples were mixed well and representative samples selected for analysis. Alfalfa hay samples were submitted to commercial laboratories for routine nutrient analysis (**Table 1**; Cumberland Analytical Services, Maugansville, MD) and Se analysis (Utah Veterinary Diagnostic Laboratory, Logan, UT). Alfalfa hay dry matter determination was completed at a temperature of 105°C for 12 to 14 h in a forced draught oven. Methods for crude protein (CP), acid detergent fiber (ADF), ash, and minerals were performed according to the Association of Official Analytical Chemists [23]. The neutral detergent fiber (NDF) was determined according to Van Soest et al. [24]. Soluble protein was determined according to Krishnamoorthy et al. [25]. Plant samples were prepared for Se analysis as previously described [26], and Se was analyzed using inductively coupled argon plasma emission spectroscopy (ICP-MS; ELAN 6000, Perkin Elmer, Shelton, CT). Quantification of Se was performed by the standard addition method, using a 4-point standard curve. A quality-control sample (in similar matrix) was analyzed after every 5 samples, and analysis was considered acceptable if the Se concentration of the quality-control sample fell within ±5% of the standard/reference value for the quality control.

Grain samples were prepared for Se analysis the same manner as alfalfa hay. Salt samples were ground using a mortar and pestle. The ground salt material (0.50 g) was placed into a labeled 30-ml digestion tube (Oak Ridge Teflon digestion tube, Nalge Nunc International, Rochester, NY). Trace metal-grade nitric acid (4.0 mL; Thermo Fisher Scientific Inc., Waltham, MA) was added to the digestion tubes. The tubes were then heated at 90°C for 1 h with the caps loose on the tubes. After digestion, tubes were allowed to cool and 5.0 mL of ultrapure water was added and the samples were again digested at 90°C for 1 h. Contents were increased to 10 mL by adding trace metal-grade nitric acid. One milliliter of the digest was transferred into another trace metal-free tube containing 9.0 mL of ultrapure water to make up a 5% (v/v) nitric acid matrix and was centrifuged at 520×g for 10 min. The supernatant was removed and the samples were analyzed to quantify Se using ICP-MS in the same manner as for plant and grain samples.

## Performance and Whole-blood Selenium Assay

Health was monitored daily during the 7-wk feeding trial. Body weights were measured at the beginning of the treatment period (baseline), and at 3 wk, 6 wk, and 7 wk (end of the Se supplementation period). To assess the effect of Se supplementation on WB-Se status, all calves were bled at 0 time (baseline) and once each week for 7 wk until study termination. Jugular venous blood was collected into evacuated ethylenediaminetetraacetic acid (EDTA) tubes (2 mL; final EDTA concentration 2 g/L; Becton Dickinson, Franklin Lakes, NJ) and stored on ice until they were frozen at −20°C. Whole-blood Se concentrations were determined by a commercial laboratory (Center for Nutrition, Diagnostic Center for Population and Animal Health, Michigan State University, E. Lansing, MI) using an ICP-MS method with modifications as previously described [15].

## Statistical Analyses

Statistical analyses were performed using SAS version 9.2 [27]. Whole-blood Se concentrations and BW were analyzed as repeated-measures-in-time using PROC MIXED. Fixed effects in the model were Se application rate (0, 22.5, 45.0, and 89.9 g Se/ha), sex of calf, (male, female), BW block, time (wk 0, 1, 2, 3, 4,

5, 6, and 7 of feeding experiment for WB-Se concentrations and wk 0, 3, 6, and 7 of feeding experiment for BW), and the interaction between Se application rate and time. Fixed effects in the model for BW were Se application rate, sex of calf, BW block, time (wk 0, 3, 6, and 7 of feeding experiment), and the interaction between Se application rate and time. An unstructured variance-covariance matrix was used to account for variation of measures within calves. The unstructured variance-covariance matrix provided the most parsimonious variance-covariance matrix based on the lowest value by the Aikaike Information Criterion.To evaluate the effect of Se application rate, linear, quadratic, and cubic contrasts were constructed. In addition, the linear response of the dependent variables Se forage content or WB-Se concentrations of beef calves to the independent variable Se fertilization rate were evaluated using univariate regression in PROC REG. Data are reported as least square means ± SEM. Statistical significance was declared at $P \leq 0.05$ and a tendency at $0.05 < P \leq 0.10$.

## Results

### Effect of Soil-Se Fertilization on Se Concentrations in Alfalfa Hay

Fertilizing fields with increasing amounts of sodium-selenate increased in a dose-dependent manner the Se-content of second-cutting alfalfa hay from 0.07 to 0.95, 1.55, and 3.26 mg Se/kg dry matter for sodium-selenate application rates of 0 (non-fertilized control), 22.5, 45.0, or 89.9 g Se/ha, respectively (**Figure 1**). Subsequent third-cutting alfalfa hay had carry-over Se concentrations of 0.16, 0.28, and 0.60 mg Se/kg dry matter for sodium-selenate application rates of 22.5, 45.0, or 89.9 g Se/ha, which is equivalent to 16.8%, 18.1%, and 18.4% carry over, respectively.

### Effect of Supranutritional Se-supplementation to Se-replete Weaned Beef Calves on Whole- blood Se Concentrations and Performance during the 7-wk Feeding Period

Based on the total amount of alfalfa hay and grain concentrate fed to each group of calves, average dry matter intake per head was calculated at 5.59 kg/head/d for alfalfa hay starting on day 12, and 0.20 kg/head/d for grain concentrate for the first 5 wk and 0.40 kg/head/d for the last 2 wk. Using measured alfalfa hay and grain concentrate values for CP, net energy for gain ($NE_g$), net energy for maintenance ($NE_m$), and total digestible nutrients (TDN), we calculated CP (1.09 kg/head/d), $NE_g$ (4.83 Mcal/head/d), $NE_m$ (8.29 Mcal/head/d), and TDN consumption (3.7 kg/head/d) and compared them to National Research Council (NRC) [28] requirements {CP (0.44 kg/head/d), $NE_g$ (0.62 Mcal/head/d), $NE_m$ (4.50 Mcal/head/d), and TDN (2.9 kg/head/d)} for growing beef calves in this weight range to verify adequate nutrient intake for growth.

Calculated Se intake from alfalfa hay was 0.4, 5.3, 8.7, and 18.2 mg Se/head/d for calves consuming hay with Se concentrations of 0.07, 0.95, 1.55, and 3.26 mg Se/kg dry matter. The measured Se concentration of the grain concentrate was 1.41 mg Se/kg dry matter. Calculated Se intake from grain concentrate was 0.28 mg Se/head/d (first 5 wk) and 0.56 mg Se/head/d (last 2 wk). The average intake of mineral supplement was 17.5 mg/head/d. The measured Se concentration of the mineral supplement without added Se was 0.10 mg Se/kg dry matter. Calculated Se intake from the mineral supplement was 0.002 mg Se/head/d. The average intake of bloat block was 120 g/head/d during the first 10 d. The reported Se concentration of the bloat block offered during the first 10 d was 13 mg Se/kg. Calculated Se intake from the bloat block fed during the first 10 d was 1.55 mg Se/head/d.

Feeding Se-fertilized alfalfa hay was effective at increasing WB-Se concentrations in weaned beef calves ($P_{Treatment}$, $P_{Time}$, and $P_{Treatment \times Time}$: *all* $P < 0.001$; **Figure 2**). The normal reference interval for WB-Se concentrations of adult cows is 120–300 ng/mL [15]. Heifer calves had greater WB-Se concentrations than male calves ($P = 0.03$). The WB-Se response increased with greater amounts of sodium-selenate applied to the soil ($P_{Linear} < 0.001$; **Figure 3**). No significant quadratic ($P = 0.11$) or cubic ($P = 0.30$) Se-dose response was detected. The WB-Se concentrations continued to increase throughout the 7-wk feeding period.

Feeding Se-fertilized alfalfa hay was effective at increasing BW in weaned beef calves ($P_{Treatment} = 0.002$; $P_{Time} < 0.001$; and $P_{Treatment \times Time} = 0.03$; **Figure 4**). Calf sex did not affect BW ($P = 0.98$). The BW response increased with greater amounts of sodium-selenate applied to the soil ($P_{Linear} < 0.001$; **Figure 4**). No significant quadratic ($P = 0.84$) or cubic ($P = 0.11$) Se-dose response was detected. Weight data were not confounded by adverse health events in the preconditioning period because only two calves were retreated for pink eye during the first week of the feeding period (existent before the start of the experiment). None of the calves died during the 7-wk feeding period or showed signs of Se-deficiency or Se-toxicosis.

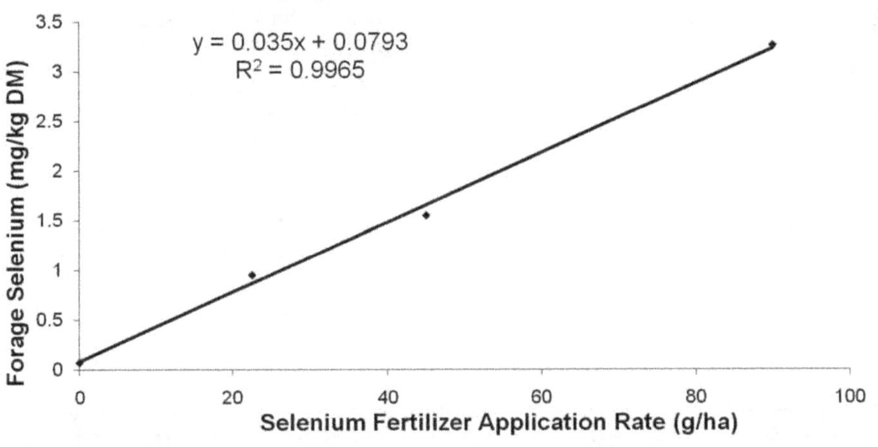


**Figure 1. Relationship between amount of Se applied by fertilization (g Se/ha) and observed forage Se content (g Se/kg DM).**

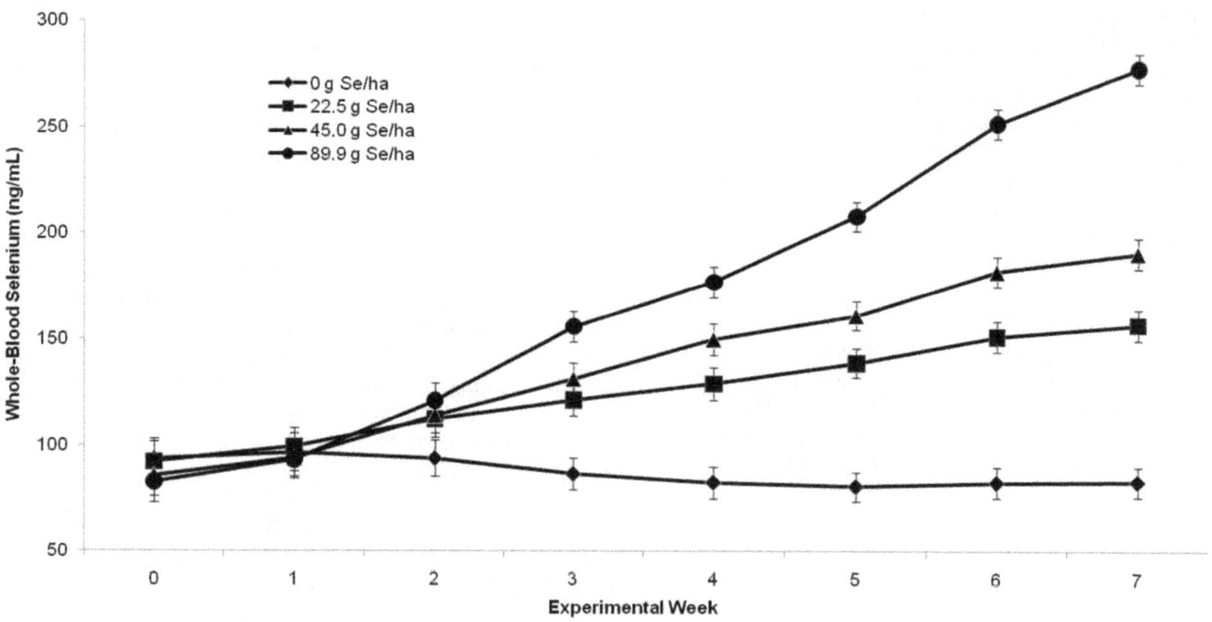


**Figure 2. Comparison of whole-blood Se concentrations (mean ± SEM) in weaned beef calves consuming alfalfa hay grown in fields not fertilized with Se (0 g Se/ha), or harvested from fields fertilized with sodium-selenate at an application rate of 22.5, 45.0, or 89.9 g Se/ha for 7 wk (n = 15 calves per group).** The normal reference interval for whole-blood Se concentrations of beef cattle is 120 to 300 ng/mL.

## Discussion

The objectives of this study were to evaluate whether fertilizing the soil of alfalfa hay fields with increasing amounts of sodium-selenate, and subsequent feeding of Se-fertilized alfalfa hay to recently weaned beef calves would improve in a dose-dependent manner WB-Se status and, consequently, increase growth rate in the preconditioning period prior to entering the feedlot. Fertilizing alfalfa hay fields with sodium-selenate increased Se content of alfalfa hay in a dose-dependent manner. Feeding Se-fertilized alfalfa hay during the 7-wk preconditioning program increased WB-Se concentrations and BW in Se-replete calves in a dose-dependent manner. Our results suggest that fertilization of alfalfa fields with sodium-selenate is a potential management tool to improve Se status and performance in weaned beef calves.

### Effect of Fertilizing Soil with Increasing Application Rates of Sodium-selenate on Se Concentrations in Alfalfa Hay

Agronomic Se-biofortification has been used in several countries with regions of low soil-Se concentrations including Finland, Denmark, New Zealand, and the United Kingdom to increase Se concentrations in the food chain [10,29,30]. In the United States, the use of feedstuffs that are naturally high in Se content is not regulated; Se fertilization, however, is not allowed in any state except for Oregon, where the Department of Agriculture does not control the use of Se as a plant fertilizer. Therefore, in Oregon it is

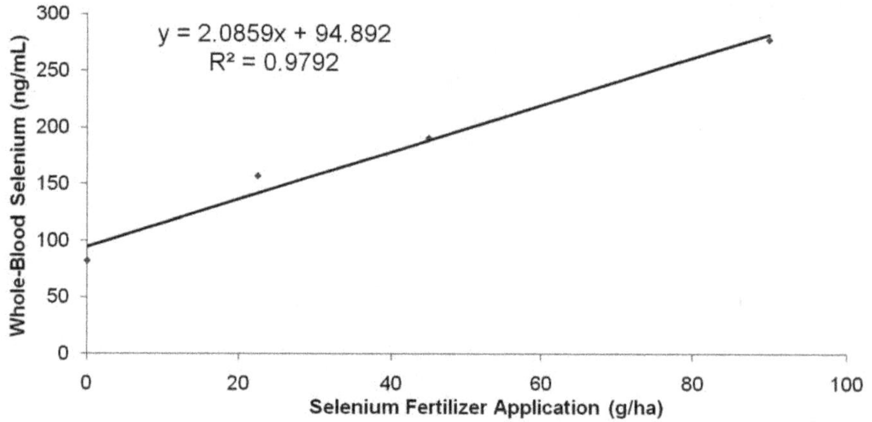


**Figure 3. Relationship between amount of Se applied by fertilization (g Se/ha) and observed WB-Se content (ng/mL) in weaned beef calves consuming alfalfa hay grown in fields not fertilized with Se (0 g Se/ha), or harvested from fields fertilized with sodium-selenate at an application rate of 22.5, 45.0, or 89.9 g Se/ha for 7 wk (n = 15 calves per group).** The normal reference interval for whole-blood Se concentrations of beef cattle is 120 to 300 ng/mL.

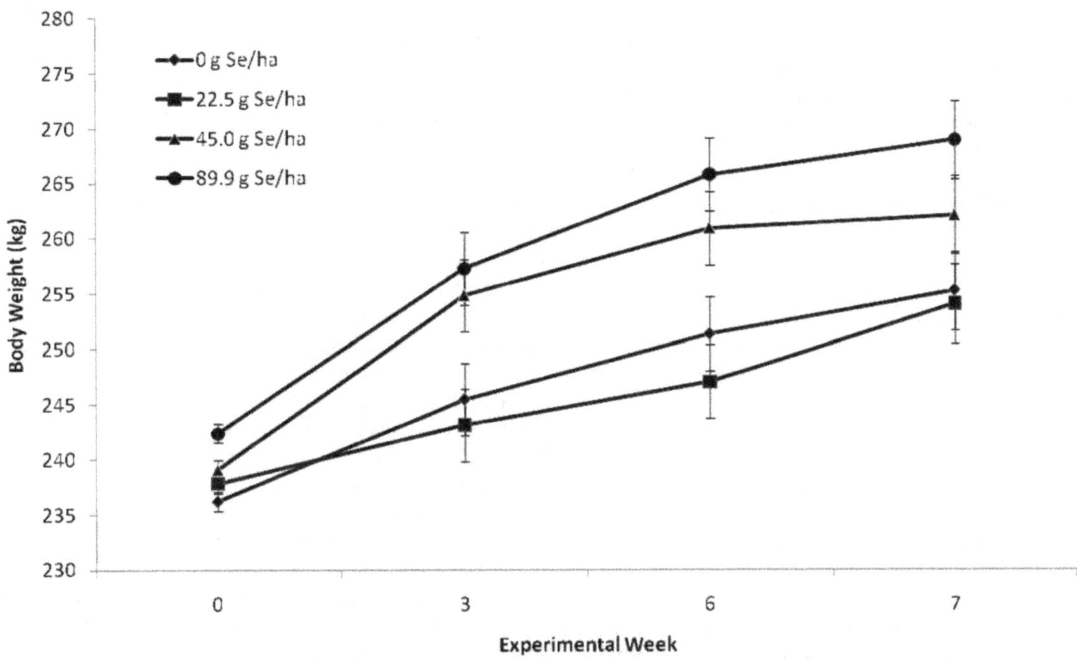


**Figure 4. Comparison of block-adjusted BW (kg; mean ± SEM) of weaned beef calves (primarily of Angus breeding and ranging in age from 4.5-to-6-mo) after consuming alfalfa hay grown in fields not fertilized with Se (0 g Se/ha), or harvested from fields fertilized with sodium-selenate at an application rate of 22.5, 45.0, or 89.9 g Se/ha for 7 wk (n = 15 calves per group).** Initial BW (baseline) ranged from 181 to 310 kg (239±3.6 kg, mean ± SEM). Final BW (7-wk) ranged from 183 to 346 kg (260±4.1 kg, mean ± SEM).

possible to produce feedstuffs with increased Se concentrations by applying Se as a fertilizer. Plants absorb Se from the soil in the form of selenate and synthesize selenoamino acids with SeMet being the major selenocompound in grassland legumes [7].

We have shown that sodium-selenate can be solubilized in water and sprayed onto soil surfaces of established alfalfa hay fields after the first cutting of alfalfa hay at three application rates in a 1×, 2×, and 4× ratio (0, 22.5, 45.0, or 89.9 g Se/ha). Hay harvested from respective field plots has a similar dose-dependent Se content (0.07, 0.95, 1.55, and 3.26 mg Se/kg dry matter) (**Figure 1**).The Se fertilizer application rates in the current study were chosen based upon WB-Se concentrations attained and absence of clinical signs of toxicosis in previous Se-fertilization studies [15,16].

In Hall et al. [15], Selcote Ultra® (10 g Se/kg as sodium selenate; Terralink, Vancouver, British Columbia, Canada) was mixed with urea-sol fertilizer and applied to pasture at a rate of 3.4 kg Selcote/ha (34 g Se/ha; 1.5× our lowest Se application rate in the current study). This resulted in forage Se of 0.11, 1.52, and 1.06 mg/kg (dry matter basis) at pre-fertilization, day 1, and day 42 of the grazing period, respectively. Beef cows grazing this forage had WB-Se concentrations of 186±5 ng/mL immediately post-grazing (day 42), which was within the normal reference interval for WB-Se concentrations of adult cows at the Michigan State University diagnostic laboratory (120–300 ng/mL). In a sheep study [16], Se applied to subclover-fescue sward type pasture at 3.4 kg Selcote/ha resulted in Se concentrations of up to 2.02 mg/kg (dry matter basis) in green forage grazed by the sheep, and WB-Se concentrations in ewes after grazing for 6 weeks of 573±20 ng/mL, with no clinical signs of Se toxicosis. When comparing the results of the current study with results from these two studies and one other that was conducted by our group that used Selcote Ultra® for Se fertilization (all at different locations in Oregon), a linear relationship between sodium-selenate application rate and Se content of forage was observed (forage selenium concencentrations in mg/kg DM = 0.0299×selenium fertilizer application rate in g/ha +0.3297; $r^2 = 0.9967$) [15,16].

The linear relationship between Se fertilizer application rate and forage Se concentration is surprising given the fact that plant species, field location, and soil differed among the studies [15,16]. It is well documented that plant species, chemical species of Se, and soil pH, sulfur, and iron content alter Se availability for plant uptake [31,32]. Based upon our results, we suggest that Se content of soil primarily determines Se content of common forage species. This is supported by the observation that, regardless of initial Se-application rate, subsequent third-cutting alfalfa hay had approximately 18% of the Se content of hay harvested after initial application of Se. The Se concentrations of third-cutting alfalfa hay were 0.16, 0.28, and 0.60 mg Se/kg (dry matter basis) for sodium-selenate application rates of 22.5, 45.0, or 89.9 g Se/ha, respectively. To our knowledge, this is the first report of Se carry-over in the soil to the next cutting of alfalfa hay.

## Effects of Feeding Alfalfa Hay Fertilized with Increasing Application Rates of Se for 7 wk on Whole-blood Se Concentrations and Performance of Weaned Beef Calves

Feeding Se-fertilized alfalfa hay was effective at increasing WB-Se concentrations in Se-replete weaned beef calves. The WB-Se concentrations increased with greater amounts of sodium-selenate applied to soil (**Figures 2 and 3**). For example, the increase in WB-Se concentrations for calves consuming alfalfa hay from fields fertilized with 89.9 vs. 45.0 g Se/ha was double the increase for calves consuming alfalfa hay from fields fertilized with 45.0 vs. 22.5 g Se/ha. Combining our results with a previous study in mature beef cows [15] indicates that there is a linear relationship between sodium-selenate application rate and WB-Se concentrations in beef cattle (WB-Se concentration in ng/mL = 1.9688×selenium fertilizer application rate in g/ha +105.67; $r^2 = 0.9594$).

The WB-Se concentrations continued to increase throughout the 7-wk feeding period.

The majority of dietary Se was supplied by the alfalfa hay, except in calves consuming non-Se fertilized alfalfa hay. In the first 5 wk, 0.28 mg Se/head/d was provided by the grain concentrate with essentially none provided by the mineral supplement. In the last 2 wk, 0.56 mg Se/head/d was provided by the grain concentrate. Thus, in the last 2 wk, total dietary Se intake was 0.96, 5.86, 9.26, and 18.76 mg Se/d for calves consuming alfalfa hay with sodium-selenite application rates of 0, 22.5, 45.0, or 89.9 g Se/ha (alfalfa hay provided 41, 90, 94, and 97% of dietary Se intake, respectively). During the first 10 d of the feeding trial, calculated Se intake from the bloat block was 1.55 mg Se/head/d. This represented the greatest source of Se for this short-term period in those calves receiving non-Se fertilized alfalfa hay.

In the United States, the FDA [33] allows Se to be added to cattle diets as sodium-selenite, sodium-selenate, or Se-yeast in complete feeds not to exceed 0.3 mg of Se/kg of diet (as fed basis), or in supplements for limit feeding not to exceed 3 mg of Se/head/day. The non-Se fortified alfalfa hay (0.07 mg Se/kg dry matter) was below this level. We were able to provide additional dietary Se using Se-fortified alfalfa hay. Our results for WB-Se concentrations in weaned beef calves fed Se-biofortified hay are similar to what we reported [15] when grazing adult beef cattle on Se-fertilized forage. No clinical signs of Se deficiency or toxicosis were observed.

Our results are consistent with SeMet from the grassland legumes being absorbed in the duodenum and incorporated into general body proteins in place of methionine. The concentration of SeMet is not regulated and ultimately reflects dietary intake [7,32]. Selenomethionine acts as a storage form of Se in body proteins, including hemoglobin and albumin in WB, from which it is slowly released by protein catabolism. The SeMet can also be trans-selenated into selenocysteine and subsequently used to provide Se requirements for selenoprotein synthesis [34].

Production benefits of agronomic Se biofortification were assessed by comparing BW gains at the end of the 7-wk Se-supplementation period. Feeding Se-fertilized alfalfa hay increased BW in a linear manner (**Figure 4**). Calves receiving Se-fertilized alfalfa hay with the highest concentration of Se had the highest BW. There results were not confounded by morbidly because none of the calves were sick during the 7-wk preconditioning period. None experienced bloating, went off feed, or showed any clinical signs of illness. These results are consistent with our previous results for lambs [17] whereby ewes raising multiples that received the highest Se-yeast dose (24.5 mg Se/wk) had heavier lambs at weaning (120 d of age) in yr 1 of supplementation and at 60 d of age in yr 2 of supplementation than ewes receiving Se-yeast at the maximum FDA-allowed level (4.9 mg Se/wk).

Selenium's role in animal performance is based upon the functions of selenoproteins, many of which have antioxidant activities [4]. Although reactive oxygen species and free radicals are a natural result of the body's normal metabolic activity, excessive stress as a result of weaning, dietary and environmental changes, comingling with other animals, and disease can lead to the over production of free radicals or accumulation of free radicals because of a lack of antioxidants. Therefore, it is important that micronutrients involved in antioxidant functions be present in tissues to provide oxidant-antioxidant balance. Although the level of Se supplementation needed for adequate performance may be less under optimum conditions, in periods of transition such as weaning, preconditioning, and shipping, where stress is a confounding factor for optimum performance, Se requirements may be increased.

Because all calves were visually healthy, it is unlikely that the observed BW response to Se supplementation is explained solely by the antioxidant activity of selenoproteins. Two selenoprotein families, the iodothyronine deiodinases, responsible for metabolism of thyroid hormones, and the thioredoxin reductases, responsible for reducing thioredoxin, are directly or indirectly through regulation of transcription factors, involved in cell growth and control of apoptosis, as well as maintenance of cellular redox status [35]. Our results suggest that supranutritional Se supplementation may have growth-promoting properties in beef cattle. Future studies are warranted to examine the effects of supranutritional Se supplementation on gene and protein expression of the iodothyronine deiodinases, thioredoxin reductases, and transcription factors regulating cell growth.

In summary, Se fertilization of alfalfa fields in a region with Se deficient soils increased in a dose-dependent manner the Se content of alfalfa hay. Supranutritional Se supplementation of recently weaned beef calves with Se-fortified alfalfa hay resulted in increased WB-Se concentrations and improved growth rates. Our results suggest that building Se-body reserves by feeding supranutritional Se levels from sodium-selenate fertilized alfalfa hay during the preconditioning program is an effective management strategy to optimize growth and health in weaned beef calves.

## Acknowledgments

Appreciation is expressed to KC Bare and Opal Springs Farms, LLC, Culver, OR for precise Se application rates to alfalfa fields and growing the alfalfa hay for the conduct of these experiments.

## Author Contributions

Conceived and designed the experiments: JAH WDM GJP. Performed the experiments: JAH JKH WRV WCS JAV CTE WDM GJP. Analyzed the data: JAH GB. Contributed reagents/materials/analysis tools: JAH GB. Wrote the paper: JAH GB CTE GJP.

## References


1. Koller LD, South PJ, Exon JH, Whitbeck GA (1983) Selenium deficiency of beef cattle in Idaho and Washington and a practical means of prevention. Cornell Vet 73: 323–332.
2. Stevens JB, Olson WG, Kraemer R, Archambeau J (1985) Serum selenium concentrations and glutathione peroxidase activities in cattle grazing forages of various selenium concentrations. Am J Vet Res 46: 1556–1560.
3. Surai PF, Fisinin VI, Papazyan TT (2008) Selenium deficiency in Europe: causes and consequences. In: Surai PF, Taylor-Pickard JA, editors. Current advances in selenium research and applications. The Netherlands: Wageningen Academic Publishers. 13–44.
4. Fairweather-Tait SJ, Collings R, Hurst R (2010) Selenium bioavailability: current knowledge and future research requirements. Am J Clin Nutr 91: 1484S–1491S.
5. Qin S, Gao J, Huang K (2007) Effects of different selenium sources on tissue selenium concentrations, blood GSH-Px activities and plasma interleukin levels in finishing lambs. Biol Trace Elem Res 116: 91–102.
6. Hall JA, Van Saun RJ, Bobe G, Stewart WC, Vorachek WR, et al. (2012) Organic and inorganic selenium: I. Oral bioavailability in ewes. J Ani Sci 90: 568–576.
7. Whanger PD (2002) Selenocompounds in plants and animals and their biological significance. J Am Col Nutr 21: 223–232.
8. Surai PF (2006) Selenium in ruminant nutrition. In: Surai PF, editor. Selenium in nutrition and health. Nottingham: Nottingham University Press. 487–587.
9. Surai PF (2006) Selenium in food and feed: selenomethionine and beyond. In: Surai PF, editor. Selenium in nutrition and health. Nottingham: Nottingham University Press. 151–212.
10. Broadley MR, White PJ, Bryson RJ, Meacham MC, Bowen HC, et al. (2006) Biofortification of UK food crops with selenium. Proc Nutr Soc 65: 169–181.
11. Makela AL, Nanto V, Makela P, Wang W (1993) The effect of nationwide selenium enrichment of fertilizers on selenium status of healthy Finnish medical students living in south western Finland. Biol Trace Elem Res 36: 151–157.

12. Whelan BR (1989) Uptake of selenite fertilizer by subterranean clover pasture in Western Australia. Australian J Experimental Agriculture 29: 517–522.
13. Whelan BR, Barrow NJ, Peter DW (1994a) Selenium fertilizers for pastures grazed by sheep. 1. Selenium concentrations in whole-blood and plasma. Australian J Agricultural Research 45: 863–875.
14. Whelan BR, Barrow NJ, Peter DW (1994b) Selenium fertilizers for pastures grazed by sheep. 2. Wool and liveweight responses to selenium. Australian J Agricultural Research 45: 877–887.
15. Hall JA, Harwell AM, Van Saun RJ, Vorachek WR, Stewart WC, et al. (2011) Agronomic biofortification with selenium: Effects on whole blood selenium and humoral immunity in beef cattle. Anim Feed Sci Technol 164: 184–190.
16. Hall JA, Van Saun RJ, Nichols T, Mosher W, Pirelli G (2009) Comparison of Se status in sheep after short-term exposure to high-Se-fertilized forage or mineral supplement. Small Ruminant Res 82: 40–45.
17. Stewart WC, Bobe G, Pirelli GJ, Mosher WD, Hall JA (2012) Organic and inorganic selenium: III. Ewe and progeny performance. J Anim Sci 90: 4536–4543.
18. Finch JM, Turner RJ (1996) Effects of selenium and vitamin E on the immune responses of domestic animals. Res Vet Sci 60: 97–106.
19. Cole NA (1985) Preconditioning calves for the feedlot. Vet Clin North Am Food Anim Pract 1: 401–411.
20. Pritchard RH, Mendez JK (1990) Effects of preconditioning on pre- and post-shipment performance of feeder calves. J Anim Sci 68: 28–34.
21. Duff GC, Galyean ML (2007) Board-invited review: recent advances in management of highly stressed, newly received feedlot cattle. J Anim Sci 85: 823–840.
22. MWPS-6 (1987) Beef Housing and Equipment Handbook, $4^{th}$ ed. Midwest Plan Service, Iowa State University, Ames, IA, USA.
23. AOAC (2000) Official Methods of Analysis, $17^{th}$ ed. Association of Official Analytical Chemists, Arlington, VA, USA.
24. Van Soest PJ, Robertson JB, Lewis BA (1991) Methods for dietary fiber, neutral detergent fiber and nonstarch polysaccharides in relation to animal nutrition. J Dairy Sci 74: 3583–3597.
25. Krishnamoorthy U, Muscato TV, Sniffen CJ, Van Soest PJ (1982) Nitrogen fractions in selected feedstuffs. J Dairy Sci 65: 217–225.
26. Davis TZ, Stegelmeier BL, Panter KE, Cook D, Gardner DR, et al. (2012) Toxicokinetics and pathology of plant-associated acute selenium toxicosis in steers. J Vet Diagn Invest 24: 319–327.
27. SAS Institute (2009) SAS User's Guide. Statistics, Version 9.2. SAS Inst Inc, Cary, NC.
28. NRC (1996) Nutrient Requirements of Beef Cattle, seventh revised edition. Natl Acad Press, Washington, DC.
29. Wang WC, Mäkelä AL, Näntö V, Mäkelä P, Lagström H (1998) The serum selenium concentrations in children and young adults: a long-term study during the Finnish selenium fertilization programme. Eur J Clin Nutr 52: 529–535.
30. Gupta UC, Gupta SC (2002) Quality of animal and human life as affected by selenium management of soils and crops. Comm Soil Sci Plant Anal 33: 2537–2555.
31. NRC (1983) Selenium in Nutrition, revised edition. Natl Acad Press, Washington, DC.
32. NRC (2007) Nutrient Requirements of Small Ruminants. Natl Acad Press, Washington, DC.
33. FDA (2009) Title 21. Food and Drugs: Food additives permitted in feed and drinking water of animals. http://www.accessdata.fda.gov/scripts/cdrh/cfdocs/cfcfr/CFRSearch.cfm?fr=573.920 Accessed 2012 July 24.
34. Rayman MP (2008) Food-chain selenium and human health: emphasis on intake. Br J Nutr 100: 254–268.
35. Rooke JA, Robinson JJ, Arthur JR (2004) Effects of vitamin E and selenium on the performance and immune status of ewes and lambs. J Agric Sci 142: 253–262.

# 11

# Establishing a Regional Nitrogen Management Approach to Mitigate Greenhouse Gas Emission Intensity from Intensive Smallholder Maize Production

**Liang Wu, Xinping Chen, Zhenling Cui*, Weifeng Zhang, Fusuo Zhang**

Center for Resources, Environment and Food Security, China Agricultural University, Beijing, People's Republic of China

## Abstract

The overuse of Nitrogen (N) fertilizers on smallholder farms in rapidly developing countries has increased greenhouse gas (GHG) emissions and accelerated global N consumption over the past 20 years. In this study, a regional N management approach was developed based on the cost of the agricultural response to N application rates from 1,726 on-farm experiments to optimize N management across 12 agroecological subregions in the intensive Chinese smallholder maize belt. The grain yield and GHG emission intensity of this regional N management approach was investigated and compared to field-specific N management and farmers' practices. The regional N rate ranged from 150 to 219 kg N $ha^{-1}$ for the 12 agroecological subregions. Grain yields and GHG emission intensities were consistent with this regional N management approach compared to field-specific N management, which indicated that this regional N rate was close to the economically optimal N application. This regional N management approach, If widely adopted in China, could reduce N fertilizer use by more than 1.4 MT per year, increase maize production by 31.9 MT annually, and reduce annual GHG emissions by 18.6 MT. This regional N management approach can minimize net N losses and reduce GHG emission intensity from over- and underapplications, and therefore can also be used as a reference point for regional agricultural extension employees where soil and/or plant N monitoring is lacking.

**Editor:** Shuijin Hu, North Carolina State University, United States of America

**Funding:** The work has been funded by the National Basic Research Program of China (973, Program: 2009CB118606) (website: http://www.973.gov.cn/AreaAppl.aspx). National Maize Production System in China (CARS-02-24)(website: http://119.253.58.231/). Special Fund for Agro-scientific Research in the Public Interest (201103003)(website: http://www.hymof.net.cn/webapp/login.asp). The funders had no role in study design, data collection and analysis, decision to publish, or preparation of the manuscript.

**Competing Interests:** The authors have declared that no competing interests exist.

* E-mail: cuizl@cau.edu.cn

## Introduction

The need to increase global food production while also increasing nitrogen (N) use efficiency and limiting environmental costs [e.g., greenhouse gas (GHG) emissions] have received increasing public and scientific attention [1–6]. Coordinated global efforts are particularly critical when dealing with N-related GHG emissions because such emissions and their impacts recognize no borders. The most rapidly developing countries, such as China and India, are becoming central to the issue, not only because these countries consume the most chemical N fertilizer [7,8], but they have also become dominating forces in the production of new N fertilizers in recent decades [7,8]. From 2001 to 2010, global N fertilizer consumption increased from 83 to 105 MT, with 83% of this global increase originating from five rapidly developing countries, specifically China (9.9 MT), India (5.2 MT), Pakistan (0.8 MT), Indonesia (1.1 MT), and Brazil (1.1 MT). In comparison, chemical N fertilizer consumption decreased by 6.5% (0.7 MT) in Western Europe and Central Europe, and increased by only 7.1% (0.8 MT) in the United States over this period [8]. Optimizing N management in these rapidly developing countries clearly has important implications worldwide.

In the past 30 years, the N application rate in many developed economies has been optimized based on recommended systems, and have included soil nitrate ($NO_3$) and plant testing [9,10], and more recently, remote sensing [11]. However, in rapidly developing countries, small-scale farming with high variability between fields and poor infrastructure in the extension service makes the use of many advanced N management technologies difficult. Fox example, the average area per farm in China is only 0.6 ha, and individually managed fields are generally 0.1–0.3 ha [12]. Therefore, the challenge is to develop agronomically effective and environmentally friendly practices that are applicable to hundreds of millions of smallholder farmers, while producing high yields and reducing N losses.

Decisions regarding the optimal N fertilizer application rate require knowledge of existing soil N supplies, crop N uptake, and the expected crop yield in response to N application [13]. Optimal N rates often vary depending on soil-specific criteria and/or crop management variables such as soil productivity, producer management level, and geographic location [14]. However, the optimal N rate will become more uniform under geographically similar soil and climatic conditions, and when the main factors causing the variation in optimal N rates are either addressed or removed [14].

Our hypothesis is that a regional N management approach could be adopted to accommodate hundreds of millions of small farmers and reduce variation among farms, increase crop yield, and lower the GHG emission intensity of maize production. In China, maize (*Zea mays* L.) is the largest food crop produced, accounting for 37% of Chinese cereal production and 22% of the global maize output in 2011 [15]. Chinese maize production results in some of the most intensive N applications globally, and the resulting enrichment of N in soil, water, and air has created serious environmental problems.

In the present study, we developed a regional N management approach across major maize agroecological regions in China. We also compared grain yield and GHG emissions between the regional N management approach and site-specific N management, and evaluated the potential for increasing grain yields and mitigating GHG emission intensity using this regional N management approach when compared to farmers' practices across each region.

## Materials and Methods

### Description of China's agroecological maize regions

In China, maize is grown primarily in 4 main agroecological regions and 12 agroecological subregions, including Northeast China (NE1, NE2, NE3, NE4), North China Plain (NCP1, NCP2), Northwest China (NW1, NW2, NW3), and Southwest China (SW1, SW2, SW3) (Fig. 1) [16]. These agroecological subregions were divided based on climatic conditions, terrains, agricultural management practices (e.g., irrigation), and soil types. Detailed information on each of these subregions is provided in Table S1 and Text S1.

### Farmers' survey

A multistage sampling technique was used to select representative farmers for a face-to-face, questionnaire-based household survey conducted once a year between 2007 and 2009 [17]. In this study, 5,406 farmers from 66 counties in 22 provinces were surveyed (Table 1). In each province, three counties were randomly selected, three townships were randomly selected in each county, two to five villages were randomly selected in each township, and 20 farmers from the villages were randomly surveyed to collect information on N fertilizer use and grain yield in each farmer's household. This study was approved by a research ethics review committee at the College of Resources and Environmental Science (CRES), China Agricultural University, Beijing, China. Data was collected through an in-house survey, which was conducted by research staff at the College of Resources and Environmental Science. Before beginning the survey, an informed consent information sheet was given to the farmer to read (or in some cases was read to the farmer), and verbal informed consent was requested. Because this study was considered anonymous and each participating household could not be identified directly or indirectly, the research ethics review committee of CRES waived the need for written informed consent from the participants.

### On-farm field experiments

In total, 1,726 on-farm maize N fertilizer experiments in 181 counties of 22 provinces were conducted from 2005 to 2010 in the NE ($n = 397$) and NW ($n = 416$) spring maize areas, and in the NCP ($n = 407$) and SW ($n = 506$) summer maize areas. All 66 counties where farm surveys were conducted were included in these 181 counties.

All experimental fields received four treatments without replication: without N fertilizer (N0), medium N rate (MN), 50% and 150% of MN. The amount of N fertilizer for the MN treatment was recommended by local agricultural extension

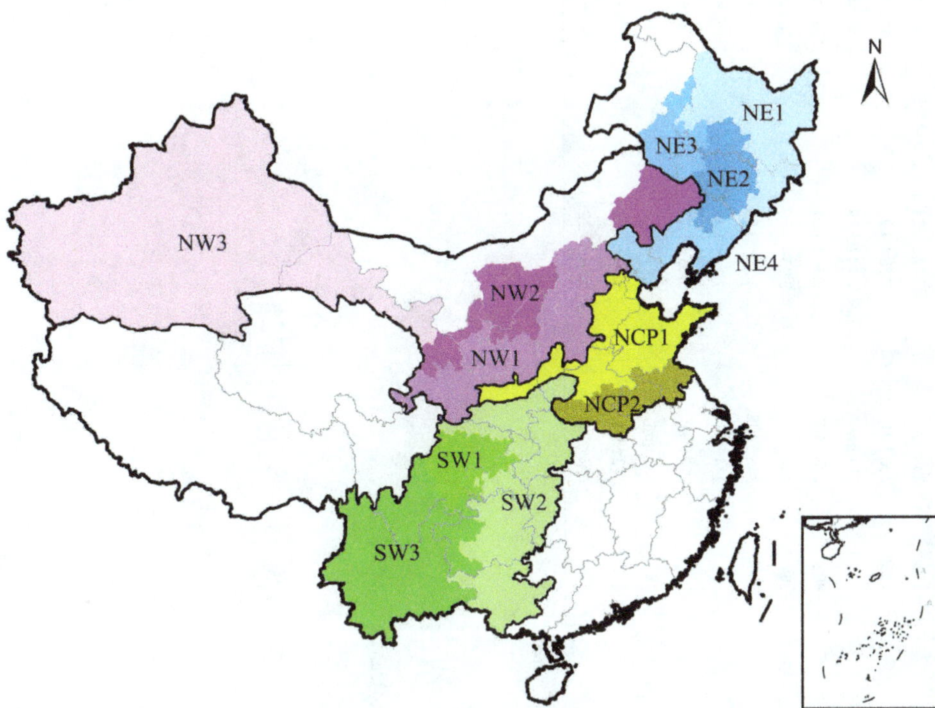


**Figure 1. Map showing the four major maize-planting agroecological regions (thick lines, NE, NCP, NW, SW) and their subregions in China (different colors).** Northeast China (NE1, NE2, NE3, NE4), North China Plain (NCP1, NCP2), Northwest China (NW1, NW2, NW3), and Southwest China (SW1, SW2, SW3). Here, we show the distribution of maize production in China; the total maize sowing area in the 12 subregions is approximately 32 million hectares, which represents 96% of the total maize production in China.

**Table 1.** N fertilizer application rate, maize grain yield, N balance, and GHG emission intensity of N fertilizer use, N fertilizer production and other sources in different agro-ecological subregions.

| Region & Subregion | n [a] | N rate (kg ha$^{-1}$) | Grain yield (Mg ha$^{-1}$) | N balance (kg N ha$^{-1}$) | GHG emission intensity (kg $CO_2$ eq Mg$^{-1}$ grain) | | | |
|---|---|---|---|---|---|---|---|---|
| | | | | | N fertilizer use | N fertilizer production | Other sources | Total |
| NE | 1263 | 195±61 | 8.91±1.19 | 32±30 | 115±38 | 182±60 | 50±17 | 347±131 |
| NE1 | 361 | 156±43 | 8.59±0.79 | 3±13 | 96±27 | 151±40 | 51±13 | 298±85 |
| NE2 | 411 | 201±59 | 9.03±1.16 | 40±28 | 116±34 | 183±54 | 49±15 | 348±124 |
| NE3 | 311 | 226±76 | 8.80±1.52 | 68±41 | 137±46 | 213±71 | 51±17 | 402±164 |
| NE4 | 180 | 205±77 | 8.68±1.50 | 49±39 | 124±51 | 196±74 | 52±21 | 373±183 |
| NCP | 1983 | 208±72 | 7.42±1.24 | 61±43 | 148±46 | 233±76 | 55±19 | 436±178 |
| NCP1 | 1460 | 206±71 | 7.68±1.27 | 54±38 | 141±49 | 223±77 | 54±19 | 418±180 |
| NCP2 | 523 | 217±66 | 7.14±1.15 | 76±46 | 161±45 | 252±75 | 57±17 | 471±174 |
| NW | 882 | 238±107 | 7.58±1.91 | 95±68 | 170±77 | 261±119 | 56±27 | 487±240 |
| NW1 | 394 | 234±103 | 6.93±1.70 | 98±73 | 182±80 | 280±123 | 58±26 | 520±282 |
| NW2 | 289 | 246±128 | 8.22±2.50 | 91±64 | 163±86 | 248±129 | 54±29 | 466±171 |
| NW3 | 199 | 234±83 | 7.15±1.48 | 106±47 | 176±62 | 272±96 | 60±20 | 508±257 |
| SW | 1278 | 250±91 | 5.45±1.17 | 144±86 | 251±93 | 381±140 | 78±30 | 710±319 |
| SW1 | 427 | 257±83 | 5.41±1.13 | 151±71 | 263±85 | 394±127 | 75±24 | 732±225 |
| SW2 | 447 | 232±90 | 5.36±1.08 | 129±93 | 232±90 | 358±134 | 84±34 | 675±359 |
| SW3 | 404 | 272±101 | 5.59±1.32 | 160±103 | 274±100 | 403±152 | 75±28 | 752±375 |

[a]n: number of observations.

employees based on experience and target yield (1.1 times the average yield of the past 5 years). The median N application rates for the 1,726 sites are shown in Table 2. Approximately one-third of the granular urea was applied by broadcasting at sowing, while the remainder was applied as a side-dressing at the six-leaf stage. All experimental fields received 30–150 kg $P_2O_5$ (P) $ha^{-1}$ as triple superphosphate and 30–135 kg $K_2O$ (K) $ha^{-1}$ as potassium chloride, based on experience and target yield. All P and K fertilizers were applied by broadcasting before sowing. No manure was used, which is common for maize production in China. Detailed information regarding the N application rate and selected soil chemical properties before maize planting at 1,726 on-farm experimental sites is provided in Table S2.

Individual plots were approximately 40 $m^2$ (5 m wide and 8 m long). All experiments were managed (including maize variety, density, planting, harvesting, herbicide and insecticide for pests, diseases, and weeds) by local farmers based on a field manual provided by local agricultural extension employees, whereas for the treatments, local agricultural extension employees conducted fertilizer applications. The time of planting and harvest were determined by farmers and differed among sites. Generally, in NE and NW, maize was planted in early May and harvested in late September. Maize was planted from June to October in NCP and from April to August in SW. Plant densities were 50,000–65,000 plants $ha^{-1}$ in NE, 70,000–75,000 plants $ha^{-1}$ in NCP, 65,000–75,000 plants $ha^{-1}$ in NW, and 45,000–50,000 plants $ha^{-1}$ in SW. The locations of the 1,726 experiments were not privately-owned or protected in any way. No specific permits were required for the field studies. The farming operations employed during the experiment were similar to the operations routinely employed on rural farms and did not involve endangered or protected species. All operations were approved by the CRES, China Agricultural University.

## Sampling and laboratory procedures

Prior to the experiments, five chemical soil properties were examined. Values were determined based on soil samples from a combined soil sample of the 10–20 cores from depths of 0–20 cm. Soil samples collected before planting were air-dried and sieved through a 0.2-mm mesh. Soil samples were used to measure organic matter content (OM) [18], alkaline hydrolyzable N (AN) [19], Olsen-P [20], $NH_4OAc$-K [21], and pH [22]. Upon harvest, approximately $2.5\times8$-$m^2$ sections of each plot were assessed, and ears were harvested from all plants by hand. The grain yield was adjusted to a moisture content of 15.5%.

## A regional N management approach

A guideline for regional N rate was calculated for each subregion through several steps. First, yield data were collected from a large number of N response trials ($n = 1{,}726$). Grain yield responses to N fertilizer curves were fit using a quadratic model with PROC NLIN (SAS Institute Inc., Cary, NC, USA) to generate yield function equations (the yield significantly ($P<0.05$) responded to N) [23,24]. Next, from the response curve equation at each experimental site, the yield increase (above the yield in the N0 treatment), gross Chinese yuan return at that yield increase (maize grain price times yield), N fertilizer cost (N fertilizer price times N fertilizer rate), and net return to N ratio (gross yuan return minus N fertilizer cost) were calculated for each 1 kg N fertilizer rate increment from 0 to 270 kg N $ha^{-1}$. Finally, for each incremental N rate, the net return was averaged across all trials in the subregional data set to generate an estimated ratio of the maximum return to N rate, and the corresponding yield across all trials at an N fertilizer:maize grain price ratio [14,25]. In recent years, the fertilizer:maize grain price ratio has remained relatively stable, and a value of 2.05 was used in this study.

## Field-specific N management

In total, grain yield responses to N fertilizer curves were fit for 1,726 on-farm sites, using a quadratic model with PROC NLIN (SAS Institute Inc.) to generate yield function equations (the yield significantly ($P<0.05$) responded to N) [23,24]. The minimum N rate for the maximum net return was calculated from the selected model based on an N:maize price ratio of 2.05.

## Nitrogen use efficiency and N balance

Nitrogen use efficiency for each treatment using the partial factor productivity ($PFP_N$) indices.

$$PFP_N = \frac{Y_N}{F_N} \tag{1}$$

Where $Y_N$ = Crop yield with N applied;

$F_N$ = Amount of N applied.

Soil surface N balance was calculated as described in the Organization for Economic Co-operation and Development (OECD) [26].

$$\text{N balance} = \text{N input} - \text{N uptake} \tag{2}$$

where N input is N applied as chemical fertilizer, and N uptake is N in the harvested yield.

$$\text{N uptake} = \text{Aboveground N uptake} \times \text{Yield} \tag{3}$$

The maize aboveground N uptake requirement per million grams (Mg) grain yield in China was determined previously; spring maize grain yield was <7.5 Mg $ha^{-1}$, 7.5–9.0 Mg $ha^{-1}$, 9.0–10.5 Mg $ha^{-1}$, and 10.5–12.0 Mg $ha^{-1}$, and N uptake requirements per Mg grain yield were 19.8, 18.1, 17.4 and 17.1 kg, respectively [27]. Summer maize N uptake requirements per Mg grain yield were 20 kg [28].

## Estimation of GHG emissions and emission intensity

Total GHG emissions during the entire life cycle of maize production, including $CO_2$, $CH_4$, and $N_2O$, consisted of three components: (1) emissions during N fertilizer application, production and transportation, (2) emissions during P and K fertilizer production and transportation, and (3) emissions from pesticide and herbicide production (delivered to the gate) and diesel fuel consumption during sowing, harvesting, and tillaging operations [29].

$$\begin{aligned} GHG = {} & (GHGm + GHGt) \times \text{N rate} + \text{total } N_2O \times 44/28 \\ & \times 298 + GHGothers \end{aligned} \tag{4}$$

where GHG (kg $CO_2$ eq $ha^{-1}$) is the total GHG emission, and GHGm is the GHG emission originating from fossil fuel mining as the industry's energy source to N product manufacturing, and was 8.21 kg $CO_2$ eq kg $N^{-1}$ (Table S3) [30]. GHGt is the N fertilizer transportation emission factor, and was 0.09 kg $CO_2$ eq kg $N^{-1}$ (Table S3) [30]. N rate is the N fertilizer application rate (kg N $ha^{-1}$). $GHG_{others}$ represents GHG emission of P and K fertilizer

**Table 2.** The number of on-farm experiments, maize yield without N, medium N rate, grain yield at the medium N rate and N rate, grain yield, GHG emission intensity of N fertilizer use, N fertilizer production and other sources for regional N management approach and field-specific N management.

| Subregion | n[a] | Yield without N (Mg ha$^{-1}$) | Medium N rate (kg ha$^{-1}$) | Yield for medium N rate (Mg ha$^{-1}$) | Regional N management approach | | | | | | Field-specific N management | | | | | |
|---|---|---|---|---|---|---|---|---|---|---|---|---|---|---|---|---|
| | | | | | N rate (kg ha$^{-1}$) | Grain yield (Mg ha$^{-1}$) | GHG emission intensity (kg $CO_2$ eq Mg$^{-1}$ grain) | | | | N rate (kg ha$^{-1}$) | Grain yield (Mg ha$^{-1}$) | GHG emission intensity (kg $CO_2$ eq Mg$^{-1}$ grain) | | | |
| | | | | | | | N fertilizer use | N fertilizer production | Other sources | Total | | | N fertilizer use | N fertilizer production | Other sources | Total |
| NE1 | 132 | 6.40±1.01[b] | 153±6 | 8.98±1.09 | 150 | 8.85 | 91 | 141 | 49 | 280 | 158±26 | 8.87±1.11 | 91±19 | 149±32 | 49±7 | 289±55 |
| NE2 | 62 | 6.82±1.23 | 147±21 | 9.05±1.51 | 150 | 9.18 | 87 | 136 | 49 | 272 | 155±25 | 9.13±1.48 | 87±14 | 143±23 | 51±8 | 281±42 |
| NE3 | 126 | 6.50±1.26 | 162±15 | 9.48±1.38 | 164 | 9.01 | 96 | 151 | 50 | 298 | 165±20 | 9.10±1.25 | 92±18 | 153±29 | 50±8 | 295±53 |
| NE4 | 77 | 6.92±1.66 | 204±28 | 8.93±1.59 | 188 | 8.76 | 113 | 178 | 55 | 346 | 191±49 | 8.84±1.48 | 117±37 | 183±53 | 57±9 | 356±94 |
| NCP1 | 348 | 6.58±1.13 | 194±22 | 8.23±1.13 | 178 | 8.13 | 115 | 182 | 58 | 355 | 179±27 | 8.14±1.19 | 113±23 | 185±36 | 59±9 | 357±64 |
| NCP2 | 59 | 6.91±1.13 | 213±30 | 8.67±0.95 | 177 | 8.37 | 111 | 176 | 55 | 342 | 185±33 | 7.59±1.12 | 129±30 | 208±45 | 62±9 | 399±80 |
| NW1 | 100 | 6.30±1.06 | 190±34 | 8.35±1.03 | 181 | 8.13 | 117 | 185 | 59 | 360 | 180±44 | 8.13±1.12 | 115±29 | 184±45 | 59±9 | 357±77 |
| NW2 | 309 | 8.12±1.83 | 190±20 | 10.53±1.71 | 176 | 10.38 | 89 | 141 | 47 | 277 | 182±34 | 10.48±1.82 | 91±25 | 148±37 | 48±9 | 288±68 |
| NW3 | 7 | 7.23±1.74 | 221±9 | 10.33±1.59 | 219 | 9.83 | 118 | 185 | 46 | 349 | 215±16 | 9.85±1.56 | 116±20 | 185±30 | 46±7 | 347±57 |
| SW1 | 78 | 5.70±1.30 | 217±22 | 7.63±1.20 | 174 | 7.46 | 123 | 194 | 63 | 379 | 191±39 | 7.56±1.19 | 134±34 | 214±53 | 63±11 | 412±93 |
| SW2 | 368 | 5.59±1.13 | 195±22 | 7.72±1.18 | 183 | 7.71 | 125 | 197 | 66 | 387 | 184±38 | 7.77±1.26 | 125±33 | 202±52 | 67±12 | 394±92 |
| SW3 | 60 | 6.00±1.09 | 207±25 | 8.29±1.33 | 186 | 8.10 | 121 | 191 | 65 | 376 | 191±37 | 8.38±1.33 | 120±28 | 192±44 | 65±11 | 376±78 |
| National[c] | - | 6.60 | 187 | 8.69 | 174 | 8.56 | 108 | 171 | 56 | 334 | 178 | 8.55 | 109 | 185 | 57 | 343 |

[a]n: number of observations.
[b]Mean ± SD.
[c]National values are computed from the regional values weighted by area. The regional weights are as follows:
NE1, 4.5%; NE2, 14.9%; NE3, 4.7%; NE4, 6.4%; NCP1, 25.6%; NCP2, 6.0%; NW1, 10.4%; NW2, 7.3%; NW3, 2.6%; SW1, 3.5%; SW2, 7.9%; SW3, 6.2%.

production and transportation, pesticide and herbicide production and transportation, and diesel fuel consumption (Table S3).

Total $N_2O$ emission included direct and indirect $N_2O$ emissions. Indirect $N_2O$ emissions were estimated with a method used by the International Panel on Climate Change [31], where 1% and 0.75% of ammonia ($NH_3$) volatilization and nitrate ($NO_3^-$) leaching, respectively, is lost as $N_2O$. $N_2O$ emission is calculated based on empirical models. Based on previous reports, the final data set consisted of 10 (30 observations) and 22 (117 observations) studies on direct $N_2O$ emissions for spring maize and summer maize, respectively. Detailed information is provided in Table S4 and Figure S1.

$$\text{Direct } N_2O \text{ emission for spring maize} = 0.576\exp(0.0049 \times \text{N rate}) \tag{5}$$

$$\text{Direct } N_2O \text{ emission for summer maize} = 0.593\exp(0.0045 \times \text{N rate}) \tag{6}$$

$NH_3$ volatilization and N leaching employs the following equation (Cui *et al* 2013, Global Change Biology, main text, Fig. 2) [6].

$$NH_3 \text{ volatilization} = 0.24 \times \text{N rate} + 1.30 \tag{7}$$

$$\text{N leaching} = 4.46\exp(0.0094 \times \text{N rate}) \tag{8}$$

The system boundaries were set as the periods of the life cycle from the production inputs (such as fertilizers, pesticides, and herbicides), delivery of the inputs to the farm gates, and farming operations. We calculated total GHG emissions expressed as kg $CO_2$ eq $ha^{-1}$ and the GHG emission intensity expressed as kg $CO_2$ eq $Mg^{-1}$ grain. The change in soil organic carbon content was also not included in our analysis, because it was difficult to detect the small magnitude of the changes that occurred over a short time [32]. The soil $CO_2$ flux as a contributor to global warming potential (GWP) was also not included in this study because the net flux was estimated to contribute less than 1% to the GWP of agriculture on a global scale [33].

To calculate total GHG emissions and emission intensity, the N rate and corresponding yield of each farm were used for farmers' N practices. The regional N rate and corresponding yield of each subregion were used for the regional N management approach, and the optimal N rate and corresponding yield of each field were used for field-specific N management.

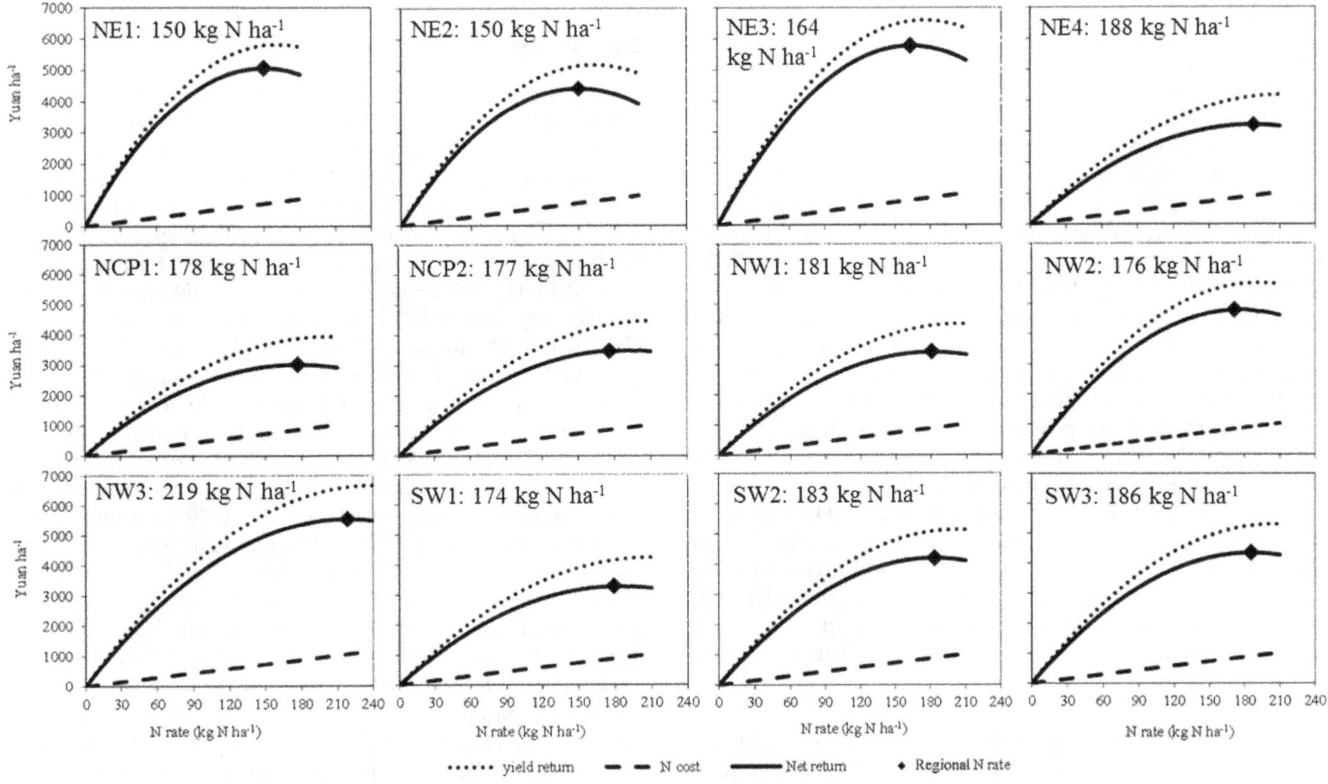


**Figure 2. Maize grain yield and fertilizer economic components of calculated net return across N rates using the regional N management approach indicated at the 2.05 price ratio (N price 4.87 yuan $kg^{-1}$ and maize price 2.37 yuan $ha^{-1}$) in the 12 agroecological subregions.** In total, 1,726 N responses trials were used to estimate the regional N rate. The net return is the increase in yield times the grain price at a particular N rate, minus the cost of that amount of N fertilizer. The maximum return is the N rate at which the net return is greatest.

## Results

### Farmers' Practice

Across all 5,406 farms, maize grain yield averaged 7.56 Mg $ha^{-1}$, the corresponding N application rate averaged 220 kg $ha^{-1}$, and the N balance averaged was 69 kg N $ha^{-1}$ (Table S5). Calculated GHG emission intensity averaged 482 kg $CO_2$ eq $Mg^{-1}$ grain (Table S5), including the contributions of 155, 242, and 85 kg $CO_2$ eq $Mg^{-1}$ grain from N fertilizer use, N fertilizer production, and other sources, respectively (data not shown).

Large variations were observed in grain yield and N fertilizer application rates across the four main agroecological regions. The N application rates followed the order SW (250 kg N $ha^{-1}$) ≈ NW (238 kg N $ha^{-1}$) > NCP (208 kg N $ha^{-1}$) ≈ NE (195 kg N $ha^{-1}$). In contrast, the maize grain yields were highest in NE (8.91 Mg $ha^{-1}$) followed by NW (7.58 Mg $ha^{-1}$), NCP (7.42 Mg $ha^{-1}$) and SW (5.45 Mg $ha^{-1}$). The GHG emission intensity averaged 347, 436, 487, and 710 kg $CO_2$ eq $Mg^{-1}$ grain for NE, NCP, NW, and SW, respectively (Table 1).

### Regional N management approach

Across all 1,726 on-farm experiments, the average grain yield under the N0 treatment, weighted by maize area in each subregion, was 6.60 Mg $ha^{-1}$ and ranged from 5.59 Mg $ha^{-1}$ (SW2) to 8.12 Mg $ha^{-1}$ (NW2) (Table 2). The average medium N rate (MN) recommended by local extension employees, weighted by maize area in each subregion, was 187 kg N $ha^{-1}$ and ranged from 147 kg N $ha^{-1}$ (NE2) to 221 kg N $ha^{-1}$ (NW3). The corresponding grain yield under MN treatment averaged 8.69 Mg $ha^{-1}$ and ranged from 7.63 Mg $ha^{-1}$ (SW1) to 10.53 Mg $ha^{-1}$ (NW2) (Table 2).

Considering all on-farm experiments, the calculated regional N rate based on the cost response to N application rate for the subregions, weighted by maize area in each subregion, averaged 174 kg N $ha^{-1}$ and ranged from 150 kg N $ha^{-1}$ (NE1 & NE2) to 219 kg N $ha^{-1}$ (NW3) (Table 2, Fig 2). The corresponding grain yield averaged 8.56 Mg $ha^{-1}$ and ranged from 7.46 Mg $ha^{-1}$ (SW1) to 10.38 Mg $ha^{-1}$ (NW2) (Table 2). Calculated GHG emission intensity, weighted by maize area in each subregion, averaged 334 kg $CO_2$ eq $Mg^{-1}$ grain and ranged from 272 kg $CO_2$ eq $Mg^{-1}$ grain (NE2) to 387 kg $CO_2$ eq $Mg^{-1}$ grain (SW2).

Based on the maize grain yield response to N application rates in all 1,726 on-farm experiments, the calculated field-specific N rate, weighted by maize area in each subregion, averaged 178 kg N $ha^{-1}$ (Table 2) and ranged from 53 kg N $ha^{-1}$ to 271 kg N $ha^{-1}$ (Table S2), with a coefficient of variation (CV) of 18% (data not shown). The corresponding grain yield averaged 8.63 Mg $ha^{-1}$ (Table 2) and ranged from 4.29 Mg $ha^{-1}$ to 14.91 Mg $ha^{-1}$ (Table S2), with a CV of 19% (data not shown). The calculated GHG emission intensity averaged 343 kg $CO_2$ eq $Mg^{-1}$ grain (Table 2). The similar N rate, grain yield and GHG emission intensity between the regional N management approach and field-specific N management supported the notion that the regional N rate was close to an economic and environmentally optimal N application (Table 2).

### Opportunities to reduce the GHG emission intensity

Compared to farmer's practices, the regional N management approach proposed reducing N fertilizer by 20.9% (220 vs. 174 kg N $ha^{-1}$). The grain yield would increase by 13.2% (7.56 vs. 8.56 Mg $ha^{-1}$). The GHG emission intensity would decrease by 30.7%, from 482 to 334 kg $CO_2$ eq $Mg^{-1}$ grain. The overuse and high variability of N use by farmers has resulted in a high variability in GHG emission intensity, ranging from 364 to 1,399 kg $CO_2$ eq $Mg^{-1}$ grain (Table S5) with a CV of 43% (data not shown).

Of the 12 agroecological subregions, NE2, NE3, NW1, NW2, SW1, SW2, and SW3 showed the highest potential for N-reduction (>20%), ranging from 21.0% to 31.5% and accounting for 55% of the total maize-sown area. Reduced N rates in other subregions ranged from 3.8% to 18.4% and accounted for 45% of the total maize-sown area. The subregions with a high yield increase potential (>15%; Fig. 3) were NCP2, NW1, NW2, NW3, SW1, SW2, and SW3, with increases ranging from 17.2% to 44.9% and accounting for 44% of the total maize-sown area. Grain yield in other regions ranged from 0.5% to 5.9%, accounting for 56% of the total maize-sown area. Subregions with a high potential to decrease GHG emission intensity (>20%) included NE2, NE3, NCP2, NW1, NW2, NW3, SW1, SW2, and SW3, ranging from 21.8% to 50.0% and accounting for 64% of the total maize-sown area. Reduced GHG emission intensity in other regions ranged from 6.0% to 15.1%, accounting for 36% of the total maize-sown area.

This regional N management approach, if widely adopted in China, regional N fertilizer consumption would be reduced by 1.4 MT (−20.3%), and 91% of this reduction would occur in the NE2, NE3, NCP1, NW1, NW2, SW1, SW2, and SW3 subregions (Table 3). At the same time, Chinese maize production could be increased by 31.9 MT (13.1%), from 244.1 MT to 276.0 MT, when undertaking this regional N management approach (Table 3). Total GHG emissions would be reduced by 18.6 MT eq $CO_2$ $year^{-1}$ (−16.9%) (from 110.2 to 91.5 MT eq $CO_2$ $year^{-1}$) (Table 3), with 91% of this reduction occurring in the NE2, NE3, NCP1, NW1, NW2, SW1, SW2, and SW3 subregions.

## Discussion

The current intensive maize system used in farmers' practices in China results in a median yield, high N application, and GHG emission intensity of 7.56 Mg $ha^{-1}$, 220 kg N $ha^{-1}$, and 482 kg $CO_2$ eq $Mg^{-1}$ grain, respectively. These yields and N application rates are higher than the reported global averages (4.81 Mg $ha^{-1}$ and 104.9 kg N $ha^{-1}$, the N rate calculated based on maize N fertilizer consumption and maize area harvested) for these crops in 2006 [8,15,34] and are similar to the previously reported Chinese averages for maize [35,36]. In comparison, grain yield in central Nebraska, USA, averaged 13.2 Mg $ha^{-1}$ with only 183 kg N $ha^{-1}$. GHG emission intensity in this region was only 231 kg $CO_2$ eq $Mg^{-1}$ grain, which was 48% lower than the average for China [37] and 109% lower than the 482 kg $CO_2$ eq $Mg^{-1}$ grain for individual farmer's practices in China. The median yield and large GHG emission intensity for Chinese maize systems were attributable to the large variation in N application rates among fields. Considering 5,406 farms, N application rates ranged from 46 (only 56% of crop N uptake) to 615 kg N $ha^{-1}$ (414% of crop N uptake). Similar results were reported by Wang *et al* (2007), showing that one-third of farmers apply too little N, while another one-third of farmers apply too much ($n = 10{,}000$) [38].

In small-scale farming, a lack of basic knowledge and information on crop responses to N fertilizer often results in the over- and underapplication of N fertilizer [39,40]. We developed and assessed regional N management approach using large pools of response trial data that have been grouped according to criteria that indicate differing N responses for regions with similar management, climates, and soil. Our guide provides a N application rate that can be used to reduce the potential for N-deficiency or N-surplus, lowers the likelihood of reduced yields and profits, and lessens GHG emissions intensity (particularly $N_2O$

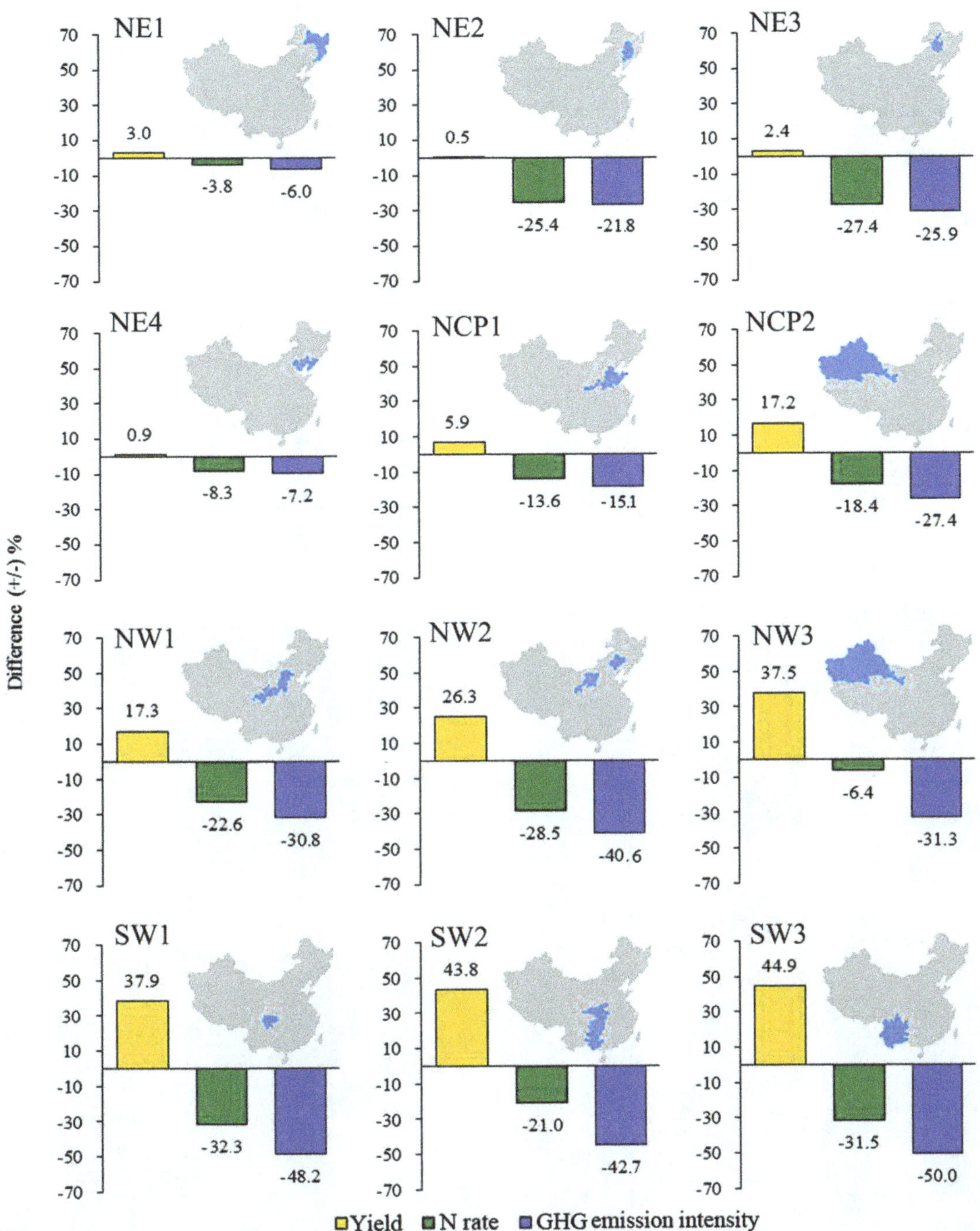


**Figure 3. Regional differences (±%) in N application rates, grain yield, and GHG emission intensity between the regional N management approach and farmers' practice in the 12 agroecological subregions.** Regional difference (±%) = (regional approach minus farmers' practice)/farmers' practice ×100.

emissions associated with N fertilization). Using a regional N management approach, potential for crop productivity increases and the mitigation of GHG emission intensity are likely to be achieved through a combination of increased N application in regions with a low N input and improved $PFP_N$ in regions where N fertilizer application is already high. Meanwhile, crop N uptake and N use efficiency can improve the ratio split application, with one-third for base dressing and two-thirds for top dressing [36]. Currently, typical farmers' practices apply 50% of the total N fertilizer before planting or at the early growth stage [36,41]. Some recent practices have indicated that the amount of basal application should be added to the ratio of the top dressing to improve N use efficiency and increase grain yield [36].

The gains in yield and reduced GHG emissions achieved using regional N management approach are significant. Moreover, we believe these benefits can be further improved by applying other best-management strategies to fertilizer (e.g., slow-release N fertilizer, N transformation inhibitors, and fertigation) [42] and related practices that enhance the crop recovery of applied N (e.g., rotation with N fixing crops, precision agriculture management

**Table 3.** Maize production, N fertilizer consumption and total GHG emission between the regional N rate and farmers' practice in 12 agro-ecological subregions.

| Subregion | Area (million ha) | N fertilizer consumption (MT) | | | Maize production (MT) | | | Total GHG emission (MT eq $CO_2$ $yr^{-1}$) | | |
|---|---|---|---|---|---|---|---|---|---|---|
| | | Farmers' practice | Regional N rate | Difference [a] | Farmers' practice | Regional N rate | Difference [a] | Farmers' practice | Regional N rate | Difference [a] |
| NE1 | 1.45 | 0.23 | 0.22 | −0.01 | 12.5 | 12.8 | 0.4 | 3.7 | 3.6 | −0.1 |
| NE2 | 4.80 | 0.96 | 0.72 | −0.24 | 43.8 | 44.1 | 0.2 | 15.2 | 12.0 | −3.3 |
| NE3 | 1.50 | 0.34 | 0.25 | −0.09 | 13.2 | 13.5 | 0.3 | 5.3 | 4.0 | −1.3 |
| NE4 | 2.06 | 0.42 | 0.39 | −0.04 | 17.9 | 18.0 | 0.2 | 6.7 | 6.2 | −0.4 |
| NCP1 | 8.24 | 1.70 | 1.47 | −0.23 | 63.3 | 67.0 | 3.7 | 26.4 | 23.8 | −2.7 |
| NCP2 | 1.94 | 0.42 | 0.34 | −0.08 | 13.9 | 16.2 | 2.4 | 6.5 | 5.6 | −1.0 |
| NW1 | 3.35 | 0.78 | 0.61 | −0.18 | 23.2 | 27.2 | 4.0 | 12.1 | 9.8 | −2.3 |
| NW2 | 2.36 | 0.58 | 0.42 | −0.17 | 19.4 | 24.5 | 5.1 | 9.0 | 6.8 | −2.2 |
| NW3 | 0.85 | 0.20 | 0.19 | −0.01 | 6.1 | 8.4 | 2.3 | 3.1 | 2.9 | −0.2 |
| SW1 | 1.14 | 0.29 | 0.20 | −0.09 | 6.2 | 8.5 | 2.3 | 4.5 | 3.2 | −1.3 |
| SW2 | 2.54 | 0.59 | 0.46 | −0.12 | 13.6 | 19.6 | 6.0 | 9.2 | 7.6 | −1.6 |
| SW3 | 1.99 | 0.54 | 0.37 | −0.17 | 11.1 | 16.1 | 5.0 | 8.4 | 6.1 | −2.3 |
| National [b] | 32.23 | 7.06 | 5.62 | −1.43 | 244.1 | 276.0 | 31.9 | 110.2 | 91.5 | −18.6 |

[a]Different mean the different of maize production, N fertilizer consumption, and total GHG emission between regional N rate and farmer's practice.
[b]National values are computed from the regional values weighted by area. The regional weights are as follows:
NE1, 4.5%; NE2, 14.9%; NE3, 4.7%; NE4, 6.4%; NCP1, 25.6%; NCP2, 6.0%; NW1, 10.4%; NW2, 7.3%; NW3, 2.6%; SW1, 3.5%; SW2, 7.9%; SW3, 6.2.

techniques) [42]. While this approach for N fertilizer management should be extended to farmers throughout the entire Chinese cereal production area, it is also relevant to other high-yield cropping systems outside of China. The economic approach to N rate recommendations based on multiple N rate trials has been applied for two to three decades in the U.S. Midwest, and has been more recently "formalized" with the Iowa State MRTN approach for seven Midwestern states [43].

This regional N management approach, if widely adopted in China, could reduce fertilizer N consumption by 20.3%, increase Chinese maize production by 13.1%, and reduce total GHG emissions by 16.9%. Moreover, the recommendations provide reasonable N rates and high net return, and can be easily adopted in rural areas of China where no available soil and/or plant N monitoring facilities exist [44]. The regional N rate can also be used as a reference point for agricultural extension employees without any soil and/or plant N monitoring. In practice, some factors also affect these suggested regional N rates, such as timing of crop rotation, tillage system, and soil productivity [14]. For example, the recommended N rate for soybean following maize rotations is lower than maize following maize rotations [14]. No-till management can delay or reduce residue breakdown, or mineralization, thereby reducing the N supplied from crop residue [14]. Soils where productivity is limited frequently require higher rates of fertilizer N to reach optimum yield. Conversely, lower rates of fertilizer N may be needed to reach optimum yield on highly productive soils [14].

Although this regional N management approach can easily be adopted in rural areas, delivering this technology to millions of farmers is challenging due to the lack of effective advisory systems and knowledgeable farmers. For example, educated young male farmers tend to leave the farming sector for more profitable jobs, leaving farmwork to the older and less educated individuals, especially in low income or remote areas [45]. In addition, adding more N fertilizer based on the regional N rate is difficult for farmers with low incomes or in remote areas. The Chinese central government has been aware of this problem and has attempted to provide agricultural technologies to these areas. For example, China has launched national programs for soil testing and fertilizer recommendations since 2005. In 2009, 2,500 counties in China were involved in the programs, receiving a total of 1.5 billion yuan from the Chinese central government [40].

Although the on-farm trials were conducted by local farmers in the same counties as the farmers' surveys (including experimental counties), the management and environment is not always the same for on-farm trials and farmers' surveys. While gains in grain yield and GHG were achieved by farmers using the trials, we believe that the majority of these gains can be realized in practice in many counties if improved agronomic and N management techniques are adopted. The management and environment differed among four maize regions; thus, N losses may also differ. For example, the annual direct $N_2O$ emission accounted for 0.92% of the applied N with an uncertainty of 29%. The highest $N_2O$ fluxes occurred in East China as compared with the lowest fluxes in West China [46]. In this study, we use the different exponential relationships of the N application rate and $N_2O$ fluxes for spring maize and summer maize, respectively. However, developing N loss models at the regional or subregional scale is difficult due to insufficient field measurement data in China. Long-term field observations covering all subregions are required to accurately assess farming potential and mitigate GHG emissions.

## Supporting Information

**Figure S1 Relationships between the N application rate and direct $N_2O$ emissions for spring maize (A) and summer maize (B) production in China based on a meta-analysis.** The direct $N_2O$ emission data was taken from Table S4.

**Table S1 The criteria and values for the sub-regional divisions.**

**Table S2 The site, year, soil type, irrigation, crop rotations, soil organic matter (SOM) content, alkaline hydrolyzable N (AN), Olsen-P (AP), $NH_4OAc$-K (AK), pH, medium N rate (MN), recommended $P_2O_5$ rate (RP), recommended $K_2O$ rate (RK), grain yield without N fertilizer, yield at 50% MN, yield at 100% MN, yield at 150% MN, economic optimal N rate (EONR), yield at EONR, and GHG emissions intensity at EONR for all 1,726 on-farm experiments.**

**Table S3 GHG emission factors of agricultural inputs.**

**Table S4 The site, year, annual mean precipitation, temperature, soil organic matter (SOM), total N content, pH, N rate, grain yield, and direct $N_2O$ emissions at different experimental sites.**

**Table S5** Descriptive statistics of the surveyed farms N fertilizer application rate, maize grain yield, $PFP_N$, N balance and GHG emission intensity for 5,406 farmed fields between 2007 and 2009 in China.

**Text S1 Detailed information for each of these regions.**

## Author Contributions

Conceived and designed the experiments: FsZ XpC. Performed the experiments: FsZ XpC. Analyzed the data: LW ZlC WfZ. Contributed reagents/materials/analysis tools: LW ZlC. Wrote the paper: LW. Designed the NH3 volatilization and N leaching models used in analysis: ZlC.

## References

1. Tilman D, Fargione J, Wolff B, D'Antonio C, Dobson A, et al. (2001) Forecasting agriculturally driven global environmental change. Science. 292: 281–284.
2. Tilman D, Cassman KG, Matson PA, Naylor R, Polasky S (2002) Agricultural sustainability and intensive production practices. Nature. 418: 671–677.
3. Conley D J, Paerl H W, Howarth RW, Boesch DF, Seitzinger SP, et al. (2009) Controlling eutrophication: nitrogen and phosphorus. Science. 323: 1014–1015.
4. Tilman D, Balzer C, Hill J, Befort BL (2011) Global food demand and the sustainable intensification of agriculture. Proc. Natl. Acad. Sci. USA.108: 20260–20264.
5. Zhang F, Cui Z, Fan M, Zhang W, Chen X, Jiang R (2011) Integrated soil-crop system management: reducing environmental risk while increasing crop productivity and improving nutrient use efficiency in China J Environ. Qual. 40: 1051–1057.
6. Cui Z, Yue S, Wang G, Meng Q, Wu L, Yang Z, et al. (2013) Closing the yield gap could reduce projected greenhouse gas emissions: a case study of maize production in China. Global Change Biol. 19: 2467–2477.
7. Zhang F, Cui Z, Chen Z, Ju X, Shen J, et al. (2012) Chapter one-Integrated nutrient management for food security and environmental quality in China. In Adv. Agron. ed Donald L S (Academic Press) 1–40 p.

8. IFA IFA Statistics (Paris: International Fertilizer Industry Association). Available at: www.fertilizer.org/ifa/HomePage/STATISTICS. Accessed 2013 Sept 6.
9. Soper R, Huang P (1963) The effect of nitrate nitrogen in the soil profile on the response of barley to fertilizer nitrogen Can. J. Soil Sci. 43: 350–358.
10. Wehrmann J, Scharpf HC, Kuhlmann H (1988) The Nmin method – an aid to improve nitrogen efficiency in plant production, In Nitrogen Efficiency in Agricultural Soils ed Jenkinson D S, Smith K A (Netherlands: Elsevier Applied Science) 38–45 p.
11. Gebbers R, Adamchuk VI (2010) Precision agriculture and food security. Science. 327: 828–831.
12. Chen X P, Cu Z L, Vitousek P M, Cassman K G, Matson P A, et al. (2011) Integrated soil-crop system management for food security. Proc. Natl. Acad. Sci. USA 108 6399–6404.
13. Dobermann A, Witt C, Abdulrachman S, Gines H, Nagarajan R, et al. (2003) Estimating indigenous nutrient supplies for site-specific nutrient management in irrigated rice. Agron. J. 95: 924–35
14. Sawyer J, Nafziger E, Randall G, Bundy L, Rehm G, Joern B (2006) Concepts and rationale for regional nitrogen rate guidelines for corn. Iowa: Iowa State University, University Extension. 15–24 p.
15. FAO FAOSTAT–Agriculture Database. Available: http://faostat.fao.org/site/339/default.aspx. Accessed 2013 Sept 6.
16. National Bureau of Statistics of China. China Statistical Yearbook. Available: http://www.stats.gov.cn/tjsj/ndsj/. Accessed 2013 Sept 6.
17. Etimi N A, Solomon VA (2010) Determinants of rural poverty among broiler farmers in Uyo, Nigeria: implications for rural household food security. J. Agric. Soc. Sci. 6: 24–28.
18. Walkley A (1947) A critical examination of a rapid method for determining organic carbon in soils-effect of variations in digestion conditions and of inorganic soil constituents. Soil Sci. 63: 251–264.
19. Khan S, Mulvaney R, Hoeft R (2001) A simple soil test for detecting sites that are nonresponsive to nitrogen fertilization. Soil Sci. Soc. Am. J. 65: 1751–1760.
20. Olsen SR (1954) Estimation of available phosphorus in soils by extraction with sodium bicarbonate (Washington, DC: US Department of Agriculture)
21. van Reeuwijk LP (1993) Procedures for soil analysis (International Soil Reference and Information Centre).
22. Richards LA (ed) (1954) Diagnosis and improvement of saline and alkali soils (Washington, DC: US USDA. U.S. Gov. Print. Office).
23. Wallach D, Loisel P (1949) Effect of parameter estimation on fertilizer optimization Appl. Stat. 641–651.
24. Magee L (1990) $R^2$ measures based on Wald and likelihood ratio joint significance tests. American Statistician. 44: 250–253.
25. Hoben J, Gehl R, Millar N, Grace P, Robertson G (2011) Nonlinear nitrous oxide ($N_2O$) response to nitrogen fertilizer in on-farm corn crops of the US Midwest. Global Change Biol. 17: 1140–1152.
26. OECE. Environmental indicators for agriculture: Methods and results (Paris: Organisation for Economic Co-operation and Development). Available: www.oecd.org/greengrowth/sustainable-agriculture/1916629.pdf. Accessed 2013 Sept 6.
27. Hou P, Gao Q, Xie R, Li S, Meng Q, et al. (2012) Grain yields in relation to N requirement: Optimizing nitrogen management for spring maize grown in China. Field Crop Res. 129: 1–6.
28. Meng Q F (2012) Strategies for achieving high yield and high nutrient use efficiency simultaneously for maize (Zea mays L.) and wheat (Triticum aestivum L.), Ph.D. Diss. China Agriculture University.
29. Forster P, Ramaswamy V, Artaxo P, Berntsen T, Betts R, et al. (2007) Changes in atmospheric constituents and in radiative forcing In Climate Change. The Physical Science Basis Contribution of Working Group I to the Fourth Assessment Report of the Intergovernmental Panel on Climate Change. ed Solomon S, Qin D, Manning M, Chen Z, Marquis M, Averyt K B, Tignor M, Miller H L 2007(Cambridge: Cambridge University Press).
30. Zhang WF, Dou ZX, He P, Ju XT, Powlson D, et al. (2013) New technologies reduce greenhouse gas emissions from nitrogenous fertilizer in China Proc. Natl. Acad. Sci. USDA 110: 8375–8380
31. Klein CD, et al. (2006) IPCC Guidelines for National Greenhouse Gas Inventories Chapter 11: $N_2O$ emissions from managed soils, and $CO_2$ emissions from lime and urea application avaluable at: www.ipcc-nggip.iges.or.jp/public/2006gl/pdf/4_Volume4/V4_11_Ch11_N2O&CO2.pdf
32. Conant RT, Ogle SM, Paul EA, Paustian K (2010) Measuring and monitoring soil organic carbon stocks in agricultural lands for climate mitigation Front. Ecol. Environ 9: 169–173.
33. IPCC 2007Climate Change 2007: Mitigation. Contribution of Working Group III to the Fourth Assessment Report of the Intergovernmental Panel on Climate Change. ed Smith P, et al(Cambridge: Cambridge University Press).
34. Heffer P (2009) Assessment of fertilizer use by crop at the global level 2006/07–2007/08. International Fertilizer Industry Association. (Paris, France).
35. Cui Z (2005) Optimization of the nitrogen fertilizer management for a winter wheat-summer maize rotation system in the North China Plain - from field to regional scale Ph.D. Diss. China Agriculture University. (Chinese with English abstract).
36. Cui Z, Chen X, Miao Y, Zhang F, Sun Q, et al. (2008) On-farm evaluation of the improved soil N-based nitrogen management for summer maize in North China. Plain Agron. J. 100: 517–525.
37. Grassini P, Cassman KG (2012) High-yield maize with large net energy yield and small global warming intensity Proc. Natl. Acad. Sci. USA 109: 1074–1079.
38. Wang JQ (2007) Analysis and evaluation of yield increase of fertilization and nutrient utilization efficiency for major cereal crops in China Ph.D. Diss. China Agriculture University.
39. Huang J, Hu R, Cao J, Rozelle S (2008) Training programs and in-the-field guidance to reduce China's overuse of fertilizer without hurting profitability J. Soil Water Conserv. 63: 165A–167A.
40. Cui Z, Chen X, Zhang F (2010) Current nitrogen management status and measures to improve the intensive wheat–maize system in China AMBIO. 39: 376–384.
41. Chen XP (2003) Optimization of the N fertilizer management of a winter wheat/summer maize rotation system in the Northern China Plain Ph.D. diss. Univ.of Hohenheim.
42. Good AG, Beatty PH (2011) Fertilizing nature: a tragedy of excess in the commons. Plos Biol. 9: e1001124.
43. Iowa State University – Agronomy Extension. Corn Nitrogen Rate Calculator. Available: extension.agron.iastate.edu/soilfertility/nrate.aspx. Accessed 2013 Sept 6.
44. Zhu Z, Chen D (2002) Nitrogen fertilizer use in China – Contributions to food production, impacts on the environment and best management strategies Nutr. Cycl. Agroecos. 63: 117–127.
45. Barning R (2008) Economic evaluation of nitrogen application in the North China Plain Ph.D. diss. Univ. of Hohenheim.
46. Lu Y, Huang Y, Zou J, Zheng X (2006) An inventory of $N_2O$ emissions from agriculture in China using precipitation-rectified emission factor and background emission. Chemosphere. 65: 1915–1924.

# 12

# Emissions of NO and $NH_3$ from a Typical Vegetable-Land Soil after the Application of Chemical N Fertilizers in the Pearl River Delta

**Dejun Li***

Department of Microbiology and Plant Biology, University of Oklahoma, Norman, Oklahoma, United States of America

## Abstract

Cropland soil is an important source of atmospheric nitric oxide (NO) and ammonia ($NH_3$). Chinese croplands are characterized by intensive management, but limited information is available with regard to NO emissions from croplands in China and $NH_3$ emissions in south China. In this study, a mesocosm experiment was conducted to measure NO and $NH_3$ emissions from a typical vegetable-land soil in the Pearl River Delta following the applications of 150 kg N $ha^{-1}$ as urea, ammonium nitrate (AN) and ammonium bicarbonate (ABC), respectively. Over the sampling period after fertilization (72 days for NO and 39 days for $NH_3$), mean NO fluxes (± standard error of three replicates) in the control and urea, AN and ABC fertilized mesocosms were 10.9±0.9, 73.1±2.9, 63.9±1.8 and 66.0±4.0 ng N $m^{-2}$ $s^{-1}$, respectively; mean $NH_3$ fluxes were 8.9±0.2, 493.6±4.4, 144.8±0.1 and 684.7±8.4 ng N $m^{-2}$ $s^{-1}$, respectively. The fertilizer-induced NO emission factors for urea, AN and ABC were 2.6±0.1%, 2.2±0.1% and 2.3±0.2%, respectively. The fertilizer-induced $NH_3$ emission factors for the three fertilizers were 10.9±0.2%, 3.1±0.1% and 15.2±0.4%, respectively. From the perspective of air quality protection, it would be better to increase the proportion of AN application due to its lower emission factors for both NO and $NH_3$.

**Editor:** Caroline P. Slomp, Utrecht University, The Netherlands

**Funding:** This work was financially supported by an innovative project of Guangzhou Institute of Geochemistry (54O7342101). The funders had no role in study design, data collection and analysis, decision to publish, or preparation of the manuscript.

**Competing Interests:** The author has declared that no competing interests exist.

* E-mail: dejunl@gmail.com

## Introduction

Increases of atmospheric trace gases are of increasing concern due to their roles in detrimental health and environmental effects. Nitrogen oxides ($NO_x = NO + NO_2$) and ammonia ($NH_3$) are among these gases. $NO_x$ catalyze the photochemical formation of ground-level ozone [1], which is a potent greenhouse gas [2] and poses a threat to human health and vegetation on regional scales [3]. The photochemical end product, $HNO_3$, is a major component of acid rain [1]. $NH_3$ enhances aerosol formation [4], and hence influences regional air quality.

Soil is a major source for atmospheric $NO_x$. According to Davidson and Kingerlee (1997), the source strength of soils is just inferior to fossil fuel combustion globally, while cropland is the second largest contributor to atmospheric $NO_x$ among all types of land uses. Current estimates of soil NO emissions from cropland range from 2.4–5.4 Tg N $yr^{-1}$ [5,6,7]. The large uncertainty is largely caused by limited data availability, especially in tropical and subtropical regions. Furthermore, fertilized cropland (12.6 Tg N $yr^{-1}$) ranks second among all the sources of atmospheric $NH_3$ and is responsible for 23.5% of the global annual emission [8]. Of the emitted NO and $NH_3$ from cropland, the major part is induced by fertilizer application [9].

Chinese croplands have been intensively managed. With the cropland area only occupying 7% of the global total, the annual consumption of synthetic N fertilizer in China accounted for ca. 30% of the total global consumption in 2004 [10]. The annual fertilizer N consumption in China has been increasing by about 6.5%, which is higher than the mean rate of 4.8% for other Asian countries [11]. Urea ($CO(NH_2)_2$) and ammonium bicarbonate ($NH_4HCO_3$, ABC) are the two most widely used synthetic N fertilizers in China, with the latter only used in China [9,12].

Most of the previous studies on $NH_3$ volatilization from Chinese croplands were carried out in North China Plain and eastern China's Yangtze River Delta [12], but, to our knowledge, little information is available for south China. NO emissions from Chinese croplands have been reported by several studies [11,13,14,15,16,17,18]. However, no study has been conducted to investigate NO emission following ABC application, although ABC is a main N fertilizer in China (Figure 1), Due the lack of a NO emission factor for ABC, one study used the emission factor for ammonium sulphate as an estimate of that for ABC [19].

Vegetable land forms an important part of China's croplands. On a national scale, vegetable land accounts for ca. 7% of the total cropland, and the ratios are typically higher in developed regions [15]. For example, in south China's Guangdong province, the ratio is 24.2% ($1.16\times10^6$ ha) in 2005 [15]; in the Yangtze Delta of east China, vegetable land (ca. $6\times10^6$ ha) also occupies one fifth of the total cropland area [14]. Compared with other cultivated lands, vegetable lands receive more fertilizer-N per unit of area. For example, N application rates in individual growing season for vegetables range from 300 to 700 kg N $ha^{-1}$, compared to 150–300 kg N $ha^{-1}$ for non-vegetable crops [20]. These higher N fertilizer application rates to vegetable lands would inevitably

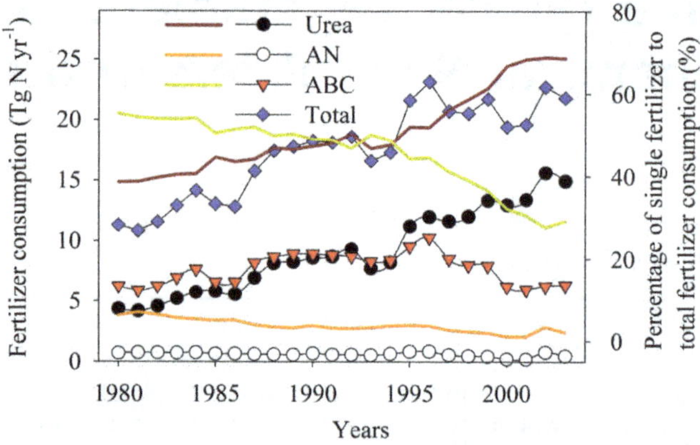


**Figure 1. Changes of fertilizer consumption (line and scatter) and percentage (bold line) of single fertilizer consumption to total consumption of urea, ammonium nitrate (AN) and ammonium bicarbonate (ABC) during 1980 to 2003 in China [10].**

induce larger amount of N loss to the environment, including leaching of nitrate to surface/ground water, and emission of nitrogen-containing gases to the atmosphere.

In the present study, we conducted a mesocosm experiment in which we measured emissions of NO and $NH_3$ from a subtropical vegetable-land soil following the application of urea, ABC, and $NH_4NO_3$ (AN), respectively. The main objective was to quantify the emission factors for the three synthetic fertilizers.

## Materials and Methods

### Ethics Statement

The vegetable field for arable soil sampling was privately owned. We collected soil under the permission of the land owner Mrs Cheng. All other necessary permits were obtained for the described field studies.

### Site Description

The experiment was conducted at a suburban site (23°10'N, 113°23'E) of Guangzhou, the capital of Guangdong province. This area enjoys a subtropical monsoon climate, with an annual mean rainfall of 1938.2 mm and annual mean air temperature of 22.5°C. Monthly mean air temperature is lowest in January (13.3°C) and highest in July (29.5°C) [21].

Flowering Chinese cabbage (*Brassica campestris* L. ssp. *Chinensis* var. *utilis* Tsen et Lee) is the most widely cultivated vegetable type in Guangdong Province [15]. Vegetables can grow under field conditions all year round in this region. Typically, a new batch of vegetables is planted approximately five days after a batch of vegetables is harvested, and thus no long fallow period exists. At the study site, a typical cultivation cycle for flowering Chinese Cabbage takes about 47 days and mainly involves five management events: sowing, harrowing and transplanting, first fertilization, second fertilization, harvesting and harrowing. The rate of fertilizer application is about 320 kg N ha$^{-1}$ yr$^{-1}$ with about 5 batches of vegetables grown in a year according to our previous field study [15].3

### Experimental Design

The arable soil (top layer of 20 cm) used in this study was collected from a vegetable field which has been cultivated for flowering Chinese cabbage for more than 10 years (Personal

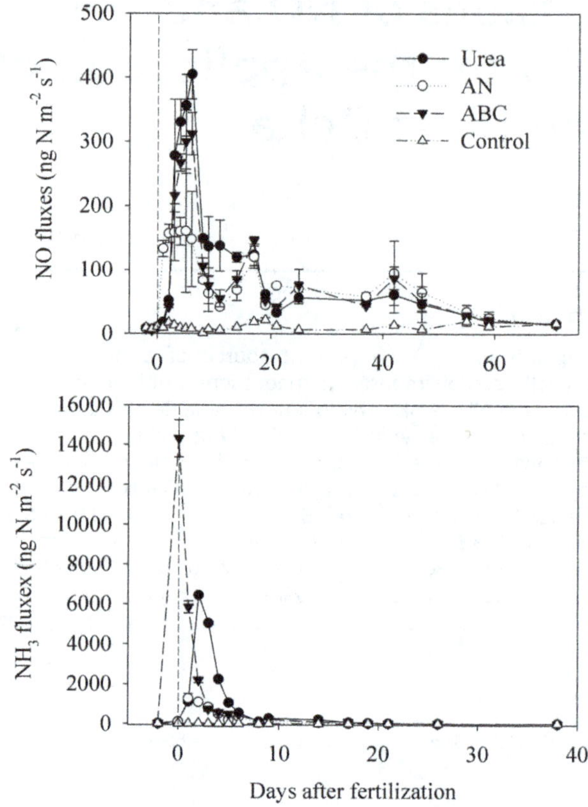


**Figure 2. NO and $NH_3$ fluxes during the sampling period.** Each value is the mean of three replicates, and error bars represent standard errors. Vertical dash line indicates the day when fertilizers were applied.

communication with the farmer). The soil type is lateritic red earth, which is typical in subtropical China. Soil properties (mean ± SE, n = 5) were analyzed after the soil was well mixed. Soil pH (extracted with KCl solution), contents of total C and total N were

**Table 1.** Fluxes of NO and $NH_3$, total emissions ($E_{total}$), net emissions ($E_{net}$), and emission factors (EF) for the three fertilizers.

| | Flux | $E_{total}$ | $E_{net}$ | EF |
|---|---|---|---|---|
| | ng N m$^{-2}$ s$^{-1}$ | mg N m$^{-2}$ | mg N m$^{-2}$ | % |
| NO | | | | |
| Urea | 73.1±2.9 a | 454.6±17.9 a | 386.7±18.7 a | 2.6±0.1 a |
| AN | 63.9±1.8 b | 397.5±10.9 b | 329.6±12.2 b | 2.2±0.1 b |
| ABC | 66.0±4.0 b | 410.3±25.1 b | 342.4±25.7 b | 2.3±0.2 b |
| Control | 10.9±0.9 c | 67.9±5.5 c | | |
| $NH_3$ | | | | |
| Urea | 493.6±4.4 b | 1663.1±27.3 b | 1633.2±27.3 b | 10.9±0.2 b |
| AN | 144.8±0.1 c | 488.0±0.5 c | 458.1±1.1 c | 3.1±0.1 c |
| ABC | 684.7±8.4 a | 2307.1±52.2 a | 2277.2±52.2 a | 15.2±0.4 a |
| Control | 8.9±0.2 d | 29.9±1.0 d | | |

Values are presented as mean ± standard error of three replicates. Different letters in a column denote that the mean values are significantly different at $P<0.05$ level.

**Table 2.** NO emissions from fertilized upland arable fields in China.

| Location | Land-use type | Fertilizer type | Duration | N rates | NO flux | Emission factors % |
|---|---|---|---|---|---|---|
| | | | | kg N ha$^{-1}$ | ng N m$^{-2}$ s$^{-1}$ | |
| 39°57′N; 116°18′E[a] | Corn | OM[j] | N/A | 88.5 | 0.14 | 0.04 |
| 39°57′N; 116°18′E[a] | Corn | Urea | N/A | 150 | 77.5 | 1.6 |
| 39°57′N; 116°18′E[a] | Corn | Urea | N/A | 300 | 76.4 | 0.75 |
| 39°57′N; 116°18′E[a] | Corn | OM+urea | N/A | 238.5 | 66.4 | 0.78 |
| 39°57′N; 116°18′E[a] | Corn | OM+urea | N/A | 238.5 | 150.8 | 1.3 |
| 34°56′N; 110°43′E[b] | Cotton | Urea+DP+PS | Jan-Dec | 66.3 | 2.4 | 0.24 |
| 32°35′N; 119°42′E[c] | Bare soil | Urea | Jan-Dec | 153 | 70.4 | 0.67 |
| 32°35′N; 119°42′E[c] | Vegetables* | | Jan-Dec | 118–548 | 2.5–142.7 | 0.05–1.24 |
| 31°16′N; 120°38′E[d] | Wheat | M+CF+urea | Nov-Jun | 191 | 21.3 | 1.75 |
| 31°16′N; 120°38′E[d] | Wheat | CF+urea | Nov-Jun | 191 | 28.5 | 2.5 |
| 31°16′N; 120°38′E[d] | Wheat | CF+urea | Nov-Jun | 191 | 22.4 | 1.87 |
| 30°50′N; 120°42′E[e] | Cabbage | CF | Mar -Jun | 45 | 11.5 | 0.6 |
| 30°50′N; 120°42′E[e] | Potato | CF+M | Mar -Jun | 45.6 | 34.2 | 3.6 |
| 30°50′N; 120°42′E[f] | Cabbage | CF | Mar -Jun | 135 | 20.9 | 1.05 |
| 30°50′N; 120°42′E[f] | Potato | CF+ M | Mar -Jun | 108 | 27.4 | 1.75 |
| 30°50′N; 120°42′E[f] | Soybean | CF+urea | Mar -Jun | 81 | 21.4 | 1.83 |
| 30°50′N; 120°42′E[g] | Cabbage | CF+M+urea | Aug-Dec | 271.2 | 33.8 | 1.2 |
| 30°50′N; 120°42′E[g] | Garlic | CF+M+urea | Aug-Dec | 267.3 | 360 | 11.56 |
| 30°50′N; 120°42′E[g] | Radish | CF+M+urea | Aug-Dec | 263.6 | 76 | 2.56 |
| 23°10′N; 113°23′E[h] | FCC | CF+M+urea | Sep-Oct | 63.5 | 47.5 | 2.4 |
| 23°10′N; 113°23′E[i] | Bare soil | Urea | Dec-Mar | 150 | 73.1 | 2.6 |
| 23°10′N; 113°23′E[i] | Bare soil | AN | Dec-Mar | 150 | 63.9 | 2.2 |
| 23°10′N; 113°23′E[i] | Bare soil | ABC | Dec-Mar | 150 | 66 | 2.3 |

[a]Walsh [37];
[b]Liu et al. [18];
[c]Mei et al., [16];
[d]Zheng et al. [11];
[e]Fang and Mu [14];
[f]Fang and Mu [38];
[g]Pang et al. [17];
[h]Li and Wang [15];
[i]This study.
[j]OM–organic matter; CF–compound fertilizer; M–Manure; FCC– Flowering Chinese cabbage; DP-diammonium phosphate; PS-potassium sulphate;
*Different vegetables were cultivated from 2004 to 2008 and various fertilizers were used.
N/A: not available.

6.99±0.02, 1.69±0.07% and 0.14±0.01%. Soil texture was sandy clay loam with a bulk density of 1.18±0.18 g cm$^{-3}$ at the mid of a cultivation cycle of the vegetable [22].

On December 10, 2005, mixed soil (after excluding coarse roots and residuals) was put into twelve plastic containers, each with a height of 40 cm and an area of 1 m$^2$. There was a 10-cm layer of gravel and 10 holes (each with a diameter of about 0.5 cm), which were used for drainage, at the bottom of each container. Each container was filled with 362 kg soil (with gravimetric soil water content of about 18.5%) to a depth of 25 cm to achieve the field bulk density measured at the mid of a cultivation cycle. Therefore, our study can be regarded as a mesocosm experiment. The mesocosms were placed on an open field in our institute, at a distance of approximately 2 km from the vegetable fields.

Urea, AN and ABC were applied at a rate of 150 kg N ha$^{-1}$. Each treatment had three replications. On December 29, 2005 (the temperature of this period was similar to the annual average, details are shown in the following section), the fertilizers were dissolved in 1 L water and applied to the soil surface with a handhold sprayer. Three mesocosms were used as control and was only treated with 1 L water. The treatment process completed within 30 min. Deionized water was sprayed to the soil to maintain the soil moisture within a range similar to that of the field (with gravimetric water content of 16.6%–22.5%, which are equivalent to volumetric soil water content of 19.6–26.6% based on the relationship between gravimetric water content and volumetric water content [23]) [15] by using moisture probe meter (MP-508B, China) to monitor the change of volumetric soil water content at the depth of 10 cm.

## Measurement of NO Fluxes

Measurement of NO fluxes started on December 25, 2005 and lasted until March 10, 2006. Measurements were conducted every one day before fertilization. Upon fertilization, NO fluxes were measured once a day for a week and then once every two days for the following two weeks. Thereafter, measurement was conducted

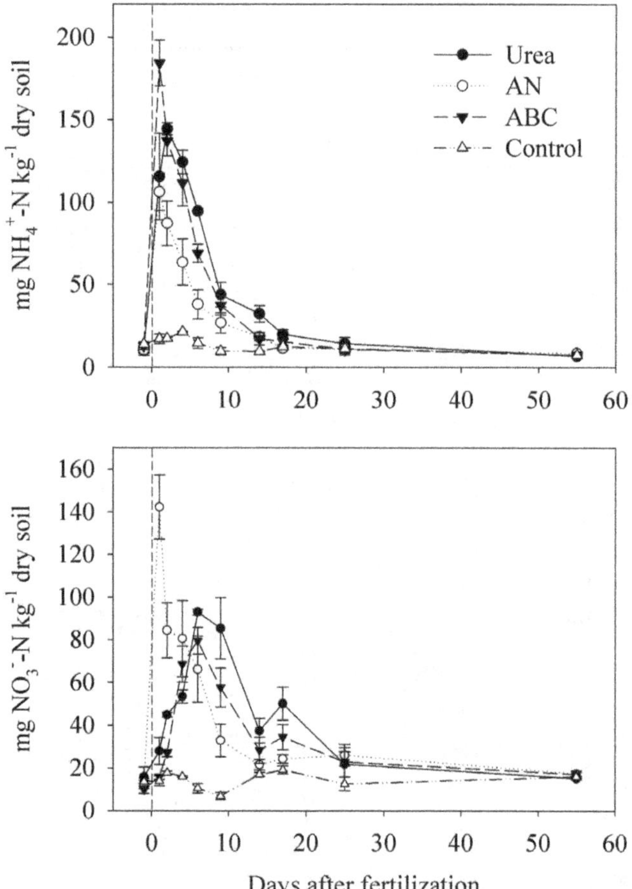


**Figure 3. Soil inorganic N pools during the sampling period.** Each value is the mean of three replicates, and error bars represent standard errors.

once every 4–12 days. On each sampling day, flux measurements were conducted in the morning (9:30–10:30), at noon (12:00–13:00), and in the afternoon (3:00–4:00).

A dynamic flow-through chamber technique was used to measure NO fluxes. The chamber system was described in detail previously [24,25]. The chambers are made of stainless steel (inner walls coated with Teflon films), each covering an area of 30 cm ×30 cm with a total volume of 9 L. Each chamber has one inlet port, one exhaust port and one outlet port for sampling. Inside each chamber, a thermo-sensor is fixed to measure air temperature, and a fan is attached to ensure sufficient mixing of air within the chamber. Different from the chamber system described previously [24,25], for this study each chamber was coupled with a steel pedestal (with a depth of 10 cm), at the top of which was a water-filled channel. One pedestal was put in the center of each container and inserted into the soil with a depth of 10 cm. During sampling, water was added to the channel of the pedestal so that the chamber and pedestal were sealed by the water. An additional reference chamber, closed at the bottom with Teflon sheet, was employed for in situ quantification of chemical reactions and chamber wall deposition effects. Ambient air was pumped into the chambers at a rate of 4 L $min^{-1}$ through 10 m long Teflon tubes with inner diameters of 4.8 mm, and the sample air was taken in through tubes of the same dimension. The residence time of air in the chambers was about 2.25 min. After about 15 min (over 5 cycles of residence time) when a steady state was reached inside the chambers, NO was analyzed by a model 42C chemiluminescence NO-$NO_2$−$NO_x$ analyzer (Thermo Environmental Instruments Inc., USA) for 3 minutes. By the difference of sampling chambers and the reference chamber, net fluxes from the soils could be calculated from equation (1) [26]:

$$F=(C_{eq}-C_0)\times \frac{M}{Vm}\times \frac{Q}{A} \quad (1)$$

where F is the net flux in ng $m^{-2}$ $s^{-1}$, M is the atomic weight of the element (N = 14.008 g $mol^{-1}$, C = 12.011 g $mol^{-1}$), $V_m$ is the standard gaseous molar volume (24.055×$10^{-3}$ $m^{-3}$ $mol^{-1}$), $C_{eq}$ is the mixing ratio (ppbv = $10^{-9}$ $m^{-3}$ $m^{-3}$) of the gas when the chamber under consideration has reached steady state, $C_0$ is the mixing ratio of the gas in the reference chamber, Q is the mass flow rate of air through the chamber (0.667×$10^{-5}$ $m^3$ $s^{-1}$), and A is the soil surface area (0.09 $m^2$) covered by the chamber.

## Measurement of $NH_3$ Fluxes

Measurement of $NH_3$ fluxes started on December 25, 2005 and lasted until February 4 2006. After sampling of NO fluxes, $NH_3$ was sampled with an acid trap method, which is commonly used for $NH_3$ measurement [27]. During sampling, the flow rate was also 4 L $min^{-1}$, and sample air was drawn at a rate of 1 L $min^{-1}$ for 30 min to the bubblers containing 20 ml of 0.7 M $H_2SO_4$. The $NH_4^+$ concentration trapped in the acid was determined by the indophenol blue method. Results gained with the acid trap method have been proven to be comparable with those by chemiluminescence $NH_3$−$NO_x$ analyzer [27]. Net fluxes were calculated by the difference of sampling chambers and the reference chamber using the equation (2):

$$F=\frac{C_s-C_r}{A\times T} \quad (2)$$

where $C_s$ and $C_r$ represent the amount of N (ng) trapped in the bubblers connected to the sampling chamber and reference chamber, respectively. A is the soil surface area (0.09 $m^2$) covered by the chamber, and T is the sampling time (1800 s).

## Auxiliary Measurements

Along with measurement of fluxes, air temperature at a height of 1.5 m, temperature inside the chambers, and soil temperature at 0–5 cm depth was measured with soil temperature probes (TES Ltd., China).

At each sampling date, three composite soil samples were collected from 0–5 cm depth. These samples were used for determinations of gravimetric soil moisture and KCl-extractable inorganic nitrogen pools. Briefly, 40 g fresh soil samples were extracted with 200 ml KCl solution (1 mol $L^{-1}$) for one hour. $NH_4^+$-N was measured by indophenol blue method and determined spectrophotometrically at 630 nm; $NO_3^-$N was determined by reduction to nitrite ($NO_2$−N) via a cadmium reactor, diazotized with sulfanilamide and is coupled to N-(1-Napthyl)-ethylenediamine dihydrochloride to form an azochromophore measured spectrophotometrically at 543 nm [22].

## Statistics and Data Analysis

Daily averaged flux of NO or $NH_3$ for each mesocosm was the average of the three measurements on each sampling day. Linear interpolation approach was used to estimate the daily fluxes for dates on which measurements were not made based on the fluxes from the dates when measurements were conducted immediately preceding and immediately following the dates that were not measured. Mean fluxes (ng N $m^{-2}$ $s^{-1}$) over the sampling period

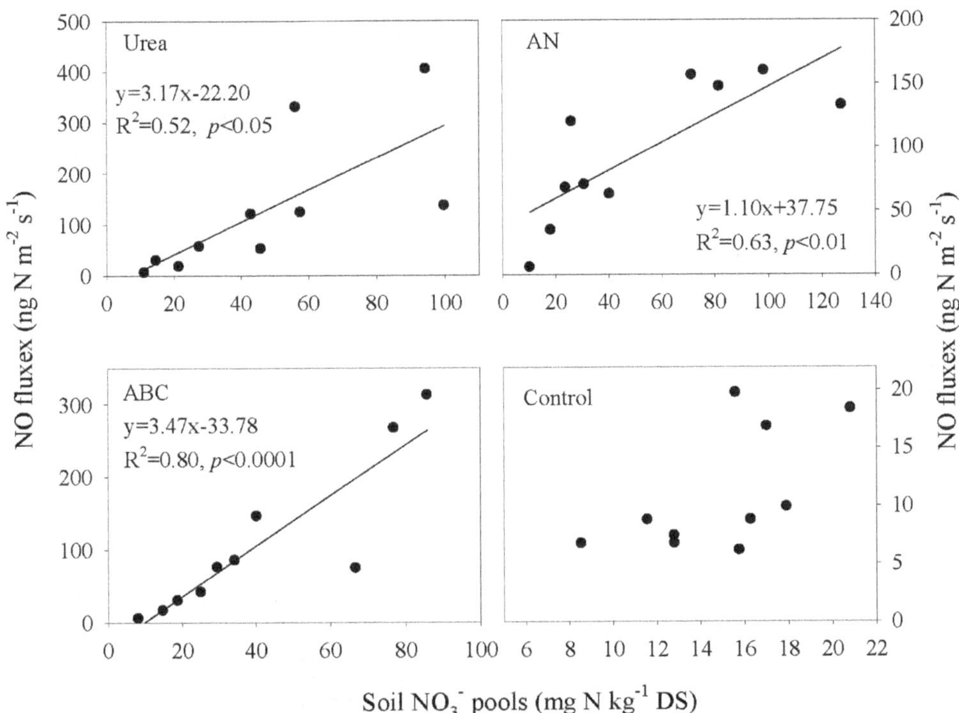


**Figure 4. Correlation between NO fluxes and soil $NO_3^-$ pools (n = 10).** Each value is the mean of three replicates.

after fertilization (72 days for NO and 39 days for $NH_3$) were calculated after data interpolation. Total emissions ($E_{total}$, mg N $m^{-2}$) were calculated by multiplying mean fluxes by sampling days after fertilization. Net emission ($E_{net}$, mg N $m^{-2}$) was obtained by subtracting $E_{total}$ of the control from that of the corresponding fertilized mesocosms. The emission factor (%) of NO ($F_{NO}$) or $NH_3$ ($F_{NH3}$) was calculated based on N atoms according to equation (3):

$$emission\,factor = \frac{total\ N\ emission_{fertilizer\ treatment} - total\ N\ emission_{control}}{applied\ N} \times 100 \quad (3)$$

Reported data of each treatment were the mean of the three replicate mesocosms on a daily basis. ANOVA analyses with post hoc LSD tests were performed using SPSS 10.0 (SPSS Ltd., USA) to identify differences among treatments. In this paper, analyses with *P* values <0.05 were considered significant.

## Results and Discussion

### NO Fluxes

Before fertilization, the mean NO flux was 6.9±0.6 (mean ± standard error of three replicates) ng N $m^{-2}$ $s^{-1}$ (Figure 2). The NO flux increased quickly in the AN fertilized mesocosms and reached 132.8±12.4 ng N $m^{-2}$ $s^{-1}$ one day after fertilization. The peaks of NO fluxes for different fertilizers occurred at different times. For AN fertilized mesocosms, the flux peaked 5 days after fertilization. For both the urea and ABC fertilized mesocosms, flux peaks were observed 6 days after fertilization. However, the peak fluxes for different fertilizers varied greatly. The averaged peak flux for urea was the greatest (404.9±38.3 ng N $m^{-2}$ $s^{-1}$), much higher (*P*<0.05) than that for AN (160.4±96.5 ng N $m^{-2}$ $s^{-1}$) or for ABC (312.1±33.0 ng N $m^{-2}$ $s^{-1}$) (Figure 2). The enhancement effects of fertilization on NO emissions lasted nearly two months.

Over the experimental period after fertilization, mean NO fluxes were 10.9±0.9, 73.1±2.9, 63.9±1.8 and 66.0±4.0 ng N $m^{-2}$ $s^{-1}$, respectively, for the control, urea, AN and ABC treatments (Table 1). The averaged NO flux for urea was significantly greater than those for AN and ABC (*P*<0.05). We previously reported that NO fluxes in a typical vegetable field (planted with flowering Chinese cabbage), where the soil for the current study was collected, varied from 20.0 to 122.1 ng N $m^{-2}$ $s^{-1}$ with an average of 47.5 ng N $m^{-2}$ $s^{-1}$ [15], which was relatively small compared with the fluxes in the present study. However, considering that only 63.5 kg N $ha^{-1}$ was applied in the field study (also see Table 2), a lower averaged NO flux was reasonable. According to a review paper, NO emissions ranged from 1.19 to 44 ng N $m^{-2}$ $s^{-1}$ from recently fertilized agricultural soils in temperate regions [5]. However, surprisingly high NO fluxes were also recorded in temperate croplands [28]. For example, Boeckx and Van Cleemput [28] observed that NO fluxes in autumn of 1997 (fertilized with 144 kg N $ha^{-1}$) and spring of 1998 (fertilized with 88 kg N $ha^{-1}$ as cattle slurry) and summer of 1998 (fertilized with 122 kg N $ha^{-1}$ as chicken manure) were19, 68, and 581 ng N $m^{-2}$ $s^{-1}$, respectively, from sandy arable soil in Belgium. Measurements of NO fluxes in tropical croplands were rather limited, but the available fluxes from tropical croplands were not significantly higher than those from temperate croplands [5], suggesting that the effect of latitude was small.

Chinese croplands are famous for relatively high N fertilization. Table 2 compiles the available data of NO fluxes from Chinese cropland soils. Averaged NO fluxes over the sampling periods varied from 0.14 to 360 ng N $m^{-2}$ $s^{-1}$. There is a strong

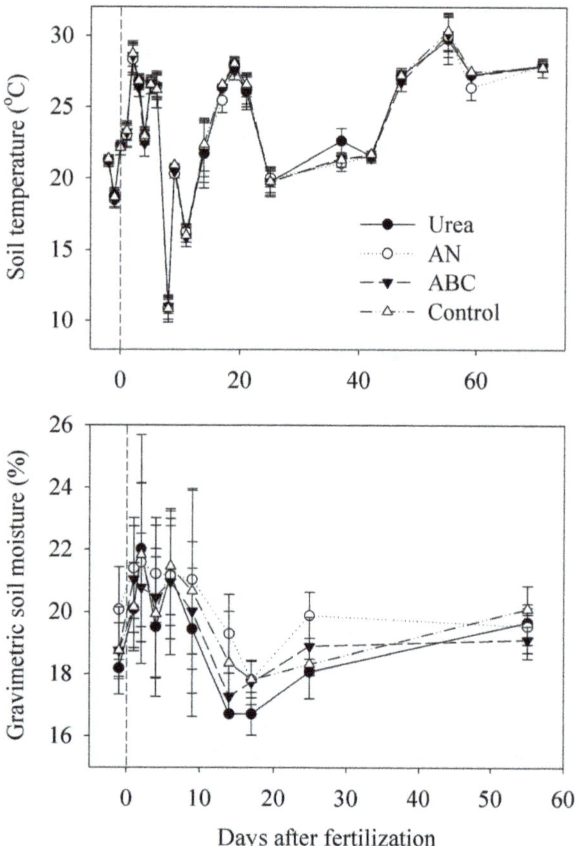


**Figure 5. Variation of soil temperature and soil moisture in different treatment mesocosms during the sampling period.** Each value is the mean of three replicates, and error bars represent standard errors. Vertical dash line indicates the day when fertilizers were applied.

correlation between N fertilizer rates and NO fluxes in an exponential way ($y = 6.7861e^{0.0094x}$, $R^2 = 0.27$, $P<0.05$, $n = 23$). However, there is no significant correlation between latitude and NO fluxes across these studies ($P>0.05$, Table 2). Our results confirm that there is no strong effect of latitude on NO fluxes.

## Influencing Factors for NO Fluxes

Many factors influence soil NO emissions, including N availability, soil moisture, soil temperature, atmospheric NO concentration, and other environmental factors that regulate the underlying processes of NO production and consumption in soils [29]. In well-drained agricultural soils, application of N fertilizers is often observed to cause large NO emissions [30]. This is because N fertilization supplies substrates ($NH_4^+$ and $NO_3^-$) to nitrifying or denitrifying bacteria, which are responsible for soil biogenic NO production [29]. In the current study, soil $NO_3^-$ pools increase after $NH_4^+$-based fertilizer application, indicating the occurrence of nitrification (Figure 3). Soil $NH_4^+$ pools were found to be significantly correlated with NO emissions only in AN fertilized mesocosms ($R^2 = 0.60$, $p<0.01$), but soil $NO_3^-$ pools were significantly correlated with NO emissions in all fertilized mesocosms (Figure 4). Since $NO_3^-$ is the product of nitrification and substrate of denitrification, it is difficult to judge which process is mainly responsible for NO production from the correlation of $NO_3^-$ and NO fluxes in the current study. Nevertheless, our field study indicated that nitrification was the dominant process [15].

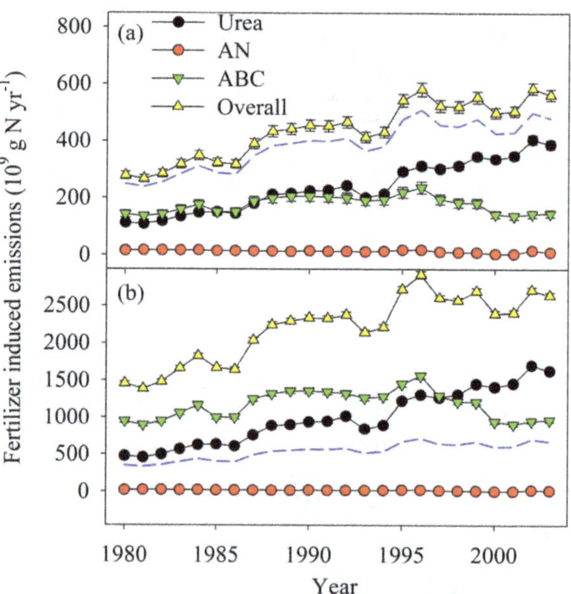


**Figure 6. Changes of fertilizer induced emissions of (a) NO and (b) $NH_3$ by consumption of urea, ammonium nitrate (AN) and ammonium bicarbonate (ABC) in China between 1980 and 2003.** Fertilizer induced emissions of NO and $NH_3$ were calculated by multiplying annual consumption of each fertilizer (Figure 1) by the corresponding emission factor obtained in the current study. The blue line in each panel represents the total emission of NO or $NH_3$ if the other two fertilizers were replaced by AN.

Since the soil used in the current study was collected from the same vegetable field where our field study was conducted, nitrification was probably also the main process responsible for NO production in the current study [15].

Soil temperature and moisture as important regulators on NO emissions have often been reported [11,15,24,29]. In the present study, soil temperature at 0–5 cm depth ranged between 10–30°C (Figure 5). A strong correlation was only found in the control ($F = 0.46\exp^{0.12T}$, $R^2 = 0.62$, $P<0.0001$, $n = 23$), probably because the effects of fertilization masked the effects of soil temperature in the fertilized mesocosms. Soil moisture (Figure 5) was maintained within a narrow range (16–25%) during the experiment, which was similar to the field soil moisture (16.6–22.5%) of typical vegetable land in this region [15]. No clear correlation was found between NO fluxes and soil moisture, probably because during the experimental period of this study, soil moisture varied within a narrow range.

## $NH_3$ Fluxes

Before fertilization, the average $NH_3$ flux was 10.7±3.8 ng N $m^{-2}$ $s^{-1}$. $NH_3$ volatilization from the ABC fertilized mesocosms increased sharply immediately after fertilization, and reached a maximum on the treatment day, then decreased rapidly (Figure 2). $NH_3$ flux from urea fertilized mesocosms peaked 3 days after fertilization (Figure 2). Peak values of $NH_3$ fluxes from AN fertilized mesocosms were observed 2 days after fertilization (Figure 2). The enhancement effect of fertilization on $NH_3$ fluxes lasted nearly three weeks. Our results were consistent with others in the pattern of changes of $NH_3$ volatilization following fertilization, i.e., ABC usually led to large $NH_3$ volatilization immediately after application, then reduced dramatically [31]. Urea is not volatile itself, but will become volatile after it is hydrolyzed into $NH_4^+$ and $CO_2$ by ureases in the soil. This process was found to be completed within 2–3 days, and $NH_3$ volatilization increases with increasing pH and temperature [8].

$NH_3$ flux measurements lasted 39 days after fertilization. During this period, mean $NH_3$ fluxes from urea, AN and ABC fertilized mesocosms were 493.6±4.4, 144.8±0.1 and 684.7±8.4 ng N $m^{-2}$ $s^{-1}$, respectively, compared to 8.9±0.2 ng N $m^{-2}$ $s^{-1}$ from the control (Table 1). The difference of averaged fluxes between any two fertilizers were significantly different ($P<0.5$).

## N loss as NO

Total NO emissions from urea, AN and ABC fertilized mesocosms were 454.6±17.9, 397.5±10.9, 410.3±25.1 mg N $m^{-2}$, respectively, significantly greater than that from the control ($P<0.05$) (Table 1).

With regard to fertilizer-induced NO emission factor (EF), there was no significant difference between AN and ABC ($P>0.05$), but EF for urea (2.6±0.1%) was significantly greater ($P<0.05$) than the other two fertilizers (Table 1). Our data supported that urea had the highest EF [32], but was in conflict with some studies which reported that a mixed form fertilizer such as AN had the greatest emission in tropical savanna soil [33]. FAO/IFA [19] used EFs for urea, AN and ABC of 0.6%, 0.5% and 0.8%, respectively. However, they used the EF for ammonium sulphate as an estimate of EF for ABC since no study was conducted to determine EF for ABC at that time.

The fertilizer-induced NO emission factors based on studies in fertilized upland arable fields in China were presented in Table 2. The field measured EFs varied from 0.04 to 11.56 with an average of 1.93 (excluding EFs from the current study). The EFs observed in the present study were within the reported range but greater than the average of all the filed observed values. However, the EFs in the current study are quite similar to those derived from the field study (EF = 2.4%) conducted at a nearby site [15]. Nevertheless, it should be noted that the EFs in the current study ware based on a mesocosm experiment, therefore limitations may be involved when making comparison with field based studies.

## N Loss as $NH_3$

Total emission of $NH_3$ from ABC fertilized mesocosms was the highest (2.31 g N $m^{-2}$), which was 1.4 and 4.7 times that from urea and AN fertilized mesocosms, respectively, and 77 times that from the control ($P<0.05$, Table 1). Total emission of $NH_3$ was significantly greater from urea than from AN fertilized mesocosms ($P<0.05$).

Fertilizer-induced $NH_3$ emission factors varied greatly for the three fertilizers (Table 1). EF for ABC was 15.2±0.4%, which was 1.4 times that for urea (10.9±0.2%), and 5 times that for AN (3.1±0.1%). Similarly, some studies in China reported that EF for ABC was about 1.5 times that for urea [34,35]. Yan et al. (2003) estimated $NH_3$ emission from dry croplands in East, South East and South Asia by adopting emission factors of 13.7% for urea and 20.5% for ABC. As no emission data for AN were available from Asian croplands, Yan et al. (2003) adopted an EF of 2.0%, which was proposed by European Environment Agency. According to the present study, the adoption of the European proposed EF for AN likely only slightly underestimated $NH_3$ emission due to AN fertilization in China. Nevertheless, more studies are needed since only one cropland soil was investigated in the current study and since our study was conducted as a mesocosm experiment.

It must be stressed that the chamber, though widely used, may lead to $NH_3$ volatilization rates somewhat different from those under field conditions. Wind speed exerts positive effects on $NH_3$ volatilization loss [36], but air currents inside the chamber by flowing through of air in addition to the mixing by fans, can not well simulate the wind above the soil under natural conditions. Thus as an approximation, $NH_3$ volatilization measured in this study might not be always consistent with that under field conditions.

## Implications for National Emissions of NO and $NH_3$

In China, urea and ABC are two main N fertilizers, while AN only accounts for a small portion in total fertilizer N consumption. During 1980 to 2003 (Figure 1), urea consumption increased continuously; ABC consumption increased until 1996 and decreased thereafter; AN consumption tended to decrease continuously [10]. The total consumption of fertilizer N (only for the three fertilizers) peaked in 1996 and then levelled off (Figure 1). The portion of urea consumption increased from 38.6% to 68.6% of the total consumption of the three fertilizers, but that of ABC and AN decreased accordingly from 55.2% to 29.2% and from 6.3% to 2.2%, respectively (Figure 1). Since the emission factors of both NO and $NH_3$ are different for the three fertilizers, changes in fertilizer consumption and proportion would have a substantial impact on national budgets of NO and $NH_3$. Since the current study compars the emission factors for the three fertilizers from a typical cropland soil in China, the impact of changes in fertilizer consumption and proportion on national budgets of NO and $NH_3$ can be roughly estimated. However, since only one soil type was studied using a mesocosm experiment design, there are limitations involved in these estimates.

Total emissions of both NO and $NH_3$ from the three fertilizers peaked in 1996 and then levelled off (Figure 6). But for different fertilizers, the changes of emission during 1980–2003 are different. Emissions from urea application increased continuously for both gases. Emissions from ABC application peaked in 1996 and then decreased. Emissions from AN application were much smaller compared to the other two fertilizers and decreased over the period. Urea application has been the dominant source since 1994 for NO and since 1998 for $NH_3$. Since EFs of both NO and $NH_3$ emissions were the lowest for AN among the three studied fertilizers, the complete replacement of the other two fertilizers by AN would greatly decrease NO and $NH_3$ emissions, especially $NH_3$ (Figure 6). Therefore, in order to reduce air pollution, it would be better to increase the proportion of AN application.

## Conclusions

In the present study, a mesocosm experiment was conducted to measure NO and $NH_3$ emissions from bare soil, which was collected from vegetable land, following the application of urea, AN and ABC. Emission factors of NO were 2.6%, 2.2% and 2.3% for the above three fertilizers, respectively; and emission factors of $NH_3$ were 10.9%, 3.1% and 15.2%, respectively. From the perspective of air quality protection, it would be better to increase the proportion of AN application due to its lower emission factors for both NO and $NH_3$.

## Author Contributions

Conceived and designed the experiments: DL. Performed the experiments: DL. Analyzed the data: DL. Contributed reagents/materials/analysis tools: DL. Wrote the paper: DL.

## References


1. Crutzen PJ (1979) The role of NO and $NO_2$ in the chemistry of the troposphere and stratosphere. Annual Reviews of Earth and Planetary Sciences 7: 443–472.
2. IPCC (2001) Climate Change 2001: The Scientific Basis. Cambridge, UK: Cambridge University Press.
3. Fowler D, Flechard C, Skiba U, Coyle M, Cape JN (1998) The atmospheric budget of oxidized nitrogen and its role in ozone formation and deposition. New Pyhtologist 139: 11–23.
4. Barthelmie RJ, Pryor SC (1998) Implications of ammonia emissions for fine aerosol formation and visibility impairment-a case study from the Lower Fraser Valley, British Columbia. Atmospheric Environment 32: 345–352.
5. Davidson EA, Kingerlee W (1997) A global inventory of nitric oxide emissions from soils. Nutrient Cycling in Agroecosystems 48: 37–50.
6. Yan XY, Ohara T, Akimoto H (2005) Statistical modeling of global soil NOx emissions. Global Biogeochemical Cycles 19: GB3019, doi:3010.1029/2004GB002276.
7. Yienger JJ, II HL (1995) Empirical model of global soil biogenic NOx emissions. Journal of Geophysical Research 100: 11447–11464.
8. Bouwman AF, Lee DS, Asman WAH, Dentener FJ, Hoek KWVD, et al. (1997) A global high-resolution emission inventory for ammonia. Global Biogeochemical Cycles 11: 561–587.
9. Yan XY, Akimoto H, Ohara T (2003) Estimation of nitrous oxide, nitric oxide and ammonia emissions from croplands in East, Southeast and South Asia. Global Change Biology 9: 1080–1096.
10. IFA (2007) Nitrogen, phosphate and potash statistics from 1973–1973/74 to 2004–2004/05. IFADATA statistics online, updated: 7 Feb. 2007. (http://www.fertilizer.org/ifa/statistics/IFADATA/dataline.asp).
11. Zheng X, Huang Y, Wang Y, Wang M (2003) Seasonal characteristics of nitric oxide emission from a typical Chinese rice-wheat rotation during the non-waterlogged period. Global Change Biology 9: 219–227.
12. Xing GX, Zhu ZL (2000) An assessment of N loss from agricultural fields to the environment in China. Nutrient Cycling in Agroecosystems 57: 67–73.
13. Walsh M (2001) $NO_x$ and $N_2O$ fluxes in an upland agroecosystem of the North China Plain: field measurements, biogeochemical simulation, and climatic sensitivity [PhD dissertation]. Fort Collins, Colorado: Colorado State University.
14. Fang SX, Mu Y (2006) Air/surface exchange of nitric oxide between two typical vegetable lands and the atmosphere in the Yangtze Delta, China. Atmospheric Environment 40: 6329–6337.
15. Li D, Wang X (2007) Nitric oxide emission from a typical vegetable field in the Pearl River Delta, China. Atmospheric Environment 41: 9498–9505.
16. Mei B, Zheng X, Xie B, Dong H, Zhou Z, et al. (2009) Nitric oxide emissions from conventional vegetable fields in southeastern China. Atmospheric Environment 43: 2762–2769.
17. Pang X, Mu Y, Lee X, Fang S, Yuan J, et al. (2009) Nitric oxides and nitrous oxide fluxes from typical vegetables cropland in China: Effects of canopy, soil properties and field management. Atmospheric Environment 43: 2571–2578.
18. Liu C, Zheng X, Zhou Z, Han S, Wang Y, et al. (2010) Nitrous oxide and nitric oxide emissions from an irrigated cotton field in Northern China. Plant and Soil 332: 123–134.
19. FAO/IFA (2001) Global estimates of gaseous emissions of $NH_3$, NO and $N_2O$ from agricultural land. Rome, Italy: Food and Agriculture Organization (FAO) of the United Nations.
20. Zheng X, Han S, Huang Y, Wang Y, Wang M (2004) Re-quantifying the emission factors based on field measurements and estimating the direct N2O emission from Chinese croplands. Global Biogeochemical Cycles 18: GB2018, doi:2010.1029/2003GB002167.
21. Guangzhou-Municipal-Statistics-Bureau (2006) Guangzhou Statistical Yearboo. Beijing: China Statistics Press.
22. National-Standard-Bureau-of-China (1987) Analytical Methods for Forest Soils (Third fascicule). Beijing: National Standard Bureau Press.
23. WMO (2008) Guide to Meteorological Instruments and Methods of Observation, WMO-No. 8, Seventh edition. Geneva: World Meteorological Organization.
24. Li DJ, Wang XM, Mo JM, Sheng GY, Fu JM (2007) Soil nitric oxide emissions from two subtropical humid forests in south China. J Geophys Res 112: D23302, doi:23310.21029/22007JD008680.
25. Li D, Wang X, Sheng G, Mo J, Fu J (2008) Soil nitric oxide emissions after nitrogen and phosphorus additions in two subtropical humid forests. J Geophys Res 113: D16301, doi:16310.11029/12007JD009375.
26. Pilegaard K, Hummelshøj P, Jensen NO (1999) Nitric oxide emission from a Norway spruce forest floor. J Geophys Res 104: 3433–3445.
27. Akiyama H, Mctaggart IP, Ball B, Scott A (2004) $N_2O$, NO, and $NH_3$ emissions from soil after the application of organic fertilizers, urea and water. Water, Air and Soil Pollution 156: 113–129.
28. Boeckx P, Cleemput OV (2006) "Forgotten" terrestrial sources of N-gases. International Congress Series 1293: 363–370.
29. Ludwig J, Meixner FX, Vogel B, Förstner J (2001) Soil-air exchange of nitric oxide: An overview of processes, environmental factors, and modeling studies. Biogeochemistry 52: 225–257.
30. Veldkamp E, Keller M (1997) Fertilizer-induced nitric oxide emissions from agricultural soils. Nutrient Cycling in Agroecosystems 48: 69–77.
31. He ZL, Alva AK, Calvert DV, Banks DJ (1999) Ammonia volatilization from different fertilizer sources and effects of temperature and soil pH. Soil Science 164: 750–758.
32. Bouwman AF, Boumans LJM, Batjes NH (2002) Modeling global annual $N_2O$ and NO emissions from fertilized fields. Global Biogeochemical Cycles 16: 1080, doi:1010.1029/2001GB001812.
33. Sanhueza E (1992) Biogenic emissions of NO and N2O from tropical savanna soils. Proceedings of International Symposium on Global Climate Change. Tokyo: Japan National Committee for the IGBC. 22–34.
34. Cai GX, Zhu ZL, Trevitt ACF, Freney JR, Simpson JR (1986) Nitrogen loss from ammonium bicarbonate and urea fertilizers applied to flooded rice. Fertilizer Research 10: 203–215.
35. Zhu ZL, Cai GX, Simpson JR, Zhang SL, Chen DL, et al. (1989) Processes of nitrogen loss from fertilizers applied to flooded rice fields on a calcareous soil in north-central China. Fertilizer Research 18: 101–115.
36. Sommer SG, Olesen JE (2000) Modelling ammonia volatilization from animal slurry applied with trail-hoses to cereals. Atmospheric Environment 34: 2361–2372.
37. Walsh M (2001) $NO_x$ and $N_2O$ Fluxes in an Upland Agroecosystem of the North China Plain: Field Measurements, Biogeochemical Simulation, and Climatic Sensitivity: Colorado State University.
38. Fang S, Mu Y (2007) NOX fluxes from three kinds of agricultural lands in the Yangtze Delta, China. Atmospheric Environment 41: 4766–4772.

# 13

# Affordable Nutrient Solutions for Improved Food Security as Evidenced by Crop Trials

**Marijn van der Velde[1]*, Linda See[1], Liangzhi You[2,3], Juraj Balkovič[1], Steffen Fritz[1], Nikolay Khabarov[1], Michael Obersteiner[1], Stanley Wood[4]**

**1** International Institute for Applied Systems Analysis (IIASA), Ecosystem Services and Management Program, Laxenburg, Austria, **2** International Food Policy Research Institute (IFPRI), Washington D.C., United States of America, **3** College of Economics and Management, Huazhong Agricultural University, Wuhan, China, **4** Global Development Program, Bill & Melinda Gates Foundation, Seattle, Washington, United States of America

## Abstract

The continuing depletion of nutrients from agricultural soils in Sub-Saharan African is accompanied by a lack of substantial progress in crop yield improvement. In this paper we investigate yield gaps for corn under two scenarios: a micro-dosing scenario with marginal increases in nitrogen (N) and phosphorus (P) of 10 kg $ha^{-1}$ and a larger yet still conservative scenario with proposed N and P applications of 80 and 20 kg $ha^{-1}$ respectively. The yield gaps are calculated from a database of historical FAO crop fertilizer trials at 1358 locations for Sub-Saharan Africa and South America. Our approach allows connecting experimental field scale data with continental policy recommendations. Two critical findings emerged from the analysis. The first is the degree to which P limits increases in corn yields. For example, under a micro-dosing scenario, in Africa, the addition of small amounts of N alone resulted in mean yield increases of 8% while the addition of only P increased mean yields by 26%, with implications for designing better balanced fertilizer distribution schemes. The second finding was the relatively large amount of yield increase possible for a small, yet affordable amount of fertilizer application. Using African and South American fertilizer prices we show that the level of investment needed to achieve these results is considerably less than 1% of Agricultural GDP for both a micro-dosing scenario and for the scenario involving higher yet still conservative fertilizer application rates. In the latter scenario realistic mean yield increases ranged between 28 to 85% in South America and 71 to 190% in Africa (mean plus one standard deviation). External investment in this low technology solution has the potential to kick start development and could complement other interventions such as better crop varieties and improved economic instruments to support farmers.

**Editor:** Luis Herrera-Estrella, Centro de Investigación y de Estudios Avanzados del IPN, Mexico

**Funding:** The Austrian Research Funding Agency (FFG) funded projects. The funders had no role in study design, data collection and analysis, decision to publish, or preparation of the manuscript. This research was supported by the Austrian Research Funding Agency (FFG) for the Project FarmSupport (No. 833421).

**Competing Interests:** The authors have declared that no competing interests exist.

* E-mail: velde@iiasa.ac.at

## Introduction

*Farming looks mighty easy when your plow is a pencil and you're a thousand miles from the corn field.* –Dwight D. Eisenhower, 1956.

The increases in global population and food demand clearly indicate that current growth in agricultural productivity is not sufficient to sustain the 9 billion people that will inhabit the Earth by 2050 [1]. Feeding the world is a multifaceted and complex challenge and a number of solutions have been offered, where closing the yield gap is one of the most frequently cited recommendations [2;3;4;5;6]. The FAO [1] suggests that 70% of the required increase in crop production in developing countries should be realized through boosting the productivity of fields already under cultivation. Without this intensification, the inevitable cropland expansion will lead to deforestation, accelerate land degradation and threaten natural habitats and biodiversity [7]. A large part of this augmentation in crop production must therefore come from soils in tropical regions which are often highly weathered, have low levels of chemical soil fertility and will need additional inputs to improve crop productivity [8]. Global fertilizer use has already increased significantly since 1960 [9] and this increase has played an important role in the Green Revolution, benefitting many developing countries in South America and Asia. Yet substantial progress is still lacking in Africa [10]. From 1960 to 2000, yields of staple crops such as wheat, rice and corn increased in South America by over 180% while African yields did not improve substantially (see Figure 1). These contrasting trajectories reflect disparities in infrastructure development, primary crop types grown, agricultural R&D and extension capacities, socio-economic conditions as well as environmental differences [10;11;12].

In many smallholder fields, fertilizer and manure inputs have been too low for too long [8]. Agricultural soils cultivated without adequate nutrient replenishment cannot reach their full crop production potential and are at risk of irreversible degradation [13]. In large parts of Africa, this has led to seemingly perpetual low per capita food production [8]. Without maintaining adequate soil fertility levels, crop yields cannot be sustained, increase over time, or respond to improved agricultural management practices.

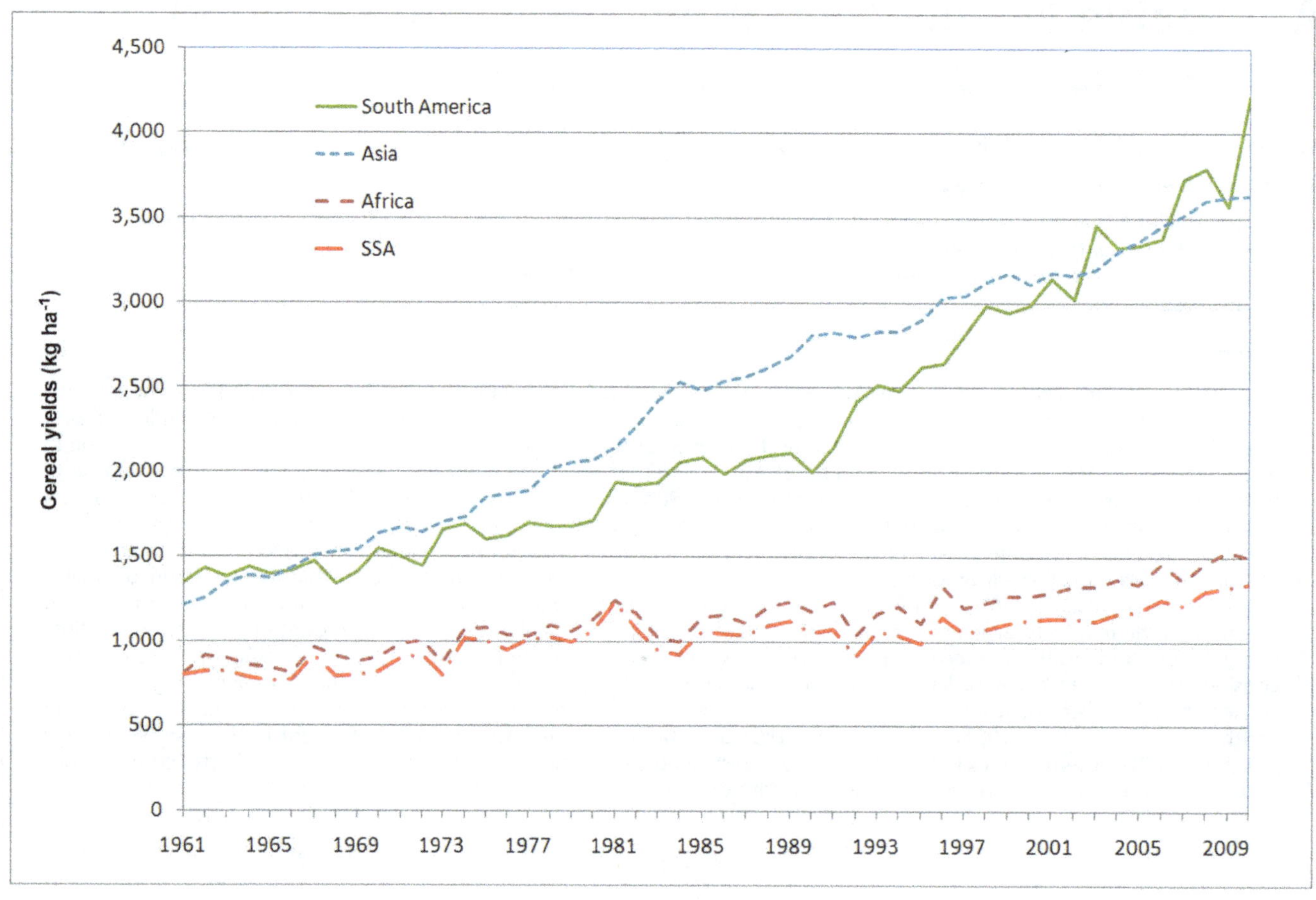


**Figure 1. Cereal yield trends since 1960 in Africa, Sub-Sahara Africa (SSA), South America and Asia.**

To improve nutrient input, smallholder farmers need actionable strategies such as micro-dosing: applications of small quantities of fertilizers. Field studies have shown that micro-dosing presents an attainable strategy for smallholder farmers in line with their financial means that can result in significant yield gains [14]. Importantly, previous higher fertilizer rate recommendations have ignored the sizeable but unlikely investment that would be required by poor and risk adverse smallholders.

Fertilizer prices in Africa are often higher relative to other developing countries. The small size of the fertilizer market, the high transportation and handling costs and the inefficient supply chain all contribute to the relatively high retail prices for fertilizers in Africa [15]. Fertilizer price also varies across regions, through different years and even among cropping seasons in the same year. In a landlocked country such as Uganda, the prices for urea in 2000 ranged from 600 shillings $kg^{-1}$ (300 US\$ $ton^{-1}$) in the central districts to over 750 shillings $kg^{-1}$ (375 US\$ $ton^{-1}$) in the eastern districts while prices for phosphate (Diammonium Phosphate, DAP) ranged from 560 shillings $kg^{-1}$ (280 US\$ $ton^{-1}$) in the long rainy season to over 700 shillings $kg^{-1}$ (350 US\$ $ton^{-1}$) in the short rainy season [16]. As a comparison, US farmers in 2000 paid from US\$80 to US\$120/ton for Urea, and US\$140 to US\$170/ton for DAP. Depending on the locations and seasons, the NPK (Nitrogen-Phosphorus-Potassium) compound fertilizer ranged from 700 shillings $kg^{-1}$ (350 US\$ $ton^{-1}$) to over 1000 shillings $kg^{-1}$ (500 US\$ $ton^{-1}$) for 1:1:2 NPK, the most common compound fertilizer in Uganda. The transportation cost in Uganda is over one third of the total fertilizer cost while it is about a fifth in Tanzania [17] due to the major sea port of Dar Es Salaam. Moreover, there are global pressures that lead to price volatility in fertilizer prices. For example, there was a fourfold increase in the price of urea from 2000 to 2008, reaching over 500 US\$ $ton^{-1}$, falling to around 200 US\$ $ton^{-1}$ in 2009 and which is currently at around 400 US\$ $ton^{-1}$.

To formulate more realistic sustainable intensification pathways, we need better estimates of smallholder yield gaps in tropical countries and to then align these with local fertilizer prices and associated investment costs. There are a number of yield gap approaches that estimate different types of attainable yield potentials across varying spatial and temporal scales [18]. Many global assessments of yield gaps use crop models or data that currently lack sufficiently detailed spatial information on soil characteristics, crop management practices and crop responses to fertilizers [19,20]. Mueller et al. [19] found that large crop production increases are possible, but will require considerable changes in nutrient and water management. Crop trials represent a valuable source of information for yield gap analysis and could be analyzed more comprehensively for this purpose, yet are rarely collected systematically. Furthermore, nutrient specific analysis of the relationships between fertilizers and crop yields has been limited, especially in the tropical and subtropical regions where crop yields are relatively low and must increase the most to meet growing demands [21].

In this paper we analyze historic data from FAO corn fertilizer trials carried out between 1969–1993 at 1358 locations in Africa and South America (Figure 2) where corn is the most commonly cultivated crop [18]. Mitscherlich-Baule crop response functions were fit to the crop trial data by optimizing the factors describing yield responses to elemental nitrogen (N) and phosphorus (P) inputs as well as an initial (residual) soil N and P. Nitrogen and Phosphorus specific fertilizer application rates from [22] were then used as inputs to the crop response functions, and the resulting yields were validated using sub-national yield statistics on corn from the International Food Policy Research Institute (IPFRI) [23] to estimate yield gaps in corn at the continental level. Only water-limited (i.e. rain fed) yield potential as opposed to irrigated yields potentials are considered here. Furthermore, we consider the importance of soil nutrient stoichiometry and the viability of micro-dosing [14,24] in order to further the African crop productivity discourse [25,26]. Yield increases associated with two different scenarios are considered here: 1) depicting a micro-dosing strategy and 2) a topping up to a conservative estimate of average nutrient fertilizer rates in the USA. Finally, the investment costs of scaling up these scenarios are calculated using the latest average fertilizer prices in Africa and South America and used to evaluate whether these approaches can function as part of an actionable development blueprint for Sub-Saharan Africa.

## Materials and Methods

### Crop Trials and Response Functions

Recently historic FAO crop fertilizer field trials have become publically accessible (http://www.fao.org/ag/agl/agll/nrdb/). These data were collected as part of FAO's Fertilizer Programme [27] that ran from 1969 until 1993. The purpose behind the programme was to undertake trials to determine suitable fertilizer application rates for locally grown crops and to demonstrate to as many farmers as possible, the positive effect of fertilizer application on crop yields and farm income. The information available from the trials includes crop yields (kg $ha^{-1}$), and application rates of the main nutrients (nitrogen (N), phosphorus (P) and potassium (K) and farmyard manure (kg $ha^{-1}$)). Unfortunately, detailed soil or meteorological information was not recorded. Nitrogen was mostly applied as urea, phosphorus (P) was mostly applied as superphosphate and potassium as part of NPK compound fertilizers. All the applied nutrients were recalculated to elemental application rates (with P calculated from $P_2O_5$ and K from $K_2O$). The application rates of farm yard manure were converted to N, P-$P_2O_5$ and K-$K_2O$ application rates following [28]. Data on corn yields from trials with at least five N and P input combinations were selected for this analysis. The Mitscherlich-Baule crop response function was used to analyze relations between nitrogen (N) fertilizer input, phosphorus-phosphate (P) fertilizer input and corn yields $y_{mb}$:

$$y_{mb} = a_1[1-\exp(-a_2(a_3+N))] * [1-\exp(-a_4(a_5+p))] \qquad (1)$$

The function allows for growth that plateaus with increasing fertilizer application, and accommodates cases of both near perfect factor substitution and near zero factor substitution, and performs superior to a quadratic and von Liebig type production function [29]. The growth plateau is represented by $a_1$, which was set equal to the maximum yield obtained in each field trial, while $a_3$ and $a_5$ represent the residual available nitrogen and phosphorus in the soil. Taking account of residual soil phosphorus is important; the cumulative cropland P surplus in certain countries in Western Europe has led to a buildup of residual soil P with expected future benefits to crop production, although lower and no effects are generally expected for Latin America and Africa [30]. The coefficients $a_2$ and $a_4$ describe the influence of the corresponding N or P fertilization on yield. The parameters $a_2$, $a_3$, $a_4$, and $a_5$ were obtained by minimizing the sum of squared errors for all applications in each experiment. The Nelder-Mead multidimensional unconstrained nonlinear minimization algorithm was used to minimize the objective function. Only those trials where a crop response function could be fit were used. This resulted in a total of 1358 unique experiments with at least five N and P input combinations; 752 in Africa and 606 in South America.

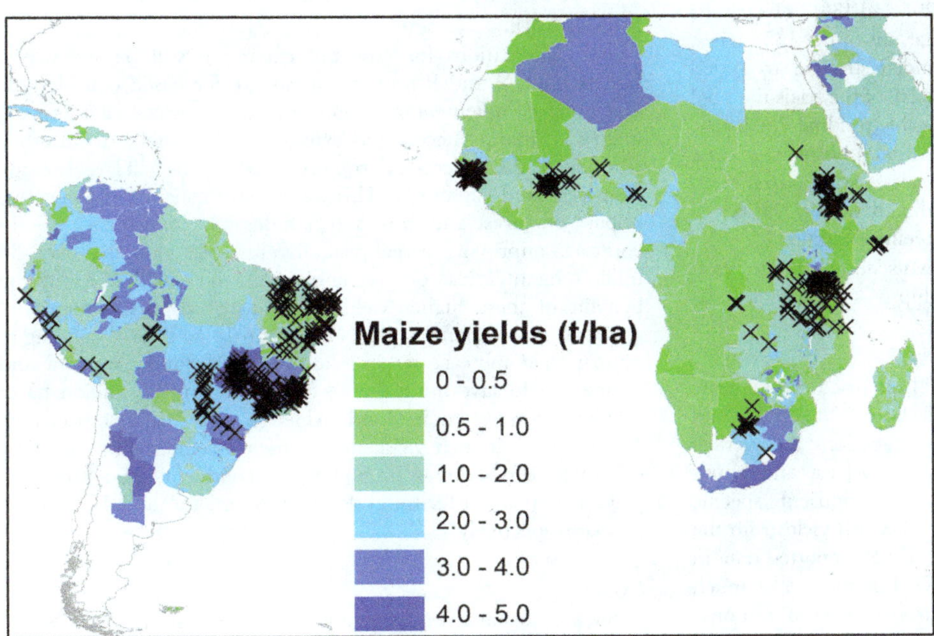


**Figure 2. Crosses indicate locations of 1358 historic FAO corn field trials with at least five N and P input combinations in Africa and South America carried out between 1969 and 1993.** Colors indicate (subnational) maize (corn) yields (ton/ha) as collected by [23].

## Current Fertilizer Application Rates

The fertilizer dataset of [22] containing crop fertilizer rates was used to assign the current N and P inputs from chemical fertilizer (Nfer, Pfer) and manure (Nman, Pman; see Figures S1, S2, S3) at the trial locations and considered representative for corn fertilizer rates [31]. Data on corn yields collected by IFPRI [23] were used as a comparison with the yields obtained from the crop response functions (Figure S4). The average cereal area, production and yield as reported by FAOSTAT for 2008–2010 were calculated for each of the scenarios. In the area scenario a constant production was assumed and an increase in yield would reduce the requirement for cropland. In the people scenario an increase in yield would produce more on the same cropland area. We assume the average cereal calorie content to be 3000 kcal $kg^{-1}$ and the average annual calorie need of a person to be 1 million kcal $year^{-1}$.

## Costs

The cost of the two proposed scenarios was calculated using the Agricultural GDP in US dollars for 2009 [32]: South America (SA) $192 billion USD; Sub-Saharan Africa (SSA) $150 billion USD. Then the costs for each region were calculated by taking the total cost, dividing by the Agricultural GDP and multiplying by 100 to arrive at the percentage of Agricultural GDP that would be required to finance the scenario.

# Results

An example of crop response trial data and the corresponding modeled Mitscherlich-Baule crop response function is shown in Figure 3. The median $r^2$ obtained by fitting the individual crop trials equaled 0.81; the 25th percentile equaled 0.66 and the 75th percentile 0.91. The resulting median, 25th and 75th percentiles for the $a_1$, $a_2$, $a_3$ and $a_4$ parameters obtained across all crop trials are presented in Table 1. Median values corresponded to 0.017 ton $kg^{-1}$, 68.4 kg N $ha^{-1}$, 0.29 ton $kg^{-1}$, 3.18 kg P $ha^{-1}$ for parameters $a_1$, $a_2$, $a_3$ and $a_4$ respectively. This is in correspondence with the parameter values reported by [29] and [33]. The results from the individual crop trials indicate that out of the 1358 trials, there were 1037 trials (76%) that responded stronger to added phosphorus than nitrogen. Similarly, for 82% of the trials the pool of residual soil N was larger than the accessible residual P. Clearly, these site-specific analyses indicate that overall, phosphorus is the nutrient most limiting crop yield. Nevertheless, at the same time, the range of parameter values obtained highlights the variety of crop yield responses depending on site-specific conditions.

Overall, the yields modeled using Mitscherlich-Baule crop response functions show a good relationship ($r^2 = 0.94$, Figure 4).

## Yield Gaps and Potentials

The fertilizer dataset of Potter et al. [22] was used to assign the current N and P inputs from chemical fertilizer (Nfer, Pfer) and manure (Nman, Pman) to the crop response functions; the average of these inputs and their distribution across Africa and South America as well as manure and fertilizer nutrient specific histograms are shown in Figures S1, S2, S3. Corn yield from the crop responses functions was compared to IFPRI reported data in Figure S4; median yields are comparable but there is a much larger variability in the yields derived from the crop response functions. This reflects both the coarser resolution of the IFPRI data [23] and the more realistic representation of the frequency distribution of yields that are attained at individual locations across both continents from the FAO crop trial data.

To indicate the potential for production increase, we calculated the average percentage yield increases resulting from an additional application of 10 kg N $ha^{-1}$, 10 kg P $ha^{-1}$, and both. Adding only N will lead to increases in crop production by ~4% and ~8%. Adding only P, on the other hand, will lead to substantially larger increases of ~12% and ~26% for South America and Africa respectively (Figure 5). This highlights the critical importance of P, of which many subsistence farmers may not be aware. The addition of both nutrients leads to increases of ~15% and ~35%, respectively, indicating that the effect of both nutrients is additive once P is applied. Thus P is clearly the limiting nutrient in improving crop yields.

In Africa, adding 10 kg $ha^{-1}$ of N or P will result in mean and median percentage increases of respectively ~5.5 and ~5.7% and ~11.7 and ~16.3%; this would thus bring a significant proportion of farmers with the lowest yields closer towards attaining average yield levels and effectively shift a bulk of smallholders out of current marginal productivity. Since these are indicative for rainfed yields, additional water resources to attain these yield increases would not be required [6]. A final experiment considers the percentage yield increase that would be obtained if 80 kg $ha^{-1}$ of N and 20 kg $ha^{-1}$ of P - a relatively conservative estimate of average rates in the USA - were applied (Figure 6). In South America this would lead to average yield increases of 30%, up to a maximum of 90% while in Africa these increases would be considerably larger, i.e. average yield increases of 70%, up to a maximum of 190%.

Even though these yield increases are considerable, they are lower than the yield potentials generally estimated in other studies. For example, yield gaps of 180 to 540% for maize have been estimated for sub-Saharan Africa [34] while a yield gap of 118% was found by Tittonell et al. [35] for western Kenya. Since most of the crop trials were done more than 20 years ago, our results will provide conservative yield estimates. In soils that have become increasingly depleted, and with new and better crop varieties that have become available since then, the response to fertilizer may be even stronger than predicted here.

## Implications

The implications for cropland expansion will be significant. Unless both N and P nutrient inputs are increased considerably and other complementary inputs and rural services such as seed, irrigation, market access and extension are available to improve crop yields, then necessary crop production gains will largely come from cropland expansion. This would have considerable negative impact on forest and grassland habitats and biodiversity [5]. In contrast, improving cereal yield by just 5% globally, over 33 million ha of forest or grassland would be saved. To put the benefits of these higher yields in context, our first scenario of applying an additional 10 kg N $ha^{-1}$ and 10 kg P $ha^{-1}$ (leading to a corn yield increase by 15% and 32% in South America and Africa) would save more than 4 million ha and 25 million ha of cropland conversion in South America and Africa respectively. Alternatively, if such yield improvement occurs in the currently cultivated cereal areas on these two continents, the improved productivity could feed an additional 64 million and 150 million people respectively.

## Costs

Such a scenario would require a total investment of US$148 million in sub-Sahara Africa (using an average of fertilizer prices paid by farmers in 2012 of Urea ($620/ton or 0.29$ $kg^{-1}$ N) and DAP ($950/ton or 0.22$ $kg^{-1}$ P) and US$79 million in South America (using an average fertilizer price paid by farmers in 2012

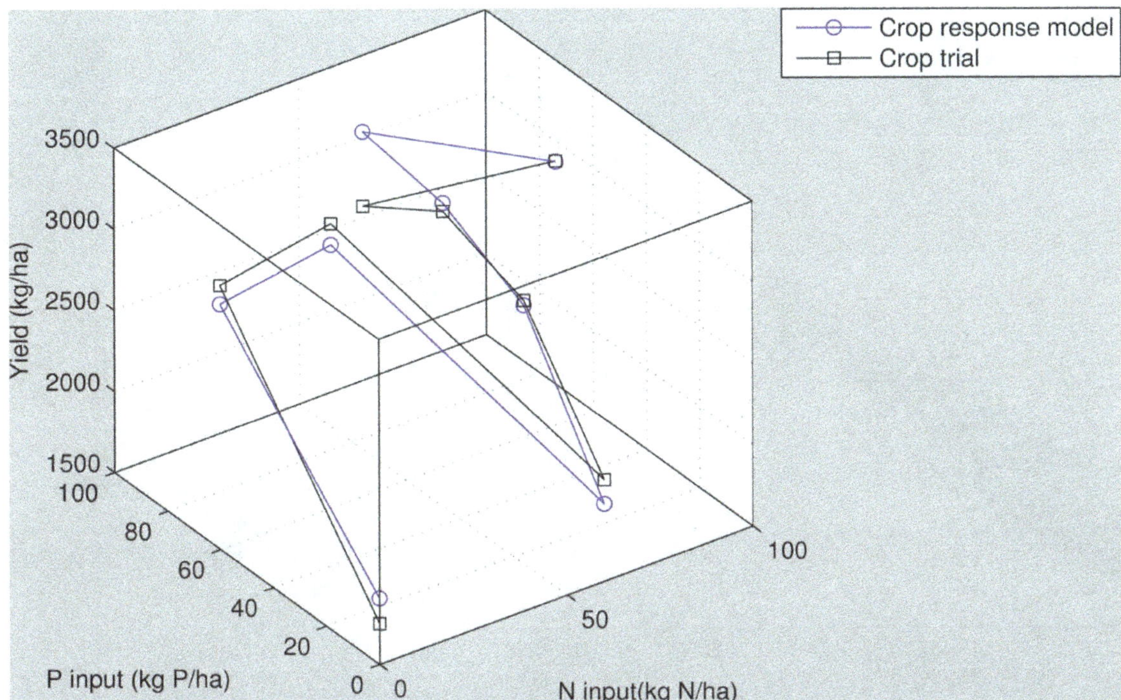


**Figure 3. A typical example of a fitted crop response trial with experimental (blue line with circles) and modeled data (black line with squares) with eight N and P input combinations and resulting yields.**

of Urea ($460/ton or 0.22$ kg$^{-1}$ N) and DAP ($680/ton or 0.16$ kg$^{-1}$ P)). The second scenario or larger nutrient inputs would amount to investments of US$798 million in sub-Sahara Africa and US$428 million in South America respectively. Maize yield increases range from 15% to over 70%, and such an investment would therefore bring considerable additional revenue to maize farmers. We acknowledge the fact that the prices of both fertilizer and maize vary with location so the actual profitability of fertilizer investment would vary spatially.

However, the direct investment in fertilizers is actually very small and is less than 1% of Agricultural GDP of both Sub-Saharan Africa and South America for the micro-dosing scenario. The calculation of the investment in terms of the percentage of Agricultural GDP that would be needed in SSA for scenario 1 equates to dividing 148 million USD by the Agricultural GDP of $150 billion USD multiplied by 100. The percentage of investment in terms of Agricultural GDP for the other scenario for SSA and the scenarios for SA were calculated in the same way (see Table 2).

**Table 1.** The median, 25th and 75th percentile values from the distributions of the Mitscherlich-Baule crop response function parameters ($a_1$, $a_2$, $a_3$ and $a_4$) fitted for the 1358 individual crop trials.

| Parameter | Median | 25th percentile | 75th percentile |
|---|---|---|---|
| a1 | 0.017321 | 0.0077335 | 0.047077 |
| a2 | 68.4297 | 21.6344 | 285.0599 |
| a3 | 0.29219 | 0.047732 | 0.29219 |
| a4 | 3.1803 | 0.99095 | 13.0806 |

## Discussion

In reality a full development blueprint would need to have a broader scope and costs would be compounded with investments in roads, agricultural extension, (local) market access, etc. [5,6]. Nevertheless, external investment in this low technology solution has the potential to kick start development and could complement other interventions such as better crop varieties, improved economic instruments to support farmers as well as new technologies involving mobile phones, crowdsourcing and data mining of internet searches [36,37]. To improve our understanding of how best to target, design and support rural development, it is insightful to compare the costs calculated here with the costs involved in a project such as the Millennium Villages (MV). The MV is an integrated approach to eradicate poverty by involving an entire community in improving their livelihoods and health in a sustainable way (http://www.millenniumvillages.org). The focused investments calculated here are significantly lower per person per year compared to the costs in the MV, which vary between 35 to 100 USD per person per year [38]. However, we clearly acknowledge that the MV has a much broader scope, engages entire communities and contributes to many other aspects of well-being and improved livelihoods as set out in the Millennium Development Goals, which would not be part of the scenarios suggested here.

We have clearly demonstrated the importance of phosphorus for closing yield gaps in Africa. The phosphorus deficiency reflects soil P supply problems that are of widespread concern in highly weathered tropical soils notorious for low levels of available P and exhibit a strong P fixation capacity [39]. For instance, approximately 82% of the land area of the American tropics is deficient in P in its natural state [40]. Combined with the fact that P reserves are likely to become exhausted during the next 30–300 years [41], this paints a bleak picture indeed. Nevertheless, if recycling programs were put in place for animal and

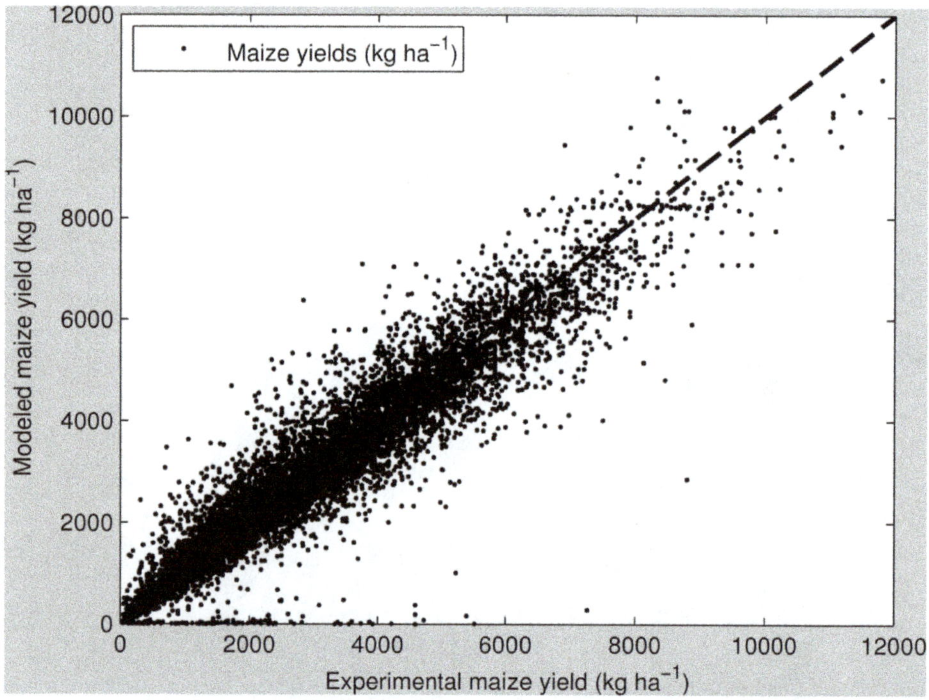


**Figure 4. Relationship between historic FAO experimental corn field trials with at least five N and P input combinations and corn yields calculated with the Mitscherlich-Baule crop response function totaling 1358 unique nutrient-yield relations ($r^2 = 0.94$).**

human excreta some of these effects might be mitigated, e.g. [42]. Raising awareness of the need to provide a more balanced stoichiometry is also a critical element in improving yields. If farmers continue to add increased supplies of N without P, they will soon reach a saturation point in yields and effectively waste valuable resources as well as contaminating groundwater due to the leaching of nitrogen. This finding also has implications for the current fertilizer subsidy programs in many developing countries. Most of these programs focus mainly on N and do not emphasize the importance of P sufficiently. Our results demonstrate that better balanced subsidy

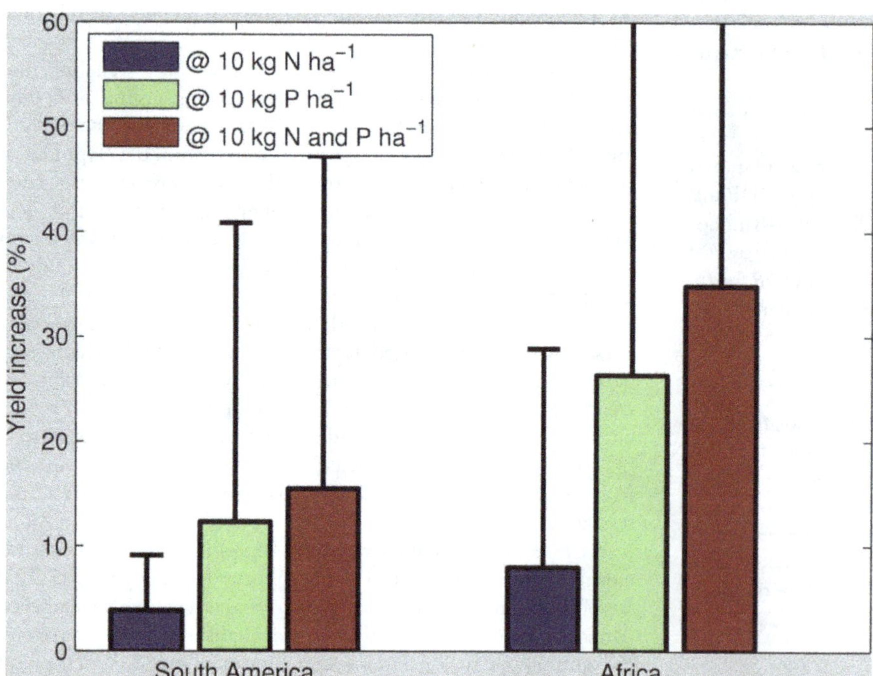


**Figure 5. Mean corn yield increase (%) across trial sites at additional applications of 10 kg N ha$^{-1}$, 10 kg P ha$^{-1}$ or 10 kg N and P ha$^{-1}$ (error bars refer to the standard deviation of the obtained yield increases observed across all trials).**

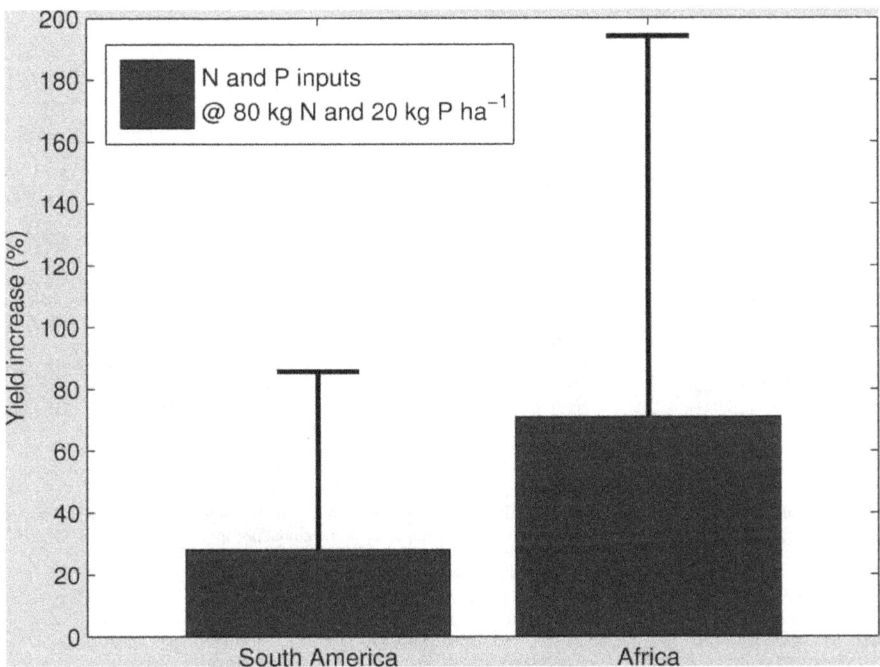


**Figure 6. Mean corn yield increase (%) across trial sites at applications of 80 kg N ha$^{-1}$ and 20 kg P ha$^{-1}$ (error bars refer to the standard deviation of the obtained yield increases observed across all trials).**

schemes taking account of both N and P would have a larger effect on crop yields.

Climate change is expected to generally have a negative impact on corn in Africa with estimates of lowering yield ranging from 3 to over 12% [43]. Corn will be the most heavily impacted crop in sub-Saharan Africa, where yield losses would occur in 65% of corn growing regions for a 1°C warming, increasing to 100% losses in areas subject to drought stress [44]. Although these pressures are considerable, and will require adaptation and fundamental changes to agricultural management, our results indicate that significant increases in yield are possible by improved nutrient management; especially during growing seasons when soil water availability is not constraining crop yields. Thus, achieving trend-growth in crop productivity through sufficient and balanced nutrient applications coupled with effective storage policies could partly offset negative climate change impacts on food security in Africa. The risk averseness of poor farmers that are prone to drought is one of the reasons why these farmers will be hesitant to invest in higher fertilizer applications. However, we have shown that a relatively small external investment would yield large improvements in both crop production and sustainable use of soils.

In contrast to previous crop productivity assessments for Africa and South America, our study allows farm input management interventions to be directly based on small scale on-the-ground observations accounting for both site-specific conditions, as well as reflecting variability in soil conditions and climates. Furthermore, the study provides yield gaps that are realistically attainable as the assessment is based on farmers' field trials and the costs associated with these interventions amount to less than 1% of Agricultural GDP in both Sub-Saharan Africa and South America. Crop field trials might be considered costly by some but provide essential and hard-won insights, and when analyzed comprehensively, have the potential - through better formulated policies and agreements - to reward global society with improved food security status for many.

**Table 2.** Cost of the proposed scenarios expressed as the percentage of Agricultural GDP.

| Region | Cost as a % of Agricultural GDP | |
|---|---|---|
| | Scenario 1 | Scenario 2 |
| SSA | 0.10% | 0.53% |
| SA | 0.04% | 0.22% |

## Supporting Information

**Figure S1 Current N and P inputs from chemical fertilizer (Nfer, Pfer) and manure (Nman, Pman) extracted and averaged from [22] for the 1358 trial locations.**

**Figure S2 Histograms of the N and P nutrient inputs from chemical fertilizer at the 1358 locations [22].**

**Figure S3 Histograms of the N and P nutrient inputs from manure at the 1358 locations [22].**

**Figure S4 Boxplots of regionally reported corn yields collected by IFPRI and corn yields obtained from the 1358 crop response functions (CRFs) with current N and P inputs from chemical fertilizer and manure (Nman, Pman) as reported by [22].**

## Author Contributions

Analyzed the data: MvdV LS LY. Wrote the paper: MvdV LS LY JB SF NK MO SW.

## References


1. FAO (2009). How to Feed the World in 2050. Report from the High-Level Expert Forum. http://www.fao.org/fileadmin/templates/wsfs/docs/expert_paper/How_to_Feed_the_World_in_2050.pdf. Accessed 2013 Feb 28.
2. Rosegrant MW, Cline SA (2003) Global food security: challenges and policies. Science 302: 1917–1918.
3. Godfray HCJ, Beddington JR, Crute IR, Haddad L, Lawrence D, et al. (2010) Food security: the challenge of feeding 9 billion people. Science 327(5967): 812–818. doi: 10.1126/science.1185383.
4. Licker R, Johnston M, Foley JA, Barford C, Kucharik CJ, et al. (2010) Mind the gap: how do climate and agricultural management explain the 'yield gap' of croplands around the world? Global Ecology and Biogeography 19: 769–782.
5. Foley JA, Ramankutty N, Brauman KA, Cassidy ES, Gerber JS, et al. (2011) Solutions for a cultivated planet. Nature 478: 337–342.
6. Mueller ND, Gerber JS, Johnston M, Ray DK, Ramankutty N, et al. (2012) Closing yield gaps through nutrient and water management. Nature. Doi:10.1038/nature11420.
7. Maitima JM, Mugatha SM, Reid RS, Gachimbi LN, Majule A, et al. (2009) The linkages between land use change, land degradation and biodiversity across East Africa. African Journal of Environmental Science and Technology 3(10): 310–325.
8. Sanchez PA (2002) Soil fertility and hunger in Africa. Science 295: 2019–2020.
9. Bumb B, Baanante C (1996) World trends in fertilizer use and projections to 2020. 2020 Brief No. 38, International Food Policy Research Institute, Washington, DC, USA.
10. Ejeta G (2010) African Green Revolution needn't be a mirage. Science 327: 831–832.
11. Thurlow R, Kilman S (2009) Enough: Why the world's poorest starve in an age of plenty. New York: Perseus Books. 302 p.
12. Nin-Pratt A, Johnson M, Magalhaes E, You L, Diao X, et al. (2011) Yield gaps and potential growth in Western and Central Africa. IFPRI Research Monograph. International Food Policy Research Institute (IFPRI), Washington, USA. DOI: 10.2499/9780896291829.
13. Lal R (2010) Managing soils for a warming earth in a food-insecure and energy-starved world. J Plant Nutr Soil Sci 173: 4–15.
14. Tmowlow S, Rohrbach D, Dimes J, Rusike J, Mupangwa W, et al. (2010) Micro-dosing as a pathway to Africa's Green Revolution: evidence from broad-scale on-farm trials. Nutr Cycl Agroecosyst 88: 3–15.
15. The World Bank (2007) Africa fertilizer policy toolkit: Promoting fertilizer use in African agriculture: Lessons learned and good practice guidelines. http://www.worldbank.org/html/extdr/fertilizeruse/about.html. Accessed 2012 August 22.
16. UBOS (2002) Uganda national household survey 1999/2000. Report on the crop survey module. Entebbe, Uganda: Uganda Bureau of Statistics: 4–30.
17. Bumb B, Johnson M, Fuente PA (2011) Policy Options for Improving Regional Fertilizer Markets in West Africa. IFPRI Discussion Paper 01084, International Food Policy Research Institute, Washington, DC, USA.
18. Lobell DB, Cassman KG, Field CB (2009) Crop yield gaps: their magnitudes, and causes. Annu Rev Environ Resourc 34: 179–204.
19. Mueller ND, Gerber JS, Johnston J, Ray DK, Ramankutty N, et al. (2012) Closing yield gaps through nutrient and water management. Nature 490: 254–257.
20. Sanchez PA, Ahamed S, Carré F, Hartemink AE, Hempel J, et al. (2009) Digital soil map of the world. Science 325(5941): 680–681.
21. FAO (2011) The state of the world's land and water resources for food and agriculture. Managing systems at risk. FAO and Earthscan.
22. Potter P, Ramankutty N, Bennett EM, Donner SD (2010) Characterizing the spatial patterns of global fertilizer application and manure production. Earth Interac 14: 1–22.
23. IFPRI (2012) Global subnational crop database. IFPRI, Washington, DC, USA.
24. Van der Velde M, See L, Fritz S (2012) Soil remedies for small-scale farming. Nature 484: 318, doi:10.1038/484318c.
25. Gilbert N (2012) Africa agriculture: dirt poor. Nature 483: 525–527, doi:10.1038/483525a.
26. Bindraban PS, Van der Velde M, Ye L, Van den Berg M, Materechera S, et al. (2012) Assessing the impact of soil degradation on food production. Current Opinion in Environmental Sustainability, 4(5), 478–488, doi:10.1016/j.cosust.2012.09.015.
27. FAO (1989) Fertilizers and food production: Summary review of trial and demonstration results 1961–1986. FAO, Rome.
28. Van Averbeke W, Yoganathan S (1997) Using Kraal Manure as a Fertiliser. Department of Agriculture and Agricultural and Rural Development Research Institute, Fort Hare.
29. Frank MD, Beattie BR, Embleton ME (1990) A comparison of alternative crop response models. Am J Agric Econ 72(3): 597–603.
30. Sattari SZ, Bouwman AF, Giller KE, van Ittersum MK (2012) Residual soil phosphorus as the missing piece in the global phosphorus crisis puzzle. Proc Natl Acad Sci U S A doi: 10.1073/pnas.1113675109.
31. IFA (2002) Fertiliser use by crop. Rome, 2002, 45 pp.
32. The World Bank (2011) Data. http://data.worldbank.org/indicator. Accessed 2013 Feb 28.
33. Finger R, Hediger W (2008) The application of robust regression to a production function comparison. The Open Agriculture Journal 2: 90–98.
34. Pingali PL, Pandey S (2001) World maize needs meeting: technological opportunities and priorities for the public sector. In: PL Pingali (Ed.) CIMMYT 1999–2000 World Maize Facts and Trends. Meeting World Maize Needs: Technological Opportunities and Priorities for the Public Sector, 1–24. Mexico: CIMMYT.
35. Tittonell P, Vanlauwe B, Corbeels M, Giller KE (2008) Yield gaps, nutrient use efficiencies and response to fertilizers by maize across heterogeneous smallholder farms of western Kenya. Plant Soil 313: 19–27.
36. Fritz S, McCallum I, Schill C, Perger C, See L, et al. (2012) Geo-Wiki: An online platform for improving global land cover. Environmental Modelling & Software 31: 110–123, doi: 10.1016/j.envsof.2011.11.015.
37. Van der Velde M, See L, Fritz S, Verheijen FGA, Khabarov N, et al. (2012) Generating crop calendars with Web search data. Environmental Research Letters 7(2): 024022, doi:10.1088/1748-9326/7/2/024022.
38. Pronyk P (2012) The Costs and Benefits of the Millennium Villages: Correcting the Center for Global Development. http://www.millenniumvillages.org/field-notes/archive/2012/4. Accessed 2013 Feb 28.
39. Buresh RJ, Smithson PC, Hellums DT (1997) Building soil phosphorus capital in Africa. In Buresh RJ, Sanchez PA and Calhoun F (eds.) Replenishing soil fertility in Africa. SSSA/ASA. Madison, WI. 111–149.
40. Sanchez PA, Salinas JG (1981) Low-input technology for managing Oxisols and Ultisols in tropical America. Advances in Agronomy 34: 279–406.
41. Cordell D, Drangeert JO, White S (2009) The story of phosphorus: Global food security and food for thought. Glob Environ Change 19: 292–305.
42. Baker LA (2011) Can urban P conservation help to avoid the brown devolution. Chemosphere 84: 779–784.
43. Nelson GC, Rosegrant M, Palazzo A, Gray I, Ingersoll C, et al. (2010) Food security, farming, and climate change to 2050. IFPRI Research Monograph. International Food Policy Research Institute (IFPRI), Washington, USA. DOI: 10.2499/9780896291867.
44. Lobell DB, Banziger M, Magorokosho C, Vivek B (2011) Nonlinear heat effects on African corn as evidenced by historical yield trials. Nat Clim Chang DOI: 10.1038/NCLIMATE1043.Dafadf.

# 14

# Eco-Stoichiometric Alterations in Paddy Soil Ecosystem Driven by Phosphorus Application

**Xia Li[1], Hang Wang[1], ShaoHua Gan[1], DaQian Jiang[2], GuangMing Tian[1], ZhiJian Zhang[1,2,3]***

**1** College of Environmental and Resource Science, Zhejiang University, Hangzhou, China, **2** Department of Earth and Environmental Engineering, Columbia University, New York, New York, United States of America, **3** China Academy of West Region Development, Zhejiang University, Hangzhou, China

## Abstract

Agricultural fertilization may change processes of elemental biogeochemical cycles and alter the ecological function. Ecoenzymatic stoichiometric feature plays a critical role in global soil carbon (C) metabolism, driving element cycles, and mediating atmospheric composition in response to agricultural nutrient management. Despite the importance on crop growth, the role of phosphorous (P) in compliance with eco-stoichiometry on soil C and nitrogen (N) sequestration in the paddy field remains poorly understood in the context of climate change. Here, we collected soil samples from a field experiment after 6 years of chemical P application at a gradient of 0 (P-0), 30 (P-30), 60 (P-60), and 90 (P-90) kg $ha^{-1}$ in order to evaluate the role of P on stoichiometric properties in terms of soil chemical, microbial biomass, and eco-enzyme activities as well as greenhouse gas (GHG: $CO_2$, $N_2O$ and $CH_4$) emissions. Continuous P input increased soil total organic C and N by 1.3–9.2% and 3%–13%, respectively. P input induced C and N limitations as indicated by the decreased ratio of C:P and N:P in the soil and microbial biomass. A synergistic mechanism among the ecoenzymatic stoichiometry, which regulated the ecological function of microbial C and N acquisition and were stoichiometrically related to P input, stimulated soil C and N sequestration in the paddy field. The lower emissions of $N_2O$ and $CH_4$ under the higher P application (P-60 and P-90) in July and the insignificant difference in $N_2O$ emission in August compared to P-30; however, continuous P input enhanced $CO_2$ fluxes for both samplings. There is a technical conflict for simultaneously regulating three types of GHGs in terms of the eco-stoichiometry mechanism under P fertilization. Thus, it is recommended that the P input in paddy fields not exceed 60 kg $ha^{-1}$ may maximize soil C sequestration, minimize P export, and guarantee grain yields.

**Editor:** Jose Luis Balcazar, Catalan Institute for Water Research (ICRA), Spain

**Funding:** This work is supported by Natural Science Foundation of China (40701162)(http://www.nsfc.gov.cn/Portal0/default152.htm). The funders had no role in study design, data collection and analysis, decision to publish, or preparation of the manuscript.

**Competing Interests:** The authors have declared that no competing interests exist.

* E-mail: zhangzhijian@zju.edu.cn

## Introduction

The balance of elements has been a main focus of global change ecology and biogeochemical cycling research. Phosphorus (P) application remains an indispensable practice for agricultural crop production. However, P export from soil to surface waters may stimulate outbreaks of water eutrophication [1]. Meanwhile, carbon (C) storage in ecosystems is controlled by the mass conservation principle and the supply of other key nutrients, such as nitrogen (N) and P [2]. Therefore, maintaining a sustainable C-N-P balance in the soil ecosystem is necessary for coping with climate change, maximizing agricultural production, and optimizing P practice.

Ecological stoichiometry (Eco-stoichiometry) is based on stoichiometric theory and the metabolic theory of ecology, which involves the balance of energy and multiple chemical elements in ecological interactions at the subcellular to ecosystem scale [3]. Eco-stoichiometry, expressed as C:N:P stoichiometric ratio, can predict nutrient cycling and microbial biomass production in ecosystems [4,5,6] and plays an important role in element regulation during biosphere-scale processes, such as soil C storage and element balance in the soil biomass [7], and also governs greenhouse gas (GHG) emissions in terrestrial ecosystems [8]. Therefore, P fertilization coupled with element eco-stoichiometry may be a determining incentive in defining the dynamics that balance C-N-P and predicting GHG emissions in the soil ecosystem.

Microorganisms drive Earth's biogeochemical cycles [9] by a "consumer-driven nutrient recycling" (CDNR-like) mechanism that determines nutrient cycling, biomass stoichiometry, and community composition [10], and mediates the global C cycle during climatic changes [11]. In turn, this influences the ecological metabolic rate [4]. Measurements of the proportion of C, N, and P in the microbial biomass may thus be a practical tool for assessing the nutrient limitations of an ecosystem. For example, a low C-to-P ratio of microorganism biomass (MBC:MBP) may stimulate soil microorganisms to release nutrients and enhance the available P pool in the environment, while a high MBC:MBP ratio could cause the microorganisms to compete for available P and enhance soil P immobilization [12]. Conceptually, plasticity and homeostasis are the fundamental mechanisms by which organisms adjust the stoichiometric equilibrium to cope with environmental disturbances [5,13]. Exogenous P input would alter the primary stoichiometric balances among the soil-microorganisms complex, which could change soil C and N storage. However, the mechanisms on interaction between the exogenous P and soil organism stoichiometry as well as the ecological feedback to dynamics of soil C and N are still unknown.

Eco-enzyme activity represents an intersection of the ecological stoichiometry, wherein eco-enzyme activity (EEA) links environmental nutrient availability with microbial production [3]. Enzyme expression is regulated by environmental signals, while ecoenzymatic activity is determined by environmental interactions [11]. This in turn mediates nutrient cycling, sequestration from soil organic matter, and decomposition biochemistry [3]. The most widely assayed eco-enzymes, β-1,4-glucosidase (BG), β,4-N-acetylglucosaminidase (NAG), leucine aminopeptidase (LAP), and acid (alkaline) phosphatase (AP), hinge functional stoichiometries in relation to organic nutrient acquisition and are used as indicators of microbial nutrient demand [3,11]. These extracellular enzymes deconstruct plant and microbial cell walls into soluble substrates for microbial assimilation, and are a measure of microbial nutrient demand [14], which reveals the rate limitations of enzymatic catalysis in relation to soil carbon storage [15]. As such, the EEA should be sensitive to the effects of P application on microbial function and provide a mechanistic indicator of P for resource acquisition in the soil ecosystem. However, to date, few studies have focused on microbial function for resource acquisition under P input in soil, and thus the underlying mechanisms are largely unknown.

The interactions among C, N, and P cycling also determine the effect of GHG emissions on the Earth's climate through their influences on C and N sequestration in soil [16,17]. Agroecosystems contribute a large percentage of global emissions, including ~60% of $N_2O$, ~39% of $CH_4$, and ~1% of $CO_2$ [18]. As one of the important cereal crops, paddy fields (approximately 28.4 M ha) in China contribute to approximately 30% of the total global rough rice yield [17,19]. Combined with $CO_2$, N fertilization may generate intermediate nitrogenous gases ($N_2O$) [20]. Better paddy fertilization management strategies are thus needed to mitigate GHG emissions [21] and preserve soil productivity [22]. Relative to N, P is a static entity, which is strongly retained in the soil matrix [23] due to its lack of a significant gaseous phase. Therefore, P management coupled with stoichiometric methodology may offer an advantageous technology for controlling GHGs emitted from paddy fields, but remains inadequately studied.

To date, few investigations have probed the role of P fertilization on soil eco-stoichiometry in paddy fields, which hinders the optimization of C-N-P biogeochemical cycles. In 2005, a paddy field experiment with annual applications of chemical P fertilizer at rates of 0.0, 30, 60, and 90 kg P $ha^{-1}.y^{-1}$ was conducted in the Yangtze River delta in southeastern China to understand P driving soil eco-stoichiometry in a paddy field ecosystem. In 2011, we collected soil samples from these experimental paddy plots to investigate the stoichiometry of soil chemical, microbial biomass, and eco-enzyme activities in response to P fertilization. Data on GHG fluxes in July and August were also probed to understand the role of P on typical GHG emissions. We hypothesized that P application could change both soil C-N-P balance and eco-stoichiometric features in paddy soil and thus stimulate soil C sequestration.

## Results

### The Effect of Phosphorus on Soil Biochemical Features and GHG Emissions

Soils collected from four treatments varied noticeably in chemical, microbial, and eco-enzymatic properties (Table 1). Continuous P application significantly enhanced total soil P by 16–75% compared to P-0. In addition, total soil C was significantly ($p<0.05$) increased by 1.3% to 9.2% for P input compared to P-0, while soil N increased from 3–13% ($p<0.05$). Similarly, soil MBC was increased by 20% to 27% under P-60 and P-90; MBN was also increased under P-60 and P-90 by 28–50%, while an increase in MBP of 52–195% was observed. Due to P application, the ratios of soil MBC:MBP, MBC:MBN, and MBN:MBP were significantly increased with increased ratios of soil C:P ($p$ = 0.001), soil C:N ($p$ = 0.039), and soil N:P ($p$ = 0.002), respectively (Fig. 1). The activities of soil eco-enzymes, such as BG, were enhanced from 39% to 75% compared to P-0, with AP decreasing by 14–33% ($p<0.01$) and NAG+LAP showing no significant changes (Table 1).

Net $CO_2$-C flux ranged from −88.7 to −204 $mg{\cdot}m^{-2}{\cdot}\ h^{-1}$ and increased significantly ($p<0.05$) with increasing P application (Fig. 2). The net emission of $CO_2$ increased by 7–45% under P input compared to P-0 in July, with increases of 17% to 40% in August. $CH_4$-C flux, which ranged from 1.7–8.8 $mg{\cdot}m^{-2}\ h^{-1}$ in July, was significantly ($p<0.05$) lower by 14%–57% under P input compared to P-0. However, the highest $CH_4$ flux occurred with P-30 in August. The $N_2O$-N flux in our study ranged from −0.02 to 0.05 $mg{\cdot}m^{-2}\ h^{-1}$. $N_2O$ emission in July was significantly ($p<0.05$) reduced by 34–75% in the tested paddy field with increasing P fertilization compared with P-0; no significant difference was found in August (Fig. 2). Analysis of variance (ANOVA) showed that the soil $CO_2$ flux significantly responded to single factors of P application and sampling time; $CH_4$ flux significantly responded to single factors of P application and interactions of treatment and sampling date, while $N_2O$ flux was significantly affected by P application and sampling date as well as interactions (Table 2).

### The Effect of Phosphorus on Soil Eco-stoichiometry

The stoichiometric ratios of soil C:P declined (from 156 to 97) significantly ($p<0.01$) with increasing P applications, and the ratio of soil N:P was reduced from 11.4 to 7.3 (Fig. 3). However, no significant difference was found for the soil C:N ratio among the four treatments. Soil MBC:MBP and MBN:MBP ratios were both found to be significantly ($p<0.05$) decreased with increases in P application, while the MBC:MBN ratio remained unchanged, except for P-90 (Fig. 3). The acquisition of C relative to organic P indicated by the ratio of ln(BG):ln(AP) was significantly increased with increasing P application (Fig 3). The ln(BG):ln(NAG+LAP), which refers to the acquisition of C relative to organic N, showed no significant differences within P application treatments, but was significantly ($p<0.01$) increased compared to P-0. However, ln(NAG+LAP):ln(AP), which refers to the acquisition of N relative to organic P, showed no significant differences with P application (Fig. 3). Scatter plots of C:N:P stoichiometry for soil chemistry, microbial biomass, and eco-enzymatic activities for the paddy system under P application compared to P-0 indicated that P application caused a C-N co-limitation on ecological stoichiometric processes (Fig. 4). This result suggests that there was a higher C and N demand for soil microorganisms relative to P, integrated soil stoichiometric ratios with biochemical properties, and eco-enzymatic activities (Table 1).

The linear relationships between soil C:P and ln(BG):ln(AP) ($p$ = 0.001), MBC:MBP and ln(BG):ln(AP) ($p$ = 0.000), and MBN:MBP and ln(NAG+LAP):ln(AP) ($p$ = 0.050) were significant and showed a negative trend (Fig. 1). However, there was no significant relationship between soil C:N and ln(BG):ln(NAG+LAP), soil N:P and ln(NAG+LAP):ln(AP), or MBC:MBN and ln(BG):ln(NAG+LAP). The significant relationships showed that soil C:P, soil C:N, soil N:P, soil C:P, MBC:MBP, and MBN:MBP substantially influenced MBC:MBP, MBC:MBN, MBN:MBP, ln(BG):ln(AP), ln(BG):ln(AP), and ln(NAG+LAP):ln(AP), respectively, and could account for the respective 69.1%, 36.2%, 64.6%,

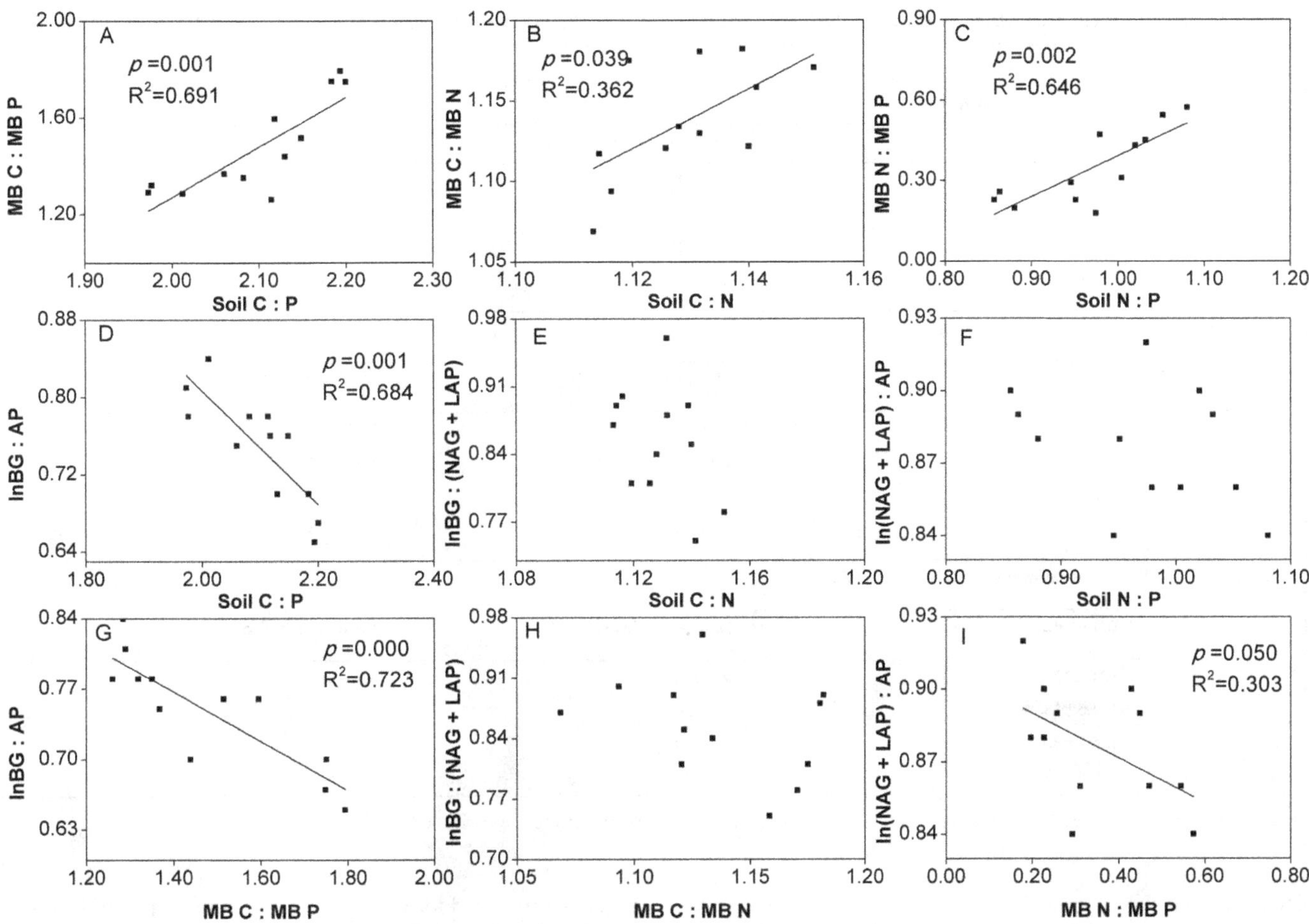


**Figure 1. Linear regression of C:N:P stoichiometry for soil chemistry, microbial biomass, and eco-enzymatic activities.** Summary of standardized major axis analysis of log10-transformed molar nutrient concentrations. Ratios of C:P, C:N, N:P acquisition activity, as indicated by ratios of ln(BG):ln(AP), ln(BG):ln(NAG+LAP), and ln(NAG+LAP):ln(AP), respectively.

68.4%, 72.3%, and 30.3% variances. Taken together, these results suggest that the microbial biomass stoichiometric ratios were positively influenced by the soil stoichiometric ratios, and both of them were negatively regulated by coenzyme activities under P application.

## Discussion

### Eco-stoichiometry upon Phosphorus Application and the Linkage to Soil Carbon and Nitrogen

Phosphorus amendment increased the decomposition of soil organic C through increased soil respiration [7]. However, the data in Table 1 indicate that paddy soil receiving P application had increased soil C and N pools by 1.3–9.2% and 3%–13%, respectively. Such percent change in soil organic C pools is consistent with a similar long-term field investigation [24], which verified that chemical fertilization (45 kg P $ha^{-1}$) could increase ~10% higher soil organic C.

The fact that the soil C:P declined (Fig. 1) while soil C and N pools increased (Table 1) upon P input indicated a higher P concentration relative to C in soils, which caused the organic matter pool to become less humic (protein-like) and provided potential energy for microbial utilization [25]. P input modified not only C:N:P ratios in the soil, but also the soil microbial biomass. The positively significant relationships between chemical and microbial stoichiometry (Fig. 1) clearly indicated that microorganisms adapted their acquisition ratios according to local resource ratios. In our study, the decreased soil MBC:MBP upon P input (Fig. 3) demonstrated that P application results in C-limitation in microbial biomass, despite the finding that soil C pools with P input were increased (Table 1). This also suggested that soil microorganisms were limited to soil C allocations and that P enrichment would improve the primary productivity, thus stimulating C sequestration in soil receiving P application, although this C-P synergistic effect [3,26] would plateau under higher P input rates at P-60 and P-90. Moreover, the distribution of ecoenzymatic C:N:P activity ratios may identify the boundaries of the microbial community response to fluctuations in nutrient availability [3,11]. The soil microbial metabolic pattern tended to be C limited (Fig. 4) with P input together with the increased ratio of lnBG:lnAP (Fig. 3), which closely responded to the decrease in soil MBC:MBP (Fig. 1). These results indicated that a greater C demand for microbial biomass due to P input enhanced the ecological function of organic C acquisition within the soil-microorganism complex. Hence, our hypothesis that soil C sequestration may be enhanced by P input was verified. Additionally, further analysis showed that MBC:MBN:MBP (49:3.3:1.0 to 21:1.6:1.0, respectively) with P application (Table 1) deviated significantly from the generally mean atomic C:N:P ratio (60:7:1) in the soil microbial biomass [27]. The C:N:P ratio in the

**Table 1.** Soil chemical, microbial, and eco-enzymatic properties in the tested paddy field after 5 years of phosphorus application.

| Treatment | Soil chemical properties | | |
|---|---|---|---|
| | Total C (mmol $kg^{-1}$) | Total N (mmol $kg^{-1}$) | Total P (mmol $kg^{-1}$) |
| P-0 | 2070±46[b] | 151±3[c] | 13.3±0.5[d] |
| P-30 | 2096±67[b] | 155±5[bc] | 15.5±1.0[c] |
| P-60 | 2185±91[ab] | 162±3[b] | 17.9±0.4[b] |
| P-90 | 2260±78[a] | 171±4[a] | 23.3±0.6[a] |
| **Treatment** | **Soil microbial properties** | | |
| | MBC (mmol $kg^{-1}$) | MBN (mmol $kg^{-1}$) | MBP (mmol $kg^{-1}$) |
| P-0 | 40.7±2.3[b] | 2.76±0.14[c] | 0.83±0.08[d] |
| P-30 | 43.9±1.4[b] | 3.14±0.22[c] | 1.26±0.30[c] |
| P-60 | 48.8±2.3[a] | 3.54±0.16[b] | 2.07±0.22[b] |
| P-90 | 51.8±1.9[a] | 4.14±0.27[a] | 2.45±0.07[a] |
| **Treatment** | **Eco-enzymatic activity properties** | | |
| | BG (nmol $h^{-1}$ $g^{-1}$) | NAG+LAP (nmol $h^{-1}$ $g^{-1}$) | AP (nmol $h^{-1}$ $g^{-1}$) |
| P-0 | 80.0±10.5[c] | 276±43[a] | 638±36[a] |
| P-30 | 111.6±18.0[b] | 259±39[a] | 576±43[b] |
| P-60 | 126.2±8.1[ab] | 256±46[a] | 538±24[b] |
| P-90 | 140.3±15.2[a] | 231±35[a] | 451±51[c] |

The different letters listed beside the data represent significant differences at $p<0.05$ (Duncan test, one-way ANOVA).

biomass reflects the physiological and biochemical constraints on the elemental composition of primary production [6,27], but differences in soil organism habitats may preclude the emergence of constrained soil microbial element ratios [13]. Our data departed from the mean ratio, which suggested that the overall investment in structural cellular material in the microbial biomass was remarkably affected by P input, indicating that no rigorous homeostasis existed for the soil microbial community in the tested paddy soil.

Soil nutrient enrichment enhances root metabolic activity and the consequent excretion of organic C, which enters the mineral soil through rhizosphere deposition [28] and simulates N fixation by promoting nutrient equilibration [29]. P enrichment would exert a synergistic function of N fixation in the soil-microorganism-plant system [27], which may contribute to soil N accumulation

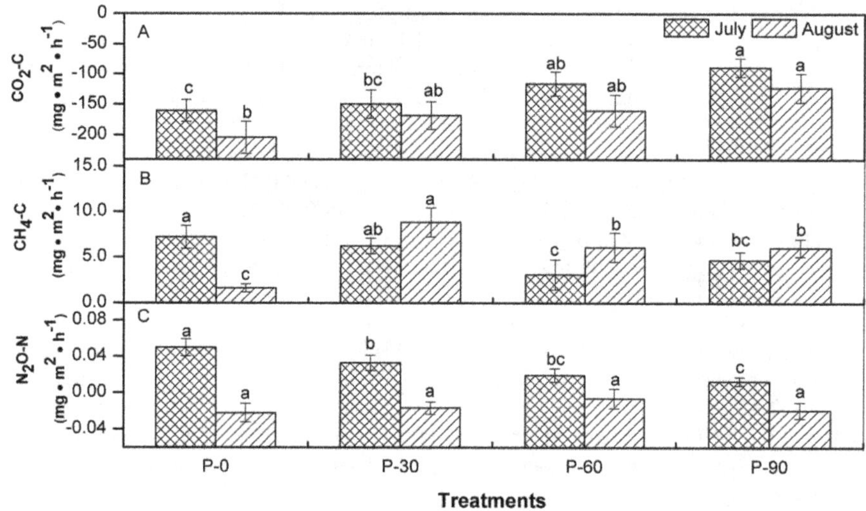


**Figure 2. Emission intensities of greenhouse gases responding to phosphorus application for samplings in July and August.** The different letters listed above bars represent significant differences at $p<0.05$ (Duncan LSD test).

**Table 2.** Results of two-way analysis of variance (repeated ANOVA) showing the *p* values for GHG emissions responding to P application and sampling time (July; August) for paddy field soil.

| Factor | $CO_2$ flux | $CH_4$ flux | $N_2O$ flux |
|---|---|---|---|
| Treatment | **<0.001** | **0.002** | **0.023** |
| Time | **0.001** | 0.510 | **<0.001** |
| Treatment×Time | 0.732 | **<0.001** | **0.001** |

(Table 1), although the soil N:P stoichiometry somewhat decreased in response to P application (Fig. 1). Moreover, the newly-fixed N might be easily stored in the soil and/or biomass under P fertilization, because no significant difference on N-related enzymatic activities was found among the four treatments (Table 1). Based on element stoichiometry [30,31], soil microorganisms would utilize more plant residues and absorb more C and N for growth with the continuous P application as well as increased soil MBC and MBN (Table 1), which might be favorable for soil C and N sequestration [28,31] in the paddy ecosystem. However, no significant difference occurred in the lnBG:ln(NAG+LAP) (with the exception of P-0) and ln(NAG+LAP):ln(AP) ratios (Fig. 3) among the P treatments, indicating that a synergistic mechanism among the ecoenzymatic stoichiometry regulates the ecological function of microbial C, N, and P acquisition. Continuous P input caused a significantly ($p<0.05$) negative relationship between the N:P eco-enzymatic and microbial stoichiometries (Fig. 1), which suggests that an abundant source of P in soil microorganisms stimulates the ecological function as a sink for soil N cycle. Together, these results support our hypothesis that P application helps to define the dynamics of the multiple balance of C-N-P in paddy soil, where the synergism of ecoenzymatic stoichiometry enhances soil C and N sequestration.

P application clearly demonstrated a predominance of C and N co-limitations for microorganism metabolic activities and C-N acquisition in our study (Fig. 4). Mechanistically, these may arise through interactions among C, N, and P in bio-molecules such as mRNA [32] and linkage to soil bacterial communities [33,34]. A biogeochemical equilibrium model verified that ecological stoichiometry constraints microbial community metabolism [11]. Further investigations are needed to determine the inter-annual variations on relationship between molecular biology and eco-enzymatic stoichiometry as well as to illustrate alterations in ecological function caused by long-term P input in paddy ecosystem. It is also an important mechanism that 30–40% of sequestered C would be transferred into sub-soil (below 30 cm) through plant roots in grassland [35]. Although a plow pan layer exists following an arable horizon in a paddy field, the vertical distribution of C sequestration over the entire soil profile at the view of C-P eco-enzymatic stoichiometry is needed to further investigate for paddy plots receiving long-term P application.

## The Role of Phosphorus on Greenhouse Gas Emission

The production of GHG emissions are regulated by soil physical properties and management practices, such as soil temperature, soil $O_2$ content, tillage, and alternations of wetting and drying

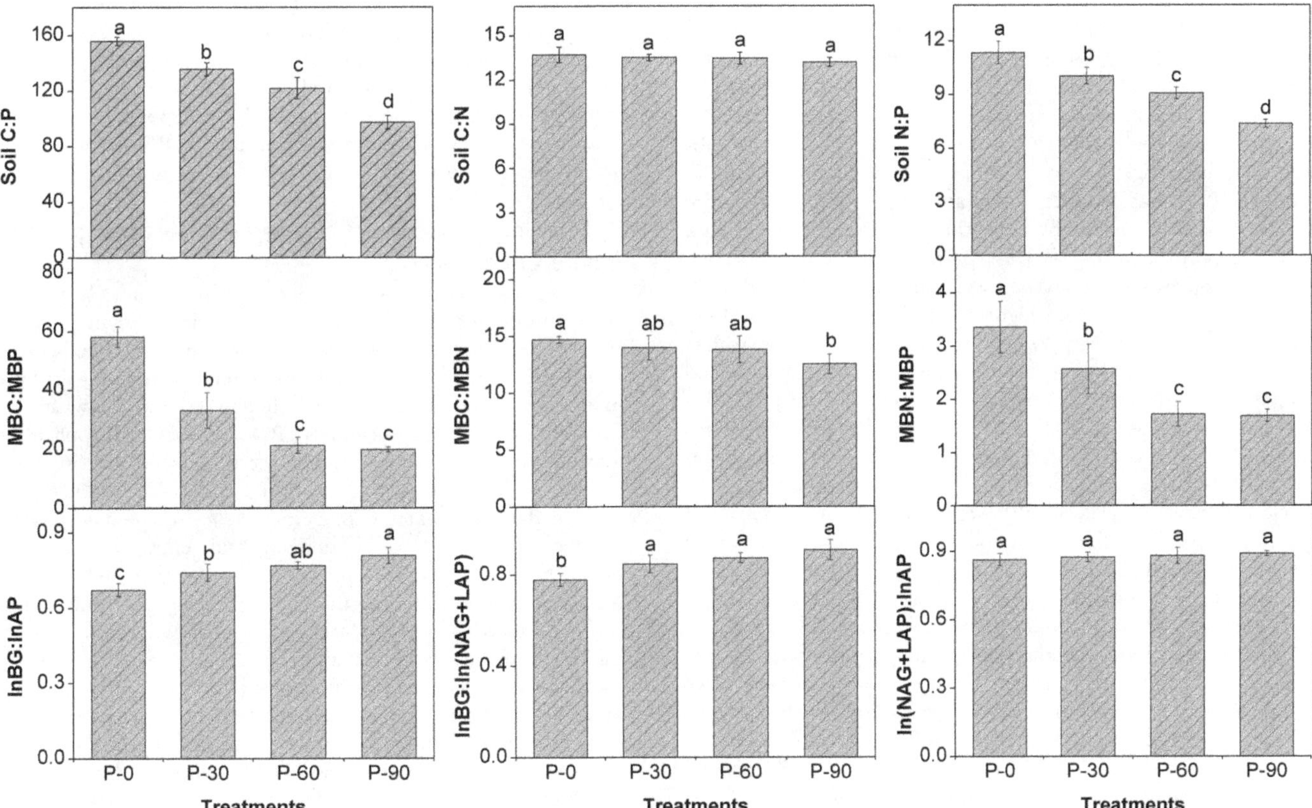


**Figure 3. Stoichiometric ratios of soil elements, soil microorganisms, and eco-enzymatic activities in the tested paddy soil under phosphorus treatments.** The different letters listed above bars represent significant differences at $p<0.05$ (Duncan test, one-way ANOVA).

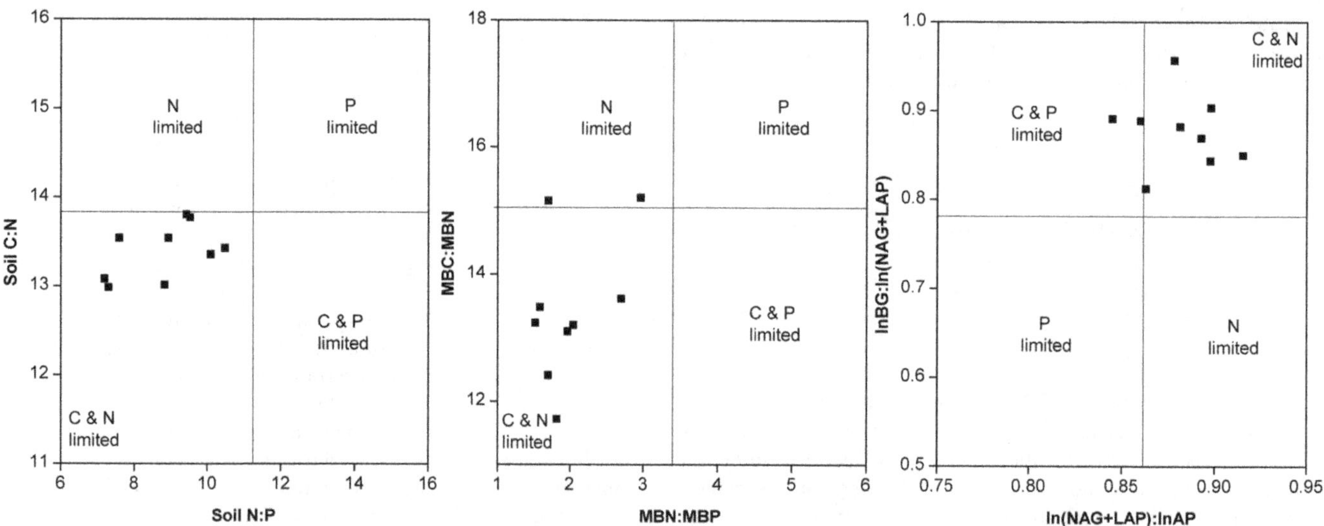


**Figure 4. Scatter plots of C:N:P stoichiometry for soil chemistry, microbial biomass, and eco-enzymatic activities for the paddy system under P application compared to P-0.**

[17,22,33]. In our study, the tested paddy soil varied only with P treatments, while the remaining environmental and agricultural factors were kept the same. Therefore, we hypothesized that the shifted GHG fluxes at two typical sampling seasons are mainly related to P treatment. Moreover, because the soil temperature is also a key factor for GHG emissions, July and August were chosen for estimating the GHG emissions subjected to P application due to the higher seasonal temperatures.

The net $CO_2$ flux is a balance of photosynthesis and respiration [17]. The negative value of the net $CO_2$ flux measured in this study was mainly due to the higher plant $CO_2$ uptake, rather than respiration emission. Increases in the soil C pool under continuous P input (Table 1) enhanced $CO_2$ fluxes over P application rates for both samplings (Fig. 2). A low P supply (fed by solution culture in 0.5 mg P $L^{-1}$) was found to stimulate root exudation and root aerenchyma development by more than two-fold compared to a high P supply (5–10 mg P $L^{-1}$) [36]. As might be expected, relatively higher $CH_4$ emissions were found for P-0 and P-30 sampling in July, while the highest emission occurred for P-30 sampling in August (Fig. 2). Shifting from continuous water-logging to midseason drainage [8,17,37] and increases of dissolved oxygen content of the rhizosphere [38,39] led to a drop in $CH_4$ flux and an increase in $N_2O$ flux in paddy soil. $CH_4$ emissions (except P-0) were commonly found to be higher for sampling in July (water-logging) than in August (oxygen secretion through developed roots); however, an overwhelmingly opposite trend occurred for $N_2O$ for the two samplings. The reason for the latter phenomenon is unclear and will require additional evaluation. The decrease ($p<0.05$) and mostly unchanged $N_2O$ flux was found upon P application in July and August, respectively (Fig. 2), which is basically consistent with the trend found in the maize system [40]. These results suggest that P input may mitigate $N_2O$ emission during the high temperature period in this study. Significant relationships between $CO_2$ flux and stoichiometric ratios (Table 3) indicated that $CO_2$ emission is not only influenced by soil chemical availability, which is basically in agreement with previous reports [7,17], but is also regulated by the co-limitation of the eco-enzymatic stoichiometric balance for paddy soil receiving P input. Although significant relationships related to stoichiometric ratios were found for both $CH_4$ and $N_2O$ emissions in July, the effect of these stoichiometric ratios for predicting both $CH_4$ and $N_2O$ is completely opposite to that of $CO_2$ (Table 3). Moreover, most of these significant relationships disappeared with increasing time (August) after P application. As such, a technical conflict would occur for simultaneously regulating three types of GHGs in terms of an eco-stoichiometry mechanism under conditions of P fertilization, and the associated strategies for such joint regulation would not be synchronous over time. GHG emissions are not only related to agricultural fertilization [38,39], but also the emission budget and associated net global warming potential, which are closely sensitive to the temporal and spatial traits [17,20,37]. Therefore, continuous probing of the eco-stoichiometric mechanisms for yearly GHG budgets responding to P input under the current field survey would be needed.

## Implication: Phosphorus Management for Coping with Climate Change

Enhanced soil C sequestration induced by P input (Table 1) is worth considering as a means of addressing climate change. However, such a practice would require technical caution, because the environmental costs of nutrient pollution from agriculture practice have been substantial, including deteriorated downstream water quality and eutrophication [1,21]. A survey of P application indicated that paddy fields may also act as an ecological sink for P dynamic with no greater than 60 kg $ha^{-1}$ of P input [23], provided that field water management (e.g., water-saving irrigation technology) [41] is optimized and takes into account physicochemical P soil absorption [29,42]. Supplemental cultivation of green manure during the fallow period (e.g., Chinese milk vetch, *Astragalus sinicus L*) or application of organic fertilizers are well-known means for enhancing soil C and N fixation [34,43]. Based on the element stoichiometric balance [3,27] and co-limitation of C and N for soil microbes (Fig. 4) in this study, excessive P in the paddy soil due to continuous P input might be effectively composited with C and N from milk vetch biomass with regard to the synergistic mechanisms among C-N-P in soil-microorganism complexes, leading to further soil C-N sequestration. Currently, rice grain yield (Table S1) among the experimental plots under P-30 (7580 kg $ha^{-1}$) was significantly higher than that of P-0 (6200 kg $ha^{-1}$), while no significant difference was

**Table 3.** Pearson correlation coefficients between $CO_2$, $CH_4$, and $N_2O$ fluxes of paddy field and soil stoichiometric ratios.

| Month | GHG flux | Chemical ratio | | | | Microbial ratio | | | Eco-enzyme ratio | |
|---|---|---|---|---|---|---|---|---|---|---|
| | | C : P | C : N | N : P | MBC: MBP | MBC: MBN | MBN: MBP | ln (BG) : ln (AP) | ln (BG) : ln (NAG+LAP) | ln (NAG+LAP) : ln (AP) |
| July | $CO_2$ flux | −0.871** | −0.578* | −0.851** | −0.732** | −0.547 | −0.675* | 0.818** | 0.786** | 0.361 |
| | $CH_4$ flux | 0.572 | 0.194 | 0.589* | 0.738** | 0.203 | 0.814** | −0.618* | −0.616* | −0.232 |
| | $N_2O$ flux | 0.841** | 0.452 | 0.839** | 0.862** | 0.549 | 0.809** | −0.740** | −0.690* | −0.372 |
| Aug | $CO_2$ flux | −0.809** | −0.446 | −0.804** | −0.658* | −0.499 | −0.555 | 0.778** | 0.822** | 0.155 |
| | $CH_4$ flux | −0.348 | −0.224 | −0.341 | −0.591* | −0.332 | −0.475 | 0.520 | 0.414 | 0.451 |
| | $N_2O$ flux | −0.195 | −0.692* | −0.099 | −0.276 | −0.216 | −0.119 | 0.093 | 0.233 | −0.308 |

Soil eco-enzymatic activities are presented as log10-transformed molar nutrient concentrations.

found between P-60 (9100 kg $ha^{-1}$) and P-90 (9070 kg $ha^{-1}$). Grain yield under P-60 is close to the highest level among a survey overview on rice yield [44] in ZheJiang, China. Thereafter, P input to paddy fields at a rate not to exceed 60 kg $ha^{-1}$ is recommended for maximizing soil C sequestration, minimizing P export, and guaranteeing grain yields.

In summary, we found that continuous P application increased the soil C and N pools. P input directly modified soil C:N:P ratios, which shifted the microbial stoichiometry of the biomass to be both C-limited and N-limited. Moreover, a synergistic mechanism among the eco-enzymatic stoichiometry regulates the ecological function of microbial C, N, and P acquisition, thus stimulating C and N sequestration in soil receiving P application in paddy fields. The P application significantly mitigated $N_2O$ and $CH_4$ emissions under P-60 and P-90 compared to P-30 or showed no difference for $N_2O$ emissions in August, while enhanced $CO_2$ fluxes were observed in both July and August samplings. Therefore, these results suggest that technical conflicts would occur for the simultaneous regulation of three major GHGs through an eco-stoichiometry mechanism under the conditions of P fertilization. The integration of soil C sequestration, minimization of P export, and guarantee of the grain yield translate into a recommended P paddy field ecosystem input rate of no greater than 60 kg $ha^{-1}$.

## Materials and Methods

(This work is unrelated to an ethics issues, and no specific permit was required for the described field study.).

### Study Site

A plot experiment on P management for paddy field was established in April 2005 at the demonstration park of YuHang County Agricultural Research Station (30°18′51.84″N, 119°54′13.37″E) in ZheJiang, China. The experimental region has a subtropical monsoon climate with an average temperature of 17.8°C and an average annual rainfall of 1450 mm. The soil type is typical clay, blue-purple paddy soil. The dominant soil type in this region is a blue-purple paddy soil (*Mollic Endoaquoll*). The soil before experimental plots construction was composed of 3% sand, 47% silt, and 50% clay in the top 150 mm, while contents of soil total P, total C and total N were found 13.7, 2087, and 148 mmol $kg^{-1}$, respectively. Local farmers in this region routinely apply approximately 25–50 kg P $ha^{-1}$ of inorganic P fertilizer or compound fertilizer in late July or early August in order to support one crop of rice and an over-wintering crop, such as wheat (*Triticum aestivum*) or rape (*Brassica Napus*).

### Rice Field Plot Experiment and Soil and Greenhouse Gases Sampling

**Rice field plot experiment.** The construction of the field experimental plots, including specific designs on plot ridges, trenches, berms and inlets/outlets, was previously described by Zhang [23]. Briefly, twelve 4 m×5 m plots were constructed in two parallel rows in 2005. In order to keep each of these plots hydrologically isolated, a high-density polyethylene impermeable membrane of 0.75 mm (thickness)×105 cm depth was first inserted between two neighboring plots, and then concrete-brick walls of 12 cm (width)×105 cm (depth) were coupled on both sides of the membrane. The experiment was conducted using a completely randomized block design with three replicates for each treatment. P was applied at rates of 0 (P-0), 30 (P-30), 60 (P-60), and 90 (P-90) kg $ha^{-1}$ in June, using superphosphate since 2005. These P treatments cover the routine rate of P application of local farmers and the excessive P rates for field experiments. All plots received 170 kg N $ha^{-1}$ (urea) and 50 kg K $ha^{-1}$ (KCl) each year. In 2011, 25-day-old rice seedlings (*Oryza sativa L.*) were transplanted at 150 mm×150 mm spacing, and rice was harvested on November 10, 2011. Details on rice grain yield and yield components are presented in Text S1.

**Soil sampling.** After application of P for 6 years from 2005 to 2011, the experimental paddy field enters a relatively stable state for soil biosphere. Therefore, soil sampling in 2011 was chosen for a comparison investigation on soil eco-stoichiometric characteristics under different P treatments. On May 21, 2011, one month before rice transplantation, each plot was divided into six subplots of grab-sampling to minimize edge effects. The arable soil samples for each plot consisted of six composited 2 cm diameter×10 cm deep cores (cultivated horizon) [45]. The samples were kept on ice and immediately transported to the laboratory. Air-dried soils were screened through 2 mm sieves and then stored in the dark at 4°C until analysis for total C, N, and P concentrations. The fresh soil samples were stored at 4°C for no longer than one week before microbial C, N, P, and enzyme activity analyses.

**Gas collection and measurements.** Sampling of GHGs was set in July and August 2011 following fertilization, representing the early-phase of rice tilling and middle-phase of rice jointing during the high temperature season. A series of lab-made gas collection apparatuses (Text S2) was constructed prior to field sampling, and then set in the middle of each plot. Gas sampling was performed between 9:00 am and 11:00 am using 10 ml vacuum glass tubes connected to the stomata sampling apparatus. For each flux determination, a series of gas samples was taken at

10 min intervals over one hour. At the same time, the air temperature within the chambers were measured, while ambient air pressure was recorded by an automatic weather station (FRT CS01A) assembled 95 m away from the experimental plots. The samples were analyzed for $CO_2$, $N_2O$, and $CH_4$ within 24 h using an Agilent 6890D GC (Agilent Technologies, Palo Alto, California, USA) equipped with an automated gas chromatographic system with an $^{63}Ni$ electron capture detector for $N_2O$ analyses and flame ionization detector for $CO_2$ and $CH_4$.

The $CO_2$, $N_2O$, and $CH_4$ fluxes were calculated using linear regression of the change in gas concentration, based on the following equation:

$$F=\sigma\frac{P}{P_0}\frac{T_0}{T}\frac{dc}{dt}H \quad (1)$$

where $F$ is the flux rate of $CO_2/CH_4/N_2O$ ($mg\cdot m^{-2}\cdot h^{-1}$), $\sigma$is the $CO_2$, $CH_4$ and $N_2O$ gas density in standard state (1.96, 0.71, and 1.96 $kg\cdot m^{-3}$, respectively), $P$ is the atmospheric pressure (kPa), $T$ the air temperature (K), $P_0$ and $T_0$is the atmospheric pressure (101.3 kPa) and the air temperature (273 K), $dc/dt$ is the unit conversion factors for calculating $CO_2$ and $N_2O$ flux rates, and H is the height of the chamber (m).

**Soil fraction properties.** Air-dried samples were analyzed for total organic C (TOC), total N, and P concentrations according to standard methods described by Bao [46]. The potassium dichromate oxidation method was used to determine total organic C contents. Total N was analyzed by the Kjeldahl method. Total P was digested by $HClO_4$-$H_2SO_4$ and measured spectrophotometrically using a continuous flow analyzer (Autoanalyzer III, BRAN+LUEBBE, Germany) set to 880 nm. The fresh soil samples used for determining microbial biomass C and N as well as the P measurements used the fumigation-extraction method as described by Wu et al. [47] and Brookes et al. [48]. Briefly, the fresh soil samples were split into two subsamples with one sample immediately extracted with 0.5 mol $L^{-1}$ $K_2SO_4$ for microbial C and N or 0.5 mol $L^{-1}$ $NaHCO_3$ for microbial P, while the other sample was fumigated with chloroform and then extracted. Following centrifugation, C, N, and P concentrations as well as microbial biomass element content were calculated from the difference between the fumigated and non-fumigated soil samples.

## Extracellular Enzyme Assays

Four eco-enzymes (i.e., BG, AP, and NAG and LAP) were selected as indicators of microbial nutrient demand in the cycles of C, N, and P, respectively (Table S2). Fluorescence-based soil assays for β-1, 4-glucosidase (BG), β-1, 4-N-acetylglucosaminidase (NAG), L-leucine aminopeptidase (LAP), and acid phosphate (AP) used MUB-linked and AMC-linked artificial substrates [15,49] (Table S2). Briefly, Sample suspensions were prepared by homogenizing 1 g (wet weight) of soil with 125 ml of 50 mmol $L^{-1}$ sodium acetate buffer using a vortex shaker for 1 min. The pH of sodium acetate buffer for the soil slurries is 6.0, which is the mean soil pH of the environmental samples [50]. Standard high throughput fluorometric enzyme assays were conducted in 96-well blank fluorescent plates (Corning Inc., costar 3603) after pipetting of buffer, slurries, references, and substrates followed a strict order and position on the well plate according to the order of Saiya-Cork [49] (Table S3). The microplates were covered and incubated in the dark at 20°C for 4 h. Then, 10 μl of 1.0 mol $L^{-1}$ NaOH were added to each well to stop the reaction and increase the fluorescence of the substrates. Different buffers respond to NaOH addition in different ways. Thus, a time frame of 1 min between NaOH addition and the reading of plates in a fluorometer was used to reduce analytical variation [50]. Following the addition of NaOH, fluorescence was measured with a Bio-Tek Synergy HT microplate reader (Bio-Tek Inc., Winooski, USA) with 365 nm excitation and 460 nm emission filters. Enzyme activities were calculated and expressed as nmol $h^{-1}$ $g^{-1}$.

## Statistical Analysis

A completely randomized design was performed to examine significant differences in soil chemical composition, microbiological properties, eco-enzymes, and their stoichiometry among the various P fertilization treatments (P-0, P-30, P-60, P-90) using two-way analysis of variance (ANOVA) at $p<0.05$ levels. All assays were conducted in triplicate. A principal component analysis (PCA) was used to transform these variables to two factors. The acquisition ratios of ln(BG):ln(AP), ln(BG):ln(NAG+LAP), and ln(NAG+LAP):ln(AP) activities were also calculated and referred to the acquisition of organic C relative to organic P and N as well as organic N relative to organic P, respectively [3]. ANOVA and Pearson correlation statistics were also performed using the SPSS statistical software version 17.0. The clustering method and PCA were performed using the Minitab statistical software version 16.0. The Origin 8.0 (Origin Lab Corporation, USA) was also used for figure preparation.

## Supporting Information

**Table S1** Rice grain yield and yield components of experimental paddy field under P fertilization.

**Table S2** Extracellular enzymes with corresponding substrate and the corresponding function.

**Table S3** The order of pipetting buffer, slurries, references, and substrates in fluorometric enzyme assays.

## Author Contributions

Conceived and designed the experiments: ZJZ DQJ. Performed the experiments: XL HW SHG. Analyzed the data: XL HW. Contributed reagents/materials/analysis tools: GMT. Wrote the paper: XL DQJ ZJZ.

## References

1. Sims JT, Simard RR, Joern BC (1998) Phosphorus loss in agricultural drainage: Historical perspective and current research. Journal of Environmental Quality 27: 277–293.
2. Hobbie SE, Nadelhoffer KJ, Hogberg P (2002) A synthesis: The role of nutrients as constraints on carbon balances in boreal and arctic regions. Plant and Soil 242: 163–170.
3. Sinsabaugh RL, Hill BH, Shah JJF (2009) Ecoenzymatic stoichiometry of microbial organic nutrient acquisition in soil and sediment. Nature 462: 795-U117.
4. Allen AP, Gillooly JF (2009) Towards an integration of ecological stoichiometry and the metabolic theory of ecology to better understand nutrient cycling. Ecology Letters 12: 369–384.

5. Hall EK, Maixner F, Franklin O, Daims H, Richter A, et al. (2011) Linking microbial and ecosystem ecology using ecological stoichiometry: A synthesis of conceptual and empirical approaches. Ecosystems 14: 261–273.
6. Yu Q, Chen QS, Elser JJ, He NP, Wu HH, et al. (2010) Linking stoichiometric homoeostasis with ecosystem structure, functioning and stability. Ecology Letters 13: 1390–1399.
7. Bradford MA, Fierer N, Reynolds JF (2008) Soil carbon stocks in experimental mesocosms are dependent on the rate of labile carbon, nitrogen and phosphorus inputs to soils. Functional Ecology 22: 964–974.
8. Zhang Y, Su SL, Zhang F, Shi RH, Gao W (2012) Characterizing spatiotemporal dynamics of methane emissions from rice paddies in northeast China from 1990 to 2010. Plos One 7.
9. Falkowski PG, Fenchel T, Delong EF (2008) The microbial engines that drive Earth's biogeochemical cycles. Science 320: 1034–1039.
10. Cherif M, Loreau M (2009) When microbes and consumers determine the limiting nutrient of autotrophs: a theoretical analysis. Proceedings of the Royal Society B-Biological Sciences 276: 487–497.
11. Sinsabaugh RL, Shah JJF (2012) Ecoenzymatic stoichiometry and ecological theory. In: Futuyma DJ, editor. Annual Review of Ecology, Evolution, and Systematics, Vol 43. Palo Alto, CA: Annual Reviews. 313–343.
12. Peng PQ, Zhang WJ, Tong CL (2005) Soil C, N and P contents and their relationships with soil physical properties in wetlands of Dongting Lake fliood plain. Chinese Journal of Applied Ecology 16 (10): 1872–1878.
13. Sterner RW, Elser JJ (2002) Ecological Stoichiometry: The Biology of Elements From Molecules to the Biosphere. Princeton, NJ: Princeton University Press.
14. Moorhead DL, Sinsabaugh RL (2006) A theoretical model of litter decay and microbial interaction. Ecological Monographs 76: 151–174.
15. Sinsabaugh RL, Lauber CL, Weintraub MN, Ahmed B, Allison SD, et al. (2008) Stoichiometry of soil enzyme activity at global scale. Ecology Letters 11: 1252–1264.
16. Falkowski P, Scholes RJ, Boyle E, Canadell J, Canfield D, et al. (2000) The global carbon cycle: A test of our knowledge of earth as a system. Science 290: 291–296.
17. Cai ZC (2012) Greenhouse gas budget for terrestrial ecosystems in China. Science China-Earth Sciences 55: 173–182.
18. OECD (2000) Environmental indicators for agriculture methods and results. Executive summary Paris.
19. IRRI (2009) World rice statistics: Rough rice production by country and geographical region-FAO1961–2007. Available: http://beta.irri.org/solutions/index.php?option = com_content&task = view&id = 250. Accessed 2009 Jul 28.
20. Davidson EA (2009) The contribution of manure and fertilizer nitrogen to atmospheric nitrous oxide since 1860. Nature Geoscience 2: 659–662.
21. Vitousek PM, Naylor R, Crews T, David MB, Drinkwater LE, et al. (2009) Nutrient imbalances in agricultural development. Science 324: 1519–1520.
22. Snyder CS, Bruulsema TW, Jensen TL, Fixen PE (2009) Review of greenhouse gas emissions from crop production systems and fertilizer management effects. Agriculture Ecosystems & Environment 133: 247–266.
23. Zhang ZJ, Zhang JY, He R, Wang ZD, Zhu YM (2007) Phosphorus interception in floodwater of paddy field during the rice-growing season in TaiHu Lake Basin. Environmental Pollution 145: 425–433.
24. Huang QR, Hu F, Huang S, Li HX, Yuan YH, et al. (2009) Effect of Long-Term Fertilization on Organic Carbon and Nitrogen in a Subtropical Paddy Soil. Pedosphere 19: 727–734.
25. Marichal R, Mathieu J, Couteaux MM, Mora P, Roy J, et al. (2011) Earthworm and microbe response to litter and soils of tropical forest plantations with contrasting C:N:P stoichiometric ratios. Soil Biology & Biochemistry 43: 1528–1535.
26. Taylor PG, Townsend AR (2010) Stoichiometric control of organic carbon-nitrate relationships from soils to the sea. Nature 464: 1178–1181.
27. Cleveland CC, Liptzin D (2007) C : N : P stoichiometry in soil: is there a "Redfield ratio" for the microbial biomass? Biogeochemistry 85: 235–252.
28. Ciampitti IA, Garcia FO, Picone LI, Rubio G (2011) Soil carbon and phosphorus pools in field crop rotations in pampean soils of Argentina. Soil Science Society of America Journal 75: 616–625.
29. Vitousek PM, Porder S, Houlton BZ, Chadwick OA (2010) Terrestrial phosphorus limitation: mechanisms, implications, and nitrogen-phosphorus interactions. Ecological Applications 20: 5–15.
30. Hill BH, Elonen CM, Seifert LR, May AA, Tarquinio E (2012) Microbial enzyme stoichiometry and nutrient limitation in US streams and rivers. Ecological Indicators 18: 540–551.
31. Hessen DO, Agren GI, Anderson TR, Elser JJ, De Ruiter PC (2004) Carbon, sequestration in ecosystems: The role of stoichiometry. Ecology 85: 1179–1192.
32. Marklein AR, Houlton BZ (2012) Nitrogen inputs accelerate phosphorus cycling rates across a wide variety of terrestrial ecosystems. New Phytologist 193: 696–704.
33. Shange RS, Ankumah RO, Ibekwe AM, Zabawa R, Dowd SE (2012) Distinct soil bacterial communities revealed under a diversely managed agroecosystem. Plos One 7.
34. Reganold JP, Andrews PK, Reeve JR, Carpenter-Boggs L, Schadt CW, et al. (2010) Fruit and soil quality of organic and conventional strawberry agroecosystems. Plos One 5.
35. Bell LW, Sparling B, Tenuta M, Entz MH (2012) Soil profile carbon and nutrient stocks under long-term conventional and organic crop and alfalfa-crop rotations and re-established grassland. Agriculture Ecosystems & Environment 158: 156–163.
36. Liu DY, Zhang RF, Wu HS, Xu DB, Tang Z, et al. (2011) Changes in biochemical and microbiological parameters during the period of rapid composting of dairy manure with rice chaff. Bioresource Technology 102: 9040–9049.
37. Shang QY, Yang XX, Gao CM, Wu PP, Liu JJ, et al. (2011) Net annual global warming potential and greenhouse gas intensity in Chinese double rice-cropping systems: a 3-year field measurement in long-term fertilizer experiments. Global Change Biology 17: 2196–2210.
38. Shrestha M, Shrestha PM, Frenzel P, Conrad R (2010) Effect of nitrogen fertilization on methane oxidation, abundance, community structure, and gene expression of methanotrophs in the rice rhizosphere. Isme Journal 4: 1545–1556.
39. Ma K, Lu YH (2011) Regulation of microbial methane production and oxidation by intermittent drainage in rice field soil. Fems Microbiology Ecology 75: 446–456.
40. Adviento-Borbe MAA, Haddix ML, Binder DL, Walters DT, Dobermann A (2007) Soil greenhouse gas fluxes and global warming potential in four high-yielding maize systems. Global Change Biology 13: 1972–1988.
41. Zhang ZJ, Yao JX, Wang ZD, Xu X, Lin XY, et al. (2011) Improving water management practices to reduce nutrient export from rice paddy fields. Environmental Technology 32: 197–209.
42. Wang K, Zhang ZJ, Zhu YM, Wang GH, Shi DC, et al. (2001) Surface water phosphorus dynamics in rice fields receiving fertiliser and manure phosphorus. Chemosphere 42: 209–214.
43. Lee CH, Do Park K, Jung KY, Ali MA, Lee D, et al. (2010) Effect of Chinese milk vetch (Astragalus sinicus L.) as a green manure on rice productivity and methane emission in paddy soil. Agriculture Ecosystems & Environment 138: 343–347.
44. Wang DY, Zhand XF, Zhou CN, Zheng GS, Zhang GX, et al. (2010) Grain yield difference investigation and reasonable planting density analysis of rice production in Zhejing Province. Acta Agricuhurae Zhejiangens 22: 330–336.
45. Sander T, Gerke HH, Rogasik H (2008) Assessment of Chinese paddy-soil structure using X-ray computed tomography. Geoderma 145: 303–314.
46. Bao SD (2000) Agro-chemical analysis of soil. Beijing: China Agricultural Press. 78–290.
47. Wu J, Joergensen RG, Pommerening B, Chaussod R, Brookes PC (1990) Measurement of soil microbial biomass C by fumigation extraction - an automated procedure. Soil Biology & Biochemistry 22: 1167–1169.
48. Brookes PC, Powlson DS, Jenkinson DS (1982) Measurement of microbial biomass phosphorus in soil. Soil Biology & Biochemistry 14: 319–329.
49. Saiya-Cork KR, Sinsabaugh RL, Zak DR (2002) The effects of long term nitrogen deposition on extracellular enzyme activity in an Acer saccharum forest soil. Soil Biology & Biochemistry 34: 1309–1315.
50. German DP, Weintraub MN, Grandy AS, Lauber CL, Rinkes ZL, et al. (2011) Optimization of hydrolytic and oxidative enzyme methods for ecosystem studies. Soil Biology & Biochemistry 43: 1387–1397.

15

# Iron: The Forgotten Driver of Nitrous Oxide Production in Agricultural Soil

**Xia Zhu[1,2,3], Lucas C. R. Silva[3], Timothy A. Doane[3]*, William R. Horwath[3]**

**1** Chengdu Institute of Biology, Chinese Academy of Sciences, Chengdu, China, **2** University of Chinese Academy of Sciences, Beijing, China, **3** Department of Land, Air, and Water Resources, University of California Davis, Davis, California, United States of America

## Abstract

In response to rising interest over the years, many experiments and several models have been devised to understand emission of nitrous oxide ($N_2O$) from agricultural soils. Notably absent from almost all of this discussion is iron, even though its role in both chemical and biochemical reactions that generate $N_2O$ was recognized well before research on $N_2O$ emission began to accelerate. We revisited iron by exploring its importance alongside other soil properties commonly believed to control $N_2O$ production in agricultural systems. A set of soils from California's main agricultural regions was used to observe $N_2O$ emission under conditions representative of typical field scenarios. Results of multivariate analysis showed that in five of the twelve different conditions studied, iron ranked higher than any other intrinsic soil property in explaining observed emissions across soils. Upcoming studies stand to gain valuable information by considering iron among the drivers of $N_2O$ emission, expanding the current framework to include coupling between biotic and abiotic reactions.

**Editor:** Ben Bond-Lamberty, DOE Pacific Northwest National Laboratory, United States of America

**Funding:** Support was provided by the California Department of Resources Recycling and Recovery (www.calrecycle.ca.gov), Agreement IWM09027; the J. G. Boswell Endowed Chair in Soil Science; and the University of California Agricultural Experiment Station (http://caes.ucdavis.edu/research/agexpstn). The funders had no role in study design, data collection and analysis, decision to publish, or preparation of the manuscript.

**Competing Interests:** The authors have declared that no competing interests exist.

* E-mail: tadoane@ucdavis.edu

## Introduction

Emission of $N_2O$ from soils is an extensively studied environmental process, given that $N_2O$ is "at the heart of debates" [1] on several prevalent current issues. Approximately two-thirds of total global emission comes from soils; most of the emission from soils is in turn attributed to agriculture [2]. The intrinsic soil properties (as opposed to temporary changes) most commonly mentioned in research studies and models as controlling emission of $N_2O$ are texture, pH, organic matter, and ability to supply inorganic nitrogen [3–12]. Production of $N_2O$ in soil is generally attributed to microbiological processes [1,2,13–17], and therefore the factors that regulate the activity of $N_2O$-producing microorganisms should be the same factors that regulate $N_2O$ production. These controlling factors are generally thought to be well recognized, but as research and related commentary on $N_2O$ emission from agricultural soils continue to accumulate, the possible role of iron is rarely considered. This is in spite of its known involvement in enzymatic reactions [2,18,19] and non-enzymatic reactions [20–23] that generate $N_2O$. The connection between iron and $N_2O$ may have been neglected because iron has never figured prominently in routine evaluations of soil for agronomic research or practical management decisions. Unlike the other soil properties cited above, iron does not have a direct and immediate bearing on the growth of most crops or on the agricultural suitability of a soil from either a physical or a chemical point of view. When it is considered, this is in instances of suspected plant deficiency or toxicity, not in the context of its potential connection with the nitrogen cycle. In addition, compared to other intrinsic properties, soil iron does not dramatically affect the short-term changes in microbiological activity generally associated with $N_2O$ production. For these reasons, once interest in $N_2O$ began to intensify, the previously reported connection with iron was already out of sight. The intent of our work was to reconsider the potential significance of iron in emission of $N_2O$ from agricultural soils.

## Materials and Methods

### Ethics statement

The soils used in this study were collected under consent of the land owners, and the compost used was collected under consent of the compost facility management.

### Soil characterization

Soils were collected from the top 15 cm in 10 agricultural fields throughout California, and were sieved to 2 mm following collection. Soil pH was measured in 1 M KCl (1:1 w:v). Percent clay, silt, and sand were determined by a modified pipet method [24]. Total carbon and nitrogen were determined on ball-milled samples by combustion-GC (Costech ECS 4010). Just prior to setting up the experiment, inorganic nitrogen (ammonium plus nitrate) was extracted by 0.5 M $K_2SO_4$ and determined colorimetrically [25,26]. Dissolved organic carbon (DOC) was determined in the same extract by UV-persulfate digestion (Teledyne-Tekmar Phoenix 8000).

We chose two commonly used, contrasting indices to characterize soil iron: that extractable by acid hydroxylamine (FeA), an index of reactive iron(III) minerals [27]; and that extractable by pyrophosphate (FeP), representing iron complexed with soil organic matter [28–30]. FeA was extracted by shaking 0.8 g soil for one hour with 40 ml 0.25 M hydroxylamine hydrochloride in 0.25 M HCl, followed by centrifugation for 30 minutes at 15600× G. FeP was extracted by shaking 1 g soil with 100 ml 0.1 M tetrasodium pyrophosphate for 16 hours, followed by centrifugation for 30 minutes at 15600× G; further centrifugation did not result in any difference in measured iron concentration, indicating that all fine iron colloids had been removed, an important consideration when using this extractant [29,30]. The concentration of iron in all extracts was determined colorimetrically [31]; pyrophosphate extracts were neutralized by a small addition of HCl prior to this determination. There was no interference from pyrophosphate in the colorimetric analysis. All analyses of soil properties were performed in duplicate. These properties are reported in Table 1.

## Experimental treatments

As stated above, the properties most commonly believed to control emission of $N_2O$ from agricultural soil include texture, pH, organic matter, and the inherent ability of the soil to release inorganic nitrogen. These are intrinsic properties which are not abruptly altered by environmental conditions; in contrast, our treatments were designed to manipulate the most common temporary extrinsic changes that influence $N_2O$ production: water content, fertilization, and organic amendments. Since these can vary across a range of values, we necessarily limited our choice of treatments. Fertilizer and compost (as a model organic amendment) were either withheld or added at a rate typical of agriculture in California, and two water contents were chosen according to the range expected in agricultural soils. Field capacity, the amount of water a soil can retain against gravity, was chosen as the upper reference point. This is not uncommon, as soil moisture can temporarily exceed field capacity following irrigation or rainfall events [32,33]. In practice we used water holding capacity (WHC) to represent field capacity. As a contrasting treatment, we chose 50% WHC. This is near the permanent wilting point of most soils [34], and it is not likely that soil moisture will fall below this in the field except during unmanaged dry seasons. Although many intermediate values could have been selected as treatments, we chose to use both ends of a typical spectrum of values in order to present a broad yet concise study.

## Experimental set-up

Prior to set-up, WHC was determined as follows: a soil sample was placed into a funnel lined with filter paper, which was then placed into a beaker of water such that just the tip of the funnel was always in contact with water; after the sample ceased to take up water, the sample was allowed to drain, and the moisture content measured. To begin the experiment, the equivalent of 50 g dry soil was placed into cups, which were themselves placed into larger jars containing a small amount of water to avoid desiccation. The larger jars were sealed with lids containing a small foam plug to allow gas exchange with the atmosphere. To imitate the timing typical of agricultural operations, 2 g finely ground finished green waste compost (corresponding approximately to a field application of 60 t $ha^{-1}$ in the top 15 cm) were mixed with the soils and incubated at 40% WHC for seven days. Treatments not receiving compost were similarly incubated. Following this preincubation, each soil received a fertilizer addition according to treatment: none, ammonium sulfate, or

**Table 1.** Characterization of the soils used in this study.

| Location | Classification[a] | FeA[b] | FeP[c] | DOC[d] | Inorganic N | Total N | Total C | Sand | Silt | Clay | pH |
|---|---|---|---|---|---|---|---|---|---|---|---|
| | | mg $kg^{-1}$ | mg $kg^{-1}$ | mg $kg^{-1}$ | mg $kg^{-1}$ | % | % | % | % | % | |
| *Sacramento Valley* | | | | | | | | | | | |
| Davis | Fine, montmorillonitic, thermic Mollic Haploxeralf | 1800 | 170 | 17 | 2 | 0.09 | 0.85 | 30 | 42 | 24 | 5.4 |
| Dixon 1 | Fine-silty, mixed, nonacid, thermic Typic Xerorthent | 2150 | 290 | 30 | 11 | 0.14 | 1.60 | 23 | 49 | 28 | 5.6 |
| Dixon 2 | Fine-silty, mixed, nonacid, thermic Typic Xerorthent | 1900 | 210 | 19 | 5 | 0.11 | 1.18 | 15 | 41 | 44 | 5.5 |
| *Salinas Valley* | | | | | | | | | | | |
| Castroville | Fine, montmorillonitic, thermic Ultic Palexerol | 710 | 550 | 88 | 32 | 0.08 | 0.75 | 72 | 15 | 13 | 6.4 |
| Salinas 1 | Fine, montmorillonitic, thermic Pachic Argixeroll | 390 | 150 | 44 | 5 | 0.07 | 0.66 | 64 | 23 | 13 | 7.2 |
| Salinas 2 | Fine, montmorillonitic, thermic Typic Pelloxerert | 1890 | 240 | 88 | 28 | 0.16 | 1.78 | 22 | 36 | 42 | 7.4 |
| Spence | Fine-loamy, mixed, thermic, Typic Argixeroll | 670 | 270 | 63 | 18 | 0.11 | 1.28 | 50 | 29 | 21 | 6.6 |
| *San Joaquin Valley* | | | | | | | | | | | |
| Five Points | Fine-loamy, mixed, superactive, thermic Typic Haplocambid | 850 | 60 | 57 | 4 | 0.08 | 0.67 | 36 | 32 | 32 | 6.8 |
| Modesto | Fine-loamy, mixed, superactive, thermic Typic Argixeroll | 410 | 240 | 164 | 130 | 0.11 | 0.97 | 72 | 18 | 10 | 6.9 |
| Sanger | Coarse-loamy, mixed, nonacid, thermic Typic Xerorthent | 390 | 260 | 28 | 4 | 0.03 | 0.30 | 61 | 32 | 7 | 4.2 |

[a]United States Department of Agriculture official soil series description, [b] acid hydroxylamine-extractable iron, [c] pyrophosphate-extractable iron, [d] dissolved organic carbon.

potassium nitrate. The amount of nitrogen added was 100 mg $kg^{-1}$ soil, corresponding approximately to a field rate of 150 kg $ha^{-1}$. Fertilizer solution was sprayed onto the soils to reach a water content of 50% or 100% WHC, depending on the treatment. For each soil there were three replicates per treatment. Samples were incubated for 14 days at 22 degrees C.

Samples for $N_2O$ analysis were taken on days 0, 1, 2, 3, 5, 9, and 14 following addition of fertilizer. The jars containing the soil cups were closed with lids containing septa and allowed to stand for one hour. Gas samples were taken at 0, 30, and 60 minutes after closure and transferred to evacuated gas sampling vials. $N_2O$ concentration was determined by gas chromatography-ECD detection (Shimadzu GC-2014). At each sampling date, the rate of $N_2O$ emission (flux) was determined by linear interpolation of the 0, 30, and 60 minute measurements. Cumulative $N_2O$ emission over the course of the incubation was calculated using these data, taking the flux measured at a given date to be the average flux for the interval represented by that date.

### Statistical analysis

To identify the soil properties that most strongly explained $N_2O$ emission in each experimental treatment, we studied the data using partial least squares (PLS) multivariate analysis, a form of structural equation modeling. This tool is particularly suitable when the number of predicting variables is greater than the number of observed variables, when multicolinearity is expected among predicting variables, and when multivariate normality can not be assumed [35–37]. PLS ranks the predicting variables by importance based on linear regression models that project the predicting variables and the observed variables to a new, multivariate space. Prior to subjecting the data to PLS analysis, predicting variables (soil properties) and the observed response (cumulative $N_2O$ emission) were standardized by centering and scaling the data to have a mean of zero and a standard deviation of one. This ensures that the predicting variables are ranked based on how much of the variation is explained when all variables have the same weight.

Although correlations among variables are possible, especially in studies that involve soil properties, this does not change the interpretation given by PLS, which depicts the relative importance of each variable separately, independently of intrinsic links between variables. Nevertheless, a correlation matrix is presented (Table 2) as an aid in understanding the relationships between the soil properties used in our study.

Following the exploratory PLS analysis, linear regressions between iron and $N_2O$ emission were calculated using unweighted, untransformed data, and were considered significant enough to report at $P<0.1$. All statistical analyses were performed using JMP 10 software.

## Results and Discussion

The results of the PLS analysis are shown in Figure 1, where each soil property is ranked according to its ability to explain cumulative $N_2O$ emission across all soils. This ranking was performed for each of the 12 different treatments studied. In five of these treatments, iron (as either FeA or FeP) ranked higher than any other measured soil characteristic in explaining observed emissions. In four additional treatments, iron was among the top four predictors.

As a complementary approach to further investigate the relationship between iron and $N_2O$ emission, simple linear regressions were calculated in which $N_2O$ data were compared against FeP and FeA. Whereas PLS was used to arrange a suite of soil properties according to their ability to explain $N_2O$ emission, regressions indicate, by the value of $r^2$, how much of the variability in $N_2O$ emission can be explained by a single property; regressions also indicate the direction of the effect (positive or negative slope) and degree of importance of the effect (absolute value of the slope). In most cases, a significant relationship between $N_2O$ emission and a given variable can be expected when that variable is ranked highly by PLS. In certain cases, however, a variable ranked highly by PLS may not necessarily yield a significant linear relationship when that variable is considered apart from the other variables; conversely, certain treatments in which a variable is not ranked highly by PLS may nonetheless yield a significant regression. The primary reason for this occasional discrepancy is the nature of the PLS procedure: by considering all predicting variables together, new predictors are generated which are composites of the original variables. Table 3 reports the results of the regressions for treatments that showed a significant relationship between $N_2O$ emission and either iron index. Despite a dataset of values for $N_2O$ emission which spanned more than three orders of magnitude across soils, several notable connections between iron and $N_2O$ emission emerged.

FeP was significantly related to $N_2O$ emission in four treatments, in which it explained between 16 and 62 percent of the variability, with a positive slope in all cases (i.e. greater

**Table 2.** Correlation matrix of the soil properties evaluated in this study.

| | FeA[a] | FeP[b] | DOC[c] | Inorganic N | Total N | Total C | Sand | Silt | Clay | pH |
|---|---|---|---|---|---|---|---|---|---|---|
| FeA | – | −0.07 | −0.41 | −0.37 | 0.68 | 0.70 | −0.91 | 0.84 | 0.79 | −0.15 |
| FeP | −0.07 | – | 0.25 | 0.05 | 0.04 | 0.08 | 0.38 | −0.37 | −0.31 | −0.10 |
| DOC | −0.41 | 0.25 | – | 0.93 | 0.25 | 0.12 | 0.53 | −0.68 | −0.29 | 0.59 |
| Inorganic N | −0.37 | 0.05 | 0.93 | – | 0.18 | 0.02 | 0.45 | −0.54 | −0.28 | 0.43 |
| Total N | 0.68 | 0.04 | 0.25 | 0.18 | – | 0.98 | −0.57 | 0.37 | 0.66 | 0.47 |
| Total C | 0.70 | 0.08 | 0.12 | 0.02 | 0.98 | – | −0.61 | 0.46 | 0.66 | 0.36 |
| Sand | −0.91 | 0.38 | 0.53 | 0.45 | −0.57 | −0.61 | – | −0.89 | −0.91 | 0.13 |
| Silt | 0.84 | −0.37 | −0.68 | −0.54 | 0.37 | 0.46 | −0.89 | – | 0.63 | −0.44 |
| Clay | 0.79 | −0.31 | −0.29 | −0.28 | 0.66 | 0.66 | −0.91 | 0.63 | – | 0.17 |
| pH | −0.15 | −0.10 | 0.59 | 0.43 | 0.47 | 0.36 | 0.13 | −0.44 | 0.17 | – |

[a]acid hydroxylamine-extractable iron, [b] pyrophosphate-extractable iron, [c] dissolved organic carbon.

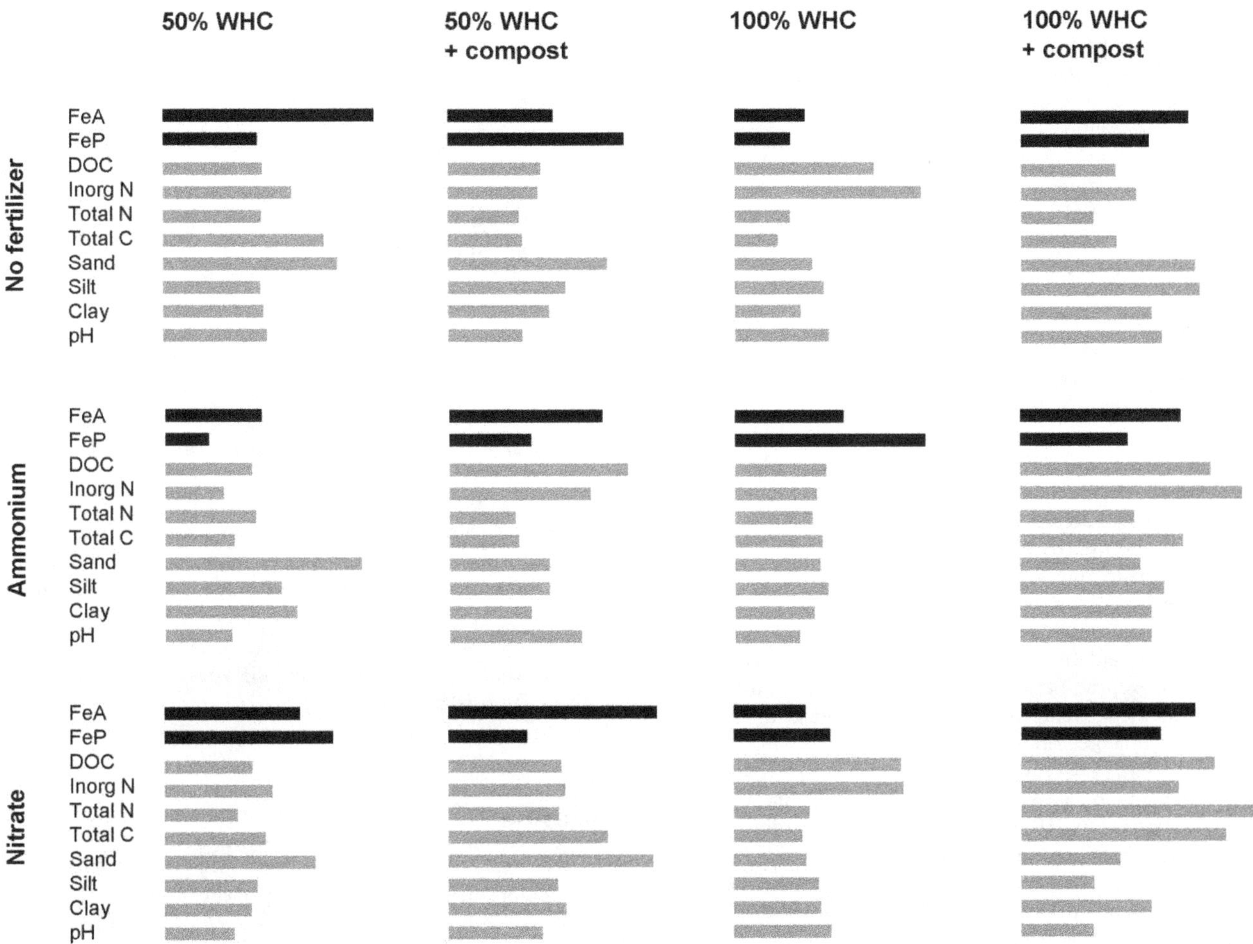


**Figure 1. Relative importance of soil properties in explaining cumulative emission of $N_2O$ under different conditions.** Result of partial least squares multivariate analysis performed across ten soils for each of 12 different treatments. Two indices of soil iron (FeA: acid hydroxylamine-extractable iron and FeP: pyrophosphate-extractable iron) were ranked alongside other soil properties commonly considered to control soil $N_2O$ emission. The size of each bar is given by the variable importance in the projection (VIP) value, and indicates the relative strength of each variable in explaining emission in that treatment. WHC = water holding capacity; DOC = dissolved organic carbon.

emission was associated with more FeP). This influence was greatest under 100% WHC when ammonium was present and compost was absent. Such a condition may be reasonably expected on occasion, since most fertilizers supply ammonium, and since this may occur close in time to irrigation or rainfall. In this treatment, an increase in FeP of 1 mg $kg^{-1}$ corresponded to an increase in cumulative emission of 11.9 ng $N_2O$-N $g^{-1}$ soil (averaged across all soils) during the course of the incubation (Table 3).

Like FeP, the connection between FeA and $N_2O$ emission was also significant under several different conditions. Unlike FeP, however, which was positively related to $N_2O$ emission, FeA was always negatively related to $N_2O$ emission. There was no treatment in which both iron indices were significantly related to $N_2O$ emission (Table 3). Considering that FeP and FeA bear almost no relationship to each other (Table 2), this difference in behavior suggests that these two indices indeed reflect two forms of iron that differ in reactivity. Also notable in Table 3 is the effect of

**Table 3.** Results of simple linear regression of cumulative $N_2O$ emission (as ng $N_2O$-N $g^{-1}$ soil) against iron, across ten soils and under 12 different conditions.

| | 50% WHC | 50% WHC + compost | 100% WHC | 100% WHC + compost |
|---|---|---|---|---|
| No fertilizer | NS | FeP: 0.37, 0.38 | FeA: 0.12, −0.09 | NS |
| Ammonium | NS | FeA: 0.28, −0.20 | FeP: 0.62, 11.9 | FeA: 0.23, −0.62 |
| Nitrate | FeP: 0.19, 0.46 | NS | FeP: 0.16, 2.1 | NS |

The first value given is that of $r^2$, and the second value is the slope of the regression. NS = regression was not significant for either iron index. WHC = water holding capacity; FeA = acid hydroxylamine-extractable iron; FeP = pyrophosphate-extractable iron.

compost in fertilized treatments: the observed negative association between $N_2O$ emission and FeA occurred only in the presence of compost, while the stimulating effect of FeP was observed only without compost.

The contrasting relationships of FeA and FeP with $N_2O$ emission could be due to differences in the reaction of either form of iron with nitrogen compounds in the soil matrix. For example, hydroxylamine is produced from biological oxidation of ammonia, and is known to generate $N_2O$ upon chemical reaction with iron(III) [20,38]. Reaction with FeP versus FeA, or locally high concentrations of either hydroxylamine or iron, could lead to more or less $N_2O$ compared to other reaction products [38]. The ability of aerobic microorganisms to acquire iron can likewise depend on its chemical nature, consequently influencing the amount of reactive compounds produced or consumed through reactions that use iron-dependent enzymes. As soil water content increases, reducing conditions may develop, especially when the depletion of oxygen is accelerated by easily metabolized organic matter. The chemical nature of existing iron(III) may determine the ease with which it is reduced to iron(II) in anaerobic microsites. This will in turn control its participation in other reactions that produce $N_2O$, such as chemodenitrification, which includes the abiotic reduction of nitrite to $N_2O$ by iron(II) [39,40]. Chemodenitrification can also produce other gases, and the relative amount of $N_2O$ released may be affected by the form of iron present. A related anaerobic process is nitrate-dependent iron(II) oxidation [41]; a recent review [42] has highlighted, in the context of this process, how the simultaneous presence of nitrate-reducing and iron(III)-reducing areas can potentially be important to nitrogen cycling. Under anaerobic conditions, iron(III) can also be linked to ammonium oxidation [43,44]. If reactions that generate $N_2O$ are active in any of the above processes, they may be stimulated or suppressed by different forms of iron, such as the two indices examined in this study. The degree of this influence under different conditions will then determine the importance of iron relative to other soil properties.

Our treatments consisted of two contrasting values for soil moisture and addition of amendments. This was done in order to explore the importance of iron across a wide range of conditions while at the same time avoiding a cumbersome dataset. It is clear from Figure 1 that the importance of iron can change between the two limits of each treatment variable. For example, between 50 and 100% WHC under ammonium fertilization, iron moves from a position of modest relevance to become the highest-ranked driver. Since our results show the importance of iron only at two distinct values, we do not know how its importance under intermediate conditions changes between the two end values. Even without such intermediate data, the differences between contrasting treatments can aid in understanding the mechanisms at work in generating $N_2O$. In the above example, the importance of iron rises markedly under ammonium fertilization as soil moisture increases from 50 to 100% WHC; FeP surpasses FeA in strength as well. As mentioned earlier, ammonia is oxidized to hydroxylamine, and this can react with iron(III) to produce $N_2O$. In a wetter soil, solutes are more mobile, which can lead to greater production of hydroxylamine as well as greater contact of hydroxylamine with iron. FeP is also likely to be more soluble than FeA. Any combination of these effects might elevate the importance of iron and change which form is more relevant in explaining the associated $N_2O$ data.

The overall position of iron among other drivers of $N_2O$ emission is determined by both its reactivity and the presence of processes subject to its influence. Ample opportunity for inquiry exists for defining the extent of the relationship between iron and $N_2O$ in managed as well as unmanaged ecosystems, and this can provide useful practical and theoretical information. For example, including iron in current models of $N_2O$ emission may strengthen their predictive ability. In addition, inasmuch as certain indices of iron can be related to its physical or chemical characteristics, observing the relationship between a given index and $N_2O$ production, and how this changes under different conditions, may provide insight into the specific reactions at work. As stated earlier, production of $N_2O$ is generally accepted to be a microbial affair, and it is logical to assume that the factors that regulate the activity of $N_2O$-producing microorganisms should be the same factors that regulate $N_2O$ production. This is not incorrect, but is perhaps a somewhat restrictive rendering; a more accurate framework might include "biotic-abiotic reaction sequences" [39] that generate $N_2O$, such as those outlined above. Indeed, "the complex interactions that occur between microorganisms and other biotic and abiotic factors" have been suggested to be a key part of further understanding greenhouse gas production and improving predictions [17].

## Conclusion

It has been recently emphasized [45] that solutions to environmental problems require explicit consideration of the couplings between element cycles. The environmental chemistry of iron has been well researched, as have many of the interrelated details of the nitrogen cycle. The specific connection between iron and $N_2O$ in soil has also been recognized in both older and recent studies. However, iron and nitrogen have yet to be brought together in agricultural systems, the foremost source of soil $N_2O$ emission. Our most important conclusion is simple: iron does indeed figure prominently among the soil properties controlling $N_2O$ emission in contrasting conditions across diverse soils. Studies concerned with the potential of agricultural soil to emit $N_2O$ will gain new momentum by remembering this "key biogeochemical engine" [46], building on a connection identified a long time ago but largely overlooked since then.

## Acknowledgments

We gratefully acknowledge the assistance of those who collected the soil samples, as well as the constructive comments received during review, which were very helpful in improving our paper.

## Author Contributions

Conceived and designed the experiments: XZ WRH. Performed the experiments: XZ. Analyzed the data: XZ LCRS. Wrote the paper: TAD.

## References

1. Reay DS, Davidson EA, Smith KA, Smith P, Melillo JM, et al. (2012) Global agricultural and nitrous oxide emissions. Nat Clim Change 2: 410-416.
2. Thomson AJ, Giannopoulos G, Pretty J, Baggs EM, Richardson DJ (2012) Biological sources and sinks of nitrous oxide and strategies to mitigate emissions. Phil Trans R Soc B 367: 1157-1168.
3. Sahrawat KL, Keeney DR (1986) Nitrous oxide emissions from soils. Advances in Soil Sci 4: 103-148.
4. Eichner M (1990) Nitrous oxide emissions from fertilized soils: summary of available data. J Environ Qual 19: 272-280.
5. Bouwman AF, Fung I, Matthews E, John J (1993) Global analysis of the potential for $N_2O$ production in natural soils. Global Biogeochem Cycles 7: 557-597.
6. Robertson K (1994) Nitrous oxide emission in relation to soil factors at low to intermediate moisture levels. J Environ Qual 23: 805-809.

7. Dobbie KE, McTaggart IP, Smith KA (1999) Nitrous oxide emissions from intensive agricultural systems: variations between crops and seasons, key driving variables, and mean emission factors. J Geophys Res 104: 26891-26899.
8. Li C, Aber J, Stange F, Butterbach-Bahl H (2000) A process-oriented model of $N_2O$ and NO emissions from forest soils: 1. Model development. J Geophys Res 105: 4369-4384.
9. Skiba U, Smith KA (2000) The control of nitrous oxide emissions from agricultural and natural soils. Chemosphere Global Change Sci 2: 379-386.
10. Bouwman AF, Boumans LJM, Batjes NH (2002) Modeling global annual $N_2O$ and NO emissions from fertilized fields. Global Biogeochem Cycles 16: 1080.
11. Freibauer A, Kaltschmitt M (2003) Controls and models for estimating direct nitrous oxide emissions from temperate and sub-boreal agricultural mineral soils in Europe. Biogeochemistry 63: 93-115.
12. Stehfest E, Bouwman L (2006) $N_2O$ and NO emission from agricultural fields and soils under natural vegetation: summarizing available measurement data and modeling of global annual emissions. Nutr Cycl Agroecosys 74: 207-228.
13. Williams EJ, Hutchinson GL, Fehsenfeld FC (1992) $NO_x$ and $N_2O$ emissions from soil. Global Biogeochem Cycles 6: 351-388.
14. Bremner JM (1997) Sources of nitrous oxide in soils. Nutr Cycl Agroecosys 49: 7-16.
15. Freney JR (1997) Emission of nitrous oxide from soils used for agriculture. Nutr Cycl Agroecosys 49: 1-6.
16. Davidson EA, Keller M, Erickson HE, Verchot LV, Veldkamp E (2000) Testing a conceptual model of soil emissions of nitrous and nitric oxides. BioScience 50: 667-680.
17. Singh BK, Bardgett RD, Smith P, Reay DS (2010) Microorganisms and climate change: terrestrial feedbacks and mitigation options. Nat Rev Microbiol 8: 779-790.
18. Meiklejohn J (1953) Iron and the nitrifying bacteria. J Gen Microbiol 8: 58-65.
19. Glass JB, Orphan VJ (2012) Trace metal requirements for microbial enzymes involved in the production and consumption of methane and nitrous oxide. Front Microbiol doi: 10.3389/fmicb.2012.00061.
20. Chao TT, Kroontje W (1966) Inorganic nitrogen transformations through the oxidation and reduction of iron. Soil Sci Soc Am Proc 30: 193-196.
21. Buresh RJ, Moraghan JT (1976) Chemical reduction of nitrate by ferrous iron. J Environ Qual 5: 320-325.
22. Chalamet A, Bardin R (1976) Action of ferrous ions on reduction of nitrous acid in hydromorphous soils. Soil Biol Biochem 9: 281-285.
23. Bremner JM, Blackmer AM, Waring SA (1980) Formation of nitrous oxide and dinitrogen by chemical decomposition of hydroxylamine in soils. Soil Biol Biochem 12: 263-269.
24. Burt R, editor (1992) Soil Survey Laboratory Methods Manual. Soil Survey Investigations Report no. 42.WashingtonDC:USDA .
25. Verdouw H, van Echteld CJA, Dekkers EMJ (1978) Ammonia determination based on indophenol formation with sodium salicylate. Water Res 12: 399-402.
26. Doane TA, Horwath WR (2003) Spectrophotometric determination of nitrate with a single reagent. Anal Lett 36: 2713-2722.
27. Lovley DR, Phillips EJP (1987) Rapid assay for microbially reducible ferric iron in aquatic sediments. Appl Environ Microbiol 53: 1536-1540.
28. Bremner JM, Heintze SG, Mann PJG, Lees H (1946) Metallo-organic complexes in soil. Nature 158: 790-791.
29. Schuppli PA, Ross GJ, McKeague JA (1983) The effective removal of suspended materials from pyrophosphate extracts of soil from tropical and temperate regions. Soil Sci Soc Am J 47: 1026-1032.
30. Loveland PJ, Digby P (1984) The extraction of Fe and Al by 0.1 M pyrophosphate solutions: a comparison of some techniques. J Soil Sci 35: 243-250.
31. Dominik P, Kaupenjohann M (2000) Simple spectrophotometric determination of Fe in oxalate and HCl soil extracts. Talanta 51: 701-707.
32. Cassel DK, Nielsen DR (1986) Field capacity and available water capacity. In: Klute A, editor. Methods of Soil Analysis.Part 1.Physical and Mineralogical Methods.Madison,WI :Soil Science Society of America. pp. 901-926.
33. Veihmeyer FJ, Hendrickson AH (1931) The moisture equivalent as a measure of the field capacity of soils. Soil Sci 32: 181-193.
34. Hendrickson AH, Veihmeyer FJ (1945). Permanent wilting percentages of soils obtained from field and laboratory trials. Plant Physiol 20: 517-539.
35. Tenenhaus M, Vinzi VE, Chatelin YM, Lauro C (2005) PLS path modeling. Comput Statd Data Anal 48: 159-205.
36. Marcoulides G, Chin WW, Saunders C (2009) A critical look at partial least squares modeling. MIS Quart 33: 171-175.
37. Vinzi V, Chim WW, Henseler J, Wang H, editors (2010) Handbook of Partial Least Squares: Concepts, Methods, and Applications. Heidelberg: Springer.
38. Bengtsson G, Fronaeus S, Bengtsson-Kloo L (2002) The kinetics and mechanism of oxidation of hydroxylamine by iron(III). J Chem Soc Dalton Trans 12: 2548-2552.
39. Burger M, Venterea RT (2011) Effects of nitrogen fertilizer types on nitrous oxide emissions. In: Guo L, Gunasekara AS, McConnel LL, editors. Understanding Greenhouse Gas Emissions from Agricultural Management.Wa-Washington DC:American Chemical Society. pp. 179-202.
40. Hanse HCB, Borggaard OK, Sorensen J (1994) Evaluation of the free energy of formation of Fe(II)-Fe(III) hydroxide-sulphate (green rust) and its reduction of nitrite. Geochim Cosmochim Acta 58: 2599-2608.
41. Straub KL, Benz M, Schink B, Widdel F (1996) Anaerobic, nitrate-dependent microbial oxidation of ferrous iron. Appl Environ Microbiol 62: 1458-1470.
42. Weber KA, Achenbach LA, Coates JD (2006) Microorganisms pumping iron: anaerobic microbial iron oxidation and reduction. Nat Rev Microbiol 4: 752-764.
43. Clement JC, Shrestha J, Ehrenfeld JG, Jaffe PR (2005) Ammonium oxidation coupled to dissimilatory reduction of iron under anaerobic conditions in wetland soils. Soil Biol Biochem 37: 2323-2328.
44. Shrestha J, Rich JJ, Ehrenfeld JG, Jaffe PR (2009) Oxidation of ammonium to nitrite under iron-reducing conditions in wetland soils. Soil Sci 174: 156-164.
45. Finzi AC, Cole JJ, Doney SC, Holland EA, Jackson RB (2011) Research frontiers in the analysis of coupled biogeochemical cycles. Front Ecol Environ 9: 74-80.
46. Silver WL, Hall SJ, Liptzin D, Yang WH (2011) The iron redox engine drives carbon, nitrogen, and phosphorus cycling in terrestrial ecosystems. American Geophysical Union Abstracts, http://adsabs.harvard.edu/abs/2011AGUFM.B11F.01S, accessed Oct.28, 2012.

# Enhanced Accumulation of Copper and Lead in Amaranth (*Amaranthus paniculatus*), Indian Mustard (*Brassica juncea*) and Sunflower (*Helianthus annuus*)

**Motior M. Rahman*, Sofian M. Azirun, Amru N. Boyce**

Institute of Biological Sciences, Faculty of Science, University of Malaya, Kuala Lumpur, Malaysia

## Abstract

***Background:*** Soil contamination by copper (Cu) and lead (Pb) is a widespread environmental problem. For phytoextraction to be successful and viable in environmental remediation, strategies that can improve plant uptake must be identified. In the present study we investigated the use of nitrogen (N) fertilizer as an efficient way to enhance accumulation of Cu and Pb from contaminated industrial soils into amaranth, Indian mustard and sunflower.

***Methods/Principal Findings:*** Plants were grown in a greenhouse and fertilized with N fertilizer at rates of 0, 190 and 380 mg $kg^{-1}$ soil. Shoots, roots and total accumulation of Cu and Pb, transfer factor (TF), translocation index were assessed to evaluate the transport and translocation ability of tested plants. Addition of N fertilizer acidified the industrial soil and caused the pH to decrease to 5.5 from an initial pH of 6.9. Industrial soil amended with N fertilizer resulted in the highest accumulation of Pb and Cu (for Pb 10.1–15.5 mg $kg^{-1}$, for Cu 11.6–16.8 mg $kg^{-1}$) in the shoots, which was two to four folds higher relative to the concentration in roots in all the three plants used. Sunflower removed significantly higher Pb (50–54%) and Cu (34–38%) followed by amaranth and Indian mustard from industrial soils with the application of N fertilizer. The TF was $<1$ while the shoot and root concentration (SC/RC) ratios of Pb and Cu were between 1.3–4.3 and 1.8–3.8, respectively, regardless of plant species.

***Conclusions:*** Sunflower is the best plant species to carry out phytoextraction of Pb and Cu. In contrast, Pb and Cu removal by Indian mustard and amaranth shows great potential as quick and short duration vegetable crops. The results suggest that the application of N fertilizer in contaminated industrial soil is an effective amendment for the phytoextraction of Pb and Cu from contaminated industrial soils.

**Editor:** Malcolm Bennett, University of Nottingham, United Kingdom

**Funding:** This research work was financially supported by the University of Malaya, Kuala Lumpur, Malaysia. The funder had no role in study design, data collection and analysis, decision to publish, or preparation of the manuscript.

**Competing Interests:** The authors have declared that no competing interests exist.

* E-mail: mmotiorrahman@gmail.com

## Introduction

A major environmental concern resulting from the dispersal of industrial and urban wastes generated by human activities is the contamination of soil and its environment [1]. Soils may become contaminated by the accumulation of toxic heavy metals and metalloids through emissions from rapidly expanding industrial areas, mine tailings, disposal of high metal wastes, leaded gasoline and paints, application of fertilizers and manures, sewage sludge, pesticides, wastewater irrigation, coal combustion residues, spillage of petrochemicals, and atmospheric deposition [2], [3]. Soils are the major sink for toxic metals released into the environment by the aforementioned anthropogenic activities and unlike organic contaminants which are oxidized to carbon (IV) oxide by microbial action, most metals do not undergo microbial or chemical degradation [4], and their total concentration in soils persists for a long time after their introduction [5], although changes in their chemical forms and bioavailability are possible. The presence of toxic metals in soil can severely inhibit the biodegradation of organic contaminants by microorganisms [6]. Among various types of soil pollutants, toxic metals pollution appears to be of great concern, especially in developing countries where it has caused soil quality deterioration either by aerial deposition or waste water discharge [7].

Regardless of the origin of the metals in the soil, high levels of many metals can result in the degradation of soil fertility and poor quality of agricultural products and poses a significant threat and hazard to human, animal and ecosystem well-being [8], [9]. Metals such as Cu, Pb and zinc (Zn) are particularly important since high quantities of these metals can decrease crop production due to the risk of biomagnifications and bioaccumulation in the food chain. There is also the risk of underground and surface water contamination [10], [11]. These cannot be removed biologically and under certain circumstances can transform from one oxidation state to another which in some metals makes them less toxic [12]. Toxic metal contamination of soil may pose risks and hazards to humans and the ecosystem through direct ingestion or contact with contaminated soil, the food chain, drinking of contaminated ground water, reduction in food quality via phytotoxicity, reduction in land usability for agricultural production causing food insecurity, and land tenure problems [13], [14].

Accumulation of toxic metals in crop plants is also of great concern because it can enter the food chain [15]. As plants acquire the necessary nutrients, such as N, phosphorus (P) and potassium (K), they also take in and accumulate toxic metals such as Pb and cadmium (Cd). Ingestion of vegetables grown in soils contaminated with toxic metals has been suggested as a possible risk to human and wildlife health. These health risks will depend on the chemical composition of the contaminated soil, its physical characteristics, the type of vegetables cultivated and the consumption rate [16]. The uptake of toxic metals by plants is often influenced by plant species, growth stage, soil type, metals and environmental factors. Toxic metal concentration in the soil solution plays a critical role in controlling metal availability to plants [17]. Many findings have shown that increasing the levels of toxic metals in the soil will bring about an increased uptake by the plants. However, the availability of the toxic metal ions are influenced by various factors including soil pH, the physical and chemical properties of soil, the clay content and manganese oxide concentration [18].

In developing countries with high population density and insufficient funds available for environmental restoration, low cost and ecologically sustainable remedial options are required to restore contaminated lands so as to reduce the associated risks, make the land resource available for agricultural production, enhance food security, and scale down land tenure problems [19]. Phytoextraction was developed as a result of efforts to find a more efficient and less hazardous technique to remediate contaminated soils [20]. It involved the removal of metals by plants through uptake and accumulation into biomass [1]. Interestingly, although phytoremediation was recognized and documented by humans more than 300 years ago, its scientific study and development was not conducted until the early 1980's [21]. However, progress in making phytoextraction a practical commercial technology has been hindered by the lack of strategies to optimize plant uptake of metals [22]. Contaminated soil can be amended by conventional ways such as physical, chemical, bioremediation and phytoremediation. Unfortunately, existing remediation methods for toxic metal removal from contaminated soils are expensive and disruptive [1]. In recent years, efforts have focused on the remediation strategies that are less expensive, less destructive and more sustainable [23−25]. Phytoremediation, defined as the use of plants to remove pollutants from the environment, is a promising technology for the remediation of contaminated soils and perhaps for the removal of metals from contaminated soil [26]. This technology can be applied to both organic and inorganic pollutants present in soil, water and air. In this respect, plants can be compared to solar driven pumps which can extract and concentrate certain elements from the environment [27]. Phytoremediation is aesthetically pleasant, soil-microorganisms friendly, enhances diversity, and derives energy from sunlight [28]. More essentially, it is able to retain the fertility status of the soil even after the removal of toxic metals. Many plants are capable of accumulating high levels of toxic metals in the soil and are potent tools to help clean up contaminated soil [29]. The success of phytoremediation depends on the solubility of heavy metals and the ability of plant to take in and partition toxic metals to the upper plant parts [30]. Several studies have been done with chelating agents for their ability to dissolve metals and enhance the uptake of metals by plants. However these chemicals have limitations due to their negative effects on the plant and soil properties [31−33]. Phytoremediation has been recommended as an alternative remediation technology for polluted soil with heavy metals but is generally perceived to be too slow. The enhance accumulation of trace pollutants in harvestable plant tissues is a prerequisite for such technology to be practical. Thus, an ideal plant selection for rehabilitation of polluted soil has to be considered based on faster and higher biomass producing plants that can tolerate and accumulate heavy metals in plant parts. Relationships between toxic metals in crops and pollution of toxic metals in soils have been well documented [34]. However, current information on the phytoremediation technology using tropical species grown with nitrogen fertilizer is limited. Taking this into account a study was undertaken to investigate the impact of N fertilizer on the accumulation of Cu and Pb in industrial soils using amaranth, Indian mustard and sunflower plants.

## Materials and Methods

### Soil Management and Experimental Design

Experiments were conducted in a greenhouse at the University of Malaya, Kuala Lumpur, Malaysia from September 2010 to December 2010. Soil samples were collected from the 0–30 cm depth randomly from 10 spots within an area of recycling electronic wastes and chemicals that contained heavy metals in Senai, Johor, (1° 28′ 0″ N, 103° 45′ 0″ E) Malaysia. No specific permits were required for the described field studies and no specific permissions were required for these activities. In addition the locations were not protected and the field studies did not involve endangered or protected species. The industrial soils were sandy loam in texture and thoroughly mixed and sieved through a 2-mm mesh to obtain homogenous soil composites before they were used as the growing medium in polyethylene pots (height 18-cm × diameter 14-cm = surface area 0.095 $m^2$). Each pot was filled with 2 kg of industrial soil. Amaranth, Indian mustard and sunflower were grown in industrial soil with N fertilizer at rates of 0, 190 and 380 mg $kg^{-1}$ soil. Each plant species was tested in an individual experiment against the three levels of N fertilizer. The experiment was conducted under completely randomized design with six replications. The seeds of amaranth, Indian mustard and sunflower were sown as per treatment. Sufficient seeds were sown to ensure healthy germination. Plants were thinned after germination and five plants per pot were kept until harvesting. Prior to sowing 0.41 g and 0.82 g urea (46%N) were applied into the soils as per treatment schedule. Continuous monitoring was done to observe growth and all plants were harvested at the age of 56 d.

### Plant and Soil Analysis

Prior to planting, the initial Pb and Cu concentrations in the industrial soil were analyzed using an atomic absorption spectrometer (AAS). In industrial soil, Cu (56.2±0.09 mg $kg^{-1}$) and Pb (40.2±0.07 mg $kg^{-1}$) were present. The initial status of soil N in industrial soil was 26.50±0.71 mg $kg^{-1}$. The soil pH was determined from the prepared soil suspension (1:2.5 soil water ratios) by using a combined pH meter model 900A (Thermo Orion, Ontario, Canada) at regular 14 d interval [35]. After harvesting all plant samples were oven-dried at 72°C for 48 h. Biomass accumulation and partitioning were determined for all the plants used. Shoot and root tissues of each plant species and soils were firstly ground by using mortar and pestle prior to toxic metal analysis. For each soil and plant parts, 800 mg of the ground particle was diluted with 1 L of distilled water to produce an 800 ppm mixture. Lead and Cu concentrations in each soil and plant parts were analyzed using a PE Analyst 400 AAS.

### Calculation

The Cu and Pb transport from soil to the shoots was evaluated using the transfer factor (TF) = WC (mg $kg^{-1}$)/TC (mg $kg^{-1}$) where, WC = elements concentration in whole plant; TC = elements soil concentration obtained with AAS [36]. The ability for Cu and Pb translocation from roots to shoots was calculated by the translocation index (TI) % = SQ (mg per pot)/WPQ (mg per pot) × 100, where, SQ = element accumulation in the shoots; WPQ = element accumulation in the whole plant (shoots+roots). Percentage removal (%) was calculated from the total concentration (TC) of elements initially present in the soil [37].

### Data Analysis

Statistical analysis was carried out by one-way ANOVA using general linear model and t-Test to evaluate significant differences between means of plant biomass, absorbed heavy metal concentrations in harvested plants at 95% level of confidence [38]. Further statistical validity of the differences among treatment means was estimated using the least significance difference (LSD) method.

## Results and Discussion

### Soil pH in Industrial Soil

Soil pH in the industrial soil was significantly affected by the application of N fertilizer regardless of plant species. From an initial pH of 6.9 the soil pH decreased significantly after the application of N fertilizer. Nitrogen fertilizer application at the rate of 380 mg $kg^{-1}$ soil developed a more acidic soil pH during experimental period. The lowest soil pH was between 5.5–5.6, with the use of N fertilizer at rates of 190 mg $kg^{-1}$ soil and 380 mg $kg^{-1}$ soil at 14 d. Thereafter the soil pH increased slightly over time in both unfertilized and fertilized industrial soils regardless of plant species (Table 1). This could possibly be due to the stable saturated moisture in the soils and plants that will result in a dilution effect of the soil pH by the presence of hydroxyl ions ($OH^-$) in water, particularly in the unfertilized industrial soil experiments. The change in soil pH with N fertilizer treatment probably involved more complex reactions due to the application of varying levels of N fertilizer. The initial acidic condition of fertilized industrial soil at 14 d is possibly due to the application of N fertilizer and the subsequent microbial stimulated aerobic conversion of $NH_4^+$ to $NO_3^-$ through nitrification, which will release $H^+$ ions into the soil. Enhanced nitrate uptake by the plants could potentially affect soil pH as a result of $OH^-$ excretion and the intake of protons ($H^+$) during uptake of soil nitrate.

Soil pH greatly influences the availability of plant nutrients [39] and controls the solubility of heavy metals and other plant nutrients in the soil that subsequently influences the effectiveness of phytoremediation. Lead is relatively unavailable to plants when the soil pH is above 6.5 and even more so when soil phosphorus levels are high [40]. The decreased soil pH can bring about an increase in heavy metal sorption, due to the increased net negative surface charge on organic matter and soil oxides at high pH levels [41]. In the present study, adding N also acidified the industrial soil, which caused the pH decrease to 5.6 from an initial pH 6.9 within 56 d in the pot experiments. The optimal soil pH for plant growth has been documented to be around 6.4 even though; sunflower can tolerate soils ranging in from pH 5.7 to over 8.0. A pH range between 6.0 to 7.2 has been reported to be optimal for many soils [42] while optimal soil pH for amaranth and Indian mustard growth has been documented to be 6.4 [43] and 6.5 [44], respectively. Thus a balance between soil pH and heavy metal availability are required for enhance plant growth. In this study the soil pH increased from 5.8 to 6.5 after 56 d when the plants were grown with N fertilizer whilst the pH increased more than 6.5 when the plants were grown in unfertilized soil. The solubility of Cu is drastically increased at pH 5.5 [32], which are close to the ideal farmland pH of 6.0−6.5 [33]. Copper is an important essential element for plants, microorganisms, animals, and humans. The connection between soil and water contamination and metal uptake by plants is determined by many chemical and physical soil factors as well as the physiological properties of the crops.

**Table 1.** Soil pH influenced by N fertilizer with amaranth, Indian mustard and sunflower grown in industrial soils over time.

| Plant species and N fertilizer (mg $kg^{-1}$ soil) | Days after emergence | | | | |
|---|---|---|---|---|---|
| | 0 | 14 | 28 | 42 | 56 |
| Amaranth: | | | | | |
| N 0 | 6.9 a | 6.8 a | 6.8 a | 6.9 a | 7.0 a |
| N 190 | 6.9 a | 5.6 b | 6.0 b | 6.2 b | 6.4 b |
| N 380 | 6.9 a | 5.5 b | 5.8 c | 6.0 c | 6.2 c |
| Indian Mustard: | | | | | |
| N 0 | 6.9 a | 6.8 a | 6.9 a | 7.0 a | 7.0 a |
| N 190 | 6.9 a | 5.6 b | 6.0 b | 6.3 b | 6.5 b |
| N 380 | 6.9 a | 5.5 b | 5.8 c | 5.9 c | 6.2 c |
| Sunflower: | | | | | |
| N 0 | 6.9 a | 6.8 a | 6.9 a | 7.0 a | 7.0 a |
| N 190 | 6.9 a | 5.6 b | 6.0 b | 6.2 b | 6.4 b |
| N 380 | 6.9 a | 5.5 b | 5.8 c | 6.0 c | 6.3 c |

Means followed by the same letters are not significantly different for each treatment means ($P<0.05$).

### Dry Matter Yield

The shoot, root and total dry matter yield was affected significantly ($P<0.001$) by application of N fertilizer in all tested plant species (Table 2). The shoot, root and total dry matter yield in the plants varied depending on the plant species. All the plant species grown with N fertilizer exhibited the highest shoot, root and total dry matter yield. Sunflower produced appreciably higher shoot, root and total dry matter than amaranth and Indian mustard. The shoot dry matter was significantly higher than roots in all tested species.

The shoot, root and whole plant accumulate content of Pb and Cu per pot was used to estimate translocation index and all the attributes were affected significantly by N fertilizer application in all tested plant species (Table 2). All the plant species grown in industrial soils with N fertilizer at the rate of 380 mg $kg^{-1}$ soil exhibited the highest shoot, root and whole plant accumulate content per pot. Sunflower produced significantly higher shoot, root and whole plant accumulate content than amaranth and Indian mustard. The shoots accumulate content was significantly higher than roots accumulate content in all the tested plant species.

Higher dry matter accumulation was possibly due to the greater vegetative growth resulting from higher photosynthetic activities influenced by the addition of N fertilizer. Sunflower and amaranth responded well to varying levels of N fertilizer in the present study.

**Table 2.** Dry matter yield, lead and copper accumulation in plant tissues.

| Plant species and N fertilizer (mg $kg^{-1}$ soil) | Dry matter yield (g $pot^{-1}$) | | | Roots accumulate content (RQ) (mg $pot^{-1}$) | | Shoots accumulate content (SQ) (mg $pot^{-1}$) | | Whole plant accumulate content (WPQ) (mg $pot^{-1}$) | |
|---|---|---|---|---|---|---|---|---|---|
| | **Root** | **Shoot** | **Total** | **Pb** | **Cu** | **Pb** | **Cu** | **Pb** | **Cu** |
| Amaranth: | | | | | | | | | |
| N 0 | 1.6 b | 10.0 b | 11.6 b | 0.007 b | 0.005 b | 0.06 c | 0.07 c | 0.07 c | 0.07 c |
| N 190 | 1.9 a | 12.1 a | 14.0 a | 0.010 a | 0.009 a | 0.12 b | 0.14 b | 0.13 b | 0.15 a |
| N 380 | 1.9 a | 12.7 a | 14.6 a | 0.011 a | 0.009 a | 0.14 a | 0.16 a | 0.15 a | 0.16 a |
| Indian Mustard: | | | | | | | | | |
| N 0 | 1.6 b | 9.8 b | 11.4 b | 0.004 b | 0.005 a | 0.05 c | 0.06 c | 0.06 c | 0.07 c |
| N 190 | 1.8 a | 11.3 a | 13.1 a | 0.005 a | 0.006 a | 0.11 b | 0.14 b | 0.12 b | 0.15 b |
| N 380 | 1.9 a | 12.0 a | 13.9 a | 0.005 a | 0.007 a | 0.14 a | 0.17 a | 0.15 a | 0.18 a |
| Sunflower: | | | | | | | | | |
| N 0 | 1.6 b | 11.0 b | 12.6 b | 0.008 c | 0.008 b | 0.09 c | 0.10 c | 0.10 c | 0.11 c |
| N 190 | 2.4 a | 13.1 a | 15.5 a | 0.013 b | 0.011 a | 0.19 b | 0.19 b | 0.20 b | 0.20 b |
| N 380 | 2.5 a | 13.6 a | 16.1 a | 0.015 a | 0.012 a | 0.21 a | 0.23 a | 0.23 a | 0.24 a |

Means followed by the same letters are not significantly different for each treatment means ($P<0.05$).

Indian mustard growth was slightly poorer compared to the other two species. The possible reason for this is the use of sandy loam soil in this study. Sunflower plants grown in industrial soils with N fertilizer exhibited significantly higher dry matter yield probably due to its broader leaves which would intercept more sunlight and enable higher rates of photosynthesis and as a result incorporate more carbon into the plants [45]. This was also observed in the case of Indian mustard grown in industrial soils with N fertilizer, where its broader leaves provide a greater surface area for photosynthesis compared to the same plant growing in unfertilized industrial soils. The results indicate that higher biomass accumulation is an important trait to enhance efficiency of phytoremediation. Soil amendments with the application of organic or inorganic fertilizer can promote higher biomass yield resulting in an increased heavy metal bioaccumulation capacity of the plants. Kenaf accumulated significantly higher biomass and absorbed higher amounts of Pb in fertilized soil compared to unfertilized soils [46]. Addition of fertilizers can significantly reduce the mobility of trace metals [47] but increases the amount of trace metals in plant shoots because of the higher total biomass [48]. The application of N and P enhanced the growth of shoot and root in Indian mustard and its subsequent ability to remove Cu from contaminated soil [49]. While N fertilizer combined with EDTA facilitated the phytoremediation efficiency of sunflower in Pb-polluted soil [50], high-dose of application of N fertilizer had an influence on the accumulation of Pb in edible parts of radish, carrot and potato. All these plants had higher concentrations of Pb when contaminated soils were treated with N fertilizer compared to untreated soils [51].

## Nitrogen, Lead and Copper Accumulation

The N uptake in all the three plant species studied were significantly higher ($P<0.01$) in industrial soil amended with N fertilizer compared to unfertilized industrial soils (Table 3). All the plant species recorded higher N uptake (64.6–75.8 mg $g^{-1}$) in industrial soil with N application compared to unfertilized industrial soils (45.8–47.8 mg $g^{-1}$). The shoot and root N were also significantly higher in industrial soil with N fertilizer. The shoot N was significantly higher than root N irrespective of plant species.

In the present study, the total extracted Pb and Cu content varied depending on the plant species. Both Pb (20.0–21.6 mg $kg^{-1}$) and Cu (19.1–21.5 mg $kg^{-1}$) uptake by sunflower grown in industrial soil with N was significantly higher than in unfertilized industrial soil (Table 3). Indian mustard and amaranth also recorded significantly higher Pb and Cu uptake when grown with N fertilizer. Both Indian mustard and amaranth recorded higher Cu (15.7–18.1 mg $kg^{-1}$ for Indian mustard; 18.7–21.4 mg $kg^{-1}$ for amaranth) than Pb uptake (12.7–14.3 mg $kg^{-1}$ for Indian mustard; 16.0–16.9 mg $kg^{-1}$ for amaranth) from industrial soil with N fertilizer. Lead and Cu uptake by the different plant species studied were in the following order: sunflower>amaranth>Indian mustard for both unfertilized and N fertilized industrial soils. Lead and Cu concentration in shoots and roots of all the plant species studied were significantly higher ($P<0.05$) in industrial soil with N application than in unfertilized industrial soil (Table 3). Regardless of plant species and soils used, Pb and Cu concentration in the shoots were two to three folds higher than in the roots. All tested parameters showed identical results between N fertilizer levels.

In all the plant species, relatively more metal accumulated in the shoots than in the roots where it was observed that the shoot and root concentration ratio (SC/RC) in all the plant species studied for Pb was >1, suggesting that Pb translocation from root to shoot was remarkably higher and shoots acted as a sink for Pb (Table 4). Indian mustard recorded remarkably higher SC/RC ratio (>3) when grown with N fertilizer. The SC/RC ratio (>1) showed that the root translocation of Pb and its partitioning from root to shoot had a positive influence on the plant species.

Copper uptake in all the plant species resulted in a relatively higher concentration of Cu in the shoots compared to the roots (Table 3). Indian mustard exhibited appreciably higher SC/RC ratio (>3) followed by sunflower when industrial soil was fertilized with N. The SC/RC ratio of all plants species was >1, suggesting that Cu translocation from root to shoot was appreciable. Considering both Pb and Cu, the higher translocation index (TI) and transfer factor (TF) were obtained, in increasing order, in amaranth, mustard and sunflower with N amended soil (Table 4). For both Pb and Cu, the highest TI was obtained in Indian

**Table 3.** Nitrogen, lead and copper concentration in plant tissues.

| Plant species and N fertilizer (mg $kg^{-1}$ soil) | N (mg $g^{-1}$) | | | Pb (mg $kg^{-1}$) | | | Cu (mg $kg^{-1}$) | | |
|---|---|---|---|---|---|---|---|---|---|
| | Root (RC) | Shoot (SC) | Total | Root (RC) | Shoot (SC) | Total | Root (RC) | Shoot (SC) | Total |
| Amaranth: | | | | | | | | | |
| N 0 | 16.8 b | 29.2 b | 46.0 b | 4.8 b | 6.2 b | 11.0 b | 3.1 b | 6.6 b | 9.7 b |
| N 190 | 21.9 a | 43.3 a | 65.2 a | 5.5 a | 10.5 a | 16.0 a | 4.5 a | 11.6 a | 16.1 a |
| N 380 | 22.6 a | 44.4 a | 67.0 a | 5.7 a | 11.2 a | 16.9 a | 4.7 a | 12.5 a | 17.2 a |
| Indian Mustard: | | | | | | | | | |
| N 0 | 16.5 b | 29.3 b | 45.8 b | 2.5 b | 5.5 b | 8.0 b | 2.9 b | 6.3 b | 9.4 b |
| N 190 | 20.6 a | 44.0 a | 64.6 a | 2.6 ab | 10.1 a | 12.7 a | 3.5 a | 12.2 a | 15.7 a |
| N 380 | 21.1 a | 45.3 a | 66.4 a | 2.7 a | 11.6 a | 14.3 a | 3.8 a | 14.3 a | 18.1 a |
| Sunflower: | | | | | | | | | |
| N 0 | 17.0 b | 30.8 b | 47.8 b | 5.2 b | 8.5 b | 13.7 b | 5.1 a | 9.2 b | 14.3 b |
| N 190 | 23.7 a | 50.7 a | 74.4 a | 5.8 a | 14.2 a | 20.0 a | 4.6 b | 14.5 a | 19.1 a |
| N 380 | 24.3 a | 51.5 a | 75.8 a | 6.1 a | 15.5 a | 21.6 a | 4.7 b | 16.8 a | 21.5 a |

Means followed by the same letters are not significantly different for each treatment means ($P<0.05$).

**Table 4.** Lead and copper removal, shoot and root concentration ratio, translocation index and transfer factor.

| Plant species and N fertilizer (mg kg$^{-1}$ soil) | Shoot and root concentration (SC/RC) ratio | | Translocation index (TI) | | Transfer Factor (TF) | | Removal (%) | |
|---|---|---|---|---|---|---|---|---|
| | Pb | Cu | Pb | Cu | Pb | Cu | Pb | Cu |
| Amaranth: | | | | | | | | |
| N 0 | 1.3 b | 2.1 b | 89.0 b | 93.0 b | 0.3 b | 0.2 b | 27.4 b | 17.2 b |
| N 190 | 1.9 a | 2.6 a | 92.0 a | 94.3 a | 0.4 a | 0.3 a | 39.8 a | 28.6 a |
| N 380 | 2.0 a | 2.7 a | 92.9 a | 94.7 a | 0.4 a | 0.4 a | 42.0 a | 30.6 a |
| Indian Mustard: | | | | | | | | |
| N 0 | 2.2 b | 2.2 b | 93.1 b | 93.0 b | 0.2 b | 0.2 b | 19.9 c | 16.4 c |
| N 190 | 3.9 a | 3.5 a | 96.1 a | 95.6 a | 0.3 ab | 0.3 a | 31.6 b | 27.9 b |
| N 380 | 4.3 a | 3.8 a | 96.4 a | 96.0 a | 0.4 a | 0.3 a | 35.6 a | 32.2 a |
| Sunflower: | | | | | | | | |
| N 0 | 1.6 b | 1.8 b | 91.8 b | 92.5 b | 0.3 b | 0.3 b | 34.1 c | 25.4 c |
| N 190 | 2.4 a | 3.2 a | 93.0 a | 94.5 a | 0.5 a | 0.3 b | 49.8 b | 34.0 b |
| N 380 | 2.5 a | 3.6 a | 93.3 a | 95.1 a | 0.5 a | 0.4 a | 53.7 a | 38.3 a |

Means followed by the same letters are not significantly different for each treatment means ($P<0.05$).

mustard and sunflower. Despite the high TI observed for Pb and Cu in all the species tested, the TF, which reflects the plant's capacity to transport metals from soil to shoots, was much higher for Pb than for Cu. Such behavior can probably be attributed to high Pb mobility in the soil due to the decreasing soil pH.

The results obtained showed that sunflower removed 49.8−53.7% of Pb 34−38.3% of Cu from industrial soil with N fertilizer compared to 34.1% of Pb and 25.4% of Cu from unfertilized industrial soil (Table 4). Amaranth removed 39.8−42% of Pb and 28.6−30.6% of Cu while Indian mustard removed 31.6−35.6% of Pb and 27.9−32.2% of Cu from

**Table 5.** Regression equation, correlation coefficient (r) and coefficients of determination ($R^2$) of different parameters.

| Regression equation | Correlation coefficient (r) | Coefficients of determination ($R^2$) |
|---|---|---|
| Relationship between N fertilizer and biomass (B): | | |
| $Y_B$ (Amaranth) = $0.011X_1+15.78$ | 0.97** | $R^2=0.939$ |
| $Y_B$ (Indian mustard) = $0.013X_1+14.857$ | 0.95** | $R^2=0.909$ |
| $Y_B$ (Sunflower) = $0.011X_1+17.044$ | 0.95** | $R^2=0.903$ |
| Relationship between biomass and Pb uptake: | | |
| $Y_{Pb}$ (Amaranth) = $1.519X_2-12.46$ | 0.99** | $R^2=0.9899$ |
| $Y_{Pb}$ (Indian mustard) = $1.504X_2-13.849$ | 0.99** | $R^2=0.9962$ |
| $Y_{Pb}$ (Sunflower) = $2.065X_2-20.798$ | 0.99** | $R^2=0.9996$ |
| Relationship between biomass and Cu uptake: | | |
| $Y_{Cu}$ (Amaranth) Cu = $1.935X_2-20.167$ | 0.99** | $R^2=0.9887$ |
| $Y_{Cu}$ (Indian mustard) = $2.112X_2-21.504$ | 0.99** | $R^2=0.9938$ |
| $Y_{Cu}$ (Sunflower) = $1.790X_2-15.705$ | 0.99** | $R^2=0.9748$ |
| Relationship between N fertilizer and Pb uptake: | | |
| $Y_{Pb}$ (Amaranth) = $0.017X_1+11.593$ | 0.94** | $R^2=0.883$ |
| $Y_{Pb}$ (Indian mustard) = $0.018X_1+8.4324$ | 0.97** | $R^2=0.941$ |
| $Y_{Pb}$ (Sunflower) = $0.022X_1+14.37$ | 0.96** | $R^2=0.913$ |
| Relationship between N fertilizer and Cu uptake: | | |
| $Y_{Cu}$ (Amaranth) = $0.021X_1+10.468$ | 0.94** | $R^2=0.879$ |
| $Y_{Cu}$ (Indian mustard) = $0.025X_1+9.767$ | 0.97** | $R^2=0.949$ |
| $Y_{Cu}$ (Sunflower) = $0.020X_1+14.614$ | 0.99** | $R^2=0.975$ |

$X_1$ = N fertilizer rate; $X_2$ = biomass;
**significant at 0.01 level of probability.

industrial soil with N fertilizer. The toxic metal removal efficiency by sunflower was 16−20% higher for Pb and 9−13% higher for Cu with the application of N fertilizer in industrial soils compared to unfertilized industrial soil. Similarly Pb removal efficiency by Indian mustard and amaranth was 12−16% and 13−15% higher, respectively. Copper removal efficiency by Indian mustard and amaranth was 11−15% and 11−14% higher in industrial soil with N fertilizer compared to unfertilized industrial soil.

A comparison of the metal content of the soil used in this study with others reported in the literature confirmed that it should be considered as a contaminated soil. The amounts of Cu ($56.2 \pm 0.09$ mg $kg^{-1}$) and Pb ($40.2 \pm 0.07$ mg $kg^{-1}$) recorded were above the maximum content commonly found in soils of industrial areas in Johor, Malaysia. Background concentrations of Pb that occur naturally in surface agricultural soils in the United States average 10 parts per million (ppm) with a range of 7 to 20 ppm [52]. Soils with Pb levels above this range are primarily the result of Pb contamination. The Canadian Council of Ministers of the Environment interim soil assessment criterion for Cu was set at 30 mg $kg^{-1}$ [53]. However there is no reference value is available for Cu or Pb from the local environmental agency in Malaysia. The shoot and root ratios were calculated in order to evaluate the translocation of the elements inside the plant, from the roots to the shoots, and its potential accumulation in the biomass (Table 4). The SC/RC ratio values are important to estimate the potential of a plant for phytoremediation purposes. One of the selection criteria for hyperaccumulator's identification is the leaf's metal concentration. Accordingly a hyperaccumulator should concentrate more than 1,000 mg $kg^{-1}$ of Ni, Pb or Cu and more than 100 mg $kg^{-1}$ of Cd [54]. Sunflower exhibited the highest concentration of Pb in shoots followed by amaranth, while Indian mustard was the lowest for both the metals considered (Table 3). However, according to the criteria already presented, none of the plants species tested would be classified as hyperaccumulator for Cu or Pb. This contradicts with the hyperaccumulator characteristic of sunflower where the SC/RC ratio should be >1. In this study all the tested plant species achieved this target although they could not achieve a leaf metal's concentration of 1,000 mg $kg^{-1}$ for Pb or Cu. A possible reason for this is the short experimental period (56 d) where sunflower did not reach maturity. As a result, the leaf metal's concentration in sunflower was <1,000 mg $kg^{-1}$ compared to that reported in other studies. Another possible explanation might be the use of pot experiments, which could have limited plant development due to the small volume of soil used. Furthermore the immobilization of elements by root adsorption may have had some influence. This mechanism prevents element translocation from roots to shoots in some species working as a defense barrier, decreasing the phytoextraction potential of such species since only a small portion of ions associated with the roots are effectively absorbed [21].

Sunflower is able to secrete organic acids which acidify the rhizosphere and increases the solubility of toxic metals like Pb and Cu [55]. The N fertilized plants showed higher biomass and potential to facilitate phytoextraction process. Sunflower absorbed appreciably higher amounts of Pb than Indian mustard and amaranth in industrial soil with N fertilizer due to its hyperaccumulator characteristic [48]. In this study all the plant species contained relatively higher concentration of Pb in the shoots compared to roots as high metal concentrations are in available form in acidic soil due to the decreased soil pH. It is well known that Pb is an immobile metal in soil, since it readily forms a precipitate because of its low aqueous solubility within the soil matrix, and in many cases it is not readily bioavailable. In addition, many plants retain Pb in their roots via sorption and precipitation with only minimal transport to the above ground harvestable plant portions [56], [57]. The leafy vegetables like Indian mustard and amaranth are expected to absorb more Cu compared to Pb due to the physiological role of Cu in plants [17]. The SC/RC ratio for Cu uptake were >1 for all the plants which proved that Cu partitioning was performed from roots to shoots. The high values of SC/RC ratio (>1) implies that plant shoots acts as sink for the absorption of Cu. The result of this study suggests that application of inorganic N fertilizer could boost the biomass accumulation of plants and subsequently phytoremediation efficiency. However, careful measures have to be taken during fertilizer application to avoid anthropogenic source of heavy metal entering the soil and the human food chain. In the literature toxicity in plants is reported in literature to occur when Cu is found in the range of 20−100 mg $kg^{-1}$, and Pb in the range of 30−300 mg $kg^{-1}$ [58], [59]. In this study the accumulations of Pb and Cu in all the tested plant species were below the toxic levels (Table 2).

The most serious source of exposure to soil Pb is through direct ingestion of contaminated soil or dust. In general, plants do not absorb or accumulate Pb. However, in soils with a high Pb content, it is possible for some Pb to be taken up. Studies have shown that Pb does not readily accumulate in the fruiting parts of vegetable and fruit crops (e.g., corn, beans, squash, tomatoes, strawberries, and apples). Higher concentrations are more likely to be found in leafy vegetables (e.g., lettuce) and on the surface of root crops (e.g., carrots). Generally, it has been considered safe to use garden produce grown in soils with total Pb levels less than 300 ppm. The risk of Pb poisoning through the food chain increases as the soil Pb level rises above this concentration. Even at soil levels above 300 ppm, most of the risk is from Pb contaminated soil or dust deposits on the plants rather than from uptake of Pb by the plant [40].

## Correlation Analysis

There were significantly positive relationships between the accumulation of Pb in the three plant species grown in the contaminated industrial soil and the dosage of the N fertilizer applied to the industrial soil (Table 5). The slopes indicated that for each 1.0 mg N application in industrial soil, 0.017 to 0.022 mg of Pb was taken up into the plant tissues (Table 5). There was also a positive correlation between N fertilizer rate and Cu uptake in the three plant species. The slopes indicated that for each 1.0 mg N application in industrial soil, 0.020 to 0.025 mg of Cu was taken up into the plant tissues (Table 5). The regression equation showed that Pb uptake was higher in sunflower while Cu uptake was higher in Indian mustard among three plant species. A strong positive relationship was observed between the accumulation of Pb and biomass ($r = 0.99$), Cu and plant biomass accumulation ($r = 0.99$) in the three plant species growing in contaminated industrial soils (Table 5).

Lead and Cu concentration were positively correlated to plant biomass due to the increased plant tissue to accumulate more heavy metals. Furthermore the higher biomass indicates that plants were bigger and probably possessed higher absorption rate so that more heavy metal uptake took place [46]. The strong positive correlation between plant biomass and plant N content was due to the positive effect of N fertilizer application. As a result, the higher biomass of plants enhanced a greater absorption of more Pb and Cu and countered the effects of heavy metal immobilization [50]. The plant N content directly or indirectly affected the heavy metal concentrations in the plants, by boosting the biomass of the plants and subsequently their ability to accumulate more heavy metals. This positive correlation suggests that inorganic or other types of fertilizers, such as organic fertilizers, can be used to increase the biomass of plants and their capacity to accumulate heavy metals from the soil.

## Conclusions

The Pb and Cu uptake ability of sunflower was appreciably greater than Indian mustard and amaranth. There was a positive relationship between N fertilizer application on plant growth and their ability to absorb Pb and Cu. Nitrogen fertilizer boosted growth parameters increasing root and shoot dry matter significantly, allowing a higher uptake of toxic metals by plants grown in fertilized industrial soil. Nevertheless, careful measures must be put in place to ensure that excessive N fertilizers are not applied to cause uncontrolled rapid above ground growth resulting in weaker plants. It is suggested that phosphorus be added to boost root growth to balance the high shoot growth in the presence of N fertilizer. In this study, sunflower was the best plant species to carry out phytoextraction of Pb and Cu due to its hyperaccumulator characteristic. Both Indian mustard and amaranth also showed Pb and Cu removal potential, as quick and short duration vegetable crops. This provides the opportunity to grow Indian mustard and amaranth several times on a piece of land to remove both Pb and Cu from contaminated soil. However, the results of the present study showed that N fertilizer application in industrial soil can lead to negative consequences on indigenous soil biota in acidic soils due to its impact on soil pH. It clearly shows the need to consider the soil characteristics of polluted soil before its remediation.

## Acknowledgments

We are grateful for the logistic support provided by the University of Malaya, Kuala Lumpur, Malaysia. We would also like to thank two anonymous reviewers for thoughtful comments and students at the Science and Environmental Management, University of Malaya.

## Author Contributions

Conceived and designed the experiments: MMR SMA ANB. Performed the experiments: MMR. Analyzed the data: MMR. Contributed reagents/materials/analysis tools: MMR SMA ANB. Wrote the paper: MMR ANB.

## References

1. Ghosh M, Singh SP (2005) A review on phytoremediation of heavy metals and utilization of it's by products. Asian J Energy Environ 6: 214–231.
2. Khan SQ, Cao Y, Zheng M, Huang YZ, Zhu YG (2008) Health risks of heavy metals in contaminated soils and food crops irrigated with wastewater in Beijing, China. Environ Poll 152(3): 686–692.
3. Zhang MK, Liu ZY, Wang H (2010) Use of single extraction methods to predict bioavail-ability of heavy metals in polluted soils to rice. Com Soil Sci Plant Anal 41(7): 820–831.
4. Kirpichtchikova TA, Manceau A, Spadini L, Panfili F, Marcus MA, et al. (2006) Speciation and solubility of heavy metals in contaminated soil using X-ray microfluorescence, EXAFS spectroscopy, chemical extraction, and thermodynamic modeling. Geochim Cosmochim Acta 70(9): 2163–2190.
5. Adriano DC (2003) Trace elements in terrestrial environments: Biogeochemistry, Bioavailability and Risks of Metals 2nd ed Springer, New York, NY, USA.
6. Maslin P, Maier RM (2000) Rhamnolipid-enhanced mineralization of phenanthrene in organic-metal co-contaminated soils. Biorem J 4(4): 295–308.
7. Demirezen D, Aksoy A, Uruc K (2007) Effect of population density on growth, biomass and nickel accumulation capacity of *Lemna gibba* (Lemnaceae). Chemos 66: 553–557.
8. Blaylock MJ, Huang JW (2000) Phytoextraction of metals. *In* Phytoremediation of toxic metals: using plants to clean-up the environment. Raskin I, Ensley BD (eds) 53–70. Wiley, NY, USA.
9. Luo C, Shen Z, Li X, (2005) Enhanced phytoextraction of Cu, Pb, Zn and Cd with EDTA and EDDS. Chemos 59: 1–11.
10. Schmidt U (2003) Enhancing phytoextraction: the effect of chemical soil manipulation on mobility, plant accumulation and leaching of heavy metals. J Environ Qual 32: 1939–1954.
11. Nowack B, Schulin R, Robinson BH (2006) Critical assessment of chelant-enhanced metal phytoextraction. Environ Sci Tech 40: 5525–5532.
12. Garbisu C, Alkorta I (2001) Phytoextraction: a cost-effective plant-based technology for the removal of metals from the environment. Biores Tech 77: 229–236.
13. McLaughlin MJ, Zarcinas BA, Stevens DP, Cook N (2000) Soil testing for heavy metals. Com Soil Sci Plant Anal 31(11–14): 1661–1700.
14. Ling W, Shen Q, Gao Y, Gu X, Yang Z (2007) Use of bentonite to control the release of copper from contaminated soils. Aust J Soil Res 45(8): 618–623.
15. Cieslinski G, Van-Rees KCJ, Huang PM, Kozak LM, Rostad HPW, et al. (1996) Cadmium uptake and bioaccumulation in selected cultivars of durum wheat and flax as affected by soil type. Plant Soil 182: 115–124.
16. Cobb GP, Sands K, Waters M, Wixson BG, Dorward-King E (2000) Accumulation of heavy metals by vegetables grown in mine wastes. Environ Toxic Chem 19: 600–607.
17. Maimon A, Khairiah J, Ahmad MR, Aminah A, Ismail BS (2009) Comparative accumulation of heavy metals in selected vegetables, their availability and correlation in lithogenic and nonlithogenic fractions of soils from some agricultural areas in Malaysia. Adv Environ Biol 3: 314–321.
18. Xian X, Shokohiford G (1989) Effects of pH on chemical forms and plant availability of cadmium, zinc and lead in polluted soils. Water Air Soil Poll 47: 265–273.
19. Raymond AW, Felix EO (2011) Heavy metals in contaminated soils: a review of sources, chemistry, risks and best available strategies for remediation. International School Res Network Ecol 2011: 1–20.
20. Nascimento CWA, Xing B (2006) Phytoextraction: a review on enhanced metal availability and plant accumulation. Sci Agric 63: 299–311.
21. Lasat MM (2000) Phytoextration of metals from contaminated soil: a review of plant/soil/metal interaction and assessment of pertinent agronomic issues. J Hazard Sub Res 2:1–25.
22. Wang AS, Angle JS, Chaney RL, Delorme TA, Reeves RD (2006) Soil pH effects on uptake of Cd and Zn by *Thlaspi caerulescens*. Plant Soil 281: 325–337.
23. Gupta AK, Sinha S (2007a) Phytoextraction capacity of the plants growing on tannery sludge dumping sites. Biores Tech 98: 1788–1794.
24. Gupta AK, Sinha S (2007b) Phytoextraction capacity of the *Chenopodium album* L grown on soil amended with tannery sludge. Biores Tech 98: 442–446.
25. Xiaomei L, Qitang W, Banks MK (2005) Effect of simultaneous establishment of *Sedum alferdii* and *Zea mays* on heavy metal accumulation in plants. Inter J Phytorem 7: 43–53.
26. Dede G, Ozdemir S, Dede OH (2012) Effect of soil amendments on phytoextraction potential of *Brassica juncea* growing on sewage sludge. Inter J Environ Sci Tech 9: 559–564.
27. Salt DE, Smith RD, Raskin I (1998) Phytoremediation. Annual Rev Plant Physio Plant Mol Bio 49: 643–668.
28. Chaney RL, Angle JS, McIntosh MS, Reeves RD, Li YM, et al. (2005) Using hyperaccu-mulator plants to phytoextract soil Ni and Cd. Zeitschrift für Naturforschung C 60: 190–198.
29. Jarup L (2003) Hazards of heavy metal contamination. British Med Bull 68: 167–182.
30. Turgut C, Pepe MK, Cutright TJ (2004) The effect of EDTA and citric acid on phytoreme-diation of Cd, Cr, and Ni from soil using *Helianthus annuus*. Environ Pollu 131: 147–154.
31. Kaplan M, Orman S, Kadar I, Koncz J (2005) Heavy metal accumulation in calcareous soil and sorghum plants after addition of sulphur-containing waste as a soil amendment in Turkey. Agric Ecosys Environ 111: 41–46.
32. Martınez CE, Motto HL (2000) Solubility of lead, zinc and copper added to mineral soils. Environ Poll 107(1): 153–158.
33. Eriksson J, Andersson A, Andersson R (1997) The state of Swedish farmlands. Tech. Rep. 4778, Swedish Environmental Protection Agency, Stockholm, Sweden.
34. Syakalima MS, Choongo KC, Chilonda P, Ahmadu B, Mwase M, et al. (2001) Bioaccumulation of lead in wildlife dependent on the contaminated environment of the Kafue flats. Bull Environ Contam Toxicol 67: 438–445.
35. Thomas GW (1996) Soil pH and soil acidity. In: Sparks DL (ed) Methods of soil analysis, Part 3-Chemical methods, Soil Sci Soc Am Book Series # 5. 475–490. Madison, USA.
36. Lubben S, Sauerbeck D (1991) The uptake and distribution of heavy metals by spring wheat. Water Air Soil Poll 57–58: 239–247.
37. Paiva HN, Carvalho JG, Siqueira JO (2002) Translocation index of nutrients in cedro (*Cedrela fissilis* Vell.) and ipe-roxo (*Tabebuia impetiginosa* (Mart.) Standl.) seedlings submitted to increasing levels of cadmium, nickel and lead. Revista Arvore 26: 467–473.
38. SAS (2003) SAS Institute, SAS Version 9.1.2(c). SAS Institute, Inc., Cary, NC 449–453.
39. Motior MR, Abdou AS, Fareed HAID, Sofian AM (2011) Responses of sulfur, nitrogen and irrigation water to the uptake of nutrients by *Zea mays* grown in sandy calcareous soil. Aust J Crop Sci 5: 347–357.
40. Rosen CJ (2002) Lead in the Home Garden and Urban Soil Environment, College of Food, Agricultural and Natural Resources Sciences, University of Minnesota, Available: http://www.extension.umn.edu/distribution/horticulture/DG2543.html.Accessed 2012 Aug 4.

41. McBride MB (1994) Environmental Chemistry of Soils. Oxford University Press, New York.
42. Putnam DH, Oplinger ES, Hicks DR, Durgan BR, Noetzel DM, et al. (1990) Sunflower. Alternative Field Crops Manual. Oplinger ES Publication A3005, University of Wisconsin-Extension. Agric Bull Available: http://www.hort.purdue.edu/newcrop/afcm/sunflower.html. Accessed 2012 Jul 31.
43. Singh BP, Whitehead WF (1992) Response of vegetable amaranth to differing soil pH and moisture regimes. II International Symposium on Specialty and Exotic Vegetable Crops, Miami, USA.
44. Zaurov DE, Perdomo P, Raskin I (1999) Optimizing soil fertility and pH to maximize cadmium removed by Indian mustard from contaminated soils. J Plant Nutr 22: 977–986.
45. Martens SN, Breshears DD, Meyer CW (2000) Spatial distributions of understory light along the grassland/forest continuum: effects of cover, height, and spatial pattern of tree canopies. Ecolog Model 126: 79–93.
46. Ho WM, Ang LH, Lee DK (2008) Assessment of Pb uptake, translocation and immobilization in kenaf (*Hibiscus cannabilus* L.) for phytoremediation of sand tailings. J Environ Sci 20: 1341–1347.
47. Madejon E, de Mora AP, Felipe E, Burgos P, Cabrera F (2006) Soil amendments reduce trace element solubility in a contaminated soil and allow re-growth of natural vegetation. Environ Poll 139: 40–52.
48. Yang B, Lan CY, Shu WS (2005) Growth and heavy metal accumulation of *Vetiveria zizanioides* grown on lead/zinc mine tailings. Acta Ecologica Sinica 25: 45–50.
49. Wu LH, Li H, Luo YM, Christie P (2004) Nutrients can enhance phytoremediation of copper-polluted soil by Indian mustard. Environ Geochem Health 26: 331–335.
50. Lin C, Liu J, Liu L, Zhu T, Sheng L, et al. (2009) Soil amendment application frequency contributes to phytoextraction of lead by sunflower at different nutrient levels. Environ Exp Bot 65: 410–416.
51. Zhou Q (2003) Interaction between heavy metals and nitrogen fertilizers applied to soil-vegetable systems. Bull Environ Contam Toxicol 71: 338–344.
52. Holmgren CGS, Meyer MW, Chaney RL, Daniels RB (1993) Cadmium, lead, zinc, copper and nickel in agricultural soils in the United States. J Environ Qual 22: 335–348.
53. BC MOE (1989) Criteria for managing contaminated soils in British Columbia-Draft. Prepared by the Waste Management Program, British Columbia Ministry of the Environment, Victoria, British Columbia.
54. Brown SL, Chaney RL, Angle JS, Baker AJM (1995) Zinc and cadmium uptake by hyperac-cumulator *Thlaspi caerulescens* grow in the nutrient solution. Soil Sci Soc Am J 59: 125–133.
55. Cunningham SD, Berti WR (2000) Phytoextraction and phytostabilization: Technical, economical and regulatory considerations of the soil-lead issue. *In:* Phytoremediation of contaminated soil and water. Terry N, Banuelos G (eds) 359–376. Lewis Publication, Florida, USA.
56. Cui Y, Dong Y, Li H, Wang Q (2004a) Effect of elemental sulphur on solubility of soil heavy metals and their uptake by maize. Environ Intern 30: 323–328.
57. Cui Y, Wang Q, Christie P (2004b) Effect of elemental sulphur on uptake of cadmium, zinc, and sulphur by oilseed rape growing in soil contaminated with zinc and cadmium. Com Soil Sci Plant Anal 35: 2905–2916.
58. Kabata-Pendias A, Pendias H (2001) Trace Elements in Soil and Plants, 3rd ed. CRC Press, Boca Raton, FL, USA.
59. Pais I, Jones Jr (2000) The Handbook of Trace Elements. CRC Press, Boca Raton, FL, USA.

# 17

# Inorganic Nitrogen Leaching from Organic and Conventional Rice Production on a Newly Claimed Calciustoll in Central Asia

**Fanqiao Meng[1]*, Jørgen E. Olesen[2], Xiangping Sun[1], Wenliang Wu[1]**

**1** College of Resources and Environmental Sciences, China Agricultural University, Beijing, China, **2** Department of Agroecology and Environment, Faculty of Agricultural Sciences, Aarhus University, Tjele, Denmark

## Abstract

Characterizing the dynamics of nitrogen (N) leaching from organic and conventional paddy fields is necessary to optimize fertilization and to evaluate the impact of these contrasting farming systems on water bodies. We assessed N leaching in organic versus conventional rice production systems of the Ili River Valley, a representative aquatic ecosystem of Central Asia. The N leaching and overall performance of these systems were measured during 2009, using a randomized block experiment with five treatments. PVC pipes were installed at soil depths of 50 and 180 cm to collect percolation water from flooded organic and conventional paddies, and inorganic N ($NH_4$-N+$NO_3$-N) was analyzed. Two high-concentration peaks of $NH_4$-N were observed in all treatments: one during early tillering and a second during flowering. A third peak at the mid-tillering stage was observed only under conventional fertilization. $NO_3$-N concentrations were highest at transplant and then declined until harvest. At the 50 cm soil depth, $NO_3$-N concentration was 21–42% higher than $NH_4$-N in percolation water from organic paddies, while $NH_4$-N and $NO_3$-N concentrations were similar for the conventional and control treatments. At the depth of 180 cm, $NH_4$-N and $NO_3$-N were the predominant inorganic N for organic and conventional paddies, respectively. Inorganic N concentrations decreased with soil depth, but this attenuation was more marked in organic than in conventional paddies. Conventional paddies leached a higher percentage of applied N (0.78%) than did organic treatments (0.32–0.60%), but the two farming systems leached a similar amount of inorganic N per unit yield (0.21–0.34 kg N $Mg^{-1}$ rice grains). Conventional production showed higher N utilization efficiency compared to fertilized organic treatments. These results suggest that organic rice production in the Ili River Valley is unlikely to reduce inorganic N leaching, if high crop yields similar to conventional rice production are to be maintained.

**Editor:** Ben Bond-Lamberty, DOE Pacific Northwest National Laboratory, United States of America

**Funding:** The study was funded by National Natural Science Foundation of China (No. 30970533) and the National Key Science and Technology Project-Organic Farming Development in the Ili River Valley (No. 2007BAC15B05). The funders had no role in study design, data collection and analysis, decision to publish, or preparation of the manuscript.

**Competing Interests:** The authors have declared that no competing interests exist.

* E-mail: mengfq@cau.edu.cn

## Introduction

Nitrogen (N) leaching is one of the primary pathways of N loss from flooded paddy farmland, representing 30–50% of total N loss from such soils [1]. The amount of leached N may equal up to 15% of the total applied N in rice (*Oryza sativa* L.) production [2,3]. Flooded rice crops typically capture only 20–40% of N applied in the harvested grain [4]. N utilization efficiency can be improved by reducing the various forms of loss, including leaching. Because reactive N is highly soluble, its subsequently high mobility in the environment makes it a major contributor to surface and groundwater contamination. Ju et al. [5] reported that high inputs (550–600 kg N $ha^{-1}$) associated with current rice production systems in China did not significantly increase crop yields, yet cause a doubling of N losses to the environment, mainly from increased denitrification in waterlogged systems. Accordingly, understanding N loss processes, particularly through leaching and $N_2O$ emission, is necessary to both improve resource use efficiency and to protect the quality of nearby water bodies. In this study, we focused on monitoring N leaching losses at different rice growth stages and soil depths in organic and conventional rice production systems.

Located in the Xinjiang Uygur Autonomous Region of northwestern China, the Ili River Valley has become an important grain production region, owing to its plentiful surface water, well-developed soils, and extensive meadow grasslands. Of the total land area (5.82 million ha), approximately 1 million ha are under cultivation, and the introduction of new irrigation technologies is facilitating the conversion of meadow grasslands to agricultural uses [6]. However, the Ili River is sensitive to pollution, which is of local and international concern as the river flows into Balkhash Lake in Kazakhstan. The river receives pollutants from both surface run-off and drainage from farming activities throughout the valley. In addition to water quality concerns, the terrestrial ecosystem may also be at risk because it contains shallow soils with a coarse texture. As such, organic farming is being promoted for the conservation of soil and water resources to reduce the overall negative environmental impact of agricultural production, as well as to increase farmer's income. However, it is unclear whether organic production can maintain high rice yields, with or without high fertilizer inputs, and what effect shifting from conventional to organic farming may have on N loading to the Ili River.

Many studies have examined N leaching from organic and conventional production [7,8]. Such research has revealed that organic farming may lower N leaching compared to conventional systems, both by reducing N inputs and by including catch crops (cover crops) in the rotation [8,9], although the magnitude of this effect is subject to high uncertainty due to variation in crop yields and input intensities [9,10]. Although rice dominates grain production in many developing countries including China, little research has been conducted in this Central Asia region about N loss in organic versus conventional rice farming, especially regarding N loss through percolation or leaching [2,4,11].

This study compared the inorganic N ($NH_4$-N and $NO_3$-N) in percolation/leachate water from organic versus conventional production systems with high yields of rice production, by the application of high-loads of animal manure. We hypothesized that organic rice farming leaches less inorganic N compared with conventional rice in the Ili River Valley. In order to calculate the amount of N leached from paddies, we had to address the technical challenge of quantifying the total volume of leachate from a single-cropping rice paddy. To do so, we designed a filtration system using PVC pipes at two depths (50 and 180 cm) to capture the leachate, and quantified the percolation rates throughout the rice growing season using a water balance based on the difference between incoming (*i.e.*, irrigation and precipitation) and outgoing (evapotranspiration and surface runoff) flows. To better understand the potential environmental effects of the treatments, we evaluated N leaching not only in absolute terms, but also in terms of leachate per unit yield of rice and as a percentage of the total N applied.

## Materials and Methods

### Experiment site

The field experiment was conducted at Chabuchaer Farm (48°65'N, 80°06'E, 634 m ASL) in the Ili River Valley of the Xinjiang Uygur Autonomous Region, China. The region has a temperate continental climate with a mean annual temperature of 9.5°C and annual precipitation of 260 mm, which falls predominantly in June, July, and August. In 2009 when monitoring was implemented, precipitation mainly happened in April to July, November and December (Fig. 1). The farm is located on an alluvial plain off the south bank of the Ili River. The land use was converted to agriculture by the local government in 2005 as part of a national grain supply project. Because this land was newly reclaimed, we could examine the contrasting effects of conventional and organic farming techniques without the usual conversion period from conventional to organic farming, which can confound the results. Our study was initiated in 2008 and the data presented here were collected in the 2009 season.

The soil was a calciustoll with salt content of 1.4%. Rice had been continuously cultivated on this land since 2005, with the aim of reducing soil salinity. The soil is a sandy loam, with 3.2% clay, 44.7% silt, and 52.1% sand. Other important characteristics of the soil (0–20 cm) are as follows: 14.0 g $kg^{-1}$ soil organic matter, 1.15 g $kg^{-1}$ total N, 11 mg $kg^{-1}$ Olsen phosphorous, 264 mg $kg^{-1}$ available potassium, and a pH of 7.9.

Permission to conduct the experiment and sampling at the site was obtained from the Ili Agricultural Technical Extension Station, and the Farmer Liu Ermao. These field studies did not involve endangered or protected species.

### Crop management

Rice (*Oryza sativa* L., cultivar Nonglin-315) was transplanted to the experimental farm on May 29, 2009. The fertilizer used was either a conventional combination of mineral fertilizers with animal manure, as used in the local production system, or only organic fertilizers (composted animal manure [B] with castor (*Ricinuscommunis* L.) bean meal [A]), as used typically in organic rice production. The level of organic fertilization was manipulated to modify the overall N input. Hence, the five treatments were 1) an unfertilized control treatment (CK), 2) low-level organic fertilization (B1A1): composted animal manure at 15,000 kg $ha^{-1}$ and castor bean meal at 2,250 kg $ha^{-1}$, 3) mid-level organic fertilization (B2A1): composted animal manure at 45,000 kg $ha^{-1}$ and castor bean meal at 2,250 kg $ha^{-1}$, 4) high-level organic fertilization (B3A1): composted animal manure at 75,000 kg $ha^{-1}$ and castor bean meal at 2,250 kg $ha^{-1}$, and 5) conventional production system (LS): composted animal manure at 11,250 kg $ha^{-1}$, urea at 277.5 kg $ha^{-1}$, diammonium phosphate at 247.5 kg $ha^{-1}$, and compounded mineral fertilizer at 75 kg $ha^{-1}$ (Table 1). The treatments were arranged in a randomized block design with three replicates.

Composted animal manure was applied on May 9 as the basal fertilizer for all treatments (except CK), 20 days before transplanting. The total amount of castor bean meal was halved and applied in two stages, first as topdressing at the early tillering stage (June 27) and then at the late tillering stage (July 24). Topdressings for LS were applied on June 27 (120 kg $ha^{-1}$ diammonium phosphate with 75 kg $ha^{-1}$ compounded fertilizer), July 6 (127.5 kg $ha^{-1}$ diammonium phosphate with 127.5 kg $ha^{-1}$ urea), and July 15 (urea, 150 kg $ha^{-1}$). The nutrient contents of the fertilizers used in the experiments are listed in Table 2.

Each paddy block had an area of $8 \times 10$ m = 80 $m^2$. Flood irrigation (about 20 cm of standing water), popular in other Central Asian countries [4], was used continuously from June 1 to September 7, except during the first week after fertilization application. Hand weeding was used in all treatments, along with the addition of herbicides to conventional replicates. The main development stages for the rice were transplant (May 29-Jun 13), tillering and elongation (~Aug 5 for organic paddies / ~Aug 10 for conventional paddies), heading (Aug 5-Sep 1 for organic / Aug 10-Sep 5 for conventional), and ripening (Sep 1-Oct 2 for organic / Sep 5-Oct 13 for conventional).

### Measurements of N cycling

Previous studies have shown that inorganic N ($NH_4$-N+$NO_3$-N), rather than particulate N, makes up 50–90% of the total N leached from the soils, and for organic fertilizers or fertile soil, this proportion maybe even higher [12,13,14]. Only inorganic N was analyzed in percolation water collected in our experiment. Percolation water was collected from flooded paddies using polyvinyl chloride (PVC) standpipes [14,15,16]. Two PVC pipes (5 cm in diameter, 150 and 250 cm in length) with sealed bottoms were installed roughly in the center of each field plot to collect drainage water from the saturated soil (Fig. 2). Both pipes were perforated 80 times (6-mm internal diameter) within a 10-cm-wide band, 30 cm from the bottom of the pipe. The porous zone of the pipe was wrapped with nylon textile to prevent sand in-filling. As the average soil layer is 2 m deep in this region, we compared the inorganic N leaching in 50 and 180 cm. At the later depth we considered that inorganic N was not usable by the rice. Thus, the pipes were installed at depths of 50 and 180 cm from the surface to the uppermost pore (Fig. 2). The gap between the collection pipe and soil wall in the porous zone was filled with quartz powder to prevent anything except soil solution from entering the pipe. At a

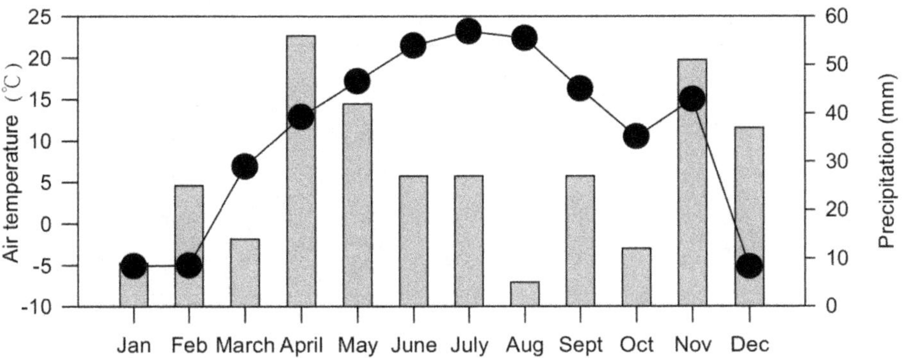


**Figure 1. Air temperature and precipitation at the experimental site in Chabuchaer County for 2009.**

depth of 20 cm, plastic film was wrapped around the PVC pipe, extending horizontally in a 30-cm radius from the pipe to reduce the preferential flow from the irrigation water.

Sampling was conducted 1 and 3 days after fertilization, and thereafter at intervals of 5 days to 2 weeks, depending on the percolate volume. Water samples were stored in plastic vials and refrigerated until analysis. Percolation and irrigation water were analyzed for inorganic N ($NH_4$-N+$NO_3$-N) content using a continuous flow N analyzer (TRAACS 2000). During the irrigation season, average $NH_4$-N and $NO_3$-N concentrations in the irrigation water were 0.11 and 0.68 mg $L^{-1}$ respectively; these values were subtracted from the respective inorganic N concentrations in percolation water to calculate net N (i.e., the amount leached).

The volume of drainage water was calculated per unit area of paddy field. We considered the spatial range of percolation into the pipe to be a half-ellipsoid, having a width of 5 cm and a depth of either 50 or 180 cm, with diameters of 27.5 cm (50/2+5/2) or 92.5 cm (180/2+5/2) [15]. The volume of each ellipsoid was calculated as $4/3*\pi*ab^2$, where $a$ and $b$ are the length of the long axis and radius of the ellipsoid, respectively. Each hectare of paddy field included 252,671 or 222,332 half-ellipsoids, at depths of 50 cm and 180 cm, respectively. Thus, total N leached per ha could be calculated by summing these half ellipsoids at each soil depth, as well as across time to calculate total N lost over one growth cycle. In total, percolation water was sampled 15 times from June 1 to September 7, the full growing season of the rice paddies.

The measured volume of percolation water in the paddy field was calibrated using the following equation [17]:

$$P+I=ET+D+R_0+P_{sd}-V_H \quad (1)$$

Where $P$ is the rainfall (mm) during the rice growth season; $I$ is irrigation (mm); $ET$ is evapotranspiration (mm), calculated using a lysimeter installed in the paddy plot; $D$ is drainage through drain pipe (mm), excluding surface runoff and percolation; $R_o$ is surface runoff (mm); $P_{sd}$ is percolation (mm); and $V_H$ is the change of the water table in the paddy field (mm). Briefly, an iron lysimeter (1 m wide×1 m long×1.2 m high), filled with original paddy soil, was installed in the paddy field. The bottom of the lysimeter was perforated to allow percolated water to drip into a storage tank (0.5 m wide×0.5 m long×0.5 m high) welded to the lysimeter for storage of the percolated water. A meter stick was also fixed inside the lysimeter to measure the water table. Evapotranspiration from the paddy field was calculated according to Equation (1). In our study, there was no artificial drainage or surface runoff, so $D$ and $R_o$ were set to 0. All of the terms are expressed in millimeters and then converted into $m^3$ $ha^{-1}$. One growth cycle (June 1 to September 7, 2009) was used to bound all calculations. The difference between $P_{sd}$ calculated from Equation (1) and from the measured percolation water volume was less than 10%; hence, here we presented percolated inorganic N as defined by its concentration ($NH_4$-N and $NO_3$-N) multiplied by measured percolate volume.

Relative N leaching loss (% of applied N) for a given plot was determined after first subtracting N leached from the nil treatment

**Table 1.** Experimental treatments in terms of total nutrient inputs for N, P, and K (kg $ha^{-1}$).

| Treatment | Fertilization | Nutrient input (kg $ha^{-1}$) | | |
|---|---|---|---|---|
| | | N | P | K |
| CK | Control with no fertilization | 0 | 0 | 0 |
| LS | Local conventional system: composted animal manure (27.6% of water content) at 11,250 kg $ha^{-1}$, urea at 277.5 kg $ha^{-1}$, diammonium phosphate at 247.5 kg $ha^{-1}$, and compounded fertilizer at 75 kg $ha^{-1}$. | 297 | 83 | 117 |
| B1A1 | Composted animal manure (27.6% of water content) at 15,000 kg $ha^{-1}$ plus castor bean meal (22.5% of water content) at 2,250 kg $ha^{-1}$ | 193 | 52 | 163 |
| B2A1 | Composted animal manure (27.6% of water content) at 45,000 kg $ha^{-1}$ plus castor bean meal (22.5% of water content) at 2,250 kg $ha^{-1}$ | 498 | 125 | 446 |
| B3A1 | Composted animal manure (27.6% of water content) at 75,000 kg $ha^{-1}$ plus castor bean meal (22.5% of water content) at 2,250 kg $ha^{-1}$ | 802 | 199 | 728 |

**Table 2.** Nutrient contents of fertilizers used in the experiments (% dry weight).

| | *N* | *P* | *K* |
|---|---|---|---|
| Composted animal manure | 1.40 | 0.34 | 1.30 |
| Castor bean meal | 2.37 | 0.84 | 1.10 |
| Urea | 46 | | |
| Diammonium phosphate | 18 | 20 | |
| Compounded fertilizer | 14 | 6.98 | 12.45 |

(CK) and then expressed as a percentage of the total N applied. Similarly, the environmental performance of organic and conventional fertilizer treatments, i.e., N leaching, was also evaluated based on the amount of inorganic N leached during the production of 1 Mg of rice (kg N $Mg^{-1}$ grain). N utilization efficiency (NUE) was estimated using the equation (2):

$$NUE = \left( \begin{array}{l} N\ uptake\ by\ rice\ in\ fertilized\ treatment \\ (kg\ N\ ha^{-1}) - \\ N\ uptake\ by\ rice\ in\ CK\ treatment \\ (kg\ N\ ha^{-1}) \end{array} \right) \quad (2)$$
$$/N\ input\ from\ the\ fertilizers(kg\ N\ ha^{-1}).$$

## Sampling and measurement of fertilizers, soils and plants

Manures were sampled to determine the content of dry matter (DM) and total N before spreading. Plant samples were taken in early October to obtain rice grain with husk and straw, by cutting stems at 2 cm above the ground from two 2 $m^2$ areas per plot. The plant samples were washed with distilled water, oven-dried at 65°C, and weighed to determine rice biomass. Plant samples were then digested using $H_2SO_4$-$H_2O_2$, after which N, P, and K in the digestion solution were determined using the Kjeldahl method, calorimetrically by a Techniconautoanalyzer, and via atomic emission spectrometry (AES), respectively.

Statistical analyses

Cumulative inorganic N leached, net inorganic N leached per Mg of rice produced, and the proportion of net leached inorganic N in total N input were compared among different treatments using the least-significant difference test after a one-way analysis of variance for a randomized block design at the 0.05 significance level. The PROC MIXED procedure in SAS v.9.1 (SAS Institute, Cary, NC, USA) was performed to analyze the effects of fertilization on $NH_4$-N and $NO_3$-N concentrations over the sampling period (15 samplings), with the treatment as the fixed effect and sampling time as the random factor. Differences among means were calculated using a Differences of Least Squares Means

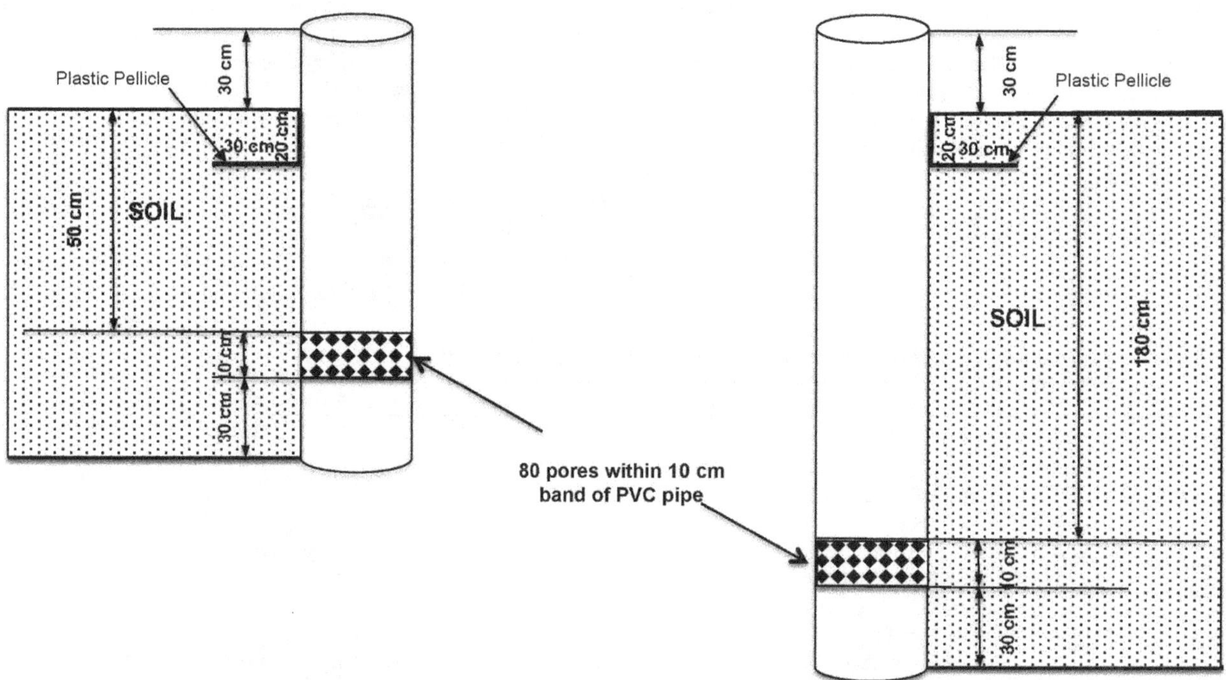


**Figure 2. Schematic of standpipes for measuring percolate inorganic N concentrations in the paddy fields.** Shaded area indicates the depth below the soil surface. Horizontal lines at 20 cm depth indicate the placement of a plastic film that extended a 30 cm-radium around the pipes.

with the PDIFF option and a Bonferroni adjustment method. The significance level was also at 0.05.

## Results

### Inorganic N concentrations in percolation water at 50 cm depth

Leaching of $NH_4$-N and $NO_3$-N during the rice growing season under different fertilization treatments is shown in Fig. 3. We observed high variance in $NH_4$-N concentration among three replicates of the same treatment, and this was also the case for $NO_3$-N (data not shown). Two peak concentrations of $NH_4$-N were observed in organic treatments, at the beginning of tillering (late June) and at heading (mid-August), and one additional $NH_4$-N peak at late-tillering (late July) in conventional treatment (LS). By PROC MIXED analysis, we found over the entire growth stage of rice, application of organic fertilizers increased $NH_4$-N concentration compared to no fertilization (CK, 0.63 mg $L^{-1}$, DF = 50.9, t = 2.76); this increase was significant for B2A1 (1.37 mg $L^{-1}$, DF = 50.9, t = 6.04) and B3A1 (1.74 mg $L^{-1}$, DF = 50.9, t = 7.65) but not for B1A1 (1.14 mg $L^{-1}$, DF = 50.9, t = 5.04). The concentration of $NH_4$-N that percolated from LS soils was significantly higher (1.20 mg $L^{-1}$, DF = 50.9, t = 5.29) than from CK soils, but not different from organic fields (B1A1, B2A1 and B3A1).

Organic and conventional fertilization affected $NO_3$-N concentrations in percolation water (Fig. 3) differently than they affected $NH_4$-N. At the 50-cm soil depth, the average $NO_3$-N concentration was lowest in CK (0.48 mg $L^{-1}$, DF = 42.2, t = 2.79) and then increased in the order of LS (1.08 mg $L^{-1}$, DF = 42.2, t = 6.27)< B1A1 (1.96 mg $L^{-1}$, DF = 42.2, t = 11.33)<B2A1 (2.16 mg $L^{-1}$, DF = 42.2, t = 12.48)<B3A1 (2.19 mg $L^{-1}$, DF = 42.2, t = 12.67). Within the percolation waters of both the CK and LS treatments, concentrations of $NH_4$-N and $NO_3$-N were similar, whereas in the organic treatments (B1A1, B2A1 and B3A1) average concentrations of $NO_3$-N were 21–42% higher than $NH_4$-N. In contrast to the two peak concentrations of $NH_4$-N, $NO_3$-N declined from high concentrations at early tillering until the end of the rice harvest

### Inorganic N concentrations in percolation water at 180 cm depth

The average $NH_4$-N concentrations at 180 cm were lowest in the CK treatment (0.41 mg $L^{-1}$, DF = 48.8, t = 1.75) and then increased as follows: LS (0.88 mg $L^{-1}$, DF = 48.8, t = 3.75) < B1A1 (1.08 mg $L^{-1}$, DF = 48.8, t = 4.61) <B2A1 (1.28 mg $L^{-1}$, DF = 48.8, t = 5.44), and B3A1 (1.57 mg $L^{-1}$, DF = 48.8, t = 6.67, Fig. 4). These results indicated that $NH_4$-N leaching increased with the increasing intensity of organic fertilization, although a high temporal variation was also observed over the monitoring period. As was found at 50 cm depth, two $NH_4$-N peaks were recorded at the beginning of tillering and at the heading stage for the organic treatments, with an additional peak for LS at mid-tilling. Different from the observations at 50 cm, $NH_4$ concentrations for all treatments increased from early heading, for about 1 month, until the start of paddy filling. $NH_4$-N concentrations in organic rice leachates at 180 cm were similar to those at 50 cm, except for LS plots, in which values at 180 cm were significantly lower than that at 50 cm depth (0.88 vs. 1.20 mg $L^{-1}$).

The temporal dynamics of $NO_3$-N concentrations at 180 cm were similar to those at 50 cm (Fig. 4), *i.e.*, as the quantity of organic fertilizer increased, leached $NO_3$-N concentrations also increased. However, significantly lower concentrations of $NO_3$-N were observed for organic treatments than for LS, and this was opposite to 50 cm depth where organic treatments percolated significantly higher concentrations of $NO_3$-N than LS. Organic treatments had much lower $NO_3$-N concentrations at 180 cm than at 50 cm (B1A1: 0.59 vs. 1.96 mg $L^{-1}$, B2A1: 0.85 vs. 2.16 mg $L^{-1}$ and B3A1: 1.17 vs. 2.19 mg $L^{-1}$) while $NO_3$-N concentrations in the LS treatment increased from 50 cm to 180 cm (1.08 vs. 1.70 mg $L^{-1}$).

### Total leached inorganic N

The rice fields were flooded during the growing period from transplant (June 1) to harvest (Sep 7). At the 50 cm depth, leached inorganic N varied from 18.5 to 23.1 kg N $ha^{-1}$ for fertilized organic treatments comparing to 12.1 kg N $ha^{-1}$ for LS (Table 3). Of the total amount of leached inorganic N, the fertilized organic treatments showed a slight higher proportion of N as $NO_3$-N (51.1~58.9%), compared to 64.2% as $NH_4$-N in LS. Higher organic fertilization rates led to higher amounts of leached $NH_4$-N, but this was not the case for $NO_3$-N, i.e., among the three fertilized organic treatments, the quantities of $NO_3$-N remained similar. For the net total inorganic N leached, B3A1 was significantly higher than other treatments, especially compared to LS.

At 180 cm, far less total inorganic N was percolated than at 50 cm for all treatments. For fertilized organic treatments (B1A1, B2A1 and B3A1), total inorganic N leached at 180 cm was only 9-13% of the amount leached at 50 cm; for the LS treatment, it was 26%. Relatively higher proportions of inorganic N were leached as $NH_4$-N (>68%) from fertilized organic production systems, whereas 61% of leached inorganic N from LS was $NO_3$-N. Among all organic treatments, B3A1 leached the highest quantity of inorganic N, significantly higher than other organic treatments, but interestingly, not higher than LS.

For the purpose of calculating N losses, we considered the inorganic N present at the depth lower than 180 cm (Table 4) to be that which could no longer be utilized by crops, and could potentially reach surface or ground water bodies. These data were used to assess the environmental performance of organic versus conventional rice production using two indicators: the amount of inorganic N leached per unit yield of rice (kg N $Mg^{-1}$ grain) and the percentage of inorganic N leached per unit applied N (% loss) at 180 cm depth. Regarding leachate per unit yield, we found that LS and organic treatments were not significantly different, i.e., a similar quantity of inorganic N was leached per production unit yield of rice (0.21~0.34 kg N $Mg^{-1}$ grain). However, in terms of percent N loss, LS leached significantly more inorganic N in total N input (0.78%) than did organic production (0.32~0.60%).

## Discussion

### Inorganic N concentration in percolation water

In flooded paddy soils, aerobic and anaerobic metabolisms occur in close proximity [18,19,20]. At the 50 cm depth in the sandy loam soil that we studied, in organic treatments, the presence of the oxidized zone close to the reduced soil zone was conductive for the transformation of $NH_4$-N mineralized from organic manure to $NO_3$-N. However, the continuous anaerobic conditions (under flood irrigation, until 1 week before harvest) at the 180 cm depth maintained $NH_4$-N as the main form of inorganic N, more so than $NO_3$-N. Particularly, in continuously flooded paddy soils with abundant organic substrate and a limited availability of electron acceptors, the reduction of $NO_3$-N to $NH_4$-N would be more efficient than the formation of $N_2$, so $NH_4$-N concentrations can be an order of magnitude higher than those of $NO_3$-N [21,22,23,24]. This situation also occurs in soils with low

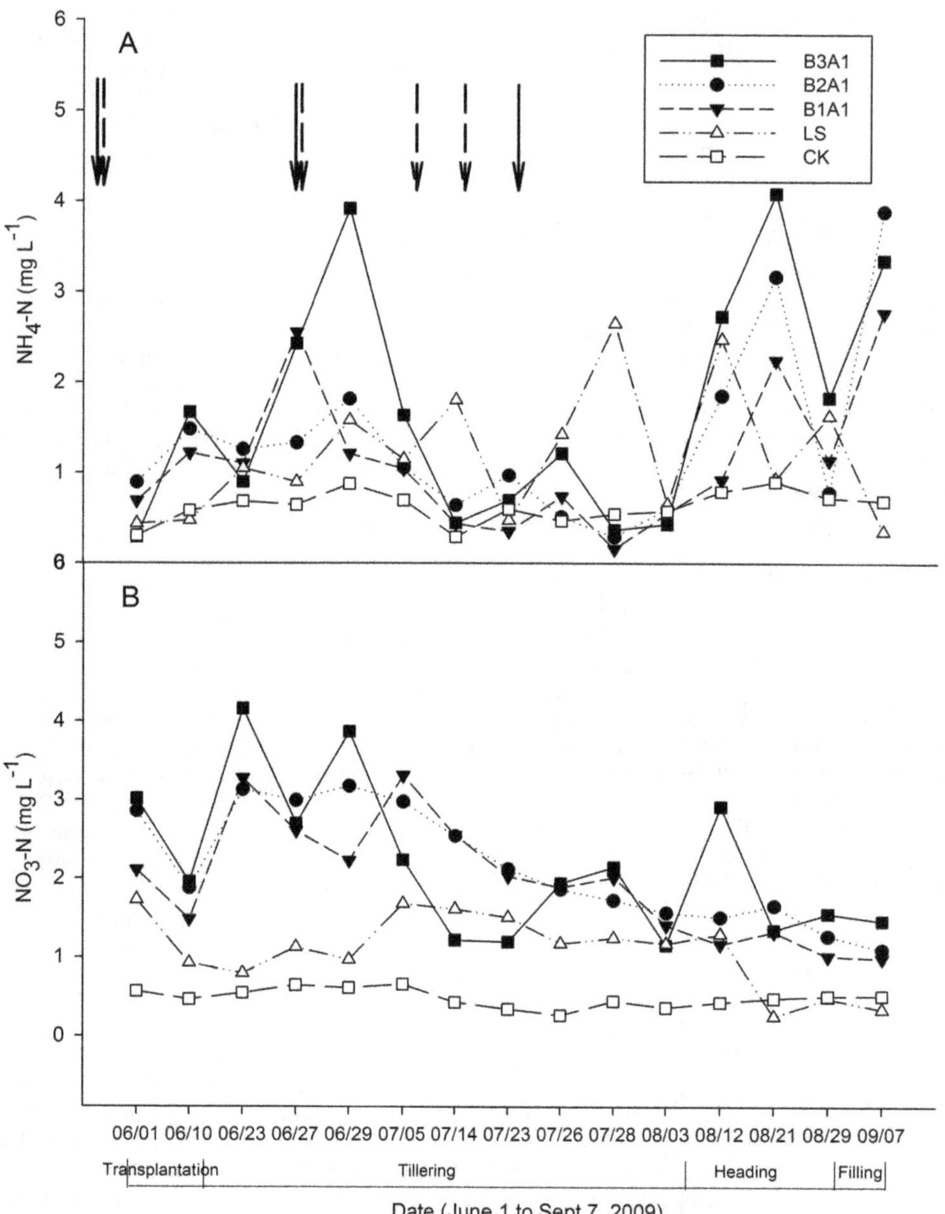


**Figure 3. Concentrations of $NH_4$-N and $NO_3$-N in leachate at the 50 cm soil depth under different fertilization treatments.** A: $NH_4$-N; B: $NO_3$-N. The solid and long-dashed arrows indicate the date of fertilizer applications for organic treatments (B1A1, B2A1, and B3A1) and for the conventional treatment (LS), respectively. Bars represent standard deviations ($n = 3$).

CEC and a low content of exchangeable base cations [19,25,26,27]. From the depth of 50 cm to 180 cm, average $NH_4$-N and $NO_3$-N concentrations decreased by about 5–10% and 47–70%, respectively, in fertilized organic treatments (B1A1, B2A1 and B3A1). The smaller decrease in $NH_4$-N concentration, compared to $NO_3$-N, may resulted from continuous decomposition of organic fertilizer and the limited soil adsorption capacity for $NH_4$-N [19,25,28]. From 50 cm to 180 cm in LS soils, there was a 27% decrease in $NH_4$-N, but a 57% increase in $NO_3$-N, indicating that mineral fertilization may have led to an increased downward movement of $NO_3$-N, compared to organic fertilizer, and the low dentrification capacity in deeper LS soils would thus have caused $NO_3$-N concentration to remain high [20]. This substantial loss of inorganic via leaching $NH_4$-N from continuous flooded rice fields has also been confirmed by other studies, which reported that $NH_4$-N might account for up to 92% of the total inorganic N in leachate - this large risk of $NH_4$-N leaching deserves more attention than the risk of $NO_3$-N loss [23,27].

During the vegetative phase of rice growth, $NO_3$-N concentrations were higher than those of $NH_4$-N, whereas the opposite was observed during the reproductive phase (Figs. 3 and 4). Rice requires more $NH_4$-N than $NO_3$-N during the vegetative stage, contributing to a suppression of $NH_4$-N concentrations in leachate [29,30]. As rice plants shifted into their reproductive stage, $NH_4$-N was continuously mineralized from the organic fertilizers, but the anaerobic conditions prevented nitrification into $NO_3$-N, both in the upper and lower soil profiles [23,31]. Unlike $NH_4$-N, the $NO_3$-N concentration in percolation water declined from transplant

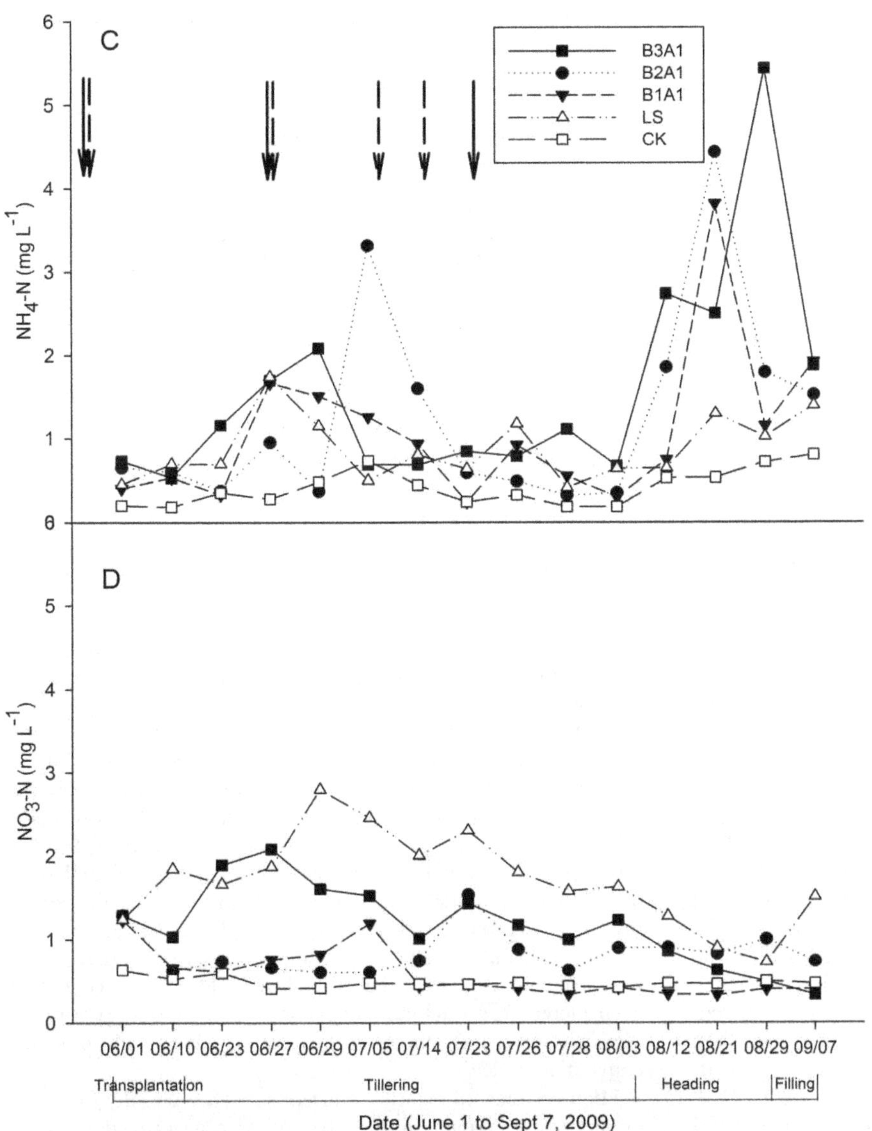


**Figure 4. Concentrations of $NH_4$-N and $NO_3$-N in leachate at the 180 cm soil depth under different fertilization treatments.** C: $NH_4$-N; D: $NO_3$-N. The solid and long-dashed arrows indicate the date of fertilizer applications for organic treatments (B1A1, B2A1, and B3A1) and for the conventional treatment (LS), respectively. Bars represent standard deviations ($n = 3$).

until harvest mainly because of continuous uptake by the plants throughout the growing season [32]. The two peaks observed in $NH_4$-N concentrations were also driven by fertilization and environmental conditions (higher temperature). At the beginning of tillering (~June 29), air temperature was close to its annual maximum (Fig. 1), which may have promoted the mineralization of organic fertilizer. However, the observed increases in $NO_3$-N concentrations lagged behind those of $NH_4$-N, because urea or organic fertilizer must first be transformed to amide N, then to ammonium and further nitrified to $NO_3$-N [28].

## Comparison of N utilization and loss from organic and conventional rice production

Organic and mineral fertilizers influenced rice yield differently, through differential effects on yield components such as the number of panicles per hill, number of grains per panicle, and grain weight. In our experiment, topdressing with castor bean meal in combination with increasing manure application resulted in higher grain weight, but mineral fertilizer still produced a larger number of panicles and grains per panicle [33]. A previous study revealed that increased mineral N supply to rice increases the amount of dry matter translocated into the grain [34]. However, if organic and mineral fertilizers are used together, higher numbers of panicles and grains per panicle can be achieved because of improved nutrient balance [35]. As the yield increased, more N was removed from the soil, but not in proportion with the increase in N input, such that N utilization efficiency (NUE) declined. Rice grown under conventional production showed the highest NUE (29.6%), which was significantly higher than that of organic treatments (8.7–17.6%, Table 5). This difference in NUE between conventional and organic production has been demonstrated in many other studies [8,32], which is attributed to slow release of N from organic fertilizers that limits N uptake by crops, and also maintains the continuous loss of N via leaching or other channels [22].

**Table 3.** Cumulative inorganic N leaching (kg N $ha^{-1}$) at 50 and 180 cm soil depths for control (CK), organic (B1A1, B2A1, and B3A1) and conventional (LS) treatments during a single rice growth cycle.

| Depth (cm) | Treatments | $NH_4$-N (kg $ha^{-1}$) | $NO_3$-N (kg $ha^{-1}$) | Total inorganic N leached (kg $ha^{-1}$) | Net inorganic N leached (kg $ha^{-1}$) [a b] |
|---|---|---|---|---|---|
| 50 | CK | 3.4 c | 0.0 c | 3.4 d | n.a. |
| | LS | 7.8 b | 4.3 b | 12.1 c | 8.7 c |
| | B1A1 | 7.6 b | 10.9 a | 18.5 b | 15.2 b |
| | B2A1 | 9.6 ab | 12.6 a | 22.2 a | 18.8 ab |
| | B3A1 | 11.3 a | 11.8 a | 23.1a | 19.8 a |
| 180 | CK | 0.4 c | 0.0 d | 0.4 d | n.a |
| | LS | 1.1 bc | 1.7 a | 2.7 ab | 2.3 ab |
| | B1A1 | 1.4 ab | 0.2 d | 1.6 c | 1.2 b |
| | B2A1 | 1.7 ab | 0.5 c | 2.2 bc | 1.8 ab |
| | B3A1 | 2.0 a | 1.0 b | 3.0 a | 2.5 a |

[a]After subtraction of the leakage quantity measured in CK.
[b]Different letters in a column denote a significant difference between treatments at the 0.05 level.

Environmental performance per se, in terms of N loss through leaching is a key concern for maintaining water quality and the integrity of organic farming. When comparing organic and conventional paddy rice production, N leaching losses are a tradeoff with rice yield; for example, if high yields are to be achieved, relatively high amounts of inorganic N must be supplied and may be lost. In our study, the highest yield of organic rice (7550 kg $ha^{-1}$) was achieved using the highest rate of N application (B3A1, 802 kg N $ha^{-1}$; Table 4); yield in these plots amounted to 80% of that achieved through conventional treatment (LS), which received inputs of only 297 kg of N $ha^{-1}$. Furthermore, organic production did not always perform well in terms of inorganic N leaching per unit yield of rice. In fact, if organic fertilization was intensified enough to obtain a "conventional" yield, even more N would be leached per unit yield than currently occurs in conventional production. Organic treatments with relatively higher N input (B2A1 and B3A1) tend to have lower overall rates of inorganic nitrogen loss than the LS treatment, but this came at the cost of a much lower rice yield. In the current study, the N input levels in organic treatments (except for B1A1) were higher than most conventional rice production in China, where for a single growth cycle N application varies from 200 to 300 kg N $ha^{-1}$ [36]. In the Ili River Valley, rice production had an average N application rate of 234 kg $ha^{-1}$, equal to the national average. This generates additional doubt that organic production can help to reduce N leaching from paddy production. In our experiments, 2.9–7.9% (8.7–19.8 kg $ha^{-1}$) of total applied N was lost through leaching at 50 cm, and 0.32–0.78% (1.2–2.5 kg $ha^{-1}$) was lost at 180 cm, whereas other studies involving organic fertilization reported losses of 0.1 to 15% [13,16,31,37,38]. Based on these findings, we believe that organic rice production in the Ili River Valley would not reduce (or may even increase) N leaching compared to conventional production, especially when organic fertilizer input is high enough to achieve conventional yields.

High reliance on external nutrient supply, especially from non-organic/conventional sources, has become a concern for large-scale organic production in recent years [39,40]. Organic

**Table 4.** Amount of inorganic N leached at 50 cm and 180 cm soil depths, defined as production of 1 Mg rice grains produced and as a % of N input.

| Depth (cm) | Treatments | Rice yield (Mg $ha^{-1}$) | N input (kg $ha^{-1}$) | Net leached inorganic N per Mg of rice grains (kg N $Mg^{-1}$) [b] | Proportion of net leached inorganic N in total N input (%) [a b] |
|---|---|---|---|---|---|
| | CK | 3.6 d | n.a. | | |
| | LS | 9.4 a | 297 | 0.93 b | 2.9 bc |
| 50 | B1A1 | 5.6bc | 193 | 2.69 a | 7.9 a |
| | B2A1 | 6.9bc | 498 | 2.72 a | 3.8 b |
| | B3A1 | 7.6 b | 802 | 2.62 a | 2.5 c |
| | CK | 3.6 | n.a. | n.a. | n.a. |
| | LS | 9.4 | 297 | 0.25 a | 0.78 a |
| 180 | B1A1 | 5.6 | 193 | 0.21 a | 0.60 ab |
| | B2A1 | 6.9 | 498 | 0.26 a | 0.36 b |
| | B3A1 | 7.6 | 802 | 0.34 a | 0.32 b |

[a]"Net N leached" was calculated by subtracting the amount of N leached in the control treatment (CK) from total treatment values.
[b]Different letters in a column denote a significant difference between treatments at the 0.05 level.

**Table 5.** Nitrogen budget for rice production.

| Treatments | N input from fertilizers (kg ha$^{-1}$) | N removed by rice (kg ha$^{-1}$) | N removed by straw (kg ha$^{-1}$) | Total N removed by rice crop (kg ha$^{-1}$) | N budget (kg ha$^{-1}$) | N utilization efficiency (%) [a] |
|---|---|---|---|---|---|---|
| CK | 0 | 40 | 13 | 53 | −53 | n.a |
| LS | 297 | 100 | 41 | 141 | 156 | 29.6 a |
| B1A1 | 193 | 59 | 28 | 87 | 106 | 17.6 b |
| B2A1 | 498 | 85 | 29 | 114 | 384 | 12.2 c |
| B3A1 | 802 | 87 | 36 | 123 | 679 | 8.7 d |

[a]Different letters in a column denote a significant difference between treatments at the 0.05 level.

fertilizers are able to meet crop N requirements, but this is achieved through high application rates, which could be 1) mineralized to release a sufficient amount of N for rice growth, and 2) synchronized between N mineralization and crop uptake [10]. However, higher organic N input can cause higher N losses through leaching, as shown in our experiments. In addition, N in paddy soils can also be denitrified to the powerful greenhouse gas $N_2O$ [14], an additional significant environmental impact that was not measured in our study. Ju et al. [5] summarized that rice production in China using 300 kg N ha$^{-1}$ per season contributed denitrification N losses of approximately 36%, which are markedly higher than losses from volatilization (12%) or leaching (0.5%). Here, we assume that denitrification and volatilization losses for mineral fertilizer in our study would be roughly comparable to those reported by Ju et al. [5] and Guan et al. [41]. Besides inorganic N, organic N in the leachate was not measured in our study, but it could eventually enter water bodies and cause of eutrophication. In addition, given that conventional rice has a higher NUE (29.6%), lower N input, and higher crop yield, N leaching from conventional rice production would not pose as high risk of contamination to local water systems as from organic rice production [10].

## Conclusions

This study indicates that inorganic N concentrations in leachate decreased as soil depth increased, but the decrease was significantly larger in organic than in conventional paddies. $NO_3$-N tended to remain in the upper soil profile of organic paddies, whereas in conventional soils, $NO_3$-N migrated further downward. In organic paddy soils, $NH_4$-N accounted for a substantial portion of inorganic N in the leachate due to the organic manure decomposition and denitrification process under the continuous flooding conditions. In terms of the N leached per unit yield of rice, organic fertilization did not perform better than conventional fertilization in all cases, particularly for the organic production with higher rice yields. Consistent with this, conventional production showed higher N utilization efficiency (29.6%) compared to organic production (8.7–17.6%). We conclude that converting conventional rice production to organic production in the Ili River Valley of Central Asia will not reduce N leaching into local water systems, especially given the high-load application of organic manure to maintain high rice yields. A longer period study with integrated monitoring of N loss through leaching, volatilization, nitrification and denitrification is necessary to compare the overall performance of organic versus conventional rice production.

## Acknowledgments

We thank Luo Xinhu and Liu Fuwang of the Ili Agricultural Technical Station, Xinjiang Uygur Autonomous Region for their support during field work. The authors would also like to thank the two anonymous reviewers for their helpful remarks.

## Author Contributions

Conceived and designed the experiments: FqM XpS WlW. Performed the experiments: XpS. Analyzed the data: FqM XpS JEO. Contributed reagents/materials/analysis tools: FqM XpS WlW. Wrote the paper: FqM JEO.

## References

1. Ghosh BC, Bhat R (1998) Environmental hazards of nitrogen loading in wetland rice fields. Environmental Pollution 102 S1: 123–126.
2. Khind CS, Meelu OP, Singh Y, Singh B (1991) Leaching losses of urea-N applied to permeable soils under lowland rice. Fertilizer Research 28: 179–184.
3. Zhou S, Sugawara S, Riya S, Sagehashi M, Toyota K, et al. (2011) Effect of infiltration rate on nitrogen dynamics in paddy soil after high-load nitrogen application containing $^{15}N$ tracer. Ecological Engineering 37: 685–692.
4. Devkota KP, Manschadi A, Lamers JPA, Devkota M, Vlek PLG (2013) Mineral nitrogen dynamics in irrigated rice–wheat system under different irrigation and establishment methods and residue levels in arid drylands of Central Asia. European Journal of Agronomy 47: 65–76.
5. Ju XT, Xing GX, Chen XP, Zhang SL, Zhang LJ, et al. (2009) Reducing environmental risk by improving N management in intensive Chinese agricultural systems. Proceedings of the National Academy of Sciences 106: 3041–3046.
6. Yang L, He GH (2008) The impact of Ili irrigation region construction on environment systems. Water Saving Irrigation 3: 34–35. (In Chinese).
7. Jemison JM, Fox RH (1994) Nitrate leaching from nitrogen-fertilized and manured corn measured with zero-tension pan lysimeters. Journal of Environmental Quality 23: 337–343.
8. Askegaard M, Olesen JE, Rasmussen IA, Kristensen K (2011) Nitrate leaching from organic arable crop rotations is mostly determined by autumn field management. Agriculture, Ecosystems and Environment 142: 149–160.
9. Hansen B, Kristensen ES, Grant R, Høgh-Jensen H, Simmelsgaard SE, et al. (2000) Nitrogen leaching from conventional versus organic farming systems-a systems modelling approach. European Journal of Agronomy 13: 65–82.
10. Kirchmann H, Bergström L (2001) Do organic farming practices reduce nitrate leaching? Communications in Soil Science and Plant Analysis 32: 7–8, 997–1028.
11. Chhabra A, Manjunath KR, Panigrahy S (2010) Non-point source pollution in Indian agriculture: Estimation of nitrogen losses from rice crop using remote sensing and GIS. International Journal of Applied Earth Observation and Geoinformation 12: 190–200.
12. Xing G, Cao Y, Shi S, Sun G, Du L, et al. (2001) N pollution sources and denitrification in waterbodies in Taihu Lake region. Science in China Series B: Chemistry 44: 304–314.
13. Tian Y, Yin B, Yang L, Yin S, Zhu Z (2007) Nitrogen runoff and leaching losses during rice-wheat rotations in Taihu Lake Region, China. Pedosphere 17: 445–456.
14. Li F, Pan G, Tang C, Zhang Q, Yu J (2008) Recharge source and hydrogeochemical evolution of shallow groundwater in a complex alluvial fan

system, southwest of North China Plain. Environmental Geology 55: 1109–1122.
15. Li CF, Cao CG, Wang JP, Zhan M, Yuan WL, et al. (2008) Nitrogen losses from integrated rice-duck and rice-fish ecosystems in southern China. Plant Soil 307: 207–217.
16. Zhu JG, Han Y, Liu G, Zhang YL, Shao XH (2000) Nitrogen in percolation water in paddy fields with a rice/wheat rotation. Nutrient Cycling in Agroecosystems 57: 75–82.
17. Luo LG, Wen DZ (1999) Nutrient balance in rice field ecosystem of northern China. Chinese Journal of Applied Ecology 10: 301–304. (In Chinese).
18. Bauder JW, Montgomery BR (1980) N-source and irrigation effects on nitrate leaching. Agronomy Journal 72: 593–596.
19. Xiong Z, Huang T, Ma Y, Xing G, Zhu Z (2010) Nitrate and ammonium leaching in variable-and permanent-charge paddy soils. Pedosphere 20: 209–216.
20. Keeney DR, Sahrawat KL (1986) Nitrogen transformations in flooded rice soils, Fertilizer Research 9: 15–38
21. George T, Ladha JK, Garrity DP, Buresh RJ (1993) Nitrate dynamics during the aerobic soil phase in lowland rice-based cropping systems. Soil Science Society of America Journal 57: 1526–1532.
22. Ouyang H, Xu YC, Shen QR (2009) Effect of combined use of organic and inorganic nitrogen fertilizer on rice yield and nitrogen use efficiency. Jiangsu Journal of Agricultural Science 25: 106–111. (In Chinese).
23. Wang X, Suo Y, Feng Y, Shohag M, Gao J, et al. (2011) Recovery of $^{15}N$-labeled urea and soil nitrogen dynamics as affected by irrigation management and nitrogen application rate in a double rice cropping system. Plant Soil 343: 195–208.
24. Islam MM, Lyamuremye F, Dick RP (1998) Effect of organic residue amendment on mineralization of nitrogen in flooded rice soils under laboratory conditions. Communications in Soil Science and Plant Analysis 29: 7–8, 971–981.
25. Qian C, Cai ZZ (2007) Leaching of nitrogen from subtropical soils as affected by nitrification potential and base cations. Plant Soil 300: 197–205.
26. Weier KL, Doran JW, Power JF, Walters DT (1993) Denitrification and the dinitrogen/nitrous oxide ratio as affected by soil water, available carbon, and nitrate. Soil Science Society of America Journal 57: 66–72
27. Stanford G, Smith SJ (1972) Nitrogen mineralization potentials of soils. Soil Science Society of America Journal 36: 465–472
28. Reddy KR, Patrick WH Jr (1986) Denitrification losses in flooded rice fields. Fertilizer Research 9: 99–116.
29. Reddy KR, Patrick WH, Lindau CW (1989) Nitrification-denitrification at the plant root-sediment interface in wetlands. Limnol. Oceanogr 34: 1004–1013.
30. Uhel C, Roumet C, Salsac L (1989) Inducible nitrate reductase of rice plants as a possible indicator for nitrification in water-logged paddy soils. Plant and soil 116: 197–206.
31. Luo L, Itoh S, Zhang Q, Yang S, Zhang Q, et al. (2010) Leaching behavior of nitrogen in a long-term experiment on rice under different N management systems. Environmental monitoring and assessment 177: 141–150.
32. Cassman KG, Peng S, Olk DC, Ladha JK, Reichardt W, et al. (1998) Opportunities for increased nitrogen-use efficiency from improved resource management in irrigated rice systems. Field Crops Research 56: 7–39.
33. Sun XP, Li GX, Meng FQ, Guo YB, Wu WL, et al. (2011) Nutrients balance and nitrogen pollution risk analysis for organic rice production in Ili reclamation area of Xinjiang. Transactions of the CSAE 27: 158–162. (In Chinese).
34. Gao JS, Qin DZ, Liu GL, Xu MG (2002) The impact of long term organic fertilization on growth and yield of rice. Tillage and Cultivation 2: 31–33. (In Chinese).
35. Li X, Liu Q, Rong XM, Xie GX, Zhang YP, et al. (2010) Effect of organic fertilizers on rice yield and its components. Hunan Agricultural Sciences 5: 64–66. (In Chinese).
36. Xi YG, Qin P, Ding GH, Fan WL, Han CM (2007) The application of RCSODS model to fertilization practice of organic rice and its effect analysis. Shanghai Agricultural Sciences 23: 28–33. (In Chinese).
37. Chowdary VM, Rao NH, Sarma PBS (2004) A coupled soil water and nitrogen balance model for flooded rice fields in India. Agriculture, Ecosystems & Environment 103: 425–441.
38. Ma J, Sun W, Liu X, Chen F (2012) Variation in the stable carbon and nitrogen isotope composition of plants and soil along a precipitation gradient in Northern China. PloS one 7: e51894.
39. Kirchmann H, Tterer KT, Bergström L (2008) Nutrient supply in organic agriculture: plant availability, sources and recycling. In: Kirchmann H, Bergström L (eds) Organic crop production–ambitions and limitations. Springer, Netherlands, 89–116.
40. Rodrigues MA, Pereira A, Cabanas JE, Dias L, Pires J, et al. (2006) Crops use-efficiency of nitrogen from manures permitted in organic farming. European journal of agronomy 25: 328–335.
41. Guan JX, Wang BR, Li DC (2009) Effect of chemical fertilizer applied combined with organic manure on yield of rice and nitrogen using efficiency.Chinese Agricultural Science Bulletin 25: 88–92. (In Chinese).

# 18

# Methane, Carbon Dioxide and Nitrous Oxide Fluxes in Soil Profile under a Winter Wheat-Summer Maize Rotation in the North China Plain

**Yuying Wang[1], Chunsheng Hu[1]*, Hua Ming[1], Oene Oenema[2], Douglas A. Schaefer[3], Wenxu Dong[1], Yuming Zhang[1], Xiaoxin Li[1]**

**1** Key Laboratory of Agricultural Water Resources, Center for Agricultural Resources Research, Institute of Genetics and Developmental Biology, Chinese Academy of Sciences, Shijiazhuang, Hebei, China, **2** Department of Soil Quality, Wageningen University, Alterra, Wageningen, The Netherlands, **3** Key Lab of Tropical Forest Ecology, Xishuangbanna Tropical Botanical Garden, Chinese Academy of Sciences, Menglun, Yunnan, China

## Abstract

The production and consumption of the greenhouse gases (GHGs) methane ($CH_4$), carbon dioxide ($CO_2$) and nitrous oxide ($N_2O$) in soil profile are poorly understood. This work sought to quantify the GHG production and consumption at seven depths (0–30, 30–60, 60–90, 90–150, 150–200, 200–250 and 250–300 cm) in a long-term field experiment with a winter wheat-summer maize rotation system, and four N application rates (0; 200; 400 and 600 kg N $ha^{-1}$ $year^{-1}$) in the North China Plain. The gas samples were taken twice a week and analyzed by gas chromatography. GHG production and consumption in soil layers were inferred using Fick's law. Results showed nitrogen application significantly increased $N_2O$ fluxes in soil down to 90 cm but did not affect $CH_4$ and $CO_2$ fluxes. Soil moisture played an important role in soil profile GHG fluxes; both $CH_4$ consumption and $CO_2$ fluxes in and from soil tended to decrease with increasing soil water filled pore space (WFPS). The top 0–60 cm of soil was a sink of atmospheric $CH_4$, and a source of both $CO_2$ and $N_2O$, more than 90% of the annual cumulative GHG fluxes originated at depths shallower than 90 cm; the subsoil (>90 cm) was not a major source or sink of GHG, rather it acted as a 'reservoir'. This study provides quantitative evidence for the production and consumption of $CH_4$, $CO_2$ and $N_2O$ in the soil profile.

**Editor:** Dafeng Hui, Tennessee State University, United States of America

**Funding:** This research is supported by the "Strategic Priority Research Program of Chinese Academy of Sciences" (Grant No. XDA0505050202 and XDA05050601) and the "National Science & Technology Pillar Program" (Grant No. 2012BAD14B07-5). It is also supported by the "National Basic Research Program of China" (Grant No. 2010CB833501); the "National Natural Science Foundation of China" (Grant No. 30970534) and the "Main Direction Program of Knowledge Innovation of Chinese Academy of Sciences" (Grant No. KSCX2-EW-J-5). The funders had no role in study design, data collection and analysis, decision to publish, or preparation of the manuscript.

**Competing Interests:** The authors have declared that no competing interests exist.

* E-mail: cshu@sjziam.ac.cn

## Introduction

Atmospheric concentrations of carbon dioxide ($CO_2$), methane ($CH_4$) and nitrous oxide ($N_2O$) have increased considerably since the industrial revolution, and are still increasing annually by about 0.5%, 1.1% and 0.3%, respectively [1]. Worldwide concerns about the increased greenhouse gases (GHGs) concentrations in the atmosphere and its effects on our future environment require a better understanding of the cause of these emissions [2]. Agricultural lands occupy 37% of the earth's land surface; about 13.5% of global anthropogenic GHG was emitted from agricultural production [1]. It was estimated that 84% of $N_2O$ and 52% of $CH_4$ emitted from agriculture activities [3]. In China, agriculture tends to produce more emissions than the global average over the last 30 years due to increased chemical and manure N inputs. Gaining a better understanding of GHG production and emission processes, and developing methods for mitigating emissions from agroecosystems are essential steps in order to mitigate climate change [4].

Agricultural soils are main sources and sinks of GHG emissions, depending on their characteristics and management. Many studies have been conducted to quantify the net fluxes of $CO_2$, $CH_4$, and $N_2O$ across the soil/atmosphere interface [3,5–9]. These studies provide an integrative estimate of the net production and consumption of $CH_4$, $CO_2$ and $N_2O$ in the soil, but do not provide information on the depth-distribution of $CH_4$, $CO_2$ and $N_2O$ production-consumption patterns within soil profiles. It has been suggested that subsurface processes exert a significant control on carbon (C) and nitrogen (N) dynamics and hence on $CO_2$, $CH_4$, and $N_2O$ emissions from soil [10], but few studies have elucidated the role of the subsoil so far. Understanding these processes might also provide a better insight into the possibilities and effectiveness of measures to reduce GHG emissions. For example, a temporary accumulation of GHG in the soil profile influences GHG flux patterns at the soil surface over time, and thereby may confuse empirical relationships between agricultural activities and measured GHG emissions [11]. Thus, measurements of $CH_4$, $CO_2$ and $N_2O$ concentration profiles may be helpful for increasing the understanding of the net exchanges of these gases between soil and atmosphere.

Though few studies have examined the production and consumption of $CO_2$ [12,13], $CH_4$ [14] and $N_2O$ [15] within individual soil horizons and their transports between soil horizons so far, very few studies have made combined measurements of the dynamics of $CO_2$, $CH_4$ and $N_2O$ production and emission processes in soil profiles in agro-ecosystems, especially in China [16]. It has been well-established that N fertilizer applications increase crop growth and $N_2O$ emissions, and tend to decrease $CH_4$ emissions into the atmosphere, but there is little information about the combined effects of N fertilizer application and irrigation on subsoil $N_2O$, $CO_2$ and $CH_4$ production, consumption and transport.

Recently, Wang et al [16] presented bi-weekly measured $CH_4$, $CO_2$ and $N_2O$ concentration profiles down to a depth of 300 cm in a winter wheat-summer maize rotation in the North China Plain, with four N application rates (0; 200; 400 and 600 kg N $ha^{-1}$). Here, we build on the results of that study, and present calculated subsurface fluxes of $CH_4$, $CO_2$ and $N_2O$ over a whole-year period. The purpose of this study is to evaluate the effects of seasonal cropping, N applications, irrigation, soil temperature, and soil moisture on net subsurface transport of $CO_2$, $N_2O$ and $CH_4$.

## Materials and Methods

### Site description

The study was conducted at Luancheng Agroecosystem Experimental Station (37°53′N, 114°41′E, elevation 50 m), Chinese Academy of Sciences. This area is at the piedmont of the Taihang Mountains, in the North China Plain. Mean annual precipitation is about 480 mm, 70% of which is in the period from July to September. Annual average air temperature is 12.5 °C. The dominant cropping system in the region is a winter wheat-summer maize double-cropping system (two crops harvested in a single year) without fallow between the crops.

### Field experimental design

The field experiment with a randomized complete block design was laid down in a winter wheat (*Triticum aestivum L.* Wheat variety Kenong 199)-summer maize (*Zea mays L.* Maize variety Xianyu 335) double-cropping system in 1998. It had four N fertilizer (urea) application treatments in triplicate: 0 (N0), 200 (N200), 400 (N400) and 600 (N600) kg N $ha^{-1}$ $year^{-1}$. Plot size was 7 m×10 m. Results of the present study refer to the period March 2007 to January 2008. Details on fertilizer application and crop management activities are presented in Table 1. Crops are flood-irrigated with pumped groundwater about five times per year, depending on rainfall distribution.

### Soil sampling, analysis and climate data collection

The soil has a silt-loam texture in the upper 90 cm and clay-loam to clay texture at depth of 90–300 cm (Table 2). All soil samples were collected from different depths of the soil profile before the GHG measurements (on 5 December 2006); soil samples were mixed to make a specific representative soil sample for each depth; and all analyses of soil chemical properties in Table 2 were based on the standard methods for soil analyses described by Sparks [17]. Soil bulk density was determined using the cutting ring method. Soil particle size analysis was done by the Bouyoucos Hydrometer Method [18]. Soil pH was measured in a suspension of 5 g soil with 25 ml distilled water after 1 h after shaking.

Soil core samples were collected from different depths (0–30, 30–60, 60–90, 90–150, 150–200, 200–250, 250–300 cm) of the soil profile in the farmland described above on 5 December 2006 (before the GHG measurements), 16 June 2007 (after the winter wheat harvest) and 11 October 2007 (after the summer maize harvest), respectively. Three different sub-samples, taken from a cross-section around the soil auger (3 meter in length), were mixed to make a specific representative soil sample for each depth from each point. The soil profile samples were sealed in dark plastic bags immediately after sampling and stored at 4°C until $NO_3$ extraction. Samples of soil $NO_3$-N were extracted with 1 M KCl solution (1:5 w/v) by shaking for 1 h. The extracts were then filtered and the concentrations of $NO_3$-N in the soil extracts were measured colorimetrically using a UV spectrophotometer (UV-2450, Shimadzu, Japan). Each measurement was replicated three times.

Soil temperature was measured using seven CS107b soil temperature probes (Cambell Scientific Inc., Logan, UT) installed at depths of 30, 60, 90, 150, 200, 250 and 300 cm. Three-meter neutron access tubes were installed at each plot. Soil moisture at seven depths (30, 60, 90, 150, 200, 250 and 300 cm.) was measured using a neutron moisture meter when gas samples were collected. Soil temperature and water content were used to explore the relationships between calculated $CO_2$, $N_2O$ and $CH_4$ fluxes and soil water-filled pore space (WFPS) and soil temperature at various depths. Daily rainfall was recorded at a weather station on the experimental site.

### Soil gas sampling and measurements

Measurements of $CO_2$, $N_2O$ and $CH_4$ concentrations in soil started in March 2007, i.e., 9 years after the start of the field experiment, assuming that by then the $CO_2$, $N_2O$ and $CH_4$ production-consumption dynamics in the subsoil had been adjusted to the experimental treatments. Seven subsurface soil air equilibration tubes were installed at each site with sampling ports at 30, 60, 90, 150, 200, 250 and 300 cm in December 2006 (for more details, see reference 16). Soil-air samples were taken twice a week between 9:00 AM and 11:00 AM, using 100 ml plastic syringes connected to the tubes via the three-way stopcocks at the surface. The surface air was concurrently sampled at a height of 5 cm above the soil surface. The gas samples were analyzed by gas chromatography (Agilent GC-6820, Agilent Technologies Inc. Santa Clara, California, US) with separate electron capture detector (ECD at 330°C) for $N_2O$ determination and flame ionization detector (FID at 200°C) for CH4 and CO2 determinations.

### Calculations of gas fluxes

The basic method of our study followed that of Campbell [19]. It was assumed that the soil conditions are uniform in horizontal direction, and that the gas diffusion in soil is in one-dimensional vertical flow, that fundamentally follows Fick's law [20,21]:

$$q = -D_p \frac{\delta c}{\delta x} \tag{1}$$

Where $q$ is the gas flux density (g gas $m^{-2}$ soil $s^{-1}$), $D_p$ is the soil-gas diffusivity ($m^3$ soil air $m^{-1}$ soil $s^{-1}$), $\frac{\delta c}{\delta x}$ is the concentration gradient between two soil layers (g gas $m^{-3}$ soil air $m^{-1}$ soil).

$D_p$ was derived from the following equation [22,23]:

$$D_p = (\varepsilon^{10/3}/E^2) \cdot D_0 \tag{2}$$

Where, $D_0$ is the gas diffusivity ($m^2$ air $s^{-1}$). We estimated the diffusion coefficient $D_0$ of $CH_4$, $CO_2$ and $N_2O$ at 298 K and 1 kPa

**Table 1.** Fertilization treatments (A); and timing of crop management activities (B).

| (A) Treatments | Basal fertilization, applied at wheat sowing (kg·ha$^{-1}$) | | | Supplementary N fertilization (kg·ha$^{-1}$) | |
|---|---|---|---|---|---|
| | N | $P_2O_5$ | $K_2O$ | Wheat (in April) | Maize (in July) |
| N0 | 0 | 65 | 0 | 0 | 0 |
| N200 | 50 | 65 | 0 | 50 | 100 |
| N400 | 100 | 65 | 0 | 100 | 200 |
| N600 | 150 | 65 | 0 | 150 | 300 |
| **(B) Timing** | **Crop management activities** | | | | |
| | **Winter-wheat season** | | | | **Summer-maize season** |
| October 3, 2006 | Basal N fertilization and irrigation (60 mm) | | | | |
| October 10, 2006 | Seeding | | | | |
| April 7, 2007 | | | | Supplementary N fertilization and irrigation (94 mm) | |
| May 19, 2007 | | | | | Irrigation (60 mm) |
| June 1, 2007 | | | | | Seeding |
| June 14, 2007 | | | | Harvest | |
| June 19, 2007 | | | | | Irrigation (60 mm) |
| July 27, 2007 | | | | | Supplementary N fertilization |
| July 29, 2007 | | | | | Irrigation (60 mm) |
| October 1, 2007 | | | | | Harvest |
| October 2, 2007 | Basal N fertilization and irrigation | | | | |
| October 7, 2007 | Seeding | | | | |

at $1.79\times10^{-5}$, $1.32\times10^{-5}$, $1.29\times10^{-5}$ $m^2$ $s^{-1}$, respectively, by using a semiempirical equation by Gilliland et al [24]. Parameter $\varepsilon$ is the soil air filled porosity ($m^3$ air $m^{-3}$ soil), and $E$ is the soil porosity ($m^3$ voids $m^{-3}$ soil).

The Millington-Quirk model was used to compute $\varepsilon$ and $E$ [25]:

$$E=1-\frac{\rho_b}{\rho_s} \tag{3}$$

$$\varepsilon=E-\theta \tag{4}$$

Where $\rho_b$ is the dry bulk density (g $m^{-3}$) at each soil depth (Table 2), $\rho_s$ is the average bulk density of surface soil (2.65 g $m^{-3}$); $\theta$ is the volumetric soil water content which was measured using a neutron moisture meter at each depth.

### Calculations of annual cumulative gas fluxes

The annual cumulative emissions were obtained by multiplying the average daily flux from two consecutive measurements within a week by the number of days between the measurements, and then summing the fluxes of these periods to an accumulative flux for the whole year [26]:

$$T=\sum_{i=1}^{n}(X_i\times24) \quad (n=1,2,3...) \tag{5}$$

Where: $T$ (kg $ha^{-1}$), $X_i$ (kg $ha^{-1}$ $h^{-1}$) and $i$ are the accumulative GHG emission, the average daily GHG emission rate, and the number of days, respectively.

### Data analyses

All data were subjected to statistical analysis (SPSS 13.0). Differences between treatments were analyzed using ANOVA, followed by LSD at the 0.05 probability level. Regression analysis was used to identify relationships between $CH_4$, $CO_2$ and $N_2O$ fluxes and the climatic variables.

## Results

### Concentrations of $CH_4$, $CO_2$ and $N_2O$

Mean concentrations and its standard deviations of $CH_4$, $CO_2$ and $N_2O$ at each depth are shown in Figure 1 A, as function of N application rates. Mean $CH_4$ concentration decreased with soil depth. Ambient air $CH_4$ concentration in the area was about 2.2 ppmv. At a depth of 30 cm, $CH_4$ concentration ranged between 1.4 and 1.6 ppmv and at depth of 60 to 300 cm between 0.3 and 0.6 ppmv. There were no clear effects of N fertilizer application on the mean $CH_4$ concentration (Figure 1 A). Mean $CH_4$ concentrations decreased significantly at soil depths of 0, 30, 60 and 90 cm ($P<0.05$); changes in mean concentration below a depth of 90 cm were not significant (Figure 1 C).

Mean $CO_2$ concentration increased with soil depth. At a depth of 30 cm, $CO_2$ concentration ranged between 8400 and 9900 ppmv and at depth of 60 to 300 cm between 16000 and 21000 ppmv. Mean $CO_2$ concentrations increased significantly at soil depths of 0, 30 and 60 cm ($P<0.05$); changes in mean concentration below a depth of 60 cm were not significant

**Table 2.** Soil characteristics at the experimental site in 2007.

| Depth (cm) | pH ($H_2O$) | Sand (%) | Silt (%) | Clay (%) | Textural class[a] | Dry bulk Density (g cm$^{-3}$) | Total organic matter (g kg$^{-1}$) | Total nitrogen (g N kg$^{-1}$) | Available nitrogen (mg N kg$^{-1}$) | Available phosphorus (mg P kg$^{-1}$) | Available potassium (mg K kg$^{-1}$) |
|---|---|---|---|---|---|---|---|---|---|---|---|
| 0–30 | 8.5 | 25 | 58 | 17 | SSL | 1.47 | 16.7 | 1.40 | 148 | 4.1 | 79.3 |
| 30–60 | 7.74 | 22 | 60 | 18 | L | 1.40 | 10.9 | 0.85 | 72.0 | 2.1 | 54.4 |
| 60–90 | 7.78 | 31 | 55 | 14 | L | 1.45 | 7.8 | 0.64 | 70.1 | 0.44 | 36.9 |
| 90–150 | 7.76 | 15 | 59 | 26 | SCL | 1.57 | 6.5 | 0.38 | 49.8 | 0.17 | 27.8 |
| 150–200 | 7.74 | 18 | 47 | 35 | GCL | 1.43 | 5.4 | 0.28 | 38.9 | 0.13 | 21.9 |
| 200–250 | 7.77 | 15 | 35 | 50 | C | 1.51 | 4.2 | 0.15 | 30.1 | 0.09 | 13.9 |
| 250–300 | 7.75 | 12 | 35 | 53 | C | 1.50 | 3.0 | 0.86 | 25.9 | 0.04 | 7.1 |

[a]SSL: Silty sandy loam; L: Loam; SCL: Silty clay loam; GCL: Gravely clay loam; C = Clay.

(Figure 1 C). There were no clear effects of N fertilizer application on the mean $CO_2$ concentrations.

Concentrations of $N_2O$ were strongly influenced by agricultural management activities such as N application and irrigation. Fertilizer N applications increased the mean $N_2O$ concentrations. Mean $N_2O$ concentrations at depth of 30 to 300 cm ranged from 600 to 1500, 1100 to 1700, 1600 to 2100 and 2500 to 3000 ppbv for the N0, N200, N400 and N600 treatments, respectively (Figure 1 A). Mean $N_2O$ concentrations increased significantly from soil surface to a depth of 30 cm ($P<0.05$), but changes in mean concentration below a depth of 30 cm were not significant (Figure 1 C). Fertilizer N application increased soil $NO_3$-N content; differences in mean $N_2O$ concentrations were correlated with differences in mean $NO_3$-N contents in the four fertilizer N treatments (Figure 1 B).

## Fluxes of $CH_4$ in soil

Diffusive fluxes between soil layers and between soil and atmosphere were calculated from the concentration gradients, using equation 1. There was a net influx of atmospheric $CH_4$ into the top 0–60 cm (Figure 2), suggesting consumption of $CH_4$ by methanotropic bacteria. Interestingly, the calculated fluxes into the soil were rather similar for the 0–30 and 30–60 cm soil layers, suggesting similar $CH_4$ uptake rates. Uptake of $CH_4$ apparently also occurred in the layers 60–90, 90–150 and 150–200 cm during the first one or two months of the measurement period (Figure 2). However, we cannot exclude the possibility that this apparent uptake of $CH_4$ in the subsoil during the first two months is an artifact related to the installation of the samplers when atmospheric $CH_4$ may have diffused into the subsoil. Fluxes between soil layers were negligible small during most of the maize growing season (Figure 2).

Annual cumulative fluxes of $CH_4$ for all soil layers and N fertilizer treatments are shown in Table 3 A. Evidently, the influx of atmospheric $CH_4$ decreased with soil depth. During the study period, mean calculated uptake was about 176 g $CH_4$ per ha by the top 30 cm, 252 g $CH_4$ per ha by the soil layer 30–60 cm, 98 g $CH_4$ per ha by the layer 60–90 cm and 22 g $CH_4$ per ha below a depth of 90 cm; mean calculated uptakes in the layers 0–30 and 30–60 cm were both significantly higher than that in the layer 90–300 cm ($P<0.05$) (Figure 1 C). Annual cumulative $CH_4$ uptake in the layer 0–90 cm (526 g $CH_4$ per ha per year) contributed about 96% to that in the layer 0–300 cm (547 g $CH_4$ per ha per year). Annual cumulative uptake in 0–30 cm layer is relatively low compared to literature data [4–6,27].

## Fluxes of $CO_2$ in soil

There was a large efflux of $CO_2$ from the top 30 cm of soil to the atmosphere from March till June, i.e., during the second half of the wheat growing season, and from August till October, i.e., during the second half of the maize growing season (Figure 3). The same holds for the upward flux from the layer 30–60 to the layer 0–30 cm. These patterns were related to the crop growing seasons of wheat and maize, and to the changes in water filled pore space (WFPS) and soil temperature. There were no clear relationships between N treatments and $CO_2$ fluxes. Treatment N200 had the smallest flux from the layer 0–30 cm to the atmosphere, but the largest from 30–60 cm to 0–30 cm from April to May. Upward fluxes from the layers 60–90 cm and especially below this layer were much smaller. There were small but significant changes in fluxes in the subsoil at the transition of the winter-wheat growing season to the summer-maize growing season (Figure 3).

Surprisingly, annual cumulative fluxes of $CO_2$ tended to decrease with increasing N fertilizer application rates (Table 3

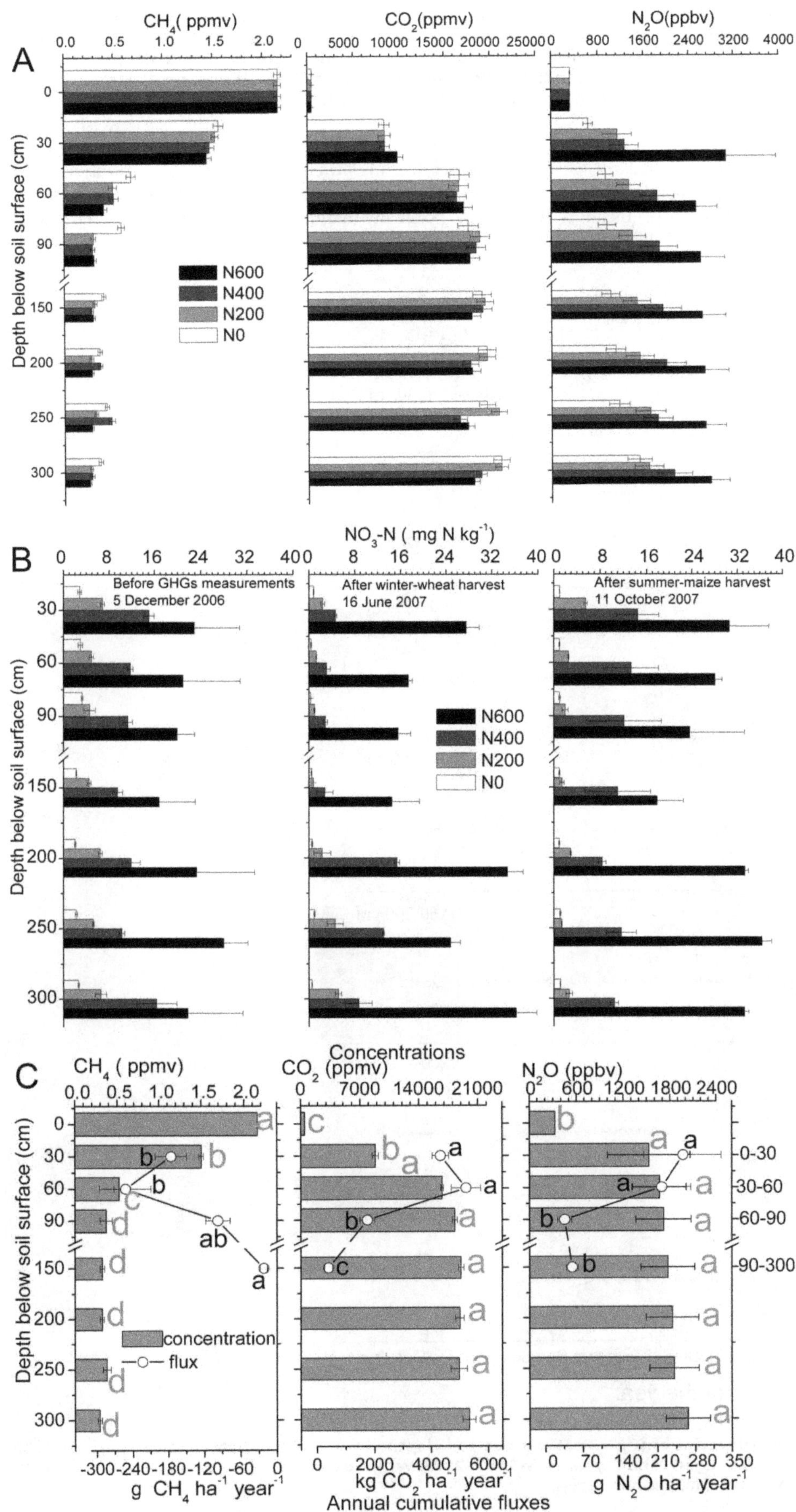

A
CH4( ppmv)
CO2(ppmv)
N2O(ppbv)
Depth below soil surface (cm)
N600
N400
N200
N0
B
NO3-N ( mg N kg-1)
Before GHGs measurements
5 December 2006
After winter-wheat harvest
16 June 2007
After summer-maize harvest
11 October 2007
C
CH4 ( ppmv)
Concentrations
CO2 (ppmv)
N2O (ppbv)
concentration
flux
0-30
30-60
60-90
90-300
g CH4 ha-1 year-1
kg CO2 ha-1 year-1
Annual cumulative fluxes
g N2O ha-1 year-1

**Figure 1. $CH_4$, $CO_2$ and $N_2O$ concentrations (mean ± standard deviations, n = 3) in soil air at various soil depths in a winter wheat-summer maize double cropping rotation receiving 0, 200, 400 and 600 kg of N $ha^{-1}$ $year^{-1}$, in 2007–2008 (A); $NO_3$-N contents (mean ± standard deviations, n = 3) at various soil depths as function of N fertilizer application rate, in 2007–2008 (B); Profiles of concentration and annual cumulative flux of $CH_4$, $CO_2$ and $N_2O$, in 2007–2008 (mean ± standard deviations, n = 4).** Same letters next to the bars indicated no significant differences between slope positions (P<0.05). (C). Note the differences in X-axes.

B). Moreover, cumulative upward fluxes were somewhat larger from the layer 30–60 cm (mean 5,227; range 3,800–6,000 kg $CO_2$ per ha) than from the layer 0–30 cm to the atmosphere (mean 4,331; range 3,800–5,000 kg $CO_2$ per ha) (Figure 1 C; Table 3 B). This suggests that a relatively large portion of total respiration in soil took place in the layer 30–60 cm. However, we can not exclude the possibility that the calculated $CO_2$ efflux from the top layer is underestimated, because the concentration gradient in the

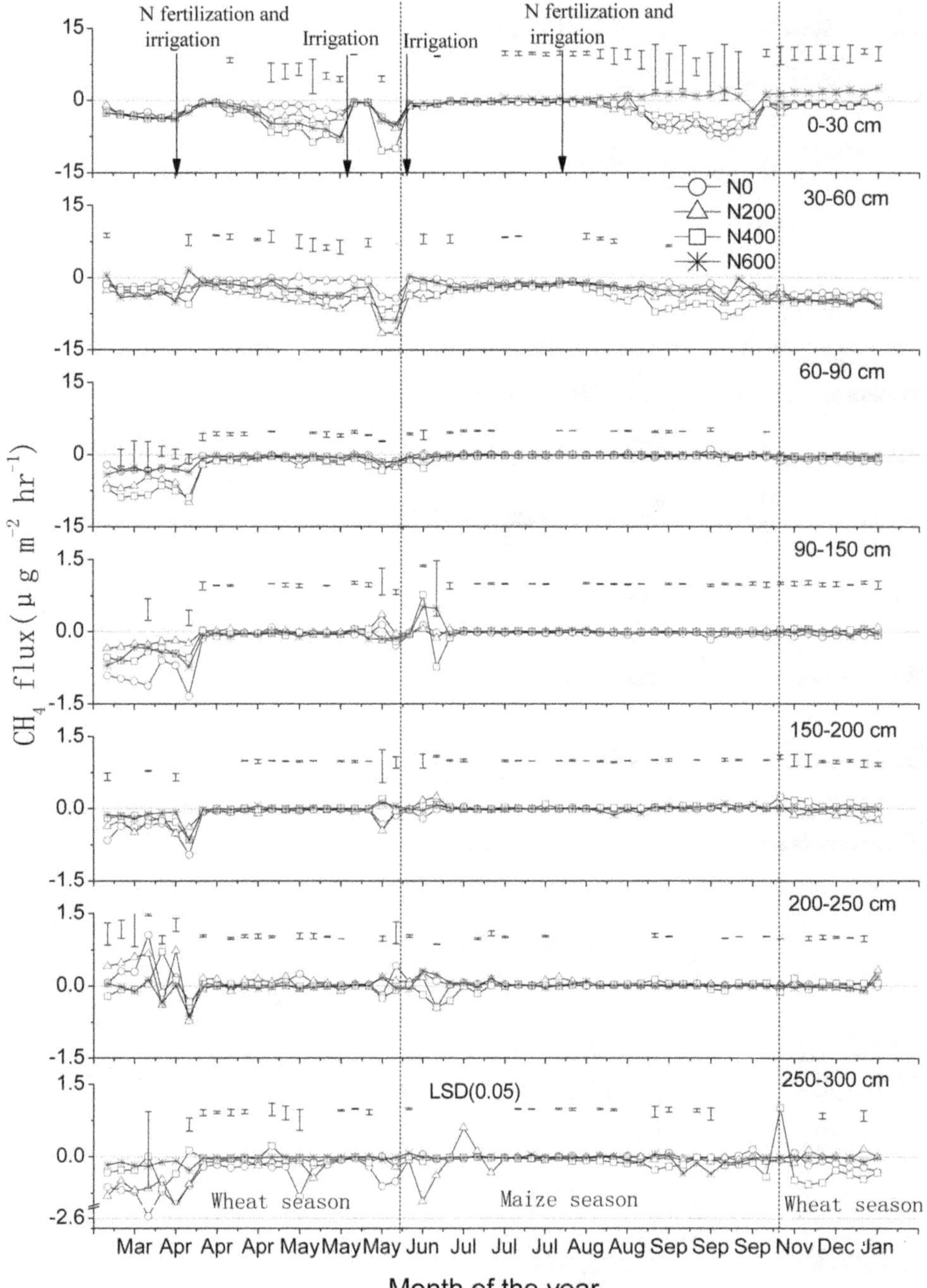


**Figure 2. $CH_4$ flux rates (means ± standard deviations, n = 3) at various soil depths in a winter wheat-summer maize double cropping rotation receiving 0, 200, 400 and 600 kg of N $ha^{-1}$ $year^{-1}$, in 2007–2008.** Vertical dashed lines indicate a change in crop. Bars in figures indicate 1 standard deviation (n = 3). Note the differences in Y-axes.

**Table 3.** Annual cumulative emissions of $CH_4$ and $N_2O$ (in g $ha^{-1}$ $yr^{-1}$) and of $CO_2$ (in kg $ha^{-1}$ $yr^{-1}$) between soil layers.

**(A) $CH_4$**

| Treatments | 0–30 cm | 30–60 cm | 60–90 cm | 90–150 cm | 150–200 cm | 200–250cm | 250–300 cm | 0–300cm |
|---|---|---|---|---|---|---|---|---|
| N0 | −167 (12) b | −138 (13) a | −63 (2) a | −15 (0.5) c | −9 (0.1) d | 5 (0.04) a | −21 (0.3) d | −408 |
| N200 | −201 (9) bc | −322 (10) c | −123 (3) b | −2 (0.07) a | −7 (0.5) c | 5 (0.08) a | −12 (0.6) c | −662 |
| N400 | −228 (3) c | −318 (8) c | −140 (10) b | −6 (0.5) b | −1 (0.1) a | −1 (0.06) b | −10 (0.4) b | −704 |
| N600 | −106 (15) a | −231 (5) b | −64 (7) a | −6 (0.5) b | −2 (0.1) b | −1 (0.05) b | −4 (0.1) a | −414 |

**(B) $CO_2$**

| Treatments | 0–30 cm | 30–60 cm | 60–90 cm | 90–150 cm | 150–200 cm | 200–250cm | 250–300 cm | 0–300cm |
|---|---|---|---|---|---|---|---|---|
| N0 | 4,986 (401) a | 6,060 (394) a | 1,730 (121) b | 163 (9) a | 92 (1) a | 12 (0.1) b | 294 (15) a | 13,337 |
| N200 | 3,819 (652) a | 6,019 (424) b | 2,194 (177) a | 194 (19) a | 19 (0.5) c | 263 (8) a | 198 (4) c | 12,706 |
| N400 | 4,640 (225) a | 4,955 (174) c | 2,100 (112) ab | 152 (15) a | −48 (1) d | −1 (0.01) c | 251 (12) b | 12,049 |
| N600 | 3,880 (383) a | 3,874 (147) c | 1,049 (120) c | 11 (2) b | 33 (0.6) b | −79 (0.5) d | 116 (4) d | 8,884 |

**(C) $N_2O$**

| Treatments | 0–30 cm | 30–60 cm | 60–90 cm | 90–150 cm | 150–200 cm | 200–250cm | 250–300 cm | 0–300cm |
|---|---|---|---|---|---|---|---|---|
| N0 | 93 (7) c | 90 (2) d | 23 (3) b | 6 (0.3) a | 9 (0.3) b | 4 (0.4) d | 62 (1) a | 287 |
| N200 | 226 (23) b | 199 (7) c | 24 (1) b | −1 (0.02) b | 12 (0.6) a | 51 (2) a | −20 (2) d | 491 |
| N400 | 263 (27) b | 358 (5) a | 23 (2) b | −6 (0.2) d | −2 (0.2) c | 20 (1) b | 35 (0.5) b | 691 |
| N600 | 447 (1) a | 222 (2) b | 74 (4) a | −3 (0.1) c | −2 (0.1) c | 11 (0.7) c | 22 (2) c | 771 |

Values (means with SE in the brackets) followed by the same letter are not significantly different within columns (one-way ANOVA with LSD; $P<0.05$)

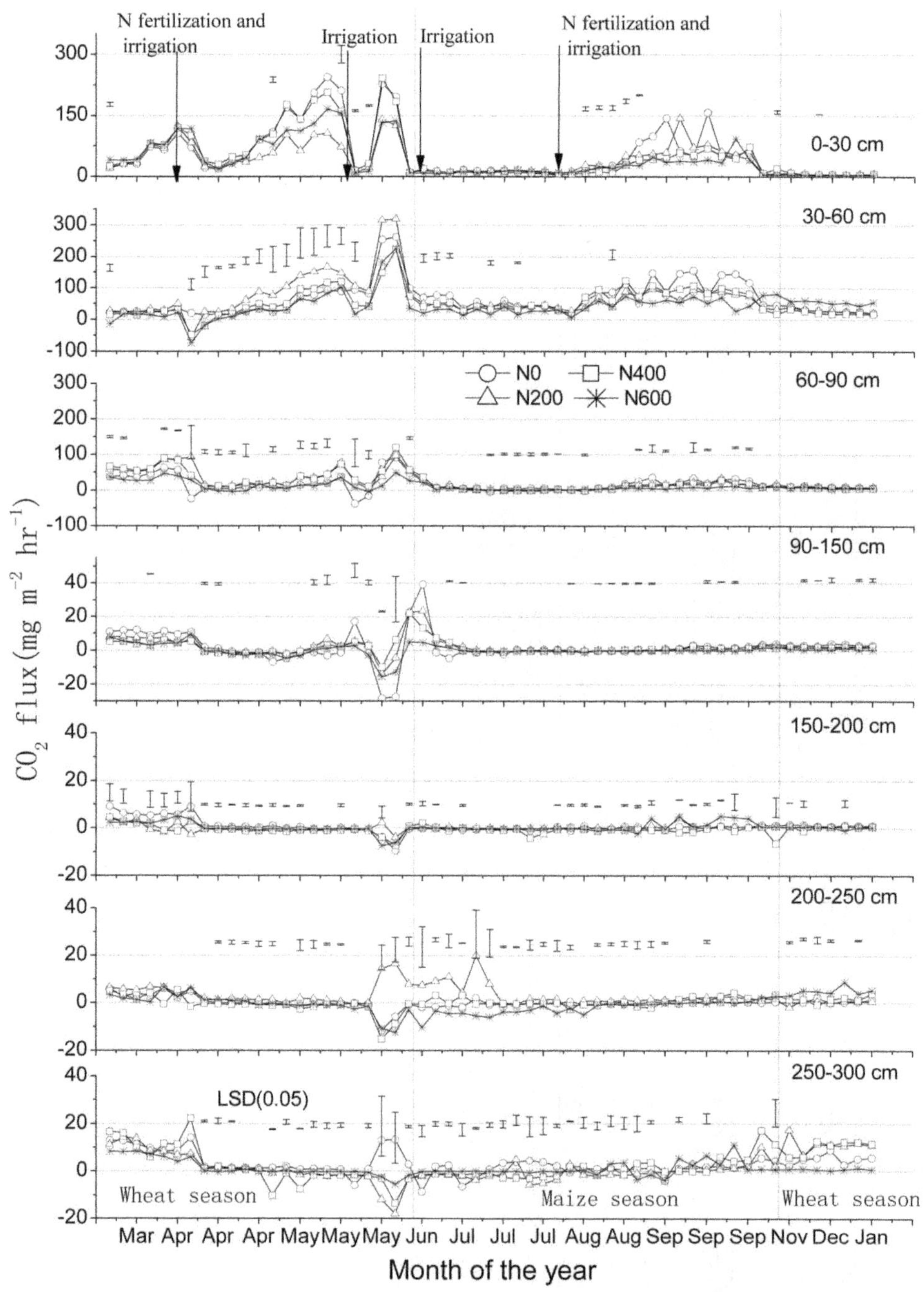


**Figure 3. $CO_2$ flux rates (means ± standard deviations, n = 3) at various soil depths in a winter wheat-summer maize double cropping rotation receiving 0, 200, 400 and 600 kg of N ha$^{-1}$ year$^{-1}$, in 2007–2008.** Vertical dashed lines indicate a change in crop. Bars in figures indicate 1 standard deviation (n = 3). Note the differences in Y-axes.

upper 0–30 cm soil layer was averaged, and soil diffusivity may be higher in the top few cm than the bulk of the top 30 cm of soil [28]. Annual cumulative $CO_2$ flux in the layer 0–90 cm (11,327 kg $CO_2$ per ha per year) contributed about 97% to that in the layer 0–300 cm (11,744 kg $CO_2$ per ha per year). Mean calculated fluxes in the layers 0–30, 30–60 and 60–90 cm were all significantly higher than that in the layer 90–300 cm ($P<0.05$); mean annual cumulative fluxes from the soil below 90 cm were very small and in upwards direction (Figure1 C; Table 3 B).

## Fluxes of $N_2O$ in soil

Fertilizer application, irrigation and precipitation events triggered an efflux of $N_2O$ from the topsoil to the atmosphere (Figure 4). The peak efflux, associated with the supplemental N fertilizer application and flooding in early April (wheat growing season), was accompanied with significant downward directed fluxes below the topsoil layer (0–30 cm). There was another relatively large efflux of $N_2O$ into the atmosphere during the relatively moist and warm August summer month (maize growing season) (Figures 4 and 5), but this peak was not accompanied with significant downward directed fluxes below the topsoil layer. In the

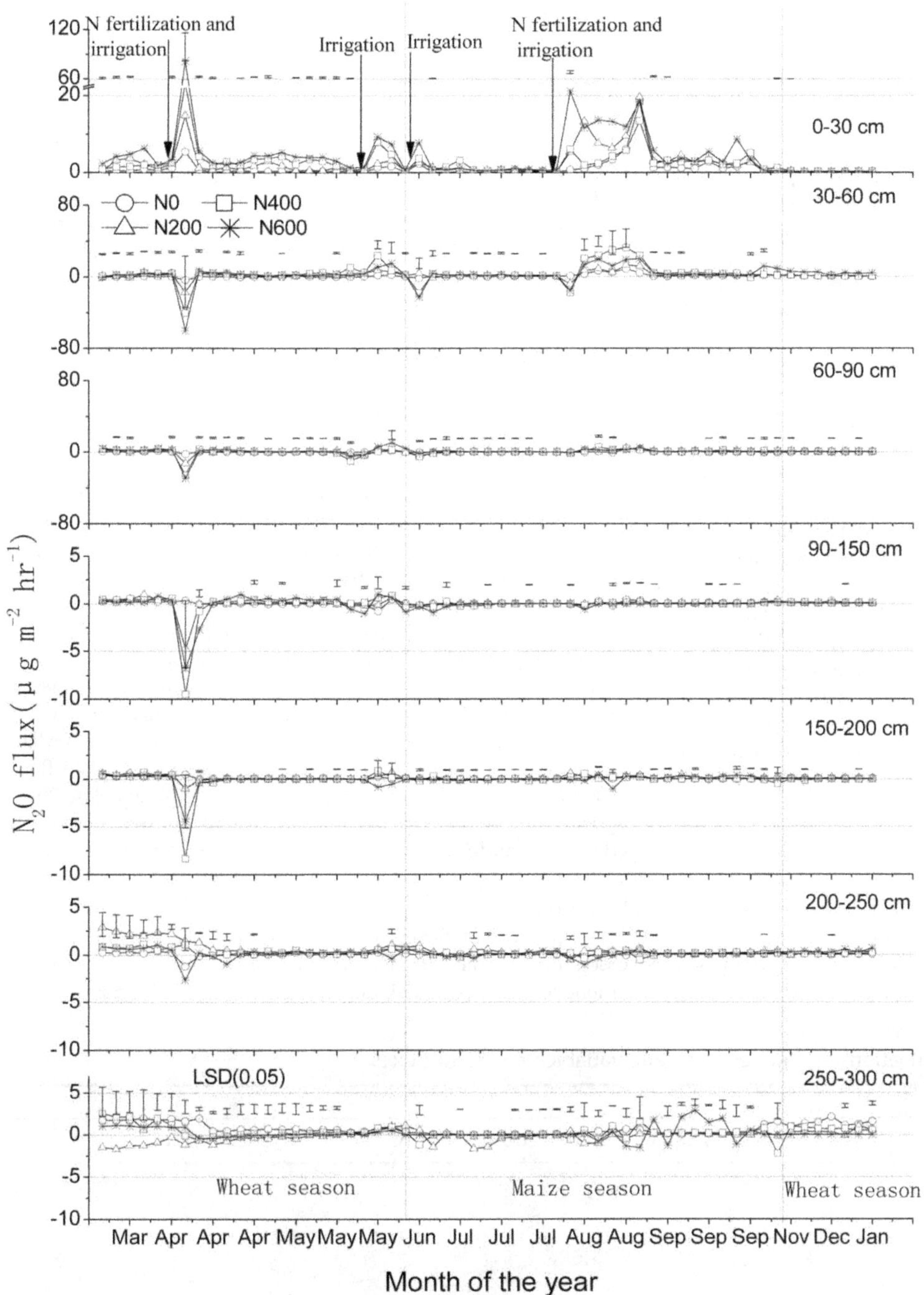


**Figure 4. $N_2O$ flux rates (means ± standard deviations, n = 3) at various soil depths in a winter wheat-summer maize double cropping rotation receiving 0, 200, 400 and 600 kg of N ha$^{-1}$ year$^{-1}$, in 2007–2008.** Vertical dashed lines indicate a change in crop. Bars in figures indicate 1 standard deviation (n = 3). Note the differences in Y-axes.

subsoil, fluxes were relatively small and directions variable (Figure 4). Essentially all seasonal fluctuations of $N_2O$ flux rates in the subsoil (60–200 cm) seem to be related to fertilizer application, irrigation and rainfall events and changes in WFPS; therefore, there was no clear evidence of $N_2O$ production in the subsoil after excluding these influence of interfering factors [16].

Annual cumulative fluxes of $N_2O$ increased with increasing N fertilizer application rates; calculated total emissions at the soil surface were 93, 226, 263 and 447 g $N_2O$ per ha for the N0, N200, N400 and N600 treatments, respectively; net upward fluxes from the 30–60 cm layer were almost as large (90, 199, 358 and 222 g $N_2O$ per ha for the N0, N200, N400 and N600 treatments, respectively) as the fluxes from the 0–30 cm layer to the atmosphere (Table 3 C). Mean calculated fluxes in the layers 0–30 and 30–60 cm were both significantly higher than those in the layers 60–90 and 90–300 cm ($P<0.05$) (Figure 1 C); mean annual cumulative fluxes from the soil below 90 cm were small but mostly in upwards direction, suggesting that the subsoil was a small source of $N_2O$, and/or that accumulated $N_2O$ from the previous season contributed to the net upward directed fluxes.

### Relations between WFPS and temperature and $CH_4$, $CO_2$ and $N_2O$ fluxes

Linear regression relationships between WFPS and $CH_4$ fluxes (positive) and between WFPS and $CO_2$ fluxes (negative) were statistically significant ($p<0.05$) for almost all soil layers (Table 4). Uptake of $CH_4$ by the soil was relatively high when WFPS was relatively low, probably because the diffusion rate of $CH_4$ into the soil was high when soil was dry, and vice versa [6]. Similarly, the upward transport of $CO_2$ was low when WFPS was high, and vice versa. This indicates that soil moisture exerted a dominant control on $CH_4$ and $CO_2$ fluxes. The linear relationship between WFPS and $N_2O$ flux was also significant for the soil layers 200–250 and 250–300 cm ($p<0.05$), but not for the other layers. Significant downward directed fluxes below the topsoil were only observed down to a deep of 200 cm (Figure 4); note that $N_2O$ fluxes were very low in 200–300 cm soil layer. Apparently, in 200–300 cm deep soil profile, soil moisture exerted a dominant control on nitrification and denitrification processes. But in 0–200 cm soil layer, WFPS was not the dominant controlling factor for the diffusive $N_2O$ flux, likely the combination of WFPS, ammonia, nitrate and metabolizable carbon, because these factors commonly control nitrification and denitrification processes.

Relationships between soil temperature and $CH_4$, $CO_2$ and $N_2O$ fluxes showed relatively large scatter (Table 4). Evidently, high temperatures are associated with the summer season, which is relatively moist (Figure 5). The significant relationships ($p<0.05$) between soil temperature and $CH_4$ fluxes at depth of 60–300 cm may be the result in part of the covariance between WFPS and soil temperature. Fluxes of $N_2O$ in soil were not significantly related to soil temperature (Table 4).

## Discussion

Fluxes of $CH_4$, $CO_2$ and $N_2O$ at the interface of soil and atmosphere are the net result of production, consumption and transport in the soil [11]. In this study, we inferred fluxes in the soil profile from changes in concentrations with depth and over time, so as to identify soil horizons of $CH_4$, $CO_2$ and $N_2O$ production and consumption, and thereby to increase the understanding of the dynamics of the net fluxes at the interface of soil and atmosphere. The study is unique in the sense that the inference of subsurface fluxes of $CH_4$, $CO_2$ and $N_2O$ in 300 cm deep soil profiles at high temporal resolution over a full year has not been reported before in such comprehensive manner.

Though the C and N cycles are intimately linked in the biosphere, there were significant differences in the dynamics of $CH_4$, $CO_2$ and $N_2O$ production, consumption and transport in the studied soil. The concentration profiles have distinct characteristics (Figure 1 A); the seasonal dynamics were much larger in the topsoil than subsoil. Moreover, the seasonal dynamics in inferred fluxes occurred during distinct periods (Figures 2, 3 and 4), and these were related to changes in WFPS (Figure 5 A), following rainfall and irrigation events. Fertilizer N application affected $N_2O$ fluxes greatly, but not those of $CH_4$ and $CO_2$. The soil under the winter wheat-summer maize double cropping system was a net sink of atmospheric $CH_4$ and a net source of $N_2O$. It was also a large source of $CO_2$ but it is unknown whether the efflux compensated the influx of C into the soil via plant growth, as the latter influx was not measured.

The inferred fluxes at the soil-atmosphere interface (Figure 1 C; Table 3) were relatively small compared to those observed in other studies [5,6,29]. Our estimated soil surface fluxes are very likely underestimates because the depth resolution of the gas samplers in the top soil was too low to capture the curvature of the concentration profile properly. Hence, our study may have underestimated the dynamics of the $CH_4$, $CO_2$ and $N_2O$ fluxes in the top soil. The depth resolution of the sampling below 90 cm appeared to be adequate. Below, we discuss the dynamics of the $CH_4$, $CO_2$ and $N_2O$ fluxes in the soil profile in more detail.

### $CH_4$ flux

Application of fertilizer N has been shown to inhibit $CH_4$ oxidation in soil [30,31], and several studies noted that non amended soils act as sink of $CH_4$ [32–34]. In our study, seasonal

**Table 4.** Linear regressions for the relationship between climatic variables and GHG fluxes.

| Climatic variable | Soil depth (cm) | $CH_4$ ($\mu g\ m^{-2}\ hr^{-1}$) | $CO_2$($mg\ m^{-2}\ hr^{-1}$) | $N_2O$ ($\mu g\ m^{-2}\ hr^{-1}$) |
|---|---|---|---|---|
| Soil water filled pore space | 0–30 | 0.853** | −0.775** | −0.149 |
| | 30–60 | 0.645** | −0.372** | 0.084 |
| | 60–90 | 0.787** | −0.852** | 0.067 |
| | 90–150 | 0.637** | −0.447** | 0.093 |
| | 150–200 | 0.771** | −0.268* | 0.084 |
| | 200–250 | −0.146 | −0.289* | −0.692** |
| | 250–300 | 0.763** | −0.532** | −0.579** |
| Soil temperature | 0–30 | −0.001 | 0.042 | 0.104 |
| | 30–60 | 0.270* | 0.356** | 0.211 |
| | 60–90 | 0.635** | −0.348** | 0.093 |
| | 90–150 | 0.620** | −0.225 | 0.062 |
| | 150–200 | 0.575** | −0.266* | −0.075 |
| | 200–250 | −0.122 | −0.299* | −0.493** |
| | 250–300 | 0.473** | −0.323* | −0.263* |

Pearson's correlation coefficient, 2-tailed tests of significance.
**Significant correlation at a <0.01.
*Significant correlation at a <0.05.

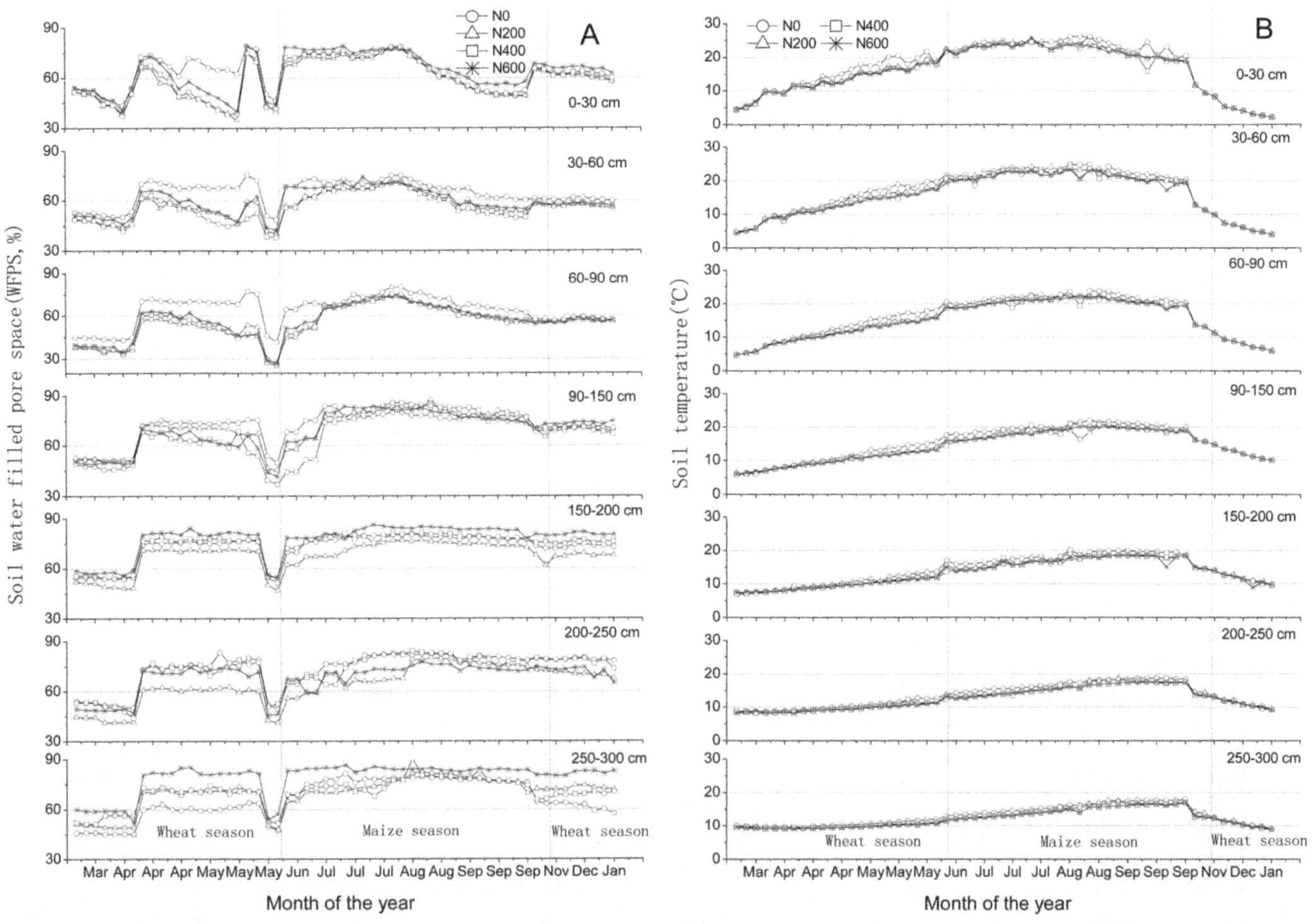


**Figure 5. Water-filled pore space (WFPS) at various soil depths in a winter wheat-summer maize double cropping rotation receiving 0, 200, 400 and 600 kg of N ha$^{-1}$ year$^{-1}$, in 2007–2008.** Bars in figures indicate 1 standard deviation (n = 3). (A); Soil temperatures at various soil depths in winter wheat-summer maize double cropping rotation receiving 0, 200, 400 and 600 kg of N ha$^{-1}$ year$^{-1}$, in 2007–2008.(B)

mean emission rates and annual cumulative fluxes of $CH_4$ for all soil layers and N fertilizer treatments were consistently directed downward (Figures 1 C and 2; Table 3 A). Though statistical significant differences in cumulative $CH_4$ fluxes between fertilizer N treatments were observed (Table 3 A), there was no clear trend that an increase in total N application decreased $CH_4$ uptake by soil. Inferred uptake was higher in the N200 and N400 treatments than in the N0 and N600 treatments at depth of 30 to 90 cm.

The magnitude of methane uptake by soils is largely controlled by diffusion of atmospheric methane into the soil [35], which in turn is strongly influenced by soil moisture [28]. The rate of diffusion of $CH_4$ in soil was high when WFPS was low. Our results showed a significant negative linear correlation between $CH_4$ uptake rate and WFPS for almost all layers; and the highest $CH_4$ uptake rates took place when WFPS was under 70% (Figures 2 and 5 A; Table 4). This is in agreement with the studies by Guo et al, Wu et al and Wang et al [4,6,16]. Inferred downward $CH_4$ fluxes decreased with depth (Figures 1 C and 2; Table 3 A). It has been reported that methanotrophic activity is most pronounced in the top soil [4,27], but our study suggests that significant uptake took place up to depths of 60 to 90 cm; below 90 cm, inferred fluxes of $CH_4$ were negligibly small (Figure 1 C).

## $CO_2$ flux

Application of 200 kg fertilizer N per ha per year and more roughly doubled grain yields relative to the control treatment [16], but did not have statistical significant effects on the $CO_2$ efflux from the soil and the diffusive flux in the soil profile (Figure 3; Table 3 B). Apparently, fertilizer N application affected predominantly aboveground biomass production, and not so much underground biomass production and respiration. Yet, we may have missed some of the topsoil dynamics, also because the incorporation of the stubbles by ploughing was in the top 15 cm of soil only. A relatively large portion of total respiration in soil took place in the layer 30–60 cm; and the total respiration in the layer was significantly higher than those in the layers 60–90 and 90–300 cm ($P<0.05$) (Figure 1 C).

When soil WFPS ranged between 40 and 70% and soil temperature was >10 °C (Figure 5), highest $CO_2$ fluxes took place at depth of 0–60 cm during the second half of the growing seasons of wheat and maize, i.e., from mid-April to mid-May and from mid-August to mid-September (Figures 3 and 5 A). These elevated $CO_2$ emissions are attributed to root respiration and to enhanced mineralization of soil organic matter by increased microbial activity [36], but also to changes in the stability and formation of soil aggregates and in the microbial community structure [37]. Sufficient soil moisture is needed to allow and support substrate diffusion to the sites of microbial activity. However, if soil moisture values exceed certain thresholds (which do depend on soil properties such as porosity, bulk density and SOC content) microbial soil respiration can get $O_2$ limited due to diffusion constrains [6]. In a saturated soil, air is pushed out of soil pore spaces and root respiration further depletes $O_2$ in the soil air [38,39]. In our study, a significant negative linear correlation was found between $CO_2$ flux rate and WFPS (40–70%) in all layers (Table 4); and low $CO_2$ fluxes took place when WFPS exceeded 70%, especially after irrigation or heavy rainfall events, i.e., from June to August (Figures 3 and 5 A). This assertion is also supported by results presented by Davidson et al, Jassal et al and Fang et al [38–40].

Generally, the $CO_2$ evolution from soil is directly correlated with soil temperature, though within a certain temperature range [41,42], and depending on the presence of active roots [43]. In our study, the relationship of soil temperature and $CO_2$ fluxes was variable, likely because of the dominant effect of WFPS (Table 4).

## $N_2O$ flux

Nitrogen application and irrigation/rainfall are main triggers for increased $N_2O$ concentrations in a soil profile and for increased emissions [15,16]. The top soil was the source of $N_2O$ production. The combined urea applications and irrigations in early April and by the end of July 2007 strongly increased $NO_3^-$ (Figure 1 B), $NH_4^+$ (data not shown) contents and WFPS (Figure 5 A) in soil, and induced large upward directed fluxes in the upper 0–30 cm soil layer; interestingly, relatively large downward directed fluxes in the subsoil only took place in April (up to the depths of 60 and 90 cm) (Figure 4), we cannot exclude the possibility that the apparent downward directed peaks during the first two months probably related to the soil structure disturbance that resulted from the installation of the samplers in December 2006.

In soil, $N_2O$ is mainly produced by nitrification and denitrification processes. The most important factors controlling these processes are $NH_4^+$ and $NO_3^-$ contents, $O_2$ partial pressure, and available carbon to fuel heterotrophic denitrification [44,45]. The rapid increases of WFPS (Figure 5 A), $NO_3$-N content (Figure 1 B) and $N_2O$ productions in the subsoil (Figures 1 A and 4) would suggest that convective transport contributed to the downward transport of water and solutes (especially in the maize growing season), which is in line with other observations [2,46]. For instance, significant upward directed fluxes of $N_2O$ took place during the relatively moist and warm August (maize growing season) than in the preceding wheat season in 30–60 cm soil layer (Figure 4). Several studies have demonstrated that higher values of soil moisture and temperature result in higher $N_2O$ fluxes [44–46] Also, Li et al found while carrying out a three-year field experiment at the same study sites that significant $NO_3$- leaching events occur predominantly during August to October (maize growing season) [47]. Zhu et al found while carrying out a four-year field experiment in a hillslope cropland that soil $NO_3$- concentrations in the subsurface soil (15–30 cm) were higher than in the topsoil (0–15 cm) during most of the maize season, indicating a rapid and effective transport of $NO_3$- to the subsurface soil following over irrigation or rainfall events [48]. Our results indicate that $NO_3$-N contents in soil layers after the maize harvest were higher than after the wheat harvest (Figure 1 B); but due to missing measurements, we can not fully rule out that high $NH_4^+$ content [47] may have contributed to relatively high $N_2O$ emissions in the warm and wet maize season. Although $N_2O$ concentration increased with soil depth, changes in inferred $N_2O$ flux below a depth of 60 cm were relatively small (Figures 1 A and 4). It may be related to the variation of the vertical $N_2O$ concentration gradient; changes in mean $N_2O$ concentration below a depth of 30 cm were not significant (Figure 1 C).

It has been frequently observed that high rates of $N_2O$ emissions take place when WFPS ranges between 30 and 70% [49]. According to Zou et al the $N_2O$ production in dry land soil of Northern China is mostly driven by nitrification [50]. Wang et al suggests that nitrification is likely a main source of the $N_2O$ production in the soil profile when WFPS varied between 45 and 70% at the study site [16]. $N_2$ starts being emitted through denitrification at a WFPS of 70%, and is the main N gas emitted when WFPS exceeds 75% [51]. This may explain the relatively high inferred $N_2O$ flux from late April to mid-May (WFPS, 40–70%) and the very low flux from late June to late July (before applying nitrogen) (WFPS, >70%) in the layer 0–30 cm (Figures 4 and 5 A).

The accumulated $N_2O$ fluxes were significantly related to N application rate. This was most apparent in the top 30 to 90 cm of soil (Table 3 C). Annual cumulative $N_2O$ flux in the layer 0–90 cm (511 g $N_2O$ per ha per year) contributed about 90% to that in the layer 0–300 cm (560 g $N_2O$ per ha per year). The 90 cm thick cinnamon top soil overlays the so-called Shajiang layer (90–140 cm) with silty clay loam texture [52]. The Shajiang layer has no crop roots, contains many iron-manganese nodules and has high bulk density (Table 2). This compacted subsoil may explain that fertilizer application and irrigation mainly affected $N_2O$ fluxes down to 90 cm (Figure 1 C; Table 3 C). In this study, calculated total emissions in the layer 0–90 cm were 206, 449, 644 and 743 g $N_2O$ per ha for the N0, N200, N400 and N600 treatments, respectively; these fluxes translate into fertilizer-derived emissions of 0.14, 0.10 and 0.07% for the N200, N400 and N600 treatments, respectively. The fertilizer induced emission factors (0.07–0.14%) were lower than the 0.30–0.39% measured by Ding et al [53] over the maize-wheat rotation year in a long-term mineral nitrogen addition field experiments (150–300 kg N $ha^{-1}$ $year^{-1}$, over 20-years) in the North China Plain. A reason for the lower fertilizer induced emission factor is probably related to the likely underestimates of soil surface $N_2O$ fluxes, because the concentration gradient in the upper 0–30 cm soil layer was averaged, and soil diffusivity may be higher in the top few cm than the bulk of the top 30 cm of soil [28]. Furthermore, due to missing measurements we can not fully rule out the indirect $N_2O$ emissions from leaching and atmospheric deposition [47,53].

## Conclusions

Our study is one of few that inferred $CH_4$, $CO_2$ and $N_2O$ transport between soil layers from changes in $CH_4$, $CO_2$ and $N_2O$ concentrations in the upper 300 cm of soil, measured at (bi)-weekly time intervals for one year in a winter wheat-summer maize double crop rotation. The top 30 to 60 cm of soil was a sink of atmospheric $CH_4$, and a source of both $CO_2$ and $N_2O$. There was little or no evidence that the subsoil (>90 cm) acted as a sink or source of GHG; rather it acted as "reservoir".

Nitrogen fertilizer application increased $N_2O$ fluxes but did not affect $CH_4$ and $CO_2$ fluxes. The fertilizer-derived $N_2O$ flux was small, likely because our sampling design may have missed $N_2O$ production in the top 15 cm of soil. This holds as well for the $CH_4$ consumption by soil and the $CO_2$ emissions from soil; both are likely underestimated. Soil moisture (WFPS) was found to play an important regulating role for $CH_4$, $CO_2$ and $N_2O$ fluxes in soil and between soil and atmosphere. Both $CH_4$ consumption and $CO_2$ fluxes in and from soil all tended to decrease with increasing WFPS.

More than 90% of the annual cumulative GHG fluxes originated at depths shallower than 90 cm. Mostly because the productive soil of our study site in the North China Plain had two distinct layers (0–90 and >90 cm), with different texture and bulk density. These differences showed up in characteristic differences in GHG concentration profiles and fluxes.

## Acknowledgments

The authors would like to thank Luancheng Agroecosystem Experimental Station, Chinese Academy of Sciences for tireless efforts with maintaining the long-term fertilizer experiments. The authors would also like to thank two anonymous reviewers, whose suggestions and comments greatly improved the manuscript.

## Author Contributions

Conceived and designed the experiments: CSH YYW HM YMZ XXL WXD OO DAS. Performed the experiments: YYW HM CSH. Analyzed the data: YYW. Wrote the paper: YYW.

## References


1. IPCC (2007) Agriculture. In: Metz, B., D.O.R., Bosch P.R. (Eds.), Climate Change 2007: Mitigation, Contribution of Working Group III to the Fourth Assessment Report of the Intergovernmental Panel on Climate Change. Cambridge University Press, Cambridge. United Kingdom and New York, NY, USA.
2. Jassal RS, Black TA, Trofymow AJ, Roy R, Nesic Z (2010) Forest-floor $CO_2$ and $N_2O$ flux dynamics in a nitrogen-fertilized Pacific Northwest Douglas-fir stand. Geoderma 157(3–4): 118–125.
3. Smith P, Martino D, Cai ZC, Gwary D, Janzen H, et al. (2008) Greenhouse gas mitigation in agriculture. Philosophical Transactions of the Royal Society B: Biological Sciences 363: 789–813.
4. Guo JP, Zhou CD (2007) Greenhouse gas emissions and mitigation measures in Chinese agroecosystems. Agricultural and Forest Meteorology 142: 270–277.
5. Kim Y, Ueyama M, Nakagawa F, Tsunogai U, Harazono Y, et al. (2007) Assessment of winter fluxes of $CO_2$ and $CH_4$ in boreal forest soils of central Alaska estimated by the profile method and the chamber method: a diagnosis of methane emission and implications for the regional carbon budget. Tellus 59B: 223–233.
6. Wu X, Yao Z, Brüggemann N, Shen ZY, Wolf B, et al. (2010) Effects of soil moisture and temperature on $CO_2$ and $CH_4$ soil-atmosphere exchange of various land use/cover types in a semi-arid grassland in Inner Mongolia, China. Soil Biology & Biochemistry 42: 773–787.
7. Banger K, Tian HQ, Lu CQ (2012) Do nitrogen fertilizers stimulate or inhibit methane emissions from rice fields? Global Change Biology 18: 3259–3267.
8. Sanz-Cobena A, Sánchez-Martín L, García-Torres L, Vallejo A (2012) Gaseous emissions of $N_2O$ and NO and $NO_3-$ leaching from urea applied with urease and nitrification inhibitors to a maize (Zea mays) crop. Agriculture, Ecosystems and Environment 149: 64–73.
9. Sanz-Cobena A, García-Marco S, Quemada M, Gabriel JL, Almendros P, et al. (2014) Do cover crops enhance $N_2O$, $CO_2$ or $CH_4$ emissions from soil in Mediterranean arable systems? Science of the Total Environment 466–467: 164–174.
10. Valentini R, Matteucci G, Dolman AJ, Schulze ED, Rebmann C, et al. (2000) Respiration as the main determinant of carbon balance in European forests. Nature 404: 861–865.
11. Bowden WB, Bormann FH (1986) Transport and loss of nitrous oxide in soil water after forest cutting. Science 233: 867–869.
12. Tang JW, Baldocchi DD, Qi Y, Xu LK (2003) Assessing soil $CO_2$ efflux using continuous measurements of $CO_2$ profiles in soils with small solid-state sensors. Agricultural and Forest Meteorology 118: 207–220.
13. Fierer N, Chadwick OA, Trumbore SE (2005) Production of $CO_2$ in Soil Profiles of a California Annual Grassland. Ecosystems 8: 412–429.
14. Gebert J, Röer IU, Scharff H, Roncato CDL, Cabral AR (2011) Can soil gas profiles be used to assess microbial $CH_4$ oxidation in landfill covers? Waste Management 31: 987–994.
15. Reth S, Graf W, Gefke O, Schilling R, Seidlitz HK, et al. (2008) Whole-year-round Observation of $N_2O$ Profiles in Soil: A Lysimeter Study.Water Air Soil Pollut: Focus 8:129–137.
16. Wang YY, Hu CS, Ming H, Zhang YM, Li XX, et al. (2013) Concentration profiles of $CH_4$, $CO_2$ and $N_2O$ in soils of a wheat–maize rotation cosystem in North China Plain, measured weekly over a whole year. Agriculture, Ecosystems & Environment 164: 260–272.
17. Sparks DL (1996) Methods of soil analysis, part 3–chemical methods. In: Sparks, D.L. (Eds.), SSSA book series, No. 5. SSSA, Inc and American Society of Agronomy, Inc. Madison, WI. pp. 475–1185.
18. Gee GW, Bauder JW (1986) Particle-size analysis. In: Klute, A. (Ed.), Methods of Soil Analysis Part 1-Physical and Mineralogical Methods. American Society of gronomy, Madison, WI. pp. 383–409.
19. Campbell GS (1985) Soil Physics with BASIC. Elsevier Science Publishers BV, Amsterdam. Soil Science Society of America Book Series: 5 Methods of Soil Analysis Part I-Physical and Mineralogical Methods.
20. Marshall TJ (1959) The diffusion of gas through porous media. J Soil Sci 10: 79–82.
21. Rolston DE (1986) 47 Gas Flux, 1103–1109. American Society of Agronomy, Inc. Soil Science Society of America, Inc. Publisher. Madison, Wisconsin USA.
22. Sallam A, Jury WA, Letey J (1984) Measurement of gas-diffusion coefficient under relatively low air-filled porosity. Soil Sci Soc Am J 48: 3–6.
23. Jury WA, Gardner WR, Gardner WH (1991) Soil Physics, fifth ed. Wiley, New York.
24. Gilliland E, Baddour R, Perkinson G, Sladek KJ (1974) Diffusion on surfaces. I. Effect of concentration on the diffusivity of physically adsorbed gases. Ind Eng Chem Fundam 13(2): 95–100.
25. Millington R, Quirk JP (1961) Permeability of porous solids. Trans Faraday Soc 57: 1200–1207.
26. Wang YY, Hu CS, Zhu B, Xiang HY, He XH (2010) Effects of wheat straw application on methane and nitrous oxide emissions from purplish paddy fields. Plant, Soil and Environ 56 (1): 16–22.
27. Stiehl-Braun PA, Powlson DS, Poulton PR, Niklaus PA (2011) Effects of N fertilizers and liming on the micro-scale distribution of soil methane assimilation in the long-term Park Grass experiment at Rothamsted. Soil Biology and Biochemistry 43(5): 1034–1041.
28. Shrestha BM, Sitaula BK, Singh BR, Bajracharya RM (2004) Fluxes of $CO_2$ and $CH_4$ in soil profiles of a mountainous watershed of Nepal as influenced by land use, temperature, moisture and substrate Addition. Nutrient Cycling in Agroecosystems 68: 155–164.
29. Chu H, Hosen Y, Yagi K (2004) Nitrogen oxide emissions and soil microbial properties as affected by N-fertilizer management in a Japanese Andisol. Soil Science and Plant Nutrition 50: 287–292.
30. Steudler PA, Bowden RD, Melillo JM, Aber JD (1989) Influence of nitrogen fertilisation on methane uptake in temperate forest soils. Nature 341: 314–316.
31. Kravchenko I, Boeckx P, Galchenko V, Van Cleemput O (2002) Shortand medium-term effects of $NH_4{}^+$on $CH_4$ and $N_2O$ fluxes in arable soils with a different texture. Soil Biology & Biochemistry 34: 669–678.
32. Flessa H, Beese F (2000) Laboratory estimates of trace gas emissions following surface application and injection of cattle slurry. J Environ Qual 29: 262–268.
33. Sherlock RR, Sommer SG, Khan RZ, Wood CW, Guertal EA, et al. (2002) Emissions of ammonia, methane and nitrous oxide from pig slurry applied to a pasture in New Zealand. J Environ Qual 31: 1491–1501.
34. Rodhe L, Pell M, Yamulki S (2006) Nitrous oxide, methane and ammonia emissions following slurry spreading on grassland. Soil Use Manage 22: 229–237.
35. Koschorreck M, Conrad R (1993) Oxidation of atmospheric methane in soil: measurements in the field, in soil cores and in soil samples. Global Biogeochemical Cycles 7: 109–121.
36. Borken W, Matzner E (2009) Reappraisal of drying and wetting effects on C and N mineralization and fluxes in soils. Global Change Biology 15: 808–824.
37. Denef K, Six J, Bossuyt H, Frey SD, Elliott ET, et al. (2001) Influence of dry-wet cycles on the interrelationship between aggregate, particulate organic matter, and microbial community dynamics. Soil Biology and Biochemistry 33: 1599–1611.
38. Davidson EA, Belk E, Boone RD (1998) Soil water content and temperature as independent or confounded factors controlling soil respiration in a temperate mixed hardwood forest. Global Change Biol 4: 217–227.
39. Jassal RS, Black TA, Drewitt GB, Novak MD, Gaumont-Guay D, et al. (2004) A model of the production and transport of $CO_2$ in soil: predicting soil $CO_2$ concentrations and $CO_2$ efflux from a forest floor. Agric For Meteorol 124: 219–236.
40. Fang YT, Gundersen P, Zhang W, Zhou GY, Christiansen JR, et al. (2009) Soil-atmosphere exchange of $N_2O$, $CO_2$ and $CH_4$ along a slope of an evergreen broad-leaved forest in southern China. Plant and Soil 319: 37–48.
41. Bajracharya RM, Lal R, Kimble JM (2000) Erosion effect on carbon dioxide concentration and carbon flux from an Ohio Alfisol. Soil Sci. Soc. Am. J 64: 694–700.
42. Fang C, Moncrieff JB (2001) The dependence of soil $CO_2$ efflux on temperature. Soil Biology & Biochemistry 33: 155–165.
43. Kelting DL, Burger JA, Edwards GS (1998) Estimating root respiration, microbial respiration in the rhizosphere, and root-free soil respiration in forest soils. Soil Biology and Biochemistry 30: 961–968.
44. Clough TJ, Sherlock RR, Kelliher FM (2003) Can liming mitigate $N_2O$ fluxes from a urine-amended soil? Aust J Soil Res 41: 439–457.
45. Clough TJ, Kelliher FM, Sherlock RR, Ford CD (2004) Lime and soil moisture effects on nitrous oxide emissions from a Urine Patch. Soil Sci Soc A J 68: 1600–1609.

46. Grandy SA, Robertson PG (2006) Initial cultivation of a temperate-region soil immediately accelerates aggregate turnover and $CO_2$ and $N_2O$ fluxes. Global Change Biology 12: 1507–1520.
47. Li XX, Hu CS, Delgado JA, Zhang YM, Ouyang ZY (2007) Increased nitrogen use efficiencies as a key mitigation alternative to reduce nitrate leaching in north China plain. Agricultural Water Management 89: 137–147.
48. Zhu B, Wang T, Kuang F, Luo Z, Tang J, et al. (2009) Measurements of nitrate leaching from a hillslope cropland in the Central Sichuan Basin, China. Soil Sci Soc Am J 73: 1419–1426.
49. Dobbie KE, Smith KA (2003) Nitrous oxide emission factors for agricultural soils in Great Britain: the impact of soil water-filled pore space and other controlling variables. Global Change Biology 9: 204–218.
50. Zou GY, Zhang FS, Chen XP, Li XH (2001) Nitrification–denitrification and $N_2O$ emission from arable soil. Soil Environ Sci 10 (4): 273–276 (in Chinese).
51. Davidson EA (1992) Sources of nitric oxide and nitrous oxide following wetting of dry soil. Soil Sci Soc Am J 56: 95–102.
52. Zhu HJ, He YG eds (1992) Soil Geography. Higher Education Press, Beijing, China (In Chinese).
53. Ding WX, Luo JF, Li J, Yu HY, Fan JL, et al. (2013) Effect of long-term compost and inorganic fertilizer application on background $N_2O$ and fertilizer-induced $N_2O$ emissions from an intensively cultivated soil. Science of the Total Environment 465: 115–124.

# 19

# Structure, Composition and Metagenomic Profile of Soil Microbiomes Associated to Agricultural Land Use and Tillage Systems in Argentine Pampas

**Belén Carbonetto[1]*, Nicolás Rascovan[1], Roberto Álvarez[2], Alejandro Mentaberry[3], Martin P. Vázquez[1]***

**1** Instituto de Agrobiotecnología de Rosario (INDEAR), Predio CCT Rosario, Santa Fe, Argentina, **2** Facultad de Agronomía, Universidad de Buenos Aires, Buenos Aires, Argentina, **3** Departamento de Fisiología y Biología Molecular y Celular, Facultad de Ciencias Exactas y Naturales, Universidad de Buenos Aires, Buenos Aires, Argentina

## Abstract

Agriculture is facing a major challenge nowadays: to increase crop production for food and energy while preserving ecosystem functioning and soil quality. Argentine Pampas is one of the main world producers of crops and one of the main adopters of conservation agriculture. Changes in soil chemical and physical properties of Pampas soils due to different tillage systems have been deeply studied. Still, not much evidence has been reported on the effects of agricultural practices on Pampas soil microbiomes. The aim of our study was to investigate the effects of agricultural land use on community structure, composition and metabolic profiles on soil microbiomes of Argentine Pampas. We also compared the effects associated to conventional practices with the effects of no-tillage systems. Our results confirmed the impact on microbiome structure and composition due to agricultural practices. The phyla *Verrucomicrobia, Plactomycetes, Actinobacteria*, and *Chloroflexi* were more abundant in non cultivated soils while *Gemmatimonadetes, Nitrospirae* and WS3 were more abundant in cultivated soils. Effects on metabolic metagenomic profiles were also observed. The relative abundance of genes assigned to transcription, protein modification, nucleotide transport and metabolism, wall and membrane biogenesis and intracellular trafficking and secretion were higher in cultivated fertilized soils than in non cultivated soils. We also observed significant differences in microbiome structure and taxonomic composition between soils under conventional and no-tillage systems. Overall, our results suggest that agronomical land use and the type of tillage system have induced microbiomes to shift their life-history strategies. Microbiomes of cultivated fertilized soils (i.e. higher nutrient amendment) presented tendencies to copiotrophy while microbiomes of non cultivated homogenous soils appeared to have a more oligotrophic life-style. Additionally, we propose that conventional tillage systems may promote copiotrophy more than no-tillage systems by decreasing soil organic matter stability and therefore increasing nutrient availability.

**Editor:** Kathleen Treseder, UC Irvine, United States of America

**Funding:** Funding for this work was provided by Agencia Nacional de Promoción Científica y Tecnológica, Argentina-PAE 37164. The funders had no role in study design, data collection and analysis, decision to publish, or preparation of the manuscript.

**Competing Interests:** The authors have declared that no competing interests exist.

* E-mail: martin.vazquez@indear.com (MPV); belen.carbonetto@indear.com (BC)

## Introduction

Agriculture is facing major challenges nowadays. Production will have to double in the next 50 years in order to face growing food demand and bioenergy needs [1,2]. This must be done without increasing environmental threats such as climate change, biodiversity loss and degradation of land and freshwater. Achieving such a goal represents one of the greatest scientific challenges ever. This is in part because of the trade-offs among economic and environmental goals and because of the insufficient knowledge about the biological, biogeochemical and ecological processes that are relevant for sustainable ecosystem functioning [3,4]. Much has been done during the last decade to gain sufficient information on agricultural ecosystem biology, still, more work needs to be done to gain deeper comprehension and to be able to reduce the negative environmental impacts of agriculture [2,5,6]. The main focus should be oriented to soil degradation. Soil fertility, as the capacity to sustain abundant crop production, needs to be preserved. Nowadays soil fertility is maintained by dependence on external inputs; with increasing water contamination [7]. In this context, the key to understand the behavior of life-supporting elements in soil, such as carbon, nitrogen, and phosphorus lies in the fluxes between their various forms in the environment, which are modulated by biology [8]. Comprehension of soil microorganism dynamics is then essential to understand soil processes that affect fertility. Ecological approaches are being taken into account in soil microbial studies trying to address these questions. These approaches involve diversity and functional analyses of soil communities [9,10]. Scholes & Scholes point out that this complex view is necessary for the comprehension of soil systems and that soil restoration of biological processes is the key to achieving lasting food and environmental security [8].

Argentine Pampas is an important player in this scenario. With a plain area of 50 million ha., nearly 50% of the whole Pampas area is devoted to crop production [11]. Cultivation began in the 19th Century in the central humid portion of the region, in soils of high fertility, and spread in last decades to the south and the semiarid west [12]. Soil degradation (i.e. intense erosion and net loss of nutrients and organic carbon) caused by the use of conventional tillage systems were reported in the Pampas [13–17].

Nowadays between 60 and 80% of production is conducted under conservational no-till practices [18]. Extensive research was done to evaluate the effects of reduced tillage and no-tillage systems on soil physical properties, water content, fertility and crop yields[19]. The main outcome of these analyses points that the adoption of limited tillage systems led to soil improvement, by augmenting organic matter content and soil structure. Still, external fertilization is needed in order to restore nutrient levels and fertility regardless the tillage system employed.

Even though the effects of different tillage practices on soil physical and chemical characteristics have been deeply studied, changes in microbial biodiversity and functioning have been poorly reported in Argentinean Pampas. Most works have studied tillage effects on microbial biomass or specific microbial activities (i.e. utilization of specific substrates, extracellular enzyme production, mineralization, etc.) rather than on full microbiome [13,20,21]. Other studies have focused on the behavior of specific bacterial taxa [22,23]. Reports with an ecological approach (i.e. microbial community analysis) have usually focused on individual effects of land use such as the application of herbicides [24,25]. In these cases, biodiversity variability has been assessed using classical fingerprinting techniques (such as RFLP and DGGE) that lack information about microbial taxonomic identity and only capture the most dominant species in the environment [26,27]. In the last few years, 16S amplicon pyrosequencing has been largely implemented to determine microbial diversity and structure of many different ecosystems worldwide [28,29]. This strategy allows a more exhaustive characterization of community patterns and composition. Moreover, some works have incorporated the use of shotgun metagenomics to study the metabolic potential of soil microbiomes [10,30]. The shotgun approach generates a massive amount of data using random high-trhoughput sequencing of soil isolated DNA. This allows the identification of functional capabilities by gene annotation and the comparison of metabolic profiles between samples. To our knowledge, Figuerola et al. [31] were the only authors studying microbial communities in agronomical soils of Argentine Pampas using high throughput sequencing approaches. They observed differences in microbial community composition of soils under no-tillage systems using 16S pyrosequencing. As a novelty, our efforts focused on assessing the impact of long-term agriculture on Pampas soil microbiomes using both shotgun metagenomics and deep 16S amplicon sequencing approaches. We evaluated the effect of more than a hundred years of agronomical land use on both community features and metabolic profiles of soil microbiomes in comparison with nearby control soils with no agricultural records. We also addressed the differences between the effects of two tillage systems: conventional tillage vs. no-tillage on microbial communities.

Several previous studies of soil microbiomes from different parts of the world showed the effects of agronomical land use on soil microbial communities [10,32–35]. Some of these studies showed differences in trophic strategies between microbial communities related to tillage; and most of them were done in experimental plots. As a novelty, we tested the impact of long term agriculture in soils sampled in production fields in the Argentine Pampas, allowing a deeper insight to the effects of intense land use on soil ecosystems functioning. We confirmed the hypothesis that agronomical practices affected Pampas soil microbiomes by promoting a shift of life-history and trophic strategies. We also showed differences in the effects of contrasting tillage systems (i.e. conventional vs. no- tillage) on community taxonomic and metabolic composition on a long term experiment.

## Materials and Methods

### Sites description and sampling

Soil samples were taken in production and experimental fields between June and August 2010. To address the effects of agricultural land use on soil microbial communities, three different production farms were sampled in the Rolling Pampas area: "La Estrella", "La Negrita" and "Criadero Klein" (See Rascovan et al and Table S1 for details). Rolling Pampas soils are classified as Typic Argiudolls [36] and mean annual rainfall and temperature were1002 mm and 16.8°C respectively. Two treatments were defined: *cultivated* for production plots, and *no cultivated* for farm-houses parks. Production plots were under cultivation for at least one century under conventional tillage systems, with a mixed rotation of pastures and annual grain crops. During the last 15 years before sampling plots were subjected to continuous crop cultivation under no-tillage systems (i.e. minimal soil disturbance, permanent soil cover, rotations and fertilization). The last crop rotation before sampling was wheat-soybean. Nitrogen and phosphorus fertilizers were applied. Samples were collected one month after soybean harvest. Soil samples were also collected nearby the farmers' houses where no agricultural land use (no tillage nor cultivation) was recorded for the last 30 years except from grass mowing. Parks around farmers' houses are usually considered as undisturbed environments in Argentine Pampas [37]. Soils under no land use were covered with grass and other herbaceous (non-woody) plants common in the region such as *Cirsium sp, Trifolium sp, Micropsis sp, Festuca sp, Dichondra sp, Cyperus sp* and *Taraxacum officinale.* For numerical analyses purposes the three farms are treated as experimental replicates. Four soil samples were taken with an auger from the upper 20 cm soil layer in each farm and treatment. A total of 24 samples were collected in Rolling Pampa soils.

In order to compare effects of contrasting tillage systems, samples were also collected in a 34-year-old experiment located in Balcarce in the Southern Pampas (See Rascovan et al and Table S1 for more details). Samples were taken in experimental plots because no production fields are using conventional tillage for crop production nowadays in the Pampas. Soils in Balcarce are a complex of Typic Argiudolls and Petrocalcic Paleudols and mean annual rainfall and temperatures were 875 mm and 13.8°C respectively. The experiment was carried out in three (175 $m^2$) experimental plots ($n = 3$). Treatments were defined as: *no tillage* (NT) and *conventional tillage* (CT). NT plots had minimal soil disturbance and permanent soil cover combined with rotations; which have included pastures and grain crops (soybean, corn, wheat) during the last 16 years. CT plots were managed with moldboard plough. Nitrogen fertilization was performed in NT and CT plots (60 kg N ha-1). Last rotation before sampling was corn-soybean. Two sub-samples were collected from all treatment and replicate plots a month after soybean harvest. A total of 12 samples were collected.

Samples were immediately sent to the lab after collection. Samples used for DNA purification were air dried and sieved through 1 mm mesh to thoroughly homogenize, break aggregates and remove roots and plant detritus, then stored at −80°C. DNA purification and library preparation was previously described in Rascovan. et al.[38].

None of the sampling sites is located in protected areas. Permissions were obtained directly from each farm owner or manager: Alejandro Cattaneo at La Negrita and La Estrella, Roberto Klein at Criadero Klein and Guillermo Studdert at Balcarce experiment.

### Soil chemical and physical measurements

Soil organic carbon was determined by wet digestion and organic matter was estimated [39]. Nitrate-nitrogen was analyzed by 2 M KCL extraction and the phenoldisulfonic acid method [40]. Extractable phosphorus was determined by the Bray method [41]. The pH was measured in a soil:water ratio 1:2.5. Salinity was estimated by the determination of electrical conductivity [42]. Texture analysis was performed by the hydrometer method [43] and nitrogen was determined by Kjeldahl method [40].

### 16S amplicon sequencing and shotgun metagenomic datasets

To analyze the effect of agronomical practices on soil microbiomes, sequence data from the previously reported Pampa dataset [38] was used. In order to evaluate microbial community structure and taxonomic composition, a total of 112,800 high-quality filtered 16S rRNA gene amplicon sequences, obtained from the 42 soil DNA samples (replicates and subsamples were included). DNA shotgun metagenomic data was used to analyze metabolic profiles. Shotgun metagenomic data completed a total of 10,445,170 sequences. In this case sequences were obtained from one subsample per sampling replicated plot.

In brief, libraries were prepared as follows: DNA was isolated from 10 g of soil of each of the 42 soil samples using the Power MaxSoil DNA Isolation Kit following the manufacturer's instructions (MO BIO Laboratories, Inc.). For amplicon libraries the V4 hyper variable region of the 16s rRNA gene was amplified. Duplicated reactions were performed using barcoded bacterial universal primers containing Roche- 454 sequencing A and B adaptors and a nucleotide multiple identifier (MID) to sort samples: 563F: 5′-CGTATCGCCTCCCTCGCGCCATCA-GACGAGTGCGTAYTGGGYDTAAAGNG -3′ (where AC-GAGTGCGT is an example, different MIDs for each sample were used) and 802R (5′-CTATGCGCCTTGCCAGCCCGCT-CAGTACCRGGGTHTCTAATCC, 5′-CTATGCGCCTTGC-CAGCCCGCTCAGTACCAGAGTATCTAATTC, 5′-CTAT-GCGCCTTGCCAGCCCGCTCAGCTACDSRGGTMTCTA-ATC, 5'-CTATGCGCCTTGCCAGCCCGCTCAGTACNVG-GGTATCTAATCC) [44]. All amplicons were cleaned using Ampure DNA capture beads (Agencourt- Beckman Coulter, Inc.) and pooled in equimolar concentrations before sequencing on a Genome Sequencer FLX (454-Roche Applied Sciences) using Titanium Chemistry according to the manufacturer's instructions.

Shotgun metagenomic libraries were prepared by nebulization, followed by tagging with GS-FLX-Titanium Rapid Library MID Adapters Kit (454-Roche Applied Sciences) and sequenced with a Genome Sequencer FLX (454-Roche Applied Sciences) using Titanium Chemistry according to the manufacturer's instructions. Sequencing runs were performed in INDEAR sequencing facility.

All the sequences used in the present study are available in The Sequence Read Archive (SRA) under accession number SRA058523 and SRA056866. See Rascovan et al. [38] for more information.

### Amplicon sequence processing, OTU classification and taxonomic assignment

Sequence data were quality controlled and denoised with the ampliconnoise.py script of QIIME [45].This script also eliminated chimeras. Sequences obtained from Rolling Pampa soil libraries and Balcarce soil libraries were processed separately. Sequences were clustered into Operational Taxonomic Units (OTUs) using the pick_otus.py script with the Uclust method [46] at 97% sequence similarity. Rolling Pampas samples yielded 2,591sequences on average (ranging from 1,455 to 3,991sequences). Balcarce samples yielded 2,329 reads on average (ranging from 1,211 to 4,755 reads).OTU representative sequences were aligned using PyNast algorithm [47] with QIIME default parameters. Phylogenetic trees containing the aligned sequences were then produced using FastTree [48]. All downstream analyses were determined after each sample was randomly rarefied to 70% the number of reads of the smallest sample (i.e. 1,080 reads for Rolling Pampa libraries and 850 reads for Balcarce libraries). Phylogenetic distances between OTUs were calculated using unweighted and weighted Unifrac [49]. Taxonomic classification of sequences was done with Ribosomal Database Project (RDP) Classifier using Greengenes database using a 50% confidence threshold [50,51].

### Microbial community analyses

Unifrac phylogenetic pairwise distances among samples were visualized with principal coordinates analysis (PCoA). Analysis of similarity statistics (ANOSIM) was calculated to test a-priori sampling groups. BIOENV analysis was performed to elucidate which soil properties correlated with community patterns. All calculations were carried out with R packages 'BiodiversityR' and 'Vegan' [52,53]. T- tests were performed with QIIME script otu_category_significance.py, and R scripts in order to elucidate differences in read abundances.

### Shotgun metagenomic sequence processing and analysis

SSF files obtained from shotgun sequencing runs were uploaded to the MG-RAST webserver [54] for sequence filter and analyses. Reads more than two standard deviations away from the mean read length were discarded. For dereplication removal MG-RAST used a simple k-mer approach to rapidly identify all 20 character prefix identical sequences. This step is required in order to remove artificial duplicate reads. We obtained an average of $1.28 \times 10^6$ filtered reads per sample for Rolling Pampa shotgun libraries, and an average of 304,258 filtered reads per sample for Balcarce libraries.

Filtered high quality sequences were assigned to Cluster of orthologous groups (COG) by the MG-RAST sever pipeline using a similarity-based approach. COGs were assigned with a maximum E value of $10^{-20}$, an average alignment of 80 amino acids length and 70% average identity.

Relative abundances were calculated by dividing the number of hits for each COG or COG-category by the total number of filtered reads in each sample. Euclidean distances based on relative abundances were calculated between sample pairs. PCoA visualizations and ANOSIM calculations were performed. All calculations were carried out with R packages 'BiodiversityR' and 'Vegan'. T- tests were performed with QIIME script otu_category_significance.py, and R scripts in order to elucidate differences in COG relative abundances between samples.

## Results

### Microbiome community changes related to agricultural land use

The PCoA visualization revealed clear differences between cultivated and noncultivated soils (ANOSIM $R = 0.8406$, $p \leq 0.001$; Figure 1A).Similar results were obtained when using Bray Curtis distance matrices (Figure S1).

The soil properties (Table S1) that best explained the phylogenetic variation observed in microbial communities were determined using Clarke and Ainsworth's BIOENV analysis. Our results showed that variables that best correlated with community

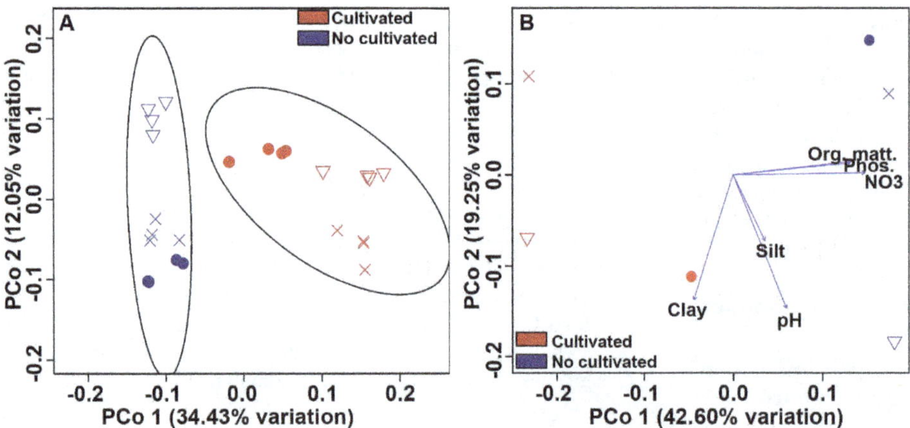


**Figure 1. PCoA plots of Pampa production field soil microbiomes based on 97% similarity Weighted Unifrac distance matrices.** A) PCoA of cultivated and non-cultivated soil microbiomes. All sub-samples are plotted. Standard error ellipses show 95% confidence areas. B) PCoA biplot of soil properties that best explained variation in community structure. Correlations were calculated using BIOENV on average data of each sampled site (Mantel r = 0.6107, p≤0.05). Circles represent samples from "La Estrella", crosses represent "Criadero Klein" samples and triangles represent "La Negrita" samples.

differences were organic matter, clay and silt content, nitrates, phosphorus and pH (Mantel r = 0.6107, p≤0.05).The PCoA biplot (Figure 1B) showed that organic matter, phosphorus and nitrate levels correlated with the first ordination axis that discriminates between cultivated and non cultivated soils. These three properties were higher in soils under no land use.

Regarding the taxonomic analyses, we observed that members of phyla *Verrucomicrobia, Planctomycetes, Actinobacteria* and *Chloroflexi* were more abundant in non cultivated soils (p≤0.05) (Figure 2). On the other hand, we found that sequences related to *Gemmatimonadetes*, candidate division WS3 and *Nitrospirae* were enriched in cultivated soils (p≤0.05) (Figure 2). No significant differences were found for *Proteobacteria* (for non of the Clases), *Acidobacteria* and *Bacteroidetes* phyla. The ten mentioned taxa represent on average 95% of total sequences of each sample.

## Microbiome metagenomic profile changes related to agricultural land use

We found that cultivated and non cultivated soils also clustered apart when metagenomic functional categories were used for the analysis. The first two components of the PCoA explained over 60% of the variability between samples (Figure 3). Standard deviation ellipses overlapped in the ordination plot, indicating that some features are shared between metagenomes. Still, a positive correlation was observed between metabolic and weighted-Unifrac distance matrices (Mantel r: 0.5036 p≤0.05). The analyses of individual COG categories revealed that the relative abundances of COG categories associated with transcription, protein modification, nucleotide transport and metabolism, wall and membrane biogenesis and intracellular trafficking and secretion were higher in cultivated soils (Figure 4, p≤0.05). A deeper analysis inside COG categories revealed that COGs related to Coenzyme A and acetyl-Coa metabolism, energy storage and starvation or quiescence such as, pantothenate kinase, phospho-transacetylase, and trehalose utilization protein were more abundant in non cultivated than in cultivated soils (Figure S2, p≤0.05). On the other hand, COGs related to rapid regulation systems, tricarboxylic acid cycle and nitrogen assimilation such as urease, citrate synthase, glutamate synthase, fumarate hydratase, S-adenosyl- homocysteine hydrolase, S-adenosyl-methionine synthetase, cobalamin biosynthesis protein and ABC transporters, were more abundant in cultivated soils (Figure S2, p≤0.05).

## Microbiome community structure and composition related to conventional tillage and no-tillage systems.

To compare the structure of microbiomes under different tillage systems, we collected samples from an experimental field located in Balcarce in the Southern Pampas. The 34-year-old experiment compared two tillage systems: no-tillage (NT) and conventional tillage (CT). Weighted Unifrac analysis showed differences in community structure associated to the tillage system employed

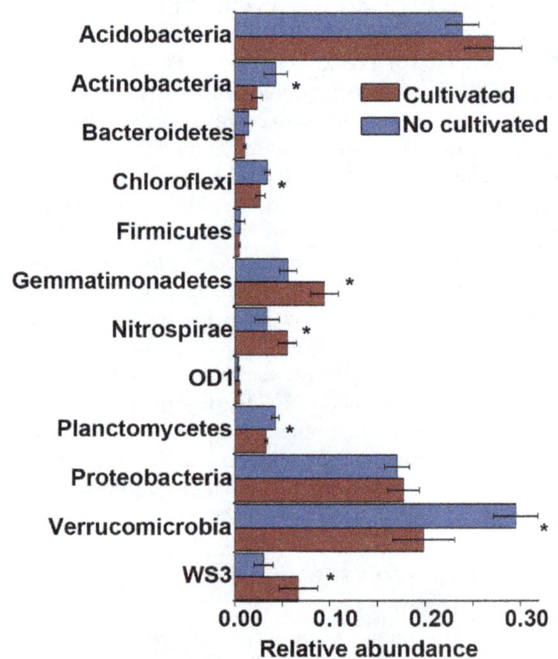


**Figure 2. Relative abundances of taxonomic groups in Pampa production field soil microbiomes.** Bars represent ± 1 standard error. (*) indicate significant differences (p≤0.05).

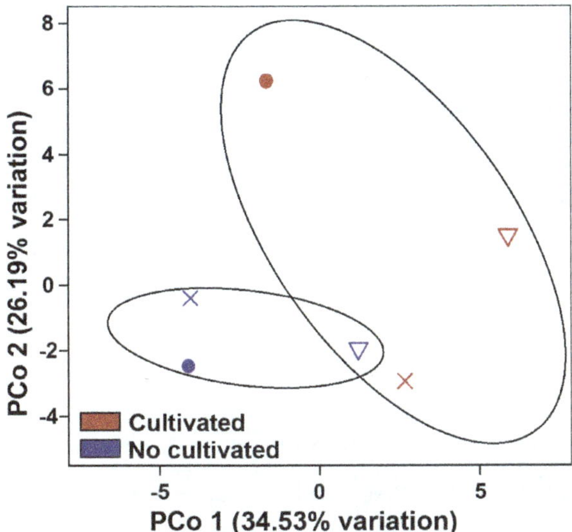


**Figure 3. PCoA plot of metagenomic data based on Euclidean distance matrices of COG categories of Pampa production field soil microbiomes.** Standard error ellipses show 95% confidence areas. Circles represent samples from "La Estrella", crosses represent "Criadero Klein" samples and triangles represent "La Negrita" samples.

(ANOSIM $R = 0.9009$, $p < 0.05$, Figure 5A). The first axis explained 31.56% of total variation and separated NT from CT. Additionally, we showed that nitrates were the only soil variable (Table S1) that significantly correlated with community structure (BIOENV analysis, Mantel $r = 0.7721$, $p \leq 0.01$, Figure 5B).

Moreover, microbiomes of CT and NT soils also differed in taxonomic composition. Members of *Acidobacteria*, *Gemmatimonadetes*, candidate division TM7 and class *Gammaproteobacteria* were more abundant in CT soils, while *Nitrospirae*, candidate divisionWS3 and *Deltaproteobacteria* were more represented in NT soils (Figure 6, $p \leq 0.05$).

## Microbiome metabolic profiles related to conventional tillage and no- tillage systems

Variation in metagenomic profiles between CT and NT microbiomes was analyzed with PCoA based on Euclidean distance matrices of COG abundances. We could not find significant differences in overall profile metabolic structure between tillage systems (Figure S3). Additionally, we did not find significant correlation between Euclidean metabolic matrices and phylogenetic matrices. However, categories related to intracellular trafficking and secretion, amino acid transport and metabolism, and energy production and conversion were shown to be more abundant in soil under CT (Figure 7).

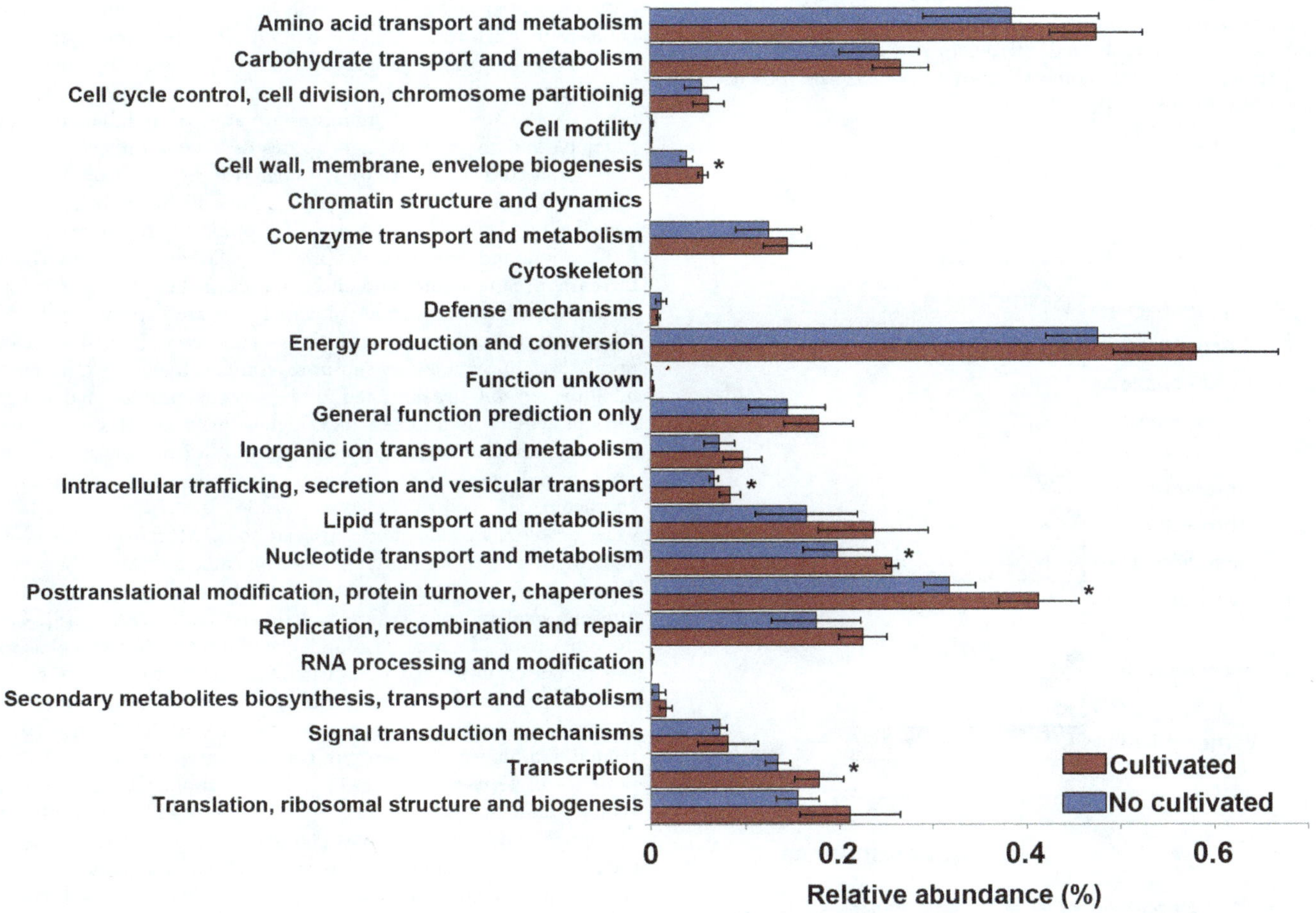

**Figure 4. Relative abundances of COG categories in Pampas production field soil microbiomes.** Bars represent ±1 standard error. (*) indicate significant differences ($p \leq 0.05$).

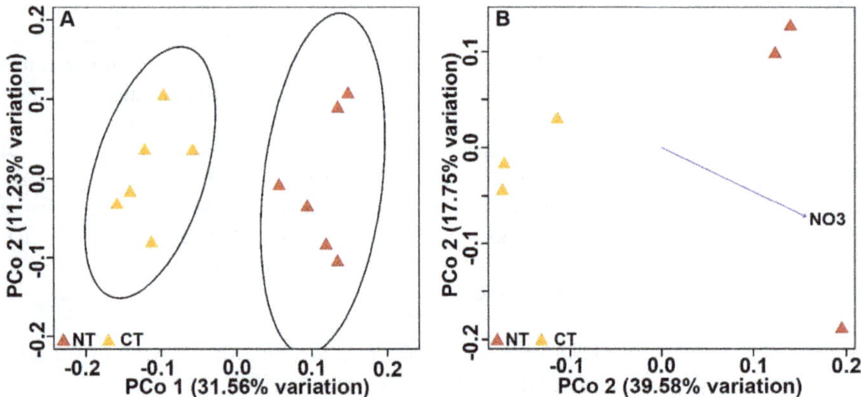


**Figure 5. PCoA plots of Balcarce experimental field soil microbiomes based on 97% similarity Weighted Unifrac distance matrices.** A) CA: conventional tillage; NT: no-tillage. All sub-samples are plotted. Standard error ellipses show 95% confidence areas. B) PCoA biplot, nitrate was the variable that best explained variation in community structure. Correlations were calculated using BIONEV on average data of each experimental plot (Mantel r = 0.7721, p≤0.01).

## Discussion

Much work still needs to be done to get a comprehensive view of the soil microbiomes. Our work is one of the firsts done in the Argentine Pampas at this resolution level, with a combination of metagenomic and phylogentic approaches; and it is aimed to contribute to the comprehension of soil microbiomes function and dynamics. In that context, our results are in agreement with previous works that showed differences in soil microbial community structure and taxonomic composition due to the presence of agricultural land use [9,10,32,33].

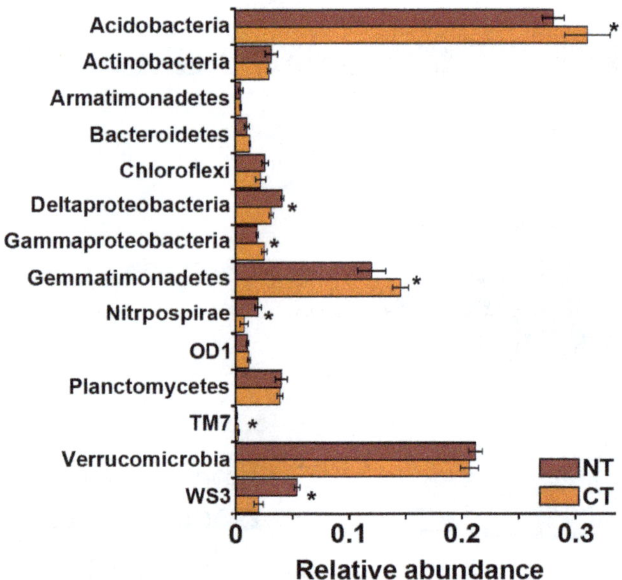


**Figure 6. Comparison of relative abundances of taxonomic groups in NT and CT soils.** CA: conventional tillage; NT: no-tillage. Bars represent ±1 standard error. (*) indicate significant differences (p≤ 0.05).

### Effects of agricultural land use on community composition of soil microbiomes in Argentine Pampas

We observed effects on taxonomic composition at the phylum level. *Verrucomicrobia, Plactomycetes, Actinobacteria* and *Chloroflexi* were more abundant in soils that were never cultivated; while *Gemmatimonadetes, Nitrospirae* and WS3 were more abundant in crop cultivated soils. These results are in agreement with the copiotroph/oligotroph hypothesis [55], that propose that high number of oligotrophic prokaryotes may be found in bogs and soils with high amounts of recalcitrant organic matter [56]. On the other hand, copiotrophic organisms are able to use labile nutrient fractions and to grow at higher rates as a consequnece. In our study, non-cultivated soils are considered to be oligotrophic since they present high levels of organic matter highly rich in humic acids [12] and cultivated soils as copiotrophic environments due to fertilization and the seasonal presence of crop residues, which increases organic matter and nitrogen accessibility [57]. Under this assumption, bacteria in non-cultivated soils are expected to be K selected and to present low growth rates and very efficient nutrient uptake systems with higher substrate affinities. In contrast, bacteria in cultivated soils are expected to be r-selected and to have higher rates of activity per biomass unit, higher turnover rates and faster growth rates. Our results showed a trend toward these statements since a reduction in the abundance of taxa with oligotrophic characteristics, such as *Verrucomicrobia* [34], and *Planctomycetes* [58] were detected in cultivated fertilized soils. Moreover, this is in agreement with recent findings that confirm a correlation between *Verrucomicrobia* abundance patterns and conditions of limited nutrient availability in Prairie Soils in the United States [9]. On the other hand, the relative abundance of phylum *Gemmatimonadetes* was increased in fertilized cultivated soils as previously described for nitrogen-fertilized forest soils [59]. Consistently with our results, these authors observed that nitrogen fertilization was related to a higher abundance of *Gemmatimonadetes* and detected no presence of *Verrucomicrobia*. Little is known about *Gemmatimonadetes* ecology and metabolism since only one representative from this phylum has been isolated and characterized [60]. Even though, their presence in environments with a wide range of nutrient concentrations and redox states suggests versatile metabolisms [61].

We can also say that cultivated soils are more heterogeneous environments than non cultivated soils. Crop rotation, periodic fertilization and pesticide application generate temporal and

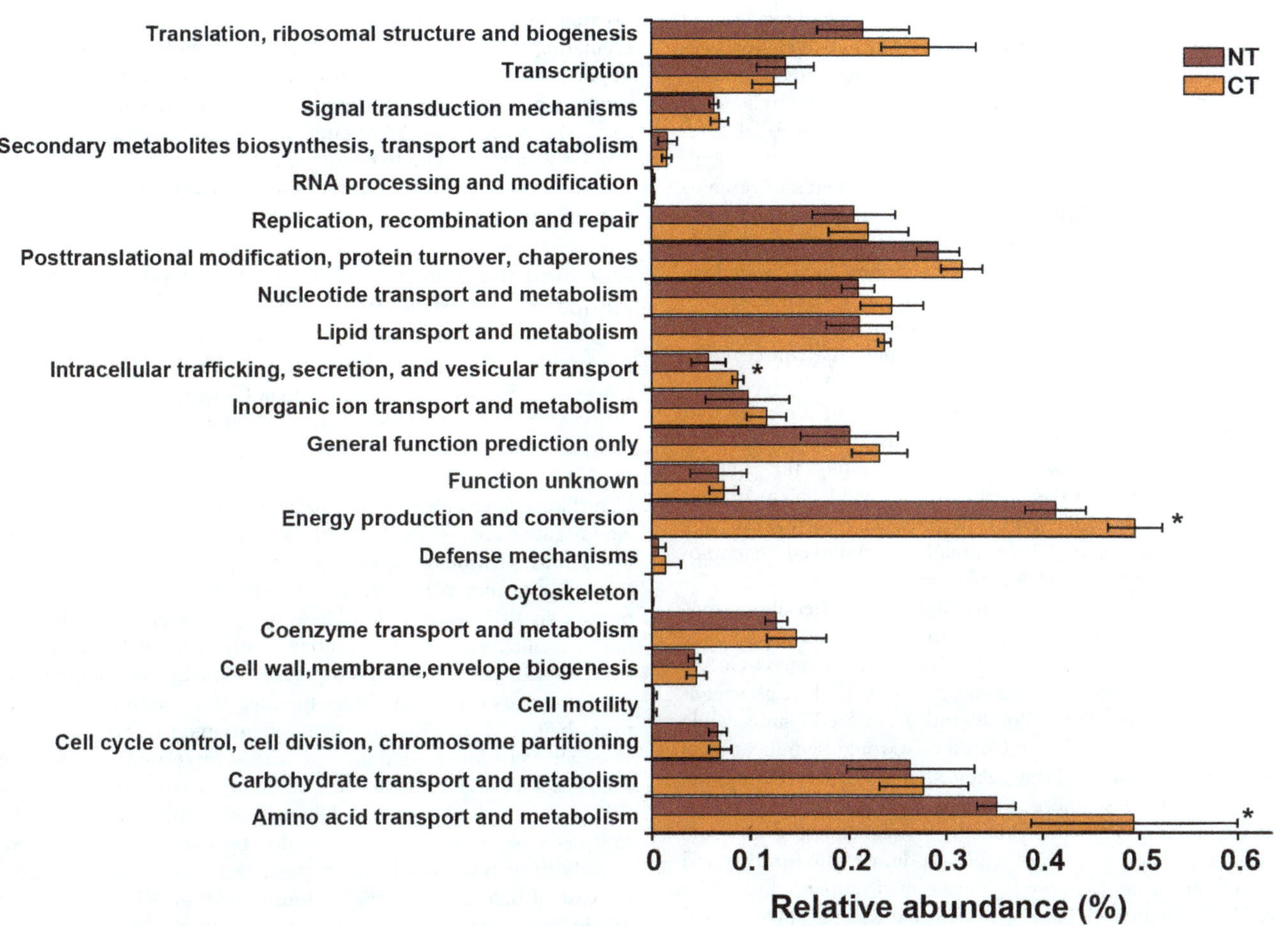


**Figure 7. Relative abundances of COG categories in soil microbiomes of tillage systems comparison experiment in Balcarce.** CA: conventional tillage; NT: no-tillage. Bars represent ±1 standard erro. (*)indicate significant differences ($p \leq 0.05$).

spatial changes in soil chemical properties, and therefore in nutrient availability and approachability for microorganisms. Bacteria dominating in this type of soils should be adapted to heterogeneity. These microorganisms should be able to fine-tune carbon and nitrogen intakes according to their metabolic needs under frequent external changes. This seems to be the case for *Gemmatimonadetes* suggesting a generalist ecological strategy. The high abundance of *Nitrospirae* may also respond to heterogeneous conditions. It has been proposed that *Nitrospirae* lineages occupy different positions on an imaginary scale reaching from K- to r-strategies [62]. We hypothesize that in cultivated soils, *Nitrospirae* K-strategists would be exploiting nitrite in high N microenvironments, while r-strategist would be mining low concentration areas in a nitrogen gradient enhanced by fertilization [63]. This competition would be less fierce in non cultivated soils due to the highest homogeneity and a less marked nitrogen gradient due to the lack of fertilization and more recalcitrant organic matter forms.

## Effects of agricultural land use on metabolic profiles of soil microbiomes in Argentine Pampas

Our results from shotgun metagenomic data also indicated a tendency for the microbiomes of cultivated soils towards adaptation to nutrient heterogeneity. The highest relative abundances of sequences assigned to COGs related to transcription, protein modifications, nucleotide transport and metabolism, wall and membrane biogenesis and intracellular trafficking and secretion in cultivated fertilized soils are consistent with a copiotrophic strategy (i.e. rapid tight metabolism regulation and fast grow rates).Moreover, some of these COG categories were previously shown to be up-represented in copiotrophic marine microorganism genomes [58]. In a deeper look we detected that the relative abundance of sequences assigned to riboswitch regulated genes was higher in cultivated soils (i.e. cobalamin biosynthesis protein, S-adenosyl-methionine synthetase, S-adenosyl- homocysteine hydrolase) [64–66]. These ancient regulators may be playing a central role in a life-style strategy adapted to nutrient heterogeneity since they were described to be the most 'economical' and fast-reacting regulatory systems (no intermediate factors involved) [67]. Moreover, the abundant riboswitch- COGs in cultivated soils were related to synthesis of B vitamins. The levels of these vitamins have already been linked to differences in community composition in marine ecosystems [68]. More studies are needed in order to address this relationship in soil environments; still, our results suggest an important role of B-vitamins in fertilized heterogeneous soils.

Another interesting observation was the highest abundance of glutamate synthase (GOGAT) related COGs in cultivated soils metagenomic profiles. The combined role of GOGAT with glutamine synthetase and glutamate dehydrogenase allows the cells to sense ammonia external levels [69]. The high abundance of this regulation system detected in cultivated soil metagenomes suggests its importance in the detection of N fluxes related with fertilization. Moreover COGs related to the tricarboxylic acid cycle (TCA), such as citrate synthase and succinate dehydrogenase, were more abundant in cultivated soil microbiomes. GOGAT nitrogen assimilation pathway and TCA are related. It has been shown that the concentration of compounds of both pathways changes considerably and rapidly upon nitrogen up shift; in contrast, the concentrations of glycolytic intermediates remains homeostatic [70]. In addition, abundance of urease related COGs were also higher in cultivated soils. The soils sampled in this study have a long history of agricultural land use and have been long fertilized with both type of nitrogen (i.e. ammonia-based and urea-based fertilizers). This kind of environmental pressure finally selected a microbiome adapted to changing N sources and availability.

It is important to mention that these results do not infer that expression or activity of these metabolisms will be necessarily increased in cultivated soil microbiomes, still, the highest abundance of these COGs could be reflected in a highest diversification and specialization. The presence of a highest copy number of strategic genes have already been linked to copiotrophic or oligotrophic life-styles [58].

Microbiomes of non cultivated soils showed a higher abundance of sequences related to Coenzyme A and acetyl-Coa metabolism than microbiomes of cultivated soils. It is known that acetyl-CoA is a fundamental building block and energy source [71]. High acetyl-CoA levels would indicate a "proliferative" or "fed" state, while low acetyl-CoA levels (and high CoA levels) would be indicative of a "quiescent" or "starved" state. Pantothenate kinase (PanK), the key enzyme in CoA syntehsis, was also highest in non cultivated soil microbiomes [72]. In addition, it was stated that some oligotrophs preferentially use lipids as immediate and stored sources of carbon and energy in marine environments [58]. The observed abundance of Pank genes may ensure a correct CoA intracellular level in fasted moments, allowing proper lipid utilization. In addition, non cultivated metagenomic profiles showed higher abundance of sequences related to trehalose utilization. Trehalose is known to serve as energy source in many microorganisms [73]. Members of *Actinobacteria* genera *Mycobacterium* and family *Frankiaceae* are known to produce and/or utilize trehalose [74–76]. As mentioned above, our results showed higher abundance of *Actinobacteria*-related sequences in non cultivated soils. Moreover, sequences classified within family *Frankiaceae* were only present in these soils and the abundance of sequences assigned as *Mycobacterium* was higher than in cultivated soils (not shown). These observations are congruent with an oligotrophic strategy based on the use of storage components for carbon and energy sources.

The responses of the phylogenetic structure and metagenomic profile to agronomic land use were significantly correlated, suggesting some degree of correspondence between these different microbiome features. These results agree with previous observations done in soils from different biomes [9,77]. Moreover, Fierer et al. found similar a similar correlation between metagenomic and phylogenetic data for microbial communities of agricultural soils under different nitrogen gradients in experimental plots [10].

### Correlation with soil properties

In addition, soil properties such as organic matter, phosphorus and nitrate levels explained most of the variability observed between cultivated and non cultivated soils. As mentioned, the highest amount of organic matter is found mostly in a recalcitrant form in these soils. Moreover, it was already established that nitrate accumulation exhibits a negative correlation with organic carbon availability [78]. In addition it has been proved that organic matter source and quality played an important role in regulating the magnitude of carbon metabolism and could be as important as nutrient abundance in water environments [79]. Our results are congruent with these observations since non cultivated soils, with highest levels of organic carbon, presented metagenomic profiles with tendencies to oligotrophy and the best explanation for this scenario is the low lability of the recalcitrant forms of carbon and nitrogen.

### Assessing the effects of different tillage systems on Pampas soil microbiomes

Our results also showed differences in the structure and composition of soil microbiomes between no-till and conventional tillage soils. Sequences related to phyla *Gemmatimonadetes*, candidate division TM7 and *Acidobacteria*, were highest in CT soils, while the abundances of *Nitrospirae* and candidate division WS3 were highest in NT soils. Tillage is the principal agent producing soil disturbance and subsequent soil structure modification [80,81]. The negative effects of CT in soil stabilization and macroaggregate losses were previously registered in Pampas soils [82]. It has also been shown that NT increases macroaggregate abundance and organic matter content [83].These increments are related with a more recalcitrant organic matter, with increased humic acid contents and nutrient retention [84–86]. The higher abundance of *Nitrospirae* in NT soils reinforces the idea of a community adapted to a better N mining in these environments. Moreover, we observed very low abundance of *Nitrobacter* related sequences in Balcarce soils and no significant difference was observed between CT and NT soils (not shown). Changes in ammonia oxidation due to a decrease in ammonia availability by humic substances have already been proposed in microcosm experiments [87]. If this is the case of Balcarce soils, higher humic acids in NT soils would be decreasing ammonia availability for oxidation into nitrite and therefore decreasing nitrate availability, compared to CT soils. The predicted highest stability of organic matter and abundance of macroaggregates in NT soils are probably generating more marked nitrite gradients than in CT soils. The highest abundance of *Nitrospirae* in NT may be reflecting a highest lineage diversity that would be better adapted to these gradients.

In addition, communities under NT showed higher abundance of sequences related to the order *Syntrophobacterales* (*Deltaproteobacteria*, Figure S4). These are known anaerobic and syntrophic organisms [88,89]. These characteristics may be an advantage in stable soils with higher number of highly humic macro aggregates since syntrophy is known to be important for community functioning in micro-environments with low nutrient levels [90]. On the other hand, CT soils presented higher relative abundance of *Gammaproteobacteria* related sequences. The order *Xanthomonadales* was the main responsible for these differences (Figure S4). This taxon has already been associated to CT practices [32].

Even though we could not find significant differences associated to tillage systems at the community structure level for metabolic profiles, COG categories related to intracellular trafficking and secretion, amino acid transport and metabolism and energy production and conversion were more abundant in CT mirobiomes. These results suggest a tendency of CT microbiomes to a more copiotrophic life-style strategy than NT microbiomes.

## Conclusion

Our results are consistent with the hypothesis that microbiomes exhibit different life- history and trophic strategies in Pampean soils under different land uses and tillage systems. Our data suggest that microbiomes of fertilized cultivated soils have more flexible metabolisms adapted to nutrient fluxes with tendencies to copiothropy while microorganisms in non cultivated soils are better adapted to lowest external nutrient availability and homogenous environment. The lowest nutrient accessibility in non cultivated soils may be explained by the higher amount of humic substances, recalcitrant organic matter and the lack of fertilizer amendments. Moreover, NT soils, with most stable structure and highest macroaggregate abundance, presented microbiomes better adapted to recalcitrant environments; while CT microbiomes presented a higher tendency to copiotrophy.

This work is of major contribution to understand how historical changes in soil properties due to agronomical land use have altered the diversity and function of below-ground communities. The importance of high-throughput characterization for the reconstruction of pre- agricultural microbiomes is being reinforced nowadays [9]. Following this direction, our findings will be very useful in future restoration and monitoring programs of Argentine Pampas ecosystems.

## Supporting Information

**Figure S1 PCoA plots of Pampa production field soil microbiomes based on average Bray Curtis distance matrices.** A) PCoA of cultivated and non cultivated soil microbiomes. Standard error ellipses show 95% confidence areas. B) PCoA biplot of soil properties that best explained variation in community structure. Correlations were calculated using BIOENV on average data of each sampled site (Mantel $r = 0.6214$, $p \leq 0.05$). Circles represent samples from "La Estrella", crosses represent "Criadero Klein" samples and triangles represent "La Negrita" samples.

**Figure S2 Relative abundances of Cluster of Orthologous groups (COGs) in Pampa production field soil microbiomes.** Bars represent ± 1 standard error. Only significant COGs are showed.

**Figure S3 Comparison of tillage systems effects on the structure of metabolic profiles. PCoA plot based on Euclidean distance matrices.** CA: conventional tillage; NT: no-tillage. Standard error ellipses show 95% confidence areas.

**Figure S4 Relative abundances of reads assigned to orders within classes *Gammaproteobacteria* and *Deltaporteobacteria* in NT and CT soils.**

## Acknowledgments

We acknowledge Germán F. Domínguez and Guillermo A. Studdert from the "Manejo Sustentable del suelo" research group (from Facultad de Ciencias Agrarias, Universidad Nacional de Mar del Plata, Unidad Integrada Balcarce) for providing samples of the tillage systems experiment. We thank Gonzalo Berhongaray, Josefina de Paepe, and María Rosa Mendoza for his help during sampling in the Rolling Pampas. We also acknowledge Roxana Colombo and Marcelo Soria for helpful discussion and advice.

## Author Contributions

Conceived and designed the experiments: BC NR RÁ AM MV. Performed the experiments: BC NR. Analyzed the data: BC. Contributed reagents/ materials/analysis tools: BC NR RÁ AM MV. Wrote the paper: BC NR MV. Obtained permission for sampling: RÁ.

## References

1. Foley JA, Ramankutty N, Brauman KA, Cassidy ES, Gerber JS, et al. (2011) Solutions for a cultivated planet. Nature 478: 337–342.
2. Tilman D, Balzer C, Hill J, Befort BL (2011) Global food demand and the sustainable intensification of agriculture. Proc Natl Acad Sci U S A 108: 20260–20264.
3. Tilman D, Cassman KG, Matson PA, Naylor R, Polasky S (2002) Agricultural sustainability and intensive production practices. Nature 418: 671–677.
4. Balmford A, Green RE, Scharlemann JPW (2005) Sparing land for nature: exploring the potential impact of changes in agricultural yield on the area needed for crop production. Global Change Biology 11: 1594–1605.
5. Power AG (2010) Ecosystem services and agriculture: tradeoffs and synergies. Philos Trans R Soc Lond B Biol Sci 365: 2959–2971.
6. Foley JA, Defries R, Asner GP, Barford C, Bonan G, et al. (2005) Global consequences of land use. Science 309: 570–574.
7. Bennett EM, Carpenter SR, Caraco NF (2001) Human Impact on Erodable Phosphorus and Eutrophication: A Global Perspective: Increasing accumulation of phosphorus in soil threatens rivers, lakes, and coastal oceans with eutrophication. BioScience 51: 227–234.
8. Scholes MC, Scholes RJ (2013) Dust Unto Dust. Science 342: 565–566.
9. Fierer N, Ladau J, Clemente JC, Leff JW, Owens SM, et al. (2013) Reconstructing the Microbial Diversity and Function of Pre-Agricultural Tallgrass Prairie Soils in the United States. Science 342: 621–624.
10. Fierer N, Lauber CL, Ramirez KS, Zaneveld J, Bradford MA, et al. (2012) Comparative metagenomic, phylogenetic and physiological analyses of soil microbial communities across nitrogen gradients. ISME J 6: 1007–1017.
11. Satorre EH, Slafer GA (1999) Wheat: Ecology and Physiology of Yield Determination: Taylor & Francis.
12. Hall AJ, Rebella CM, Ghersa CM, Culot JP (1992) Field crop systems of the Pampas. In: Pearson CJ, editor. Field Crop Ecosystems of the World. Amsterdam: Elsevier Science & Technology Books. pp. 413–450.
13. Alvarez R, Díaz RA, Barbero N, Santanatoglia OJ, Blotta L (1995) Soil organic carbon, microbial biomass and CO2-C production from three tillage systems. Soil and Tillage Research 33: 17–28.
14. Bernardos JN, Viglizzo EF, Jouvet V, Lértora FA, Pordomingo AJ, et al. (2001) The use of EPIC model to study the agroecological change during 93 years of farming transformation in the Argentine pampas. Agricultural Systems 69: 215–234.
15. Alvarez R (2001) Estimation of carbon losses by cultivation from soils of the Argentine Pampa using the Century Model. Soil Use and Management 17: 62–66.
16. Hevia GG, Buschiazzo DE, Hepper EN, Urioste AM, Antón EL (2003) Organic matter in size fractions of soils of the semiarid Argentina. Effects of climate, soil texture and management. Geoderma 116: 265–277.
17. Quiroga AR, Buschiazzo DE, Peinemann N (1996) Soil Organic Matter Particle Size Fractions in Soils of the Semiarid Argentinian Pampas. Soil Science 161.
18. Kassam A, Friedrich T, Shaxson F, Pretty J (2009) The spread of Conservation Agriculture: justification, sustainability and uptake. International Journal of Agricultural Sustainability 7: 292–320.
19. Alvarez R, Steinbach HS (2009) A review of the effects of tillage systems on some soil physical properties, water content, nitrate availability and crops yield in the Argentine Pampas. Soil and Tillage Research 104: 1–15.
20. Gomez E, Bisaro V, Conti M (2000) Potential C-source utilization patterns of bacterial communities as influenced by clearing and land use in a vertic soil of Argentina. Applied Soil Ecology 15: 273–281.
21. Aon MA, Cabello MN, Sarena DE, Colaneri AC, Franco MG, et al. (2001) I. Spatio-temporal patterns of soil microbial and enzymatic activities in an agricultural soil. Applied Soil Ecology 18: 239–254.
22. Agaras B, Wall LG, Valverde C (2012) Specific enumeration and analysis of the community structure of culturable pseudomonads in agricultural soils under no-till management in Argentina. Applied Soil Ecology 61: 305–319.
23. Nievas F, Bogino P, Nocelli N, Giordano W (2012) Genotypic analysis of isolated peanut-nodulating rhizobial strains reveals differences among populations obtained from soils with different cropping histories. Applied Soil Ecology 53: 74–82.
24. Zabaloy MC, Garland JL, Gómez MA (2008) An integrated approach to evaluate the impacts of the herbicides glyphosate, 2,4-D and metsulfuron-methyl on soil microbial communities in the Pampas region, Argentina. Applied Soil Ecology 40: 1–12.
25. Zabaloy MC, Gómez E, Garland JL, Gómez MA (2012) Assessment of microbial community function and structure in soil microcosms exposed to glyphosate. Applied Soil Ecology 61: 333–339.

26. Deng W, Xi D, Mao H, Wanapat M (2008) The use of molecular techniques based on ribosomal RNA and DNA for rumen microbial ecosystem studies: a review. Molecular Biology Reports 35: 265–274.
27. Pontes D, Lima-Bittencourt C, Chartone-Souza E, Amaral Nascimento A (2007) Molecular approaches: advantages and artifacts in assessing bacterial diversity. Journal of Industrial Microbiology & Biotechnology 34: 463–473.
28. Sogin ML, Morrison HG, Huber JA, Welch DM, Huse SM, et al. (2006) Microbial diversity in the deep sea and the underexplored "rare biosphere". Proceedings of the National Academy of Sciences 103: 12115–12120.
29. Fortunato CS, Herfort L, Zuber P, Baptista AM, Crump BC (2012) Spatial variability overwhelms seasonal patterns in bacterioplankton communities across a river to ocean gradient. ISME J 6: 554–563.
30. Delmont TO, Prestat E, Keegan KP, Faubladier M, Robe P, et al. (2012) Structure, fluctuation and magnitude of a natural grassland soil metagenome. ISME J 6: 1677–1687.
31. Figuerola ELM, Guerrero LD, Rosa SM, Simonetti L, Duval ME, et al. (2012) Bacterial Indicator of Agricultural Management for Soil under No-Till Crop Production. PLoS ONE 7: e51075.
32. Souza RC, Cantão ME, Vasconcelos ATR, Nogueira MA, Hungria M (2013) Soil metagenomics reveals differences under conventional and no-tillage with crop rotation or succession. Applied Soil Ecology 72: 49–61.
33. Lauber CL, Ramirez KS, Aanderud Z, Lennon J, Fierer N (2013) Temporal variability in soil microbial communities across land-use types. ISME J 7: 1641–1650.
34. Ramirez KS, Craine JM, Fierer N (2012) Consistent effects of nitrogen amendments on soil microbial communities and processes across biomes. Global Change Biology 18: 1918–1927.
35. Ramirez K, Lauber CL, Knight R, Bradford MF (2010) Consistent effects of nitrogen fertilization on soil bacterial communities in contrasting systems. Ecology: 3463–3470
36. Lavado RS (2008) La Región Pampeana: historia, características y uso de sus suelos. In: Alvarez R, editor. Materia Orgánica Valor agronómico y dinámica en suelos pampeanos. Buenos Aires: Editorial Facultad de Ingenieria. pp. 1–11.
37. Berhongaray G, Alvarez R, De Paepe J, Caride C, Cantet R (2013) Land use effects on soil carbon in the Argentine Pampas. Geoderma 192: 97–110.
38. Rascovan N, Carbonetto B, Revale S, Reinert M, Alvarez R, et al. (2013) The PAMPA datasets: a metagenomic survey of microbial communities in Argentinean pampean soils. Microbiome 1: 21.
39. Nelson DW, Sommers LE, Sparks DLE, Page ALE, Helmke PAE, et al. (1996) Total Carbon, Organic Carbon, and Organic Matter. Methods of Soil Analysis Part 3-Chemical Methods: Soil Science Society of America, American Society of Agronomy. pp. 961–1010.
40. Bremner JME, Sparks DLE, Page ALE, Helmke PAE, Loeppert RH (1996) Nitrogen-Total. Methods of Soil Analysis Part 3-Chemical Methods: Soil Science Society of America, American Society of Agronomy. pp. 1085–1121.
41. Kuo SE, Sparks DLE, Page ALE, Helmke PAE, H LR (1996) Phosphorus. Methods of Soil Analysis Part 3-Chemical Methods: Soil Science Society of America, American Society of Agronomy. pp. 869–919.
42. Rhoades JDE, Sparks DLE, Page ALE, Helmke PAE, H. LR (1996) Salinity: Electrical Conductivity and Total Dissolved Solids. Methods of Soil Analysis Part 3-Chemical Methods: Soil Science Society of America, American Society of Agronomy. pp. 417–435.
43. Gee GW, Bauder JWEKA (1986) Particle-size Analysis. Methods of Soil Analysis: Part 1—Physical and Mineralogical Methods: Soil Science Society of America, American Society of Agronomy. pp. 383–411.
44. Cole J, Wang Q, Cardenas E, Fish J, Chai B, et al. (2009) The Ribosomal Database Project: improved alignments and new tools for rRNA analysis. Nucleic acids research: 141–145.
45. Caporaso JG, Kuczynski J, Stombaugh J, Bittinger K, Bushman FD, et al. (2010) QIIME allows analysis of high-throughput community sequencing data. Nat Meth 7: 335–336.
46. Edgar RC (2010) Search and clustering orders of magnitude faster than BLAST. Bioinformatics 26: 2460–2461.
47. Caporaso JG, Bittinger K, Bushman FD, DeSantis TZ, Andersen GL, et al. (2010) PyNAST: a flexible tool for aligning sequences to a template alignment. Bioinformatics 26: 266–267.
48. Price MN, Dehal PS, Arkin AP (2009) FastTree: Computing Large Minimum Evolution Trees with Profiles instead of a Distance Matrix. Molecular Biology and Evolution 26: 1641–1650.
49. Lozupone C, Knight R (2005) UniFrac: a New Phylogenetic Method for Comparing Microbial Communities. Applied and Environmental Microbiology 71: 8228–8235.
50. DeSantis TZ, Hugenholtz P, Larsen N, Rojas M, Brodie EL, et al. (2006) Greengenes, a Chimera-Checked 16S rRNA Gene Database and Workbench Compatible with ARB. Applied and Environmental Microbiology 72: 5069–5072.
51. Wang Q, Garrity GM, Tiedje JM, Cole JR (2007) Naïve Bayesian Classifier for Rapid Assignment of rRNA Sequences into the New Bacterial Taxonomy. Applied and Environmental Microbiology 73: 5261–5267.
52. Kindt R, Coe R (2005) Tree Diversity Analysis: A Manual and Software for Common Statistical Methods for Ecological and Biodiversity Studies: World Agrofirestry Centre.
53. Dixon P (2003) VEGAN, a package of R functions for community ecology. Journal of Vegetation Science 14: 927–930.
54. Meyer F, Paarmann D, D'Souza M, Olson R, Glass EM, et al. (2008) The metagenomics RAST server - a public resource for the automatic phylogenetic and functional analysis of metagenomes. BMC Bioinformatics 9: 386.
55. Fierer N, Bradford MA, Jackson RB (2007) Toward an ecological classification of soil bacteria. Ecology 88: 1354–1364.
56. Dion P, Nautiyal CS (2008) Microbiology of Extreme Soils: Springer.
57. Galantini J, Rosell R (2006) Long-term fertilization effects on soil organic matter quality and dynamics under different production systems in semiarid Pampean soils. Soil and Tillage Research 87: 72–79.
58. Lauro FM, McDougald D, Thomas T, Williams TJ, Egan S, et al. (2009) The genomic basis of trophic strategy in marine bacteria. Proceedings of the National Academy of Sciences 106: 15527–15533.
59. Nemergut DR, Townsend AR, Sattin SR, Freeman KR, Fierer N, et al. (2008) The effects of chronic nitrogen fertilization on alpine tundra soil microbial communities: implications for carbon and nitrogen cycling. Environmental Microbiology 10: 3093–3105.
60. Zhang H, Sekiguchi Y, Hanada S, Hugenholtz P, Kim H, et al. (2003) Gemmatimonas aurantiaca gen. nov., sp. nov., a Gram-negative, aerobic, polyphosphate-accumulating micro-organism, the first cultured representative of the new bacterial phylum Gemmatimonadetes phyl. nov. International Journal of Systematic and Evolutionary Microbiology 53: 1155–1163.
61. DeBruyn JM, Nixon LT, Fawaz MN, Johnson AM, Radosevich M (2011) Global Biogeography and Quantitative Seasonal Dynamics of Gemmatimonadetes in Soil. Applied and Environmental Microbiology 77: 6295–6300.
62. Maixner F, Noguera DR, Anneser B, Stoecker K, Wegl G, et al. (2006) Nitrite concentration influences the population structure of Nitrospira-like bacteria. Environmental Microbiology 8: 1487–1495.
63. Attard E, Poly F, Commeaux C, Laurent F, Terada A, et al. (2010) Shifts between Nitrospira- and Nitrobacter-like nitrite oxidizers underlie the response of soil potential nitrite oxidation to changes in tillage practices. Environmental Microbiology 12: 315–326.
64. Edwards AL, Reyes FE, Héroux A, Batey RT (2010) Structural basis for recognition of S-adenosylhomocysteine by riboswitches. RNA 16: 2144–2155.
65. Loenen WA (2006) S-adenosylmethionine: jack of all trades and master of everything? Biochem Soc Trans 34: 330–333.
66. Winkler WC, Breaker RR (2005) Regulation of bacterial gene expression by riboswitches. Annual Review of Microbiology 59: 487–517.
67. Nudler E, Mironov AS (2004) The riboswitch control of bacterial metabolism. Trends in biochemical sciences 29: 11–17.
68. Sañudo-Wilhelmy SA, Cutter LS, Durazo R, Smail EA, Gómez-Consarnau L, et al. (2012) Multiple B-vitamin depletion in large areas of the coastal ocean. Proceedings of the National Academy of Sciences.
69. Yan D (2007) Protection of the glutamate pool concentration in enteric bacteria. Proceedings of the National Academy of Sciences 104: 9475–9480.
70. Doucette CD, Schwab DJ, Wingreen NS, Rabinowitz JD (2011) α-ketoglutarate coordinates carbon and nitrogen utilization via enzyme I inhibition. Nat Chem Biol 7: 894–901.
71. Cai L, Tu BP (2011) On Acetyl-CoA as a Gauge of Cellular Metabolic State. Cold Spring Harbor Symposia on Quantitative Biology 76: 195–202.
72. Leonardi R, Rehg JE, Rock CO, Jackowski S (2010) Pantothenate Kinase 1 Is Required to Support the Metabolic Transition from the Fed to the Fasted State. PLoS ONE 5: e11107.
73. Elbein AD, Pan YT, Pastuszak I, Carroll D (2003) New insights on trehalose: a multifunctional molecule. Glycobiology 13: 17R–27R.
74. Barabote RD, Xie G, Leu DH, Normand P, Necsulea A, et al. (2009) Complete genome of the cellulolytic thermophile Acidothermus cellulolyticus 11B provides insights into its ecophysiological and evolutionary adaptations. Genome Research 19: 1033–1043.
75. Lopez MF, Fontaine MS, Torrey JG (1984) Levels of trehalose and glycogen in Frankia sp. HFPArI3 (Actinomycetales). Canadian Journal of Microbiology 30: 746–752.
76. Tropis M, Meniche X, Wolf A, Gebhardt H, Strelkov S, et al. (2005) The Crucial Role of Trehalose and Structurally Related Oligosaccharides in the Biosynthesis and Transfer of Mycolic Acids in Corynebacterineae. Journal of Biological Chemistry 280: 26573–26585.
77. Fierer N, Leff JW, Adams BJ, Nielsen UN, Bates ST, et al. (2012) Cross-biome metagenomic analyses of soil microbial communities and their functional attributes. Proceedings of the National Academy of Sciences 109: 21390–21395.
78. Taylor PG, Townsend AR (2010) Stoichiometric control of organic carbon-nitrate relationships from soils to the sea. Nature 464: 1178–1181.
79. Apple JK, del Giorgio PA (2007) Organic substrate quality as the link between bacterioplankton carbon demand and growth efficiency in a temperate salt-marsh estuary. ISME J 1: 729–742.
80. Bayer C, Mielniczuk J, Amado TJC, Martin-Neto L, Fernandes SV (2000) Organic matter storage in a sandy clay loam Acrisol affected by tillage and cropping systems in southern Brazil. Soil and Tillage Research 54: 101–109.
81. Langdale GW, West LT, Bruce RR, Miller WP, Thomas AW (1992) Restoration of eroded soil with conservation tillage. Soil Technology 5: 81–90.
82. Bongiovanni MD, Lobartini JC (2006) Particulate organic matter, carbohydrate, humic acid contents in soil macro- and microaggregates as affected by cultivation. Geoderma 136: 660–665.
83. Plaza-Bonilla D, Cantero-Martínez C, Viñas P, Álvaro-Fuentes J (2013) Soil aggregation and organic carbon protection in a no-tillage chronosequence under Mediterranean conditions. Geoderma 193–194: 76–82.

84. Jiao Y, Whalen JK, Hendershot WH (2006) No-tillage and manure applications increase aggregation and improve nutrient retention in a sandy-loam soil. Geoderma 134: 24–33.
85. Tivet F, de Moraes Sá JC, Lal R, Borszowskei PR, Briedis C, et al. (2013) Soil organic carbon fraction losses upon continuous plow-based tillage and its restoration by diverse biomass-C inputs under no-till in sub-tropical and tropical regions of Brazil. Geoderma 209–210: 214–225.
86. Slepetiene A, Slepetys J (2005) Status of humus in soil under various long-term tillage systems. Geoderma 127: 207–215.
87. Dong L, Córdova-Kreylos AL, Yang J, Yuan H, Scow KM (2009) Humic acids buffer the effects of urea on soil ammonia oxidizers and potential nitrification. Soil Biology and Biochemistry 41: 1612–1621.
88. Sieber JR, McInerney MJ, Gunsalus RP (2012) Genomic Insights into Syntrophy: The Paradigm for Anaerobic Metabolic Cooperation. Annual Review of Microbiology 66: 429–452.
89. McInerney MJ, Sieber JR, Gunsalus RP (2009) Syntrophy in anaerobic global carbon cycles. Current Opinion in Biotechnology 20: 623–632.
90. Kim HJ, Boedicker JQ, Choi JW, Ismagilov RF (2008) Defined spatial structure stabilizes a synthetic multispecies bacterial community. Proceedings of the National Academy of Sciences 105: 18188–18193.

# 20

# Deep 16S rRNA Pyrosequencing Reveals a Bacterial Community Associated with Banana *Fusarium* Wilt Disease Suppression Induced by Bio-Organic Fertilizer Application

**Zongzhuan Shen[1]⁹, Dongsheng Wang[3]⁹, Yunze Ruan[2], Chao Xue[1], Jian Zhang[1], Rong Li[1], Qirong Shen[1]***

**1** National Engineering Research Center for Organic-based Fertilizers, Key Laboratory of Plant Nutrition and Fertilization in Low-Middle Reaches of the Yangtze River, Ministry of Agriculture, Jiangsu Key Lab and Engineering Center for Solid Organic Waste Utilization, Jiangsu Collaborative Innovation Center for Solid Organic Waste Resource Utilization, Nanjing Agricultural University, Nanjing, China, **2** Hainan key Laboratory for Sustainable Utilization of Tropical Bio-resources, College of Agriculture, Hainan University, Haikou, China, **3** Nanjing Institute of Vegetable Science, Nanjing, China

## Abstract

Our previous work demonstrated that application of a bio-organic fertilizer (BIO) to a banana mono-culture orchard with serious *Fusarium* wilt disease effectively decreased the number of soil *Fusarium* sp. and controlled the soil-borne disease. Because bacteria are an abundant and diverse group of soil organisms that responds to soil health, deep 16 S rRNA pyrosequencing was employed to characterize the composition of the bacterial community to investigate how it responded to BIO or the application of other common composts and to explore the potential correlation between bacterial community, BIO application and *Fusarium* wilt disease suppression. After basal quality control, 137,646 sequences and 9,388 operational taxonomic units (OTUs) were obtained from the 15 soil samples. *Proteobacteria, Acidobacteria, Bacteroidetes, Gemmatimonadetes* and *Actinobacteria* were the most frequent phyla and comprised up to 75.3% of the total sequences. Compared to the other soil samples, BIO-treated soil revealed higher abundances of *Gemmatimonadetes* and *Acidobacteria*, while *Bacteroidetes* were found in lower abundance. Meanwhile, on genus level, higher abundances compared to other treatments were observed for *Gemmatimonas* and *Gp4*. Correlation and redundancy analysis showed that the abundance of *Gemmatimonas* and *Sphingomonas* and the soil total nitrogen and ammonium nitrogen content were higher after BIO application, and they were all positively correlated with disease suppression. Cumulatively, the reduced *Fusarium* wilt disease incidence that was seen after BIO was applied for 1-year might be attributed to the general suppression based on a shift within the bacteria soil community, including specific enrichment of *Gemmatimonas* and *Sphingomonas*.

**Editor:** Gabriele Berg, Graz University of Technology (TU Graz), Austria

**Funding:** This work was supported by the National Natural Science Foundation of China (41101231 and 31372142), Natural Science Foundation of Hainan province (313045), the Priority Academic Program Development of Jiangsu Higher Education Institutions (PAPD), 111 project (B12009), the Agricultural Ministry of China (201103004), the Innovative Research Team Development Plan of the Ministry of Education of China (IRT1256), the China Postdoctoral Science Foundation (2011M501248 and 2012T50479), and the (KJ2011007) and The Central Financial Support to the Central and Western Nniversities to Specially Enhance the Comprehensive Strength (ZDZX2013023). The funders had no role in study design, data collection and analysis, decision to publish, or preparation of the manuscript.

**Competing Interests:** The authors have declared that no competing interests exist.

* E-mail: shenqirong@njau.edu.cn

⁹ These authors contributed equally to this work.

## Introduction

Banana *Fusarium* wilt disease, which is caused by *Fusarium oxysporum* f. sp. *cubense* race 4 (FOC) and reported to be the most limiting factor in banana production worldwide, has spread quickly in *Cavendish*-production areas since 1996, and it affects approximately 90% of the banana industry in China [1–3]. Among the managements for controlling the disease, such as crop rotation, biocontrol, application of chemical fungicides and cropping of resistant banana cultivars [4–8], biocontrol is the most promising technique for disease prevention because of owning the advantages of environmental protection, safety, high economic benefits and longevity at the same time [9]. However, direct inoculation of functional microorganisms into the soil without a suitable organic substrate cannot be expected to be successful due to the absence of nutrients [10]. Many reports have demonstrated that biocontrol agents combined with organic materials to create novel bio-organic fertilizers (BIOs) can enhance the suppression of *Fusarium* wilt disease in the soil by ameliorating the structure of the microbial community [11–14].

The composition of the soil microbial community and induced changes caused by its amendment, provide useful information on soil health and quality [15]. Maintaining biodiversity of soil microbes is crucial to soil health because a decrease in soil microbial diversity is responsible for the development of soil-borne diseases [16]. Determining the responses of soil bacterial communities to different organic amendments is particularly important because the bacterial community is one of the main components that determine soil health and is believed to be one of the main drivers in disease suppression [17]. Despite the known key roles of bacteria in soil health and the significant change in soil bacterial composition and activity after BIO application, information regarding the variation of soil bacterial communities that are affected by different organic amendments is still lacking. More importantly, understanding soil microbial community structure shifts following implementation of various organic amendments is an important component when selecting fertilizer types to improve soil function and health.

As described in our previous work, *Fusarium* wilt disease was more effectively controlled by a 1-year application of BIO than by the other composts in a field experiment [12]. In that study, the effects of different types of composts on soil bacterial communities were mainly assessed using traditional PCR-DGGE fingerprinting and culture-dependent methods. Taking into account the large size of the bacterial community and the heterogeneity of the soils, only a tiny fraction of the bacterial diversity was unraveled by that study. Recently, pyrosequencing of 16 S rRNA gene fragments has been applied for in-depth analysis of soil bacterial communities [18,19]. This method could provide a large number of parallel reads to characterize the unseen majority of the soil microbial community and offer an opportunity to achieve a high throughput and deeper insight into the effects of different types of composts on soil bacterial communities [20], thus it is an improvement over previous fingerprinting techniques, such as PCR-DGGE or T-RFLP, which are not entirely specific and do not result in many sequences [15].

We used a deep 16 S rRNA pyrosequencing approach to further investigate how the soil bacteria community responded to the application of BIO or other common composts and to explore the potential correlation between bacterial community, BIO application and *Fusarium* wilt disease suppression. This study was the first to provide information on the banana soil bacterial community in a single soil type that was exposed to different organic amendments using deep 16 S rRNA pyrosequencing. Therefore, the aims of this study were to answer the following questions: (1) Does the soil bacteria community that is amended with BIO differ from that exposed to other common composts? (2) Does the *Fusarium* wilt disease incidence correlate with the bacterial community? (3) Does the disease suppression after BIO application correlate with the physicochemical properties of the soil?

## Materials and Methods

### Ethics statement

Our study was carried out on the farmers' land (18°23′ N, 108°44′ E) with property rights in China (1996-2035) and farmer Yusheng Li should be contacted for future permissions. No specific permits were required for the described field studies and the locations are not protected. The field studied did not involve endangered or protected species.

### Field experiment

Five treatments were established as randomized, complete block designs with three replicates at the "Wan Zhong" banana orchard in Hainan, China and included a general operation control (GCK) and soil that was amended with four different types of organic amendments: bio-organic fertilizer (BIO), cattle manure compost (CM), Chinese medicine residue compost (CMR) and pig manure compost (PM). And each replicate was planted with 170 banana tissue culture plantlets (*Musa acuminate* AAA *Cavendish* cv. Brazil) with an area of 667 $m^2$. Worthy to notify, the bio-organic fertilizer (BIO) contained a biocontrol agent *Bacillus* sp. and was prepared by a solid fermentation method according to Chen et al. [21]. The orchard has been continuously cropped banana for more than 10 years and was abandoned by farmers to growing banana for high *Fusarium* wilt disease incidence (50%). The detailed information regarding the field experiment setting and amendments were described in our previous report [12].

### Soil sample collection and DNA extraction

The soil sample collection and DNA extraction methods were described in detail as supplementary information to our previous study [12]. Five individual, healthy banana trees that were at least 5 m apart in each treatment plot were randomly selected for sample collection, and the collected soil samples from each tree were mixed as a composite soil sample for each replicate plot. For each tree, composite soil from 4 random sites of the trunk base was collected using a 25-mm soil auger at a depth of 20 cm. All soil samples were transported to the laboratory and stored at −70°C for subsequent DNA extraction after sifting through a 2-mm sieve. Total soil DNA was extracted using PowerSoil DNA Isolation Kits (MoBio Laboratories Inc., Carlsbad, USA) according to the manufacturer's protocol. The concentration and quality (ratio of A260/A280) of the DNA were determined using a spectrophotometer (NanoDrop 2000, ThermoScientific, USA).

### Polymerase chain reaction amplification and deep 16 S rRNA pyrosequencing

PCR reactions for each sample were performed in triplicate (including two negative control reactions) with 2 μM of each primer, 0.25 μM of dNTPs, 4 μL of 5 × FastPfu Buffer, 1 U of FastPfu DNA polymerase (2.5 U/μL, TransGen Biotech Co., Ltd., Beijing, China) and approximately 20 ng of soil DNA template at a final volume of 20 μL. The forward primer consisted of the 25-bp 454 adapter A, 2-bp linker A and 15-bp universal bacterial primer 27F [22], and the reverse primer consisted of the 25-bp 454 adapter B, 2-bp linker B, a 10-bp barcode and the 19-bp universal bacterial primer 533R [23]. Detailed information regarding the primer sequence is shown in Table S1. These primers target an approximately 500-bp region of the 16 S rRNA gene that contains variable regions 1 to 3 (V1–V3), which is well-suited for accurate phylogenetic placement of bacterial sequences [24].

Amplifications were performed using an Eppendorf Mastercycler thermocycler (Eppendorf North America, Hauppauge, NY) with the following temperature program: an initial denaturation step of 95°C for 4 min, followed by 25 cycles of denaturation at 95°C for 30 s, annealing at 55°C for 30 s, extension at 72°C for 30 s and a final elongation at 72°C for 5 min. PCR amplicon libraries were purified from a 1.2% agarose gel and quantified using the PicoGreen dsDNA reagent (Promega, USA). Equal amplicons from each sample were then pooled in equimolar concentrations into a single aliquot. After cleaning, precipitating, and re-suspending the amplicons in nuclease-free water, an emPCR was carried out to attach the single strands onto beads for further 454 pyrosequenicng. Pyrosequencing was performed on a Roche 454 GS-FLX Titanium System at Majorbio Bio-pharm Technology Co., Ltd (Shanghai, China).

### Bioinformatic analysis

After pyrosequencing, raw sequences were analyzed using the Mothur software following the Schloss standard operating procedure [25]. Briefly, sequences with a minimum flow length of 450 flows were denoised using the Mothur-based reimplementation of the PyroNoise algorithm with the default parameters [26]. Sequences with more than 1 mismatch to the barcode, 2 mismatches to the primer, any ambiguous base call, homopolymers longer than 8 bases and reads shorter than 250 bp were eliminated, and the filtered sequences were then trimmed and assigned to soil samples based on unique 10-base barcodes. After removing the barcode and primer sequences, the unique sequences were aligned against the Silva bacteria database [27]. After screening, filtering, preclustering, and chimera removal, the retained sequences were used to build a distance matrix with a distance threshold of 0.2. Using the average neighbor algorithm with a cut-off of 97% similarity, bacterial sequences were clustered to operational taxonomic units (OTU), and the representative sequence for each OTU was picked and classified using a Ribosomal Database Project naive Bayesian rRNA classifier with a confidence threshold of 80% [28]. Lastly, the resulting matches for each set of sequence data were summarized at various levels of taxonomic hierarchal structure (e.g., phylum and genera). All raw sequences have been deposited in DDBJ SRA under the accession number DRA001282.

To correct for sampling effects, we used a randomly selected subset of 7,817 sequences per sample to further analyze the richness and diversity of the bacterial community. All analyses were based on the OTU clusters with a cut-off of 3% dissimilarity. The richness index of the Chao1 estimator (Chao1) [29] and the abundance-based Coverage estimator (ACE) [30] was calculated to estimate the number of observed OTUs that were present in the sampling assemblage. The diversity within each individual sample was estimated using the nonparametric Shannon diversity index [31]. Good's nonparametric Coverage estimator was used to estimate the percentage of the total species that were sequenced in each sample [32], and a rarefaction curve generated using the Mothur software was used to compare the relative levels of bacterial OTU diversity across all soil samples.

To compare bacterial community structures across all samples, a heat map based on the abundant phyla were performed in R (Version 3.0.2) with the gplots package [33,34], and principal coordinates analysis (PCoA) based on the OTU composition was performed using the Mothur software. To examine the relationship between the frequencies of abundant phyla, samples and environmental variables, redundancy analysis (RDA) was carried out using CANOCO for Windows [35].

### Statistical analysis

The relationships between the selected taxonomy group (abundant phyla or genera) or bacterial community indices (Chao1, ACE and Shannon) and *Fusarium* wilt disease incidence (DI) were calculated using the SPSS 13.0 software program. For all parameters, data were compared using a one-way analysis of variance (ANOVA) at the end of each bioassay. Mean comparison was performed using Fisher's least significant difference test (LSD) and the Duncan multiple range test with a significance level of $p < 0.05$.

## Results

After filtering the reads based on basal quality control, 137,646 sequences with an average length of 254 bases were obtained from 15 soil samples when using Mothur flowgrams strategy to analyze sequences. The number of high-quality sequences per sample varied from 7,817 to 11,234 (Table 1). Based on 97% species similarity, in total 9,388 OTUs were found, and 12,845 sequences (9.3% of the total sequences) were returned as unclassified.

**Table 1.** Good quality sequences that were used to further analysis after basic quality control for treatments: bio-organic fertilizer (BIO), cattle manure compost (CM), Chinese medicine residue compost (CMR), general operation control (GCK) and pig manure compost (PM).

| Treatments | Good quality sequences |
|---|---|
| BIO1 | 9,382 |
| BIO2 | 9,666 |
| BIO3 | 7,817 |
| CM1 | 9,937 |
| CM2 | 8,521 |
| CM3 | 8,459 |
| CMR1 | 8,736 |
| CMR2 | 9,280 |
| CMR3 | 8,614 |
| GCK1 | 8,192 |
| GCK2 | 8,473 |
| GCK3 | 11,234 |
| PM1 | 8,695 |
| PM2 | 11,185 |
| PM3 | 9,455 |
| Total | 137,646 |

### Bacterial community composition

As shown in Fig. 1, although the phyla compositions of the different soil samples were similar, some obvious variations in the relative abundances of phyla between different fertilizer treatments were still observed. The classified sequences for each sample were affiliated with 19 bacterial phyla, and the remaining sequences were unclassified. The most abundant phyla of *Proteobacteria*, *Acidobacteria* and *Bacteroidetes* were found in all treatments at a relative abundance of approximately 35%, 15% and 10%, respectively, and 9 phyla (*Actinobacteria*, *Gemmatimonadetes*, *Nitrospirae*, *Firmicutes*, *Chloroflexi*, *Verrucomicrobia*, *TM7*, *Armatimonadetes* and *Planctomycetes*) were found in all samples at a relative abundance of higher than 1%, but lower than 6%, with some obvious variations. The relative abundances of *Acidobacteria* and *Gemmatimonadetes* were highest, while those of *Bacteroidetes* were lowest, in the BIO-treated soil sample compared with the other treatments (CM, CMR, GCK and PM).

The most abundant classified genera (>1%) for each treatment are shown in Table 2, which shows 12, 16, 14, 12 and 15 most frequently classified genera for the BIO, CM, CMR, GCK and PM treatments, respectively. Among the most frequent genera, only 10, including *Gemmatimonas*, *Gp1*, *Gp4*, *Gp6*, *Burkholderia*, *Gp3*, *Nitrospira*, *Ohtaekwangia*, *TM7_genus_incertae_sedis* and *3_genus_incertae_sedis* were represented in all treatments. Moreover, in comparison to other treatments, significantly higher abundances of the genera *Gemmatimonas* and *Gp4* were observed in BIO-treated soil among the most 10 abundant genera.

### Bacterial α-diversity

The bacterial richness and diversity of the different fertilizer treatments were calculated based on 7,817 randomly selected sequences (Table 3). The richness index, Chao1 and ACE showed that the CM-treated soil exhibited the lowest number of OTUs,

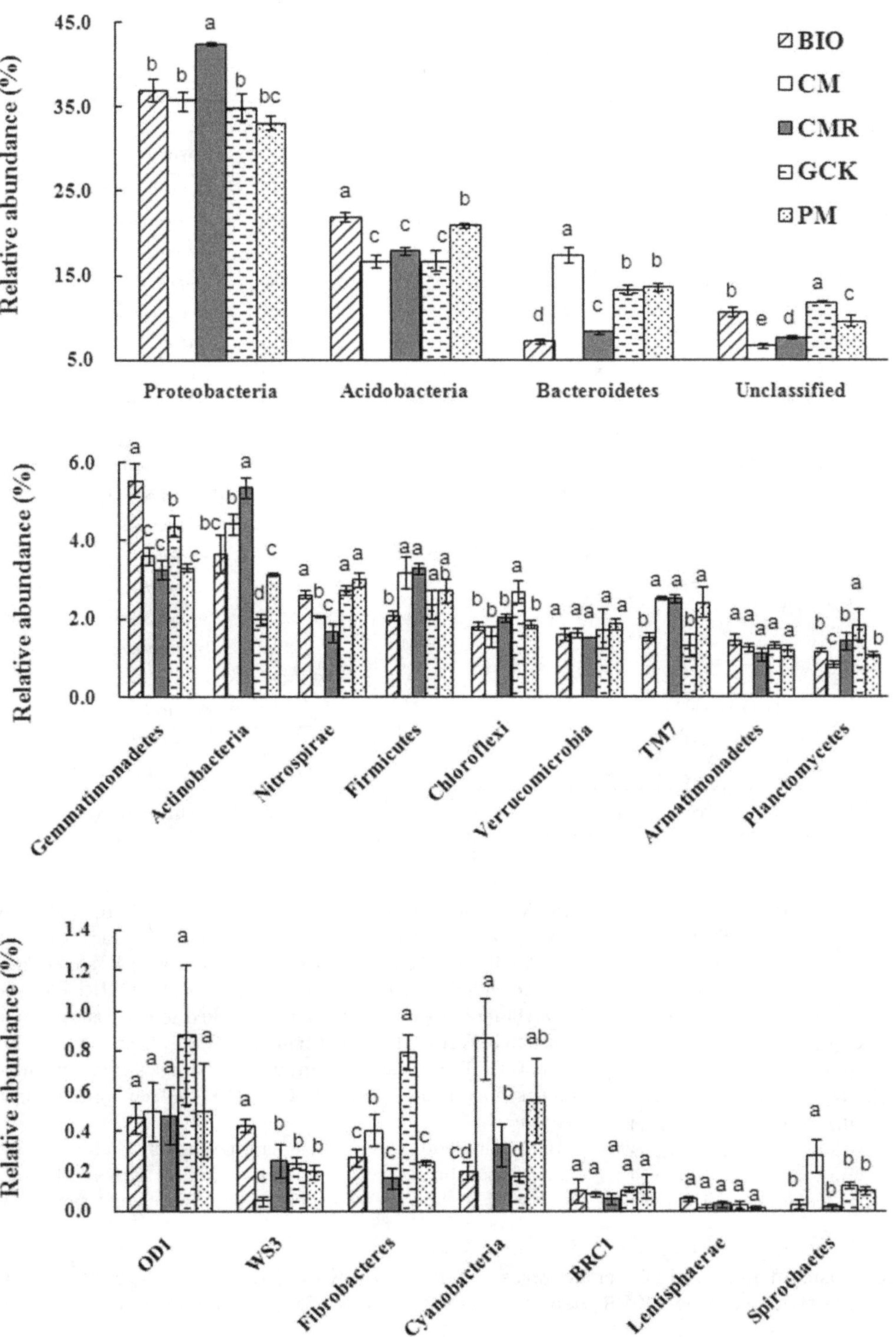


**Figure 1. The relative abundance of the phyla for treatments with bio-organic fertilizer (BIO), cattle manure compost (CM), Chinese medicine residue compost (CMR), general operation control (GCK) and pig manure compost (PM).** Bars represent the standard error of the three replicates and different letters above each phylum indicate significantly difference at 0.05 probability level according to the Duncan test.

while the BIO-treated soil showed the highest number with no significant difference between the CMR, PM and GCK treatments. The CM treatment had the lowest Shannon diversity index value ($H'$), while the highest values were of the GCK and PM treatments. CM treatment showed the highest Good's query Coverage (ranging from 0.87 to 0.90 for all treatments), and no significant difference was observed for the other treatments.

Similar results were observed with 3% dissimilarity after comparing the rarefaction curves of the mean pooled sequences of 3 replicates of each treatment, with the GCK treatment showing the highest OTU number and CM treatment showing the lowest

**Table 2.** Frequency of the most abundant bacterial genera, indicated in % of all classified sequences, within each treatment of bio-organic fertilizer (BIO), cattle manure compost (CM), Chinese medicine residue compost (CMR), general operation control (GCK) and pig manure compost (PM).

| % | BIO | CM | CMR | GCK | PM | Phylum |
|---|---|---|---|---|---|---|
| *Gemmatimonas* | 5.56±0.42a | 3.62±0.22c | 3.27±0.25c | 4.38±0.25b | 3.33±0.10c | *Gemmatimonadetes* |
| *Gp1* | 5.49±0.31c | 6.54±0.17b | 7.07±0.39a | 2.43±0.07d | 6.59±0.75b | *Acidobacteria* |
| *Gp4* | 4.62±0.27a | 2.15±0.08d | 2.19±0.35d | 3.55±0.46b | 2.72±0.14c | *Acidobacteria* |
| *Gp6* | 4.49±0.19a | 1.49±0.07c | 2.38±0.16b | 5.31±1.07a | 2.73±0.60b | *Acidobacteria* |
| *Burkholderia* | 3.76±1.00d | 8.68±0.77b | 10.79±2.02a | 1.46±0.43e | 6.51±1.90c | *Proteobacteria* |
| *Gp3* | 2.90±0.19a | 2.84±0.45a | 2.10±0.23b | 2.35±0.68a | 2.77±0.10a | *Acidobacteria* |
| *Nitrospira* | 2.64±0.10b | 2.07±0.01c | 1.66±0.23d | 2.73±0.12b | 3.01±0.17a | *Nitrospirae* |
| *Ohtaekwangia* | 1.70±0.19d | 2.18±0.13c | 1.32±0.06e | 3.31±0.11a | 2.83±0.16b | *Bacteroidetes* |
| *TM7_genus_incertae_sedis* | 1.55±0.09b | 2.54±0.04a | 2.52±0.11a | 1.33±0.30b | 2.44±0.38a | *TM7* |
| *3_genus_incertae_sedis* | 1.07±0.19a | 1.13±0.12a | 1.11±0.07a | 1.17±0.42a | 1.39±0.10a | *Verrucomicrobia* |
| *Sphingomonas* | 1.71±0.49a | 1.10±0.05b | 1.47±0.05a | | | *Proteobacteria* |
| *Gp5* | 1.17±0.12a | | | 1.12±0.12a | 1.12±0.17a | *Acidobacteria* |
| *Bacillus* | | 1.67±0.11a | 1.78±0.06a | | 1.44±0.12b | *Firmicutes* |
| *Niastella* | | 2.96±0.23a | | | 1.55±0.20b | *Bacteroidetes* |
| *Gp2* | | 1.48±0.19b | 1.09±0.05c | | 1.68±0.05a | *Acidobacteria* |
| *Beggiatoa* | | | | 1.46±0.16 | | *Proteobacteria* |
| *Gp13* | | | | | 1.49±0.06 | *Acidobacteria* |
| *Segetibacter* | | 1.87±0.12 | | | | *Bacteroidetes* |
| *Chitinophaga* | | 1.36±0.04 | | | | *Bacteroidetes* |
| *Frateuria* | | | 1.06±0.13 | | | *Proteobacteria* |

Only the genera frequency higher than 1% was listed in the table. Values are the means followed by standard error of the mean. Different letters indicate statistically significant differences at the 0.05 probability level according to Fisher's least significant difference test (LSD) and the Duncan test.

OTU number, However, the rarefaction curves did not reach saturation, which indicated that more sequencing efforts were needed (Fig. 2).

## Bacterial community structure

The analysis of microbial communities using hierarchical cluster analysis showed that the bacterial communities from the same treatment were more similar to each other than those from different treatments, as observed for the 5 highly supported clusters that were made up of samples from different fertilizer-treated soils (Fig. 3). Bacterial community structure from soil samples that were amended with common composts (CM, CMR, and PM) clustered together while soil samples from BIO and GCK were clustered together based on weighted UniFrac algorithm (Fig. 3a). Bacterial community membership from soil samples that were amended with organic amendments (CM, CMR, PM and BIO) clustered together and were separated to general operation control (GCK) based on unweighted UniFrac algorithm (Fig. 3b). Moreover, BIO-treated soil grouped separately from common compost treatments (CM, CMR and PM), which were grouped together.

Heat map analysis of the abundant phyla within a hierarchical cluster based on Bray–Curtis distance indices showed different patterns of community structure among the different treatments

**Table 3.** Calculations of Chao1, ACE, Shannon and Good's Coverage indices for treatments with bio-organic fertilizer (BIO), cattle manure compost (CM), Chinese medicine residue compost (CMR), general operation control (GCK) and pig manure compost (PM) at a 97% similarity threshold.

| Treatments | Chao1 | ACE | Shannon | Coverage |
|---|---|---|---|---|
| BIO | 3,751±220a | 5,398±292a | 6.60±0.04b | 0.87±0.01b |
| CM | 3,105±75b | 4,085±91b | 6.38±0.05c | 0.90±0.01a |
| CMR | 3,477±174a | 4,904±216ab | 6.46±0.03c | 0.88±0.01b |
| GCK | 3,588±173a | 5,112±395a | 6.76±0.05a | 0.88±0.01b |
| PM | 3,724±236a | 5,573±108a | 6.70±0.04a | 0.88±0.01b |

Values indicate the means followed by standard error of the mean. Different letters indicate statistically significant differences at the 0.05 probability level according to Fisher's least significant difference test (LSD) and the Duncan test.

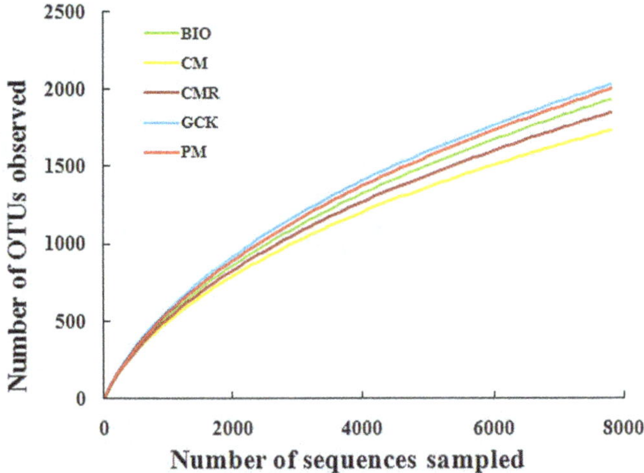


**Figure 2. Rarefaction analysis at different 3% dissimilarity levels for treatments with bio-organic fertilizer (BIO), cattle manure compost (CM), Chinese medicine residue compost (CMR), general operation control (GCK) and pig manure compost (PM).**

and similar patterns for the same treatment in triplicate (Fig. 4a). Moreover, BIO treatment showed a different pattern of community structure from those of other soil samples and enriched phyla of *Acidobacteria*, *Gemmatimonadetes*, *WS3* and *Lentisphaerae*, as shown in blue. Principal coordinates analysis (PCoA) based on the OTU composition also clearly showed variations among these different fertilizer treatments (Fig. 4b). The first two principal components could explain 83.1% of the variation of the individual samples of the total bacterial community. The bacterial community of the BIO-treated soil was well-separated from that of common compost-treated soils (CM, PM and CMR) along the first component (PCoA1) and was separated from the general control (GCK) along the second component (PCoA2).

## Relationship between disease incidence and the selected parameters

According to the disease incidence reported in our previous paper [12] and based on line regression analysis, a significant correlation between the abundance of the *Gemmatimonadetes*, *Bacteroidetes*, *Lentisphaerae* and *SR1* phyla and *Fusarium* wilt disease incidence was found (Table S2). Among these phyla, *Lentisphaerae* and *SR1* were not considered further due to their low abundance and random distribution. A clear negative correlation between *Gemmatimonadetes* ($r = -0.579$, $p = 0.024$) and the disease incidence and a clear positive correlation between *Bacteroidetes* ($r = 0.600$, $p = 0.018$) and the disease incidence were observed (Fig. 5a).

Line regression analysis between the 20 most-abundant classified genera and disease incidence showed that *Gemmatimonas*, *Ohtaekwangia* and *Sphingomonas* were significantly correlated to disease incidence (Table S3). A strong negative correlation between disease incidence and *Gemmatimonas* ($r = -0.579$, $p = 0.024$) and *Sphingomonas* ($r = -0.689$, $p = 0.005$) and a positive correlation with *Ohtaekwangia* ($r = 0.764$, $p = 0.001$) were observed (Fig. 5b). Unfortunately, some classified genera that were generally considered to contain plant growth-promoting rhizobacteria (PGPR) strains, which can suppress soil-borne fungi or promote plant growth, were only present in limited amounts, and their presence was not correlated with disease incidence (Table S4). Furthermore, in our research, no significant correlation was found between the whole bacteria community indices (richness and diversity) and disease incidence (Table S5).

The RDA that was performed on the phyla data and soil chemical properties showed that the first two RDA components could explain 88.6% of the total variation (Fig. 6). The first component (RDA1) separated the BIO and CMR treatments from the other fertilizer treatments and explained 61.1% of the variation, and the second component (RDA2), which separated the BIO from the CMR treatment, explained 27.5% of the variation. All soil chemical properties sufficiently explained the variation in phyla data ($p = 0.002$, Monte Carlo test). Ammonium nitrogen (NH4-N) and electricity conductivity (EC) accounted for a large amount of the variation in the distribution of the BIO treatment from other treatments along the RDA1 and RDA2 axes.

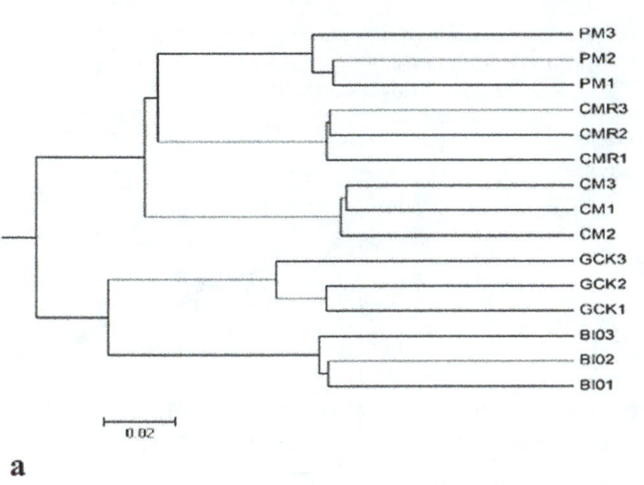


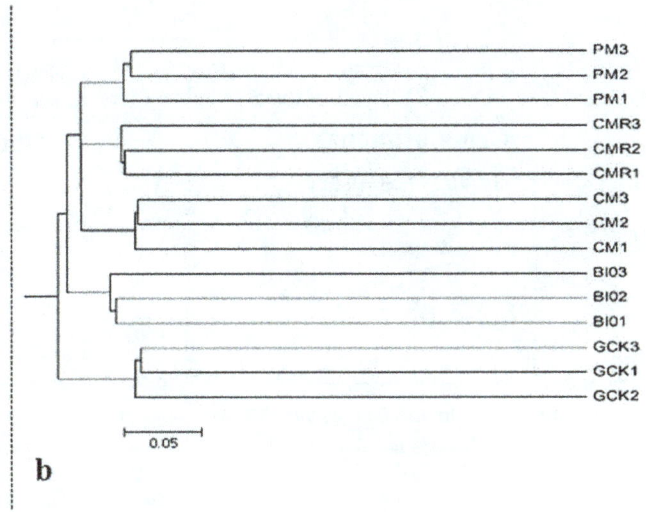


**Figure 3. Hierarchical cluster tree constructed based on the distance matrix that was calculated using the (a) weighted UniFrac algorithm and (b) unweighted UniFrac algorithm for treatments with bio-organic fertilizer (BIO), cattle manure compost (CM), Chinese medicine residue compost (CMR), general operation control (GCK) and pig manure compost (PM).**

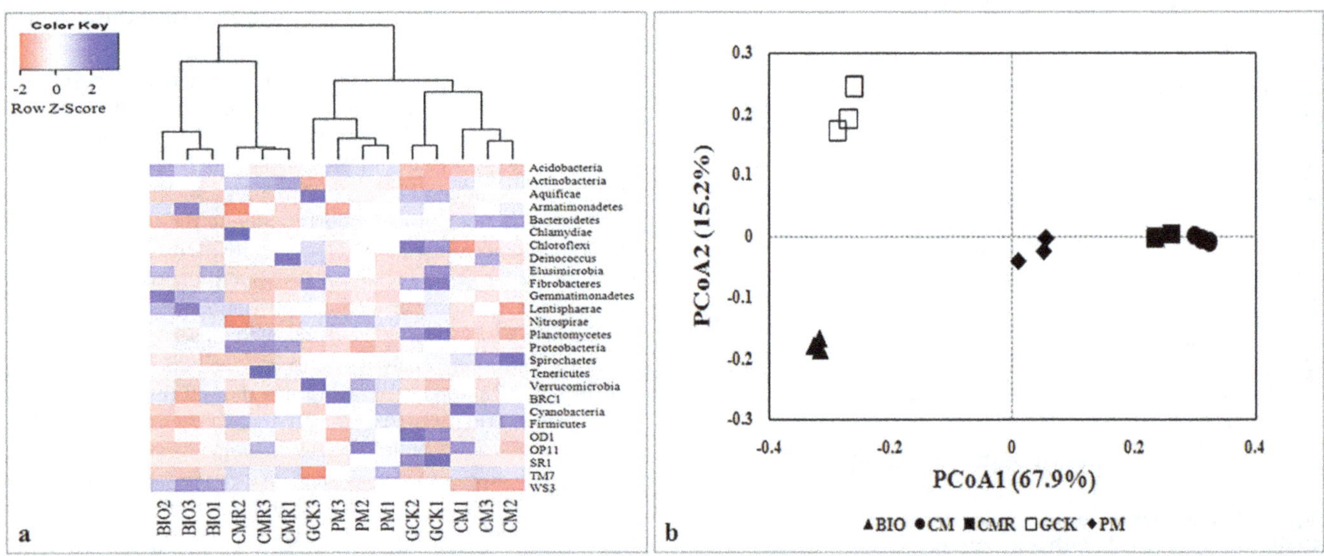


**Figure 4. Heat map of the bacterial communities based on abundance of phyla (a) and Jackknifed principal coordination analysis (PCoA) plots with unweighted UniFrac distance metric (b) from treatments with bio-organic fertilizer (BIO), cattle manure compost (CM), Chinese medicine residue compost (CMR), general operation control (GCK) and pig manure compost (PM).** Color from pink to blue indicates increasing abundance.

As shown by their close grouping and by the vectors, BIO-treated soil with the lowest disease incidence was positively related to the higher relative abundant phyla of *Gemmatimonadetes* and *Lentisphaerae*, the higher content of NH4-N and the EC, and it was negatively related to *Bacteroidetes*, a higher content of soil nitrate nitrogen (NO3-N) and higher total carbon to nitrogen ratio (C/N). Furthermore, the relative abundance of *Gemmatimonadetes* was positively correlated with soil pH, EC and NH4-N contents and negatively correlated with the soil total carbon (TOC) and C/N ratio. Moreover, the relative abundance of *Lentisphaerae* was

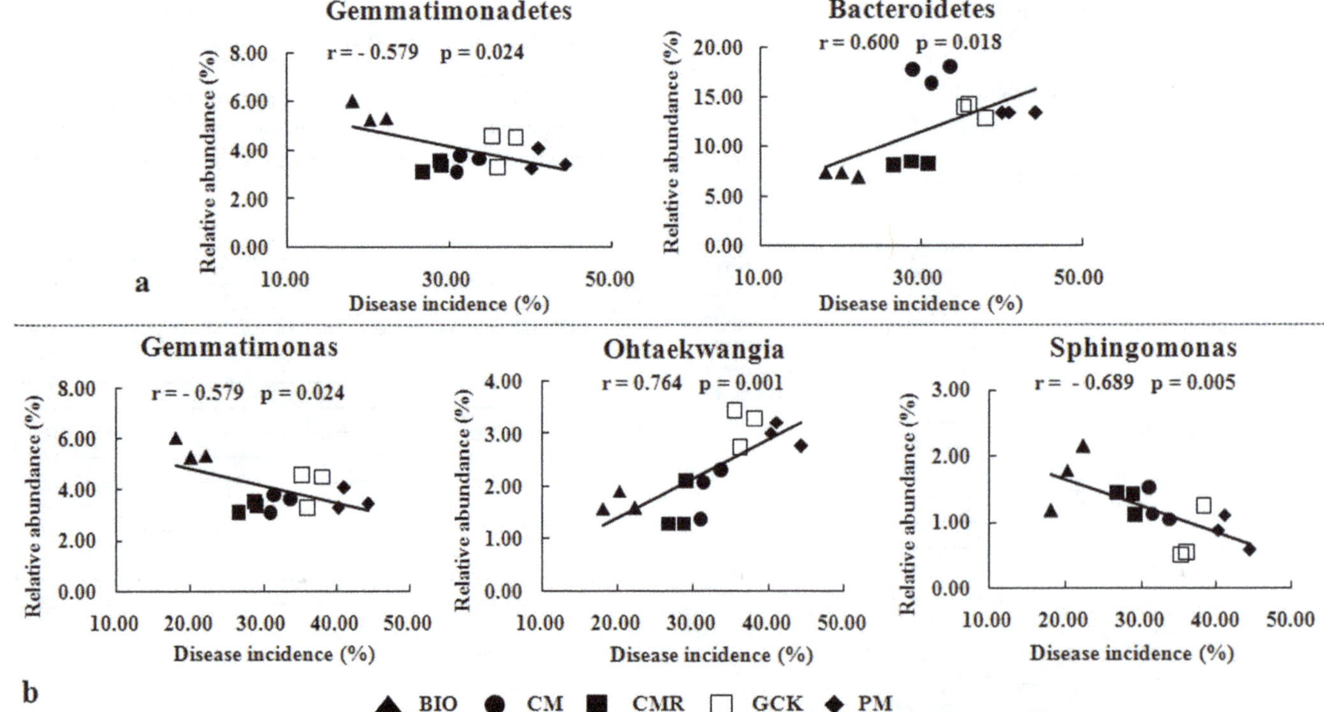


**Figure 5. Correlation analysis between the relative abundance of two bacteria phyla (a), three of the most classified bacteria genera (b) and banana *Fusarium* wilt disease incidence for treatments with bio-organic fertilizer (BIO), cattle manure compost (CM), Chinese medicine residue compost (CMR), general operation control (GCK) and pig manure compost (PM).**

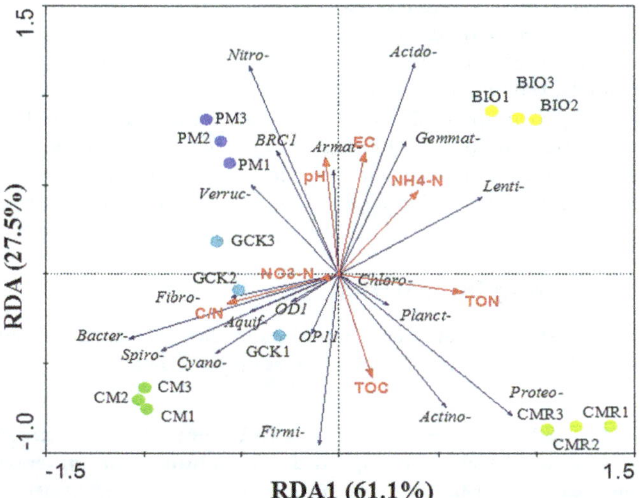


**Figure 6. Redundancy analysis (RDA) of the abundant phyla and soil properties for soil samples from treatments with bio-organic fertilizer (BIO), cattle manure compost (CM), Chinese medicine residue compost (CMR), general operation control (GCK) and pig manure compost (PM).**

positively correlated with the total nitrogen (TON) and NH4-N contents of the soil and negatively correlated with the soil C/N ratio. In contrast, the relative abundance of *Bacteroidetes* was positively correlated with the soil C/N ratio and negatively correlated with the soil TON and NH4-N contents (Fig. 6 and Table S6).

## Discussion

In our previous study, the main potential mechanism by which the BIO application reduced the *Fusarium* population has been revealed by culture-depended and PCR-DGGE methods [12]. However, deeper research should be done to further explore the potential mechanism. To our knowledge, this detailed comparison of the soil bacteria community after the application of BIO or other common composts in a banana orchard with serious *Fusarium* wilt disease was the first to be assessed using deep 16 S rRNA pyrosequencing, although this method has been used to study the long-term effects of selected, common composts on the soil bacteria community composition or structure [15,36]. The obtained results supported the hypothesis that soil amended with different organic materials showed different responses by the bacterial community or suppression of *Fusarium* wilt disease [15,37–39].

Phyla analysis revealed that *Proteobacteria*, *Acidobacteria*, *Bacteroidetes*, *Gemmatimonadetes*, *Actinobacteria* and *Firmicutes* were the most common phyla, but with some variety in relative abundance. This finding roughly corresponded with those of previous articles that investigated agricultural or other type soils in which these phyla accounted for more than 74.0% of the sequences that were examined using deep 16 S rRNA pyrosequencing [18,19]. The relative abundance of *Acidobacteria* was relatively high in our study due to the experiment being conducted in acidic soil [40,41]. However, in our study, BIO and PM treatments with the higher pH showed the higher relative abundance of *Acidobacteria*. This finding was contrary to a previous study that showed that pH had a negative relation to *Acidobacteria* abundance [40,42]. The reason for this phenomenon is still unclear and may be due to the narrow pH value range of the treated soil; however, a few articles have shown no obvious correlation between pH and abundance of *Acidobacteria* [15,36]. Analysis of the most abundant genera (>1%) also revealed significant differences between the bacterial communities of different treatments, a higher abundance of *Gemmatimonas* and *Gp4* in BIO-treated soil compared to other soil samples.

These changes could correspond to the decline of *Fusarium* wilt disease incidence. Thus, further correlation analysis was performed. Interestingly, the results showed that *Fusarium* wilt disease incidence might be related to the *Gemmatimonadetes* and *Bacteroidetes* phyla and/or *Gemmatimonas* genus, which belongs to *Gemmatimonadetes*, *Ohtaekwangia*, which belongs to *Bacteroidetes*, and *Sphingomonas*, which belongs to *Proteobacteria*. The high abundance of the *Bacteroidetes* phylum and the *Ohtaekwangia* genus that was observed in this study might positively correspond to *Fusarium* wilt disease incidence because this finding is in accordance with the report that the relative abundance of *Bacteroidetes* was similar between the initial and disease stages and followed by a significant decrease when suppressiveness was reached, as investigated using a 16 S rRNA-based microarray method [43], although, *Bacteroidetes* was also reported to possess the potential ability for biocontrol [44]. Moreover, we found the *Gemmatimonadetes* phylum and *Gemmatimonas* and *Sphingomonas* genera might respond to the suppression of *Fusarium* wilt disease via BIO application. *Gemmatimonas* and *Gemmatimonadetes* are a recently proposed genus and phylum, respectively, and they widely exist in multiple terrestrial and aquatic habitats. However, little is known about the ecological functions of this genus/phylum, except that Yin et al. [45] reported that the *Gemmatimonas* genus was found at a higher frequency in the rhizosphere of healthy plants using 454 pyrosequencing. *Sphingomonas*, which belongs to the *Sphingomonadaceae* order and *Proteobacteria* phylum, is widely distributed in natural habitats and is utilized for a wide range of biotechnological applications due to its remarkable biodegradative and biosynthetic capabilities [46]. Kyselková et al. [44] reported that bacteria affiliated with *Sphingomonadaceae* were more prevalent in tobacco-suppressive rhizosphere soil. Wachowska et al. [47] also reported that *Sphingomonas* could be used as biological agents to control winter wheat pathogens, such as *Fusarium*, under greenhouse conditions.

Analysis using rarefaction, Chao1 and ACE showed that the OTU numbers for BIO treatment were not significantly higher than for the other treatments. Furthermore, the diversity for BIO treatment that was estimated by the Shannon index and Coverage was also not the highest. All of the results indicated that a 1-year application of BIO could not significantly increase the bacteria community richness and diversity at the whole-community-structure level, which was in accordance with results of a previous study that used pyrosequencing to show that soil bacterial community richness and diversity were similar after a 5-year application of different organic amendments [15]. Although many previous articles indicated that the richness and/or diversity of the soil microbial community may respond to disease incidence [12,38], this phenomenon was not observed in this study because no obvious correlation between the indices and *Fusarium* wilt disease was observed (Table S5). This may be due to all 1-year treatments being performed on the same soil, which possessed similar bacteria community indices at the beginning.

In our study, the results of phylogenetic structure analyzed using the hierarchical cluster tree, heat map analysis based on the phyla frequency and PCoA analysis based on the OTU composition all showed that the bacterial community of BIO-treated soil differed from the common compost treatments (CM, CMR, and PM) and the control (GCK). All of the results confirmed that BIO application altered the bacterial community, which was roughly similar to the results of our previous investigation using PCR-DGGE that showed that BIO-treated soil grouped away from other soil samples [12]. Poulsen et al. [15] also reported similar results suggesting that soil amended with MSW-compost was separate from other amendments or the control, which indicated that the soil bacterial community responds differently to different compost amendments.

It has been reported that the chemical properties of soil can influence the suppressiveness of soil towards diseases [48]. In our RDA analysis, the BIO treatment with lowest *Fusarium* wilt disease incidence was highly correlated with the highest proportion of *Gemmatimonadetes* and lowest proportion of *Bacteroidetes*. Furthermore, the proportion of *Gemmatimonadetes* was positively correlated with soil pH, EC and NH4-N and negatively correlated with TOC and the C/N ratio. However, *Bacteroidetes* was positively correlated with the soil C/N ratio and negatively correlated with TON and NH4-N (Fig. 5, Table S6). Therefore, suppression of *Fusarium* wilt disease might be highly correlated with soil properties because *Fusarium* wilt disease incidence was positively correlated with the C/N ratio and negatively correlated to NH4-N and TON (Table S7), which was in agreement with reports from several previous studies. For example, Hamel et al. [49] reported a positive association between the TON content of the soil and the suppressiveness towards *Fusarium* spp. on asparagus. However, the form of N, either as NO3-N or NH4-N, is also important for disease suppression. Pérez-Piqueres et al. [50] reported that suppressive soil contained higher rates of NH4–N than conductive soil when studying the effect of compost amendment on soil suppressiveness toward *Rhizoctonia solani* disease, and Mallett and Maynard [51] reported that the incidence of *Armillaria* root disease significantly increased with decreasing NH4-N concentration on the organic surface horizon. In contrast, Oyarzun et al. [52] reported that the disease suppression ability of *Thielaviopsis basicola* was positively associated with a decreased C/N ratio.

In this study, after analyzing all of the data, the abundance of *Bacillus* was not enriched after BIO application. This finding combined with our previous results, the main mechanism reduced the *Fusarium* population for BIO application might be attributed to a general suppression that the BIO application altered the soil microbial composition and stimulated the population of soil bacteria, actinomycetes and some beneficial microorganisms [12], indicated that the genus might not necessarily reflect the individual species that has functional importance in suppressing endemic soil disease and all the results revealed by further deep 16S rRNA pyrosequencing confirmed that the main potential mechanism by which the BIO application reduced the *Fusarium* population was deduced to the fact that the specific bio-organic fertilizer containing functional microbes altered the soil microbial composition and stimulated the population of some beneficial microorganisms, thus resulting in a general suppression.

## Conclusions

Deep 16 S rRNA pyrosequencing assessment of soil bacterial communities from different compost-treated soil in a monoculture banana orchard revealed significant differences among all treatments, including differences in community structure, composition, richness, diversity and bacterial phylogeny. Phyla of *Gemmatimonadetes* and *Acidobacteria* were significantly elevated in BIO treatment in comparison to other treatments. A decrease was also found for *Bacteroidetes* in BIO treatment. Moreover, genera of *Gemmatimonas* and *Gp4* were significantly elevated in BIO treatment in comparison to other treatments. Additionally, the enrichment of *Gemmatimonas* and *Sphingomonas* and the TON and NH4-N soil content was positively correlated with disease suppression. Cumulatively, the reduction of the *Fusarium* wilt disease incidence after a 1-year application of BIO might be attributed to the fact that application of a BIO fertilizer containing *Bacillus* sp. induced general suppression in the soil by modulating the bacterial community and specific suppression by enriching *Gemmatimonas* and *Sphingomonas*.

## Supporting Information

**Table S1 Primer sequences used for preparation of samples for deep 16S rRNA pyrosequencing.**

**Table S2 Line regression coefficient of the most abundant phyla (>1%) and Fusarium wilt disease incidence.** * in the table means correlation is significant at the 0.05 level, ** in the table means correlation is significant at the 0.01 level.

**Table S3 Line regression coefficient of the most frequent classified genera (>1%) and Fusarium wilt disease incidence.** * in the table means correlation is significant at the 0.05 level, ** in the table means correlation is significant at the 0.01 level.

**Table S4 Line regression coefficient of selected bacteria genera and Fusarium wilt disease incidence.** * in the table means correlation is significant at the 0.05 level, ** in the table means correlation is significant at the 0.01 level.

**Table S5 Line regression coefficient of the bacteria community indices and Fusarium wilt disease incidence.** * in the table means correlation is significant at the 0.05 level, ** in the table means correlation is significant at the 0.01 level.

**Table S6 Line regression coefficient (r) between selected phyla in all samples and soil properties.** * in the table means correlation is significant at the 0.05 level, ** in the table means correlation is significant at the 0.01 level.

**Table S7 Line regression coefficient (r) between Fusarium wilt disease incidence in all samples and soil properties.** * in the table means correlation is significant at the 0.05 level, ** in the table means correlation is significant at the 0.01 level.

## Acknowledgments

We thank Majorbio Bio-pharm Biotech Company (Shanghai, China) for deep 16 S rRNA barcode pyrosequencing and Hainan Wanzhong Agriculture Company for huge help to us in banana planting.

## Author Contributions

Conceived and designed the experiments: QS RL. Performed the experiments: ZS DW YR CX JZ. Analyzed the data: ZS DW YR. Contributed reagents/materials/analysis tools: CX JZ. Wrote the paper: ZS QS RL.

## References


1. O'Donnell K, Kistler HC, Cigelnik E, Ploetz RC (1998) Multiple evolutionary origins of the fungus causing Panama disease of banana: concordant evidence from nuclear and mitochondrial gene genealogies. P Natl Acad Sci USA 95: 2044–2049.
2. Butler D (2013) Fungus threatens top banana. Nature 504: 195–196.
3. Chen YF, Chen W, Huang X, Hu X, Zhao JT, et al. (2013) Fusarium wilt-resistant lines of Brazil banana (*Musa* spp., AAA) obtained by EMS-induced mutation in a micro-cross-section cultural system. Plant Pathol 62: 112–119.
4. Getha K, Vikineswary S (2002) Antagonistic effects of *Streptomyces violaceusniger* strain G10 on *Fusarium oxysporum* f. sp. *cubense* race 4: Indirect evidence for the role of antibiosis in the antagonistic process. J Ind Microbiol Biot 28: 303–310.
5. Getha K, Vikineswary S, Wong W, Seki T, Ward A, et al. (2005) Evaluation of *Streptomyces* sp. strain g10 for suppression of *Fusarium* wilt and rhizosphere colonization in pot-grown banana plantlets. J Ind Microbiol Biot 32: 24–32.
6. Raguchander T, Jayashree K, Samiyappan R (1997) Management of Fusarium wilt of banana using antagonistic microorganisms. J Biol Control 11: 101–105.
7. Saravanan T, Muthusamy M, Marimuthu T (2003) Development of integrated approach to manage the fusarial wilt of banana. Crop Prot 22: 1117–1123.
8. Sivamani E, Gnanamanickam S (1988) Biological control of *Fusarium oxysporum* f. sp. *cubense* in banana by inoculation with *Pseudomonas fluorescens*. Plant Soil 107: 3–9.
9. Wang BB, Yuan J, Zhang J, Shen ZZ, Zhang MX, et al. (2012) Effects of novel bioorganic fertilizer produced by *Bacillus amyloliquefaciens* W19 on antagonism of *Fusarium* wilt of banana. Biol Fertil Soils 49: 435–446.
10. El-Hassan S, Gowen S (2006) Formulation and delivery of the bacterial antagonist *Bacillus subtilis* for management of lentil vascular wilt caused by *Fusarium oxysporum* f. sp. *lentis*. J Phytopathol 154: 148–155.
11. Kavino M, Harish S, Kumar N, Saravanakumar D, Samiyappan R (2010) Effect of chitinolytic PGPR on growth, yield and physiological attributes of banana (*Musa* spp.) under field conditions. Appl Soil Ecol 45: 71–77.
12. Shen ZZ, Zhong ST, Wang YG, Wang BB, Mei XL, et al. (2013) Induced soil microbial suppression of banana fusarium wilt disease using compost and biofertilizers to improve yield and quality. Eur J Soil Biol 57: 1–8.
13. Cotxarrera L, Trillas-Gay MI, Steinberg C, Alabouvette C (2002) Use of sewage sludge compost and *Trichoderma asperellum* isolates to suppress Fusarium wilt of tomato. Soil Biol Biochem 34: 467–476.
14. Zhao QY, Dong CX, Yang XM, Mei XL, Ran W, et al. (2011) Biocontrol of *Fusarium* wilt disease for *Cucumis melo* melon using bio-organic fertilizer. Appl Soil Ecol 47: 67–75.
15. Poulsen PHB, Al-Soud WA, Bergmark L, Magid J, Hansen LH, et al. (2013) Effects of fertilization with urban and agricultural organic wastes in a field trial-Prokaryotic diversity investigated by pyrosequencing. Soil Biol Biochem 57: 784–793.
16. Mazzola M (2004) Assessment and mangement of soil microbial community structre for disease suppression. Annu Rev Phytopathol 42: 35–59.
17. Garbeva P, Van VJ, Van EJ (2004) Microbial diversity in soil: selection of microbial populations by plant and soil type and implications for disease suppressiveness. Annu Rev Phytopathol 42: 243–270.
18. Acosta-Martinez V, Dowd S, Sun Y, Allen V (2008) Tag-encoded pyrosequencing analysis of bacterial diversity in a single soil type as affected by management and land use. Soil Biol Biochem 40: 2762–2770.
19. Roesch LF, Fulthorpe RR, Riva A, Casella G, Hadwin AK, et al. (2007) Pyrosequencing enumerates and contrasts soil microbial diversity. ISME J 1: 283–290.
20. Binladen J, Gilbert MTP, Bollback JP, Panitz F, Bendixen C, et al. (2007) The use of coded PCR primers enables high-throughput sequencing of multiple homolog amplification products by 454 parallel sequencing. PLoS One 2: e197.
21. Chen LH, Yang XM, Raza W, Luo J, Zhang FG, et al. (2011) Solid-state fermentation of agro-industrial wastes to produce bioorganic fertilizer for the biocontrol of *Fusarium* wilt of cucumber in continuously cropped soil. Bioresour Technol 102: 3900–3910.
22. Dethlefsen L, Huse S, Sogin ML, Relman DA (2008) The pervasive effects of an antibiotic on the human gut microbiota, as revealed by deep 16 S rRNA sequencing. PLoS Biol 6: e280.
23. Huse SM, Dethlefsen L, Huber JA, Welch DM, Relman DA, et al. (2008) Exploring microbial diversity and taxonomy using SSU rRNA hypervariable tag sequencing. PLoS Genet 4: e1000255.
24. Liu Z, Lozupone C, Hamady M, Bushman FD, Knight R (2007) Short pyrosequencing reads suffice for accurate microbial community analysis. Nucleic Acids Res 35: e120.
25. Schloss PD, Westcott SL, Ryabin T, Hall JR, Hartmann M, et al. (2009) Introducing mothur: open-source, platform-independent, community-supported software for describing and comparing microbial communities. Appl Environ Microb 75: 7537–7541.
26. Quince C, Lanzen A, Davenport RJ, Turnbaugh PJ (2011) Removing noise from pyrosequenced amplicons. BMC Bioinformatics 12: 38.
27. Pruesse E, Quast C, Knittel K, Fuchs BM, Ludwig W, et al. (2007) SILVA: a comprehensive online resource for quality checked and aligned ribosomal RNA sequence data compatible with ARB. Nucleic Acids Res 35: 7188–7196.
28. Wang Q, Garrity GM, Tiedje JM, Cole JR (2007) Naive Bayesian classifier for rapid assignment of rRNA sequences into the new bacterial taxonomy. Appl Environ Microb 73: 5261–5267.
29. Chao A (1984) Nonparametric estimation of the number of classes in a population. Scand J Stat 11: 265–270.
30. Eckburg PB, Bik EM, Bernstein CN, Purdom E, Dethlefsen L, et al. (2005) Diversity of the human intestinal microbial flora. Science 308: 1635–1638.
31. Washington H (1984) Diversity, biotic and similarity indices: a review with special relevance to aquatic ecosystems. Water Res 18: 653–694.
32. Bunge J, Fitzpatrick M (1993) Estimating the number of species: a review. J Am Stat Assoc 88: 364–373.
33. Warnes GR, Bolker B, Bonebakker L, Gentleman R, Huber W, et al. (2011) gplots: Various R programming tools for plotting data. R package version 2.
34. R Development Core Team (2012) R: A language and environment for statistical computing. Vienna, Austria, http://www.r-project.org.
35. Etten EV (2005) Multivariate analysis of ecological data using CANOCO. Austral Eco 30: 486–487.
36. Chaudhry V, Rehman A, Mishra A, Chauhan PS, Nautiyal CS (2012) Changes in bacterial community structure of agricultural land due to long-term organic and chemical amendments. Microbial Ecol 64: 450–460.
37. Bonanomi G, Antignani V, Pane C, Scala F (2007) Suppression of soilborne fungal diseases with organic amendments. J Plant Pathol 89: 311–324.
38. Qiu MH, Zhang RF, Xue C, Zhang SS, Li SQ, et al. (2012) Application of bio-organic fertilizer can control *Fusarium* wilt of cucumber plants by regulating microbial community of rhizosphere soil. Biol Fert Soils 48: 807–816.
39. Sun H, Deng S, Raun W (2004) Bacterial community structure and diversity in a century-old manure-treated agroecosystem. Appl Environ Microbiol 70: 5868–5874.
40. Lauber CL, Hamady M, Knight R, Fierer N (2009) Pyrosequencing-based assessment of soil pH as a predictor of soil bacterial community structure at the continental scale. Appl Environ Microb 75: 5111–5120.
41. Rousk J, Bååth E, Brookes PC, Lauber CL, Lozupone C, et al. (2010) Soil bacterial and fungal communities across a pH gradient in an arable soil. ISME J 4: 1340–1351.
42. Shen CC, Xiong JB, Zhang HY, Feng YZ, Lin XG, et al. (2012) Soil pH drives the spatial distribution of bacterial communities along elevation on Changbai Mountain. Soil Biol Biochem 57: 204–211.
43. Sanguin H, Sarniguet A, Gazengel K, Moënne-Loccoz Y, Grundmann G (2009) Rhizosphere bacterial communities associated with disease suppressiveness stages of take-all decline in wheat monoculture. New Phytol 184: 694–707.
44. Kyselková M, Kopecký J, Frapolli M, Défago G, Ságová-Marečková M, et al. (2009) Comparison of rhizobacterial community composition in soil suppressive or conducive to tobacco black root rot disease. ISME J 3: 1127–1138.
45. Yin C, Hulbert SH, Schroeder KL, Mavrodi O, Mavrodi D, et al. (2013) Role of bacterial communities in the natural suppression of *Rhizoctonia solani* bare patch of wheat (*Triticum aestivum* L.). Appl Environ Microb 79: 7428–7438.
46. Balkwill DL, Fredrickson JK, Romine MF (2006) Sphingomonas and related genera. In Dworkin M, Falkow S, Rosenberg E, Schleifer K, Stackebrandt E, editors. The Prokaryotes: Delta, Epsilon Subclass. Springer, New York. pp. 605–629.
47. Wachowska U, Irzykowski W, Jędryczka M, Stasiulewicz-Paluch AD, Głowacka K (2013) Biological control of winter wheat pathogens with the use of antagonistic *Sphingomonas* bacteria under greenhouse conditions. Biocontrol Sci Tech 23: 1110–1122.
48. Höper H, Alabouvette C (1996) Importance of physical and chemical soil properties in the suppressiveness of soils to plant diseases. Eur J Soil Biol 32: 41–58.
49. Hamel C, Vujanovic V, Jeannotte R, Nakano-Hylander A, St-Arnaud M (2005) Negative feedback on a perennial crop: Fusarium crown and root rot of asparagus is related to changes in soil microbial community structure. Plant Soil 268: 75–87.
50. Pérez-Piqueres A, Edel-Hermann V, Alabouvette C, Steinberg C (2006) Response of soil microbial communities to compost amendments. Soil Biol Biochem 38: 460–470.
51. Mallett K, Maynard D (1998) Armillaria root disease, stand characteristics, and soil properties in young lodgepole pine. Forest Eco Manag 105: 37–44.
52. Oyarzun P, Gerlagh M, Zadoks J (1998) Factors associated with soil receptivity to some fungal root rot pathogens of peas. Appl Soil Ecol 10: 151–169.

# 21

# Effects of Biochar on Soil Microbial Biomass after Four Years of Consecutive Application in the North China Plain

**Qing-zhong Zhang[1]*, Feike A. Dijkstra[2], Xing-ren Liu[1], Yi-ding Wang[1], Jian Huang[1], Ning Lu[1]**

**1** Key Laboratory of Agricultural Environment, Ministry of Agriculture, Sino-Australian Joint Laboratory For Sustainable Agro-Ecosystems, Institute of Environment and Sustainable Development in Agriculture, Chinese Academy of Agricultural Sciences, Beijing, China, **2** Centre for Carbon, Water and Food, Department of Environmental Sciences, The University of Sydney, Camden, New South Wales, Australia

**Abstract**

The long term effect of biochar application on soil microbial biomass is not well understood. We measured soil microbial biomass carbon (MBC) and nitrogen (MBN) in a field experiment during a winter wheat growing season after four consecutive years of no (CK), 4.5 (B4.5) and 9.0 t biochar $ha^{-1}$ $yr^{-1}$ (B9.0) applied. For comparison, a treatment with wheat straw residue incorporation (SR) was also included. Results showed that biochar application increased soil MBC significantly compared to the CK treatment, and that the effect size increased with biochar application rate. The B9.0 treatment showed the same effect on MBC as the SR treatment. Treatments effects on soil MBN were less strong than for MBC. The microbial biomass C:N ratio was significantly increased by biochar. Biochar might decrease the fraction of biomass N mineralized ($K_N$), which would make the soil MBN for biochar treatments underestimated, and microbial biomass C:N ratios overestimated. Seasonal fluctuation in MBC was less for biochar amended soils than for CK and SR treatments, suggesting that biochar induced a less extreme environment for microorganisms throughout the season. There was a significant positive correlation between MBC and soil water content (SWC), but there was no significant correlation between MBC and soil temperature. Biochar amendments may therefore reduce temporal variability in environmental conditions for microbial growth in this system thereby reducing temporal fluctuations in C and N dynamics.

**Editor:** Andrew C. Singer, NERC Centre for Ecology & Hydrology, United Kingdom

**Funding:** This work was supported by the S & T Innovation Program of Chinese Academy of Agricultural Sciences. The funders had no role in study design, data collection and analysis, decision to publish, or preparation of the manuscript.

**Competing Interests:** The authors have declared that no competing interests exist.

* Email: ecologyouth@126.com

## Introduction

Biochar is the product of the thermal degradation of organic materials in the absence of air (pyrolysis) and is distinguished from charcoal by its use as a soil amendment [1]. Application of biochar has been proposed as a novel approach to improve soil fertility, increase soil carbon sequestration and mitigate the greenhouse effect [2]. Nevertheless, the turnover of soil organic matter and biochar and their interaction in soils remain poorly understood. Positive and negative priming effects on native organic carbon mineralization in biochar-amended soils have been reported [3–6], and the priming direction is thought to be controlled by the biochar type (determined by production conditions and sources used) and soil pH. Moreover, the presence of soil organic matter can stimulate the mineralization of the more labile components of biochar over the short term, but over the long term, the biochar-soil interaction may enhance soil C storage via processes of organic matter sorption to biochar and physical protection [3,5,6]. There is strong evidence that priming effect on soil organic matter decomposition relies on microbial biomass [7]. Research on soil microbial biomass dynamics with biochar application will help the understanding of priming effect on soil organic matter and biochar decomposition. Biochar addition can increase soil microbial biomass, and may also affect the soil biological community composition, which in turn will affect nutrient cycling, plant growth, and greenhouse gas emission, as well as soil organic carbon mineralization mentioned above [1].

However, the present knowledge on soil microbial biomass dynamics due to biochar application is mainly based on the comparison between biochar application and no biochar treatment without considering crop residue return practice, and lacks long-term field experimental data [8–12]. Our field experiment with consecutive biochar amendments and crop residue return lasted nearly 4 years in a winter wheat-maize relay cropping system until October 2010, with total inputs of biochar ranging from 18.0 to 36.0 t $ha^{-1}$. The aim of this paper is to examine effects of biochar and wheat straw residue incorporation on the temporal and spatial variation in microbial biomass carbon (MBC) and microbial biomass nitrogen (MBN) measured within a wheat growing season under different soil depths after four years of consecutive application.

## Materials and Methods

### Site description

The field experiment was conducted at an experimental station (36°58′N, 117°59′E, elevation 17 m) for ecological and sustainability research in Huantai County, Shandong Province, China, and was begun in 2007 [13]. This site has a warm, temperate, continental monsoon climate with a mean annual temperature of 12.4°C. The mean annual precipitation was 600 mm, with most rainfall occurring in June, July, and August. The soil is classified to Fluvic Cambisol according to the USDA system. The soil was a sandy loam, and the proportion of sand, silt, and clay particles in the top 20 cm of soil was 70.8, 26.9, and 2.3%, respectively. The soil bulk density of the top 20 cm of soil before the biochar amendment was 1.52 g $cm^{-3}$, and the soil organic matter (SOM) content was 15 g $kg^{-1}$ of soil. The soil pH was 8.1 before experiment, and was not significantly changed due to biochar amendment [13].

### Biochar

Milled biochar was purchased from a local company (Jinfu Biochar Company, Liaoning Province), with a density of 0.297 g $cm^{-3}$ and a pH of 8.2 [14]. The biochar was made by incomplete self-combustion of crushed corncob in an open-top concrete tank with an igniting apparatus at the bottom for about 24 hours at about 360°C. The incomplete self-combustion process did not need an external energy source. The biochar contained 65.7% carbon and 0.909% N (analyzed with an elemental analysis apparatus, Flash EA 2000, Thermo Electron Corporation, Italy). Available phosphorus content was 0.08% (extracted with 0.5 M $NaHCO_3$ at a pH of 8.5, and analyzed with a colorimetric method), and available potassium content was 1.60% (extracted with 2.0 M $HNO_3$, and analyzed with a flare photometer, FP640, Cany, China). Ash content of the biochar was 72.0% (determined by dry combustion in a muffle furnace at 550°C for 2 h).

### Experimental design

The experimental crops were winter wheat (*Triticum aestivum* L.) and maize (*Zea mays* L.) in relay cropping. Generally, the maize was sown in early June and matured in late September. The winter wheat was then sown in early October and harvested in early June of the next year. The field experiment was a randomized complete block design and each of the experimental plots was 6 m×6 m = 36 $m^2$, four treatments (CK, B4.5, B9.0, SR) with three replications. The experiment included a control treatment with no biochar addition (CK), two biochar treatments with 4.5 and 9.0 t $ha^{-1}$ $yr^{-1}$ biochar applied (B4.5 and B9.0, respectively), and a treatment where all wheat and maize straw residue produced in the plot was returned to the field and incorporated into the soil after harvest by a straw returning machine (SR), which allowed us to compare biochar vs. fresh litter effects on microbial biomass. Based on the estimated total amount of aboveground crop residues of about 15 t $ha^{-1}$ $yr^{-1}$ for local normal croplands and the empirical value of about 30% in weight of crop residues left as biochar from the local biochar company, biochar amount can be obtained from the field crop residues is equivalent to that used in the B4.5 treatment.

The biochar was distributed equally to each crop (half applied before sowing of wheat and the other half before sowing of maize). Inorganic basal N fertilizers as urea of 200 kg N $ha^{-1}$ $yr^{-1}$ (165 kg N $ha^{-1}$ $yr^{-1}$ was used before 2009), P as superphosphate of 52.5 kg $P_2O_5$ $ha^{-1}$ $yr^{-1}$, and K as potassium sulfate of 37.5 kg $K_2O$ $ha^{-1}$ $yr^{-1}$ were applied in all treatments. Before 2009, all fertilizers were used as base fertilizer. From 2009, half of nitrogen fertilizer was applied as base fertilizer, and the other half was applied as topdressing. Biochar and the base fertilizers were broadcast on the soil surface and incorporated into the soil by rotary tillage to depth of 15 cm before seeding.

### Soil sampling

Soil samples were collected during the winter wheat growing period (from October 2010 to June 2011) divided into five functional stages: (1) wheat post-seeding stage; (2) before freezing stage, (3) reviving stage; (4) shooting stage, and (5) harvest stage. Soil samples were collected at depths of 0–5, 5–10, 10–20, and 20–30 cm in each experimental plot. The soil samples were collected from five points randomly, and mixed into one sample, each mixed soil sample was divided into two parts. One part of the soil sample was determined for soil water content (SWC) and the other part was prepared for microbial analysis. All samples were immediately stored in sealed plastic bags in a cooler and transported to laboratory and stored in refrigerator at 4°C. All microbiological determinations were performed within one week of sampling.

### Measurement and monitoring

Microbial biomass carbon (MBC) and nitrogen (MBN) in soil were determined by fumigation extraction method [15,16], and the value of 0.45 was used for both the fraction of biomass C mineralized ($K_C$) and the fraction of biomass N mineralized ($K_N$). Soil samples were thoroughly mixed and ground to pass through a 2-mm sieve, and then the soil moisture was adjusted to about 40% water holding capacity. We fumigated 20.0 g (dry weight equivalent) of soil with ethanol-free chloroform for 24 h. Both fumigated and non-fumigated soils were extracted with 80 ml of 0.5 M $K_2SO_4$ by shaking for 30 min on a reciprocating shaker at 40 cycles per minute and then filtered (soil:water = 1:4). The TOC analyzer (Multi N/C 2100, Jena, Germany) was used to determine the C and N in the extracts.

SWC was determined gravimetrically by oven-drying at 105°C for 48 h. Soil temperature at 5-cm depth in each replication was monitored hourly by a temperature probe (tolerance: ±0.2°C over the 0–70°C range; temperature measurement range: −50 to +70°C; model 109, Campbell Scientific, Logan, UT, USA) connected to a datalogger (model U12-006, Jimuduoli, Beijing, China) from 22 October 2010, except during harvesting and sowing (June 10–July 2) of each year.

### Data analyses

We used repeated measures ANOVA to test for main effects of treatment (between-subjects factor), soil depth (within-subjects factor) and date (within-subjects factor), and their interactive effects on MBC, MBN and microbial biomass C:N ratio. When necessary, data were log-transformed to reduce heteroscedasticity and improve assumptions of normality. We used Pearson's test to determine whether there was a significant correlation between microbial and environmental soil properties (moisture and temperature) and Fisher's least significant difference (LSD) test to determine the significant difference in coefficient of seasonal variability (CV) of MBC and MBN between different treatments. All statistical analyses were performed with JMP (version 4.0.4; SAS Institute, Cary, NC, USA).

## Results

### Microbial biomass C (MBC)

Variability of MBC under different rates of biochar and straw residue addition was large (Fig. 1). Treatment effects on MBC were significant (Table 1), and on average, largest in the B9.0 treatment, followed by the SR treatment, and lowest in CK. There were significant treatment*date and treatment*date*depth interactive effects ($P = 0.02$ and $P < 0.0001$ respectively, Table 1). The greatest MBC was found at depth of 10–20 cm treated with SR on 20 October of 2010 (post-seeding stage), followed by the same treatment in 5 November of 2010 (before freezing stage) among all treatments and both soil layers during the whole experimental period. At depth of 0–5 and 5–10 cm, the MBC content in the treatments of B9.0 and SR in general showed the largest increases compared to CK. The MBC in the B9.0 treatment increased by 118% to 763%, while the MBC in the SR treatment was 2% lower to 722% higher compared to CK, depending on the time of year (Fig. 1a, b). The MBC at 0–5 and 5–10 cm in the B4.5 treatment also increased by 6% to 246%, depending on the time of year.

The MBC in the B9.0 treatment increased by 45% to 294% at soil depths of 10–20 and 20–30 cm compared to CK (Fig. 1c, d). The MBC in the SR treatment decreased by 62% during the shooting stage (25 April) at 20–30 cm soil depth, but showed some of the largest increases (between 50% and 408%) at other times at this depth. The MBC in the B4.5 treatment at 20–30 cm soil depth increased in most of the stages, but decreased at 10–20 cm soil depth by 19%, 40%, 6%, and 10% in the winter wheat post-seeding stage (20 October 2010), before freezing stage (5 November 2010), reviving stage (26 March 2011), and harvest stage (5 June 2011), respectively.

Table 2 presents the coefficients of seasonal variation of the MBC under different treatments. The CV of CK varied between 20% and 38%, and the overall CV was 25%. The overall CV of B4.5 and B9.0 treatments was 11% and 16%, respectively. The CVs were significantly smaller than that of CK. But the CV of SR treatment varied from 40% to 63%, which was significantly higher than that of CK. The CV of the surface soil tended to be higher than that of the subsoil in the same treatment, but not for the SR treatment.

## Microbial biomass N (MBN)

As with MBC, there was large variability in MBN among the different biochar and SR treatments, dates, and depths (Fig. 2, Table 1). However, unlike the MBC results, across date and depth the largest MBN was observed in the SR treatment (on average 35, 106, and 31% higher than the CK, B4.5 and B9.0 treatment respectively). The MBN in the B4.5 treatment decreased at most dates and soil depths, and across all dates compared to the CK treatment. Across date, the MBN in the B4.5 treatment decreased by 40, 47, 3, and 30% at 0–5, 5–10, 10–20, and 20–30 cm soil depth. The MBN in the B9.0 treatment often increased at 0–5, 5–10, and 20–30 cm soil depth during the stages of wheat post-seeding stage, before freezing stage, and reviving stage compared to the CK treatment. Large MBN pools were particularly observed in the SR treatment at 10–20 and 20–30 cm soil depth. At the

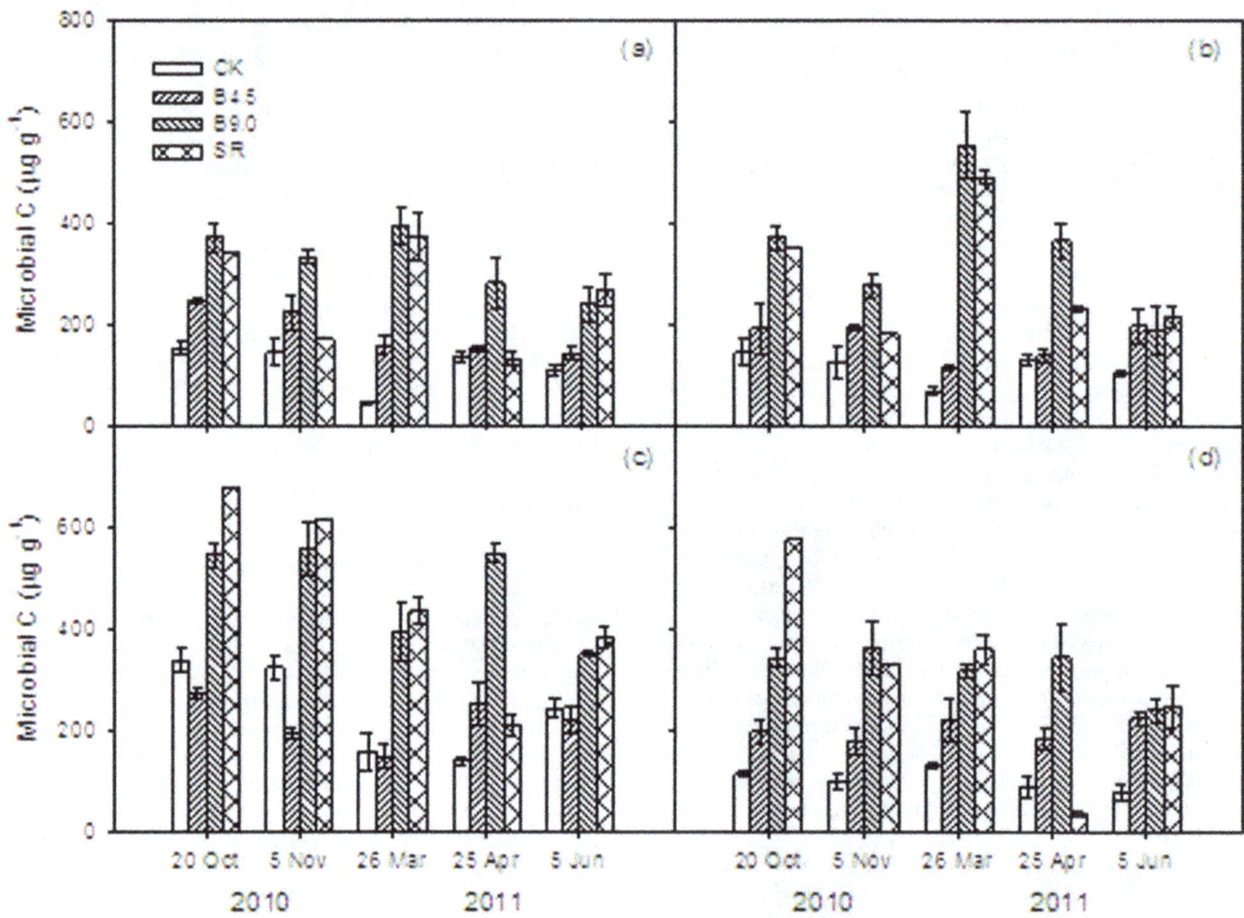


**Figure 1. Microbial biomass carbon (MBC) under different treatments (CK: control, B4.5: 4.5 t ha$^{-1}$ yr$^{-1}$ biochar addition, B9.0: 9.0 t ha$^{-1}$ yr$^{-1}$ biochar addition, SR: incorporation of wheat straw) and different soil depths (a: 0–5, b: 5–10, c: 10–20, and d: 20–30 cm) during winter wheat season.** Vertical bars represent the standard errors for means of each treatment ($n = 3$).

**Table 1.** Repeated Measures ANOVA P values.

| Variable | Microbial C | Microbial N | Microbial C:N |
|---|---|---|---|
| Treatm | 0.0003 | <0.0001 | <0.0001 |
| Date | <0.0001 | <0.0001 | 0.002 |
| Depth | <0.0001 | 0.0001 | <0.0001 |
| Treatm*date | 0.02 | <0.0001 | 0.0005 |
| Treatm*depth | 0.12 | <0.0001 | 0.0004 |
| Date*depth | <0.0001 | 0.0001 | 0.005 |
| Treatm*date*depth | <0.0001 | 0.0003 | 0.0007 |

shooting stage, the MBN of the four treatments were all relatively low.

The CV of MBN in the CK treatment varied from 24% to 44%, and the overall CV was 32% (Table 3). The overall CV of B4.5 was 19%, which was significantly lower than that of other treatments. The overall CV of B9.0 and SR treatments were 49% and 65% respectively, which were significantly larger than that of CK. From this, it can be seen that addition of 4.5 t $ha^{-1}$ of biochar decreased the temporal and spatial fluctuation of MBN, but that a higher biochar addition increased the fluctuation of MBN.

### Microbial biomass C:N ratio

Variability in microbial biomass C:N ratio was large among the different treatments. We observed significant treatment effects on microbial biomass C:N ratio (Table 1, Fig. 3), with largest ratios in the B9.0 treatment (on average 181% greater than CK) and B4.5 treatment (on average 93% higher than CK), suggesting that the increase in MBC with biochar addition was larger compared to the increase in MBN. The greatest microbial biomass C:N ratio ratios were found at 10–20 cm and 20–30 cm soil depth in the B9.0 treatment.

## Discussion

### Soil microbial biomass

The change in MBC reflects the process of microbial growth, death and organic matter degradation. Our results showed that MBC increased with biochar amendment compared to CK, which suggested that microbial growth could be accelerated by biochar addition. Reported biochar effects on soil MBC are quite inconsistent. Several studies found that there was no significant effect of biochar amendment on soil MBC [8,11]. Dempster et al. [12] found that MBC significantly decreased with biochar addition while MBN was unaltered in a coarse textured soil, and others observed the same positive effects of biochar addition on microbial biomass as ours [1,17–19]. Moreover, a positive linear relationship between microbial biomass and biochar concentration was also observed in a highly weathered soil [20]. Biochar type was thought as the driving parameter for any effects on soil microbial biomass, community, and activity [9,21].

Compared to wheat straw residue return, after 4 years of annual application, biochar increased MBC in 0–30 cm soil when applied at a high rate (9.0 t $ha^{-1}$ $yr^{-1}$, $p<0.05$), but decreased MBC when applied at a low rate (4.5 t $ha^{-1}$ $yr^{-1}$, $p<0.01$). The amount of C added in the B4.5 treatment most likely did not induce a similar increase in MBC as in the SR treatment, while the amount of C added in the B9.0 treatment did.

The MBN at 0–30 cm soil depth significantly decreased in the B4.5 treatment, but increased in the B9.0 treatment compared to CK ($p<0.01$), whereas the MBN at 0–30 cm soil depth in both biochar treatments were lower compared to SR treatment ($p<0.05$). Zavalloni et al. [8] found that biochar amendment at a rate of 5% had no significant influence on soil MBN in an incubation experiment. The increase in MBC and the decrease in MBN for biochar treatments indicate that biochar in soil acted as a carbon source rather than a nitrogen source for soil microbes. In return, this could have consequences for N cycling (e.g., increased microbial N immobilization with biochar addition). Our previous study in the field showed that biochar addition decreased soil available N by 7–10% compared to CK ($p>0.05$), but increased total N by 14–21% ($p<0.05$) in the 0–15 cm soil [22].

In the relatively unfertile and coarse-textured soil of our study, microbial biomass was probably limited by both a suitable growth environment and by C availability [17]. The labile fraction of biochar has been shown to stimulate microbial activity and abundance in some cases. Sorption of comparatively polar organic matter and nutrients could support energy for microorganisms, while macro-and micropores of biochar, which hold air and water, likely support microorganisms' livable habitat [1].Our results

**Table 2.** Coefficients of seasonal variation of MBC under different treatments (%).

| Treatment | 0–5 cm | 5–10 cm | 10–20 cm | 20–30 cm | 0–30 cm |
|---|---|---|---|---|---|
| CK | 37.0ab | 25.4ab | 38.1ab | 20.3b | 25.0b |
| B4.5 | 25.7bc | 22.9b | 22.6bc | 10.4c | 11.2c |
| B9.0 | 19.4c | 34.2ab | 20.5c | 15.3b | 15.6c |
| SR | 40.6ab | 43.1a | 40.4a | 63.3a | 41.6a |

Note: Different small letters indicate significant differences among treatments (CK: control, B4.5: 4.5 t $ha^{-1}$ $yr^{-1}$ biochar addition, B9.0: 9.0 t $ha^{-1}$ $yr^{-1}$ biochar addition, SR: incorporation of wheat straw) at the $P<0.05$ level (L.S.D.).

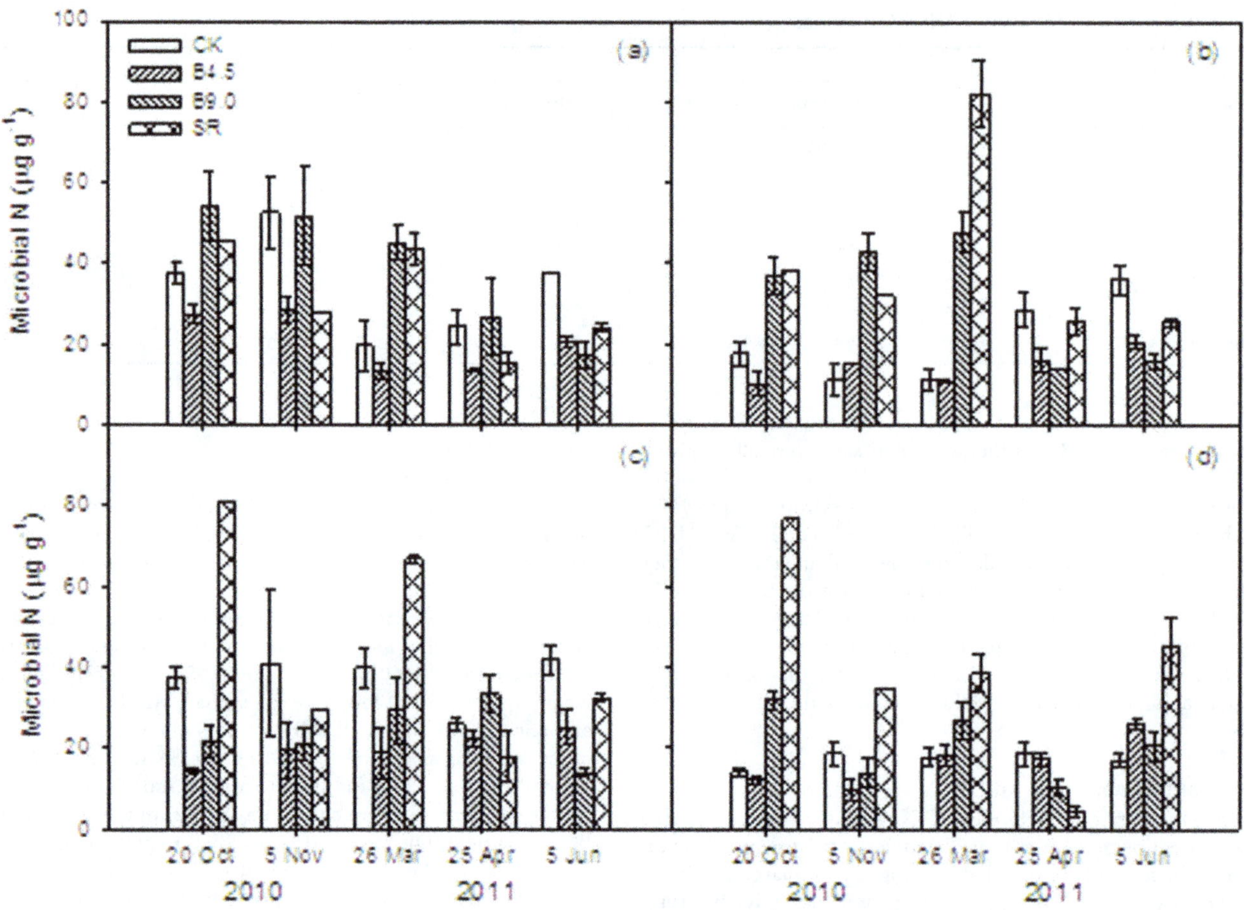


**Figure 2. Microbial biomass nitrogen (MBN) under different treatments (CK: control, B4.5: 4.5 t ha$^{-1}$ yr$^{-1}$ biochar addition, B9.0: 9.0 t ha$^{-1}$ yr$^{-1}$ biochar addition, SR: incorporation of wheat straw) and different soil depths (a: 0–5, b: 5–10, c: 10–20, and d: 20–30 cm) during winter wheat season.** Vertical bars represent the standard errors for means of each treatment ($n = 3$).

suggest that biochar supplied livable habitat for microorganisms stimulating microorganisms which can use the carbon source from the labile fraction of biochar.

## Soil microbial biomass C:N

Changes in the microbial biomass C:N ratio can reflect changes in the relative availability of C and N to microbes, but could also reflect changes in the microbial community structure. In our study, the microbial biomass C:N ratio significantly increased with increasing biochar application and the microbial biomass C:N ratio more than doubled in the B9.0 treatment compared to the control (Fig. 3). A pot experiment showed that there was no significant effect on the microbial biomass C:N ratio of biochar addition at a rate of 25 t ha$^{-1}$ [12]. Nicolardot et al. [23] determined that there was a significantly positive correlation between soil microbial biomass C:N ratio and the C:N ratio of the added organic matter. In contrast, Kushwaha et al. [24] reported a decrease in microbial biomass C:N ratio after straw return, while Kallenbach and Grandy [25] found no effect on microbial biomass C:N ratio after application of organic carbon, suggesting that changes in microbial biomass C:N ratio cannot solely be explained by organic amendments with relatively high C content. Several studies observed a change in soil microbial structure and abundance with biochar application [1]. In our study the

**Table 3.** Coefficients of seasonal variation of MBN under different treatments (%).

| Treatment | 0–5 cm | 5–10 cm | 10–20 cm | 20–30 cm | 0–30 cm |
|---|---|---|---|---|---|
| CK | 38.1ab | 43.6ab | 23.7b | 31.4a | 31.7b |
| B4.5 | 34.9b | 23.9b | 6.9c | 34.9c | 18.6c |
| B9.0 | 49.5a | 52.5a | 43.7ab | 58.8b | 49.1ab |
| SR | 42.0ab | 55.3a | 66.9a | 93.6a | 65.4a |

Note: Different small letters indicate significant differences among treatments (CK: control, B4.5: 4.5 t ha$^{-1}$ yr$^{-1}$ biochar addition, B9.0: 9.0 t ha$^{-1}$ yr$^{-1}$ biochar addition, SR: incorporation of wheat straw) at the $P<0.05$ level (L.S.D.).

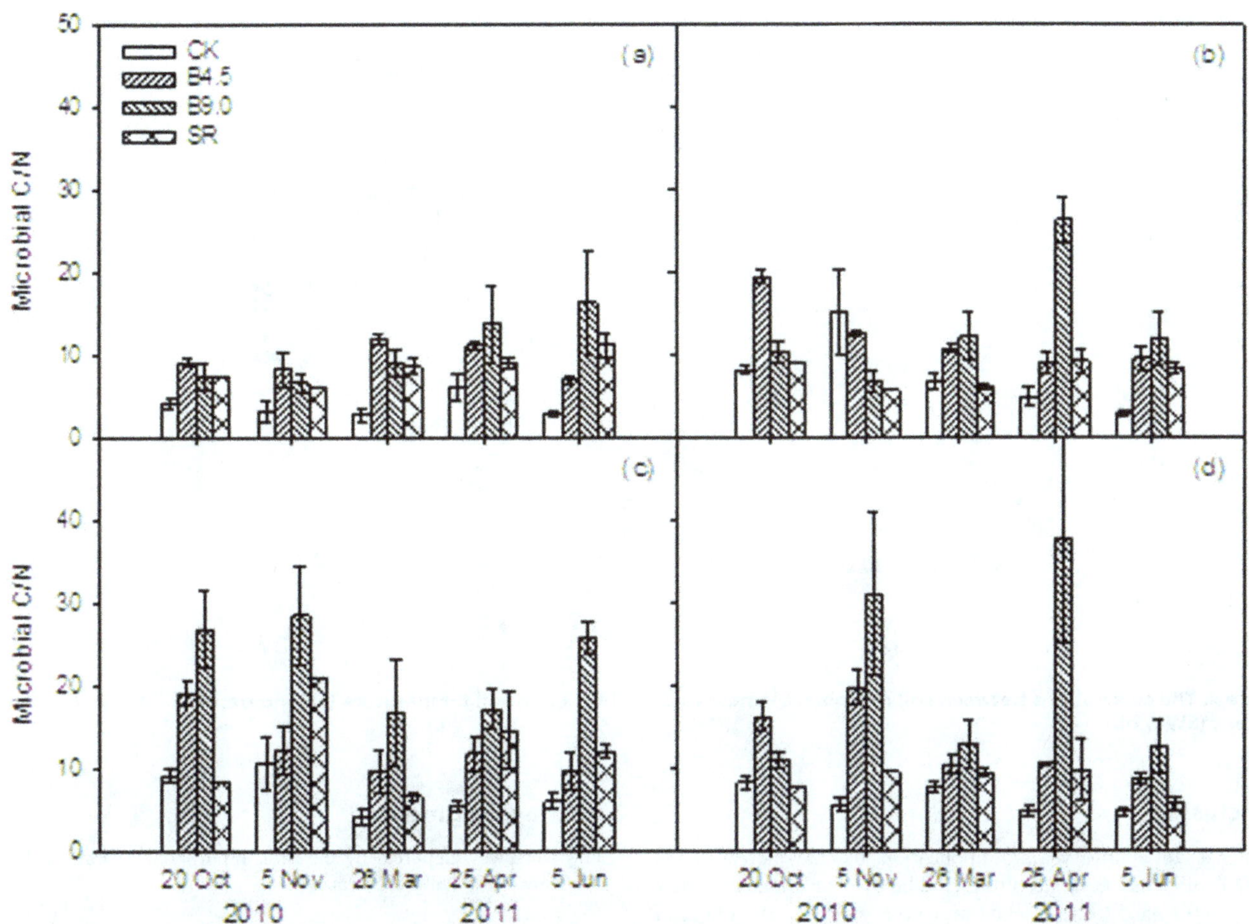


**Figure 3. The ratio of soil microbial biomass carbon (MBC) to soil microbial biomass nitrogen (MBC/MBN) under different treatments (CK: control, B4.5: 4.5 t ha$^{-1}$ yr$^{-1}$ biochar addition, B9.0: 9.0 t ha$^{-1}$ yr$^{-1}$ biochar addition, SR: incorporation of wheat straw) and different soil depths (a: 0–5, b: 5–10, c: 10–20, and d: 20–30 cm) during winter wheat season.** Vertical bars represent the standard errors for means of each treatment ($n = 3$).

increased microbial C:N ratio with biochar suggests increased microbial N limitation, while it is unclear if the increased microbial C:N ratio was also due to changes in microbial community structure. Moreover, biochar application may alter the release of MBN or the fraction of biomass N mineralized ($K_N$). The value of $K_N$ had a wide range from 0.28–0.81 [26–28], and it was difficult to be measured precisely. The high values of microbial C:N ratio for biochar treatments especially at 10–20 and 20–30 cm soil depths might be due to the same value of $K_N$ used for all treatments herein. It is worthy of further study on the effect of biochar on the fraction of biomass N mineralized in the fumigation-extraction method.

## Relationship between MBC and environmental factors

Data analysis showed that there was a significant positive correlation between SWC and MBC (**Fig. 4**, $R^2 = 0.172$, $P < 0.001$). The positive relationship between SWC and MBC suggests that SWC may have limited microbial activity at the experimental site. Biochar addition can increase soil water content [29,30]. Therefore, biochar application could influence soil microbial biomass via variation in soil water content.

However, the increases in MBC with biochar addition did not always coincide with an increase in SWC. For instance, MBC increased sharply in the reviving stage (March 2011) in the B9.0 treatment while SWC did not. Moreover, our study showed that biochar application showed a limited effect on soil water content [13,14], which also did not coincide with the higher MBC in the biochar treatments than in the CK treatment.

There was no significant correlation between MBC and soil temperature, and our result was supported by other studies [31,32]. For instance, Contin et al. [32] found that MBC did not change significantly at different incubation temperatures in arable and grassland soils. But we should point out that the spike in MBC at 5–10 cm depth of soil on March 26 may be related to the soil temperature increase that occurred during the same time.

The response of microbial biomass to seasonal changes is important in regulating microbial turnover, which may in turn influence nutrient availability and ultimately plant nutrition [33]. Seasonal fluctuations in MBC in the biochar treatments were smaller than in the CK and SR treatments (Table 2), whereas, seasonal fluctuations in MBN showed no obvious differences among the CK, B9.0 and SR treatments, and only the seasonal fluctuations in the B4.5 treatment was lower than in the other treatments. Biochar may have reduced extreme environmental conditions for microbial growth, which could have important implications for C and N dynamics, and crop yield [34,35].

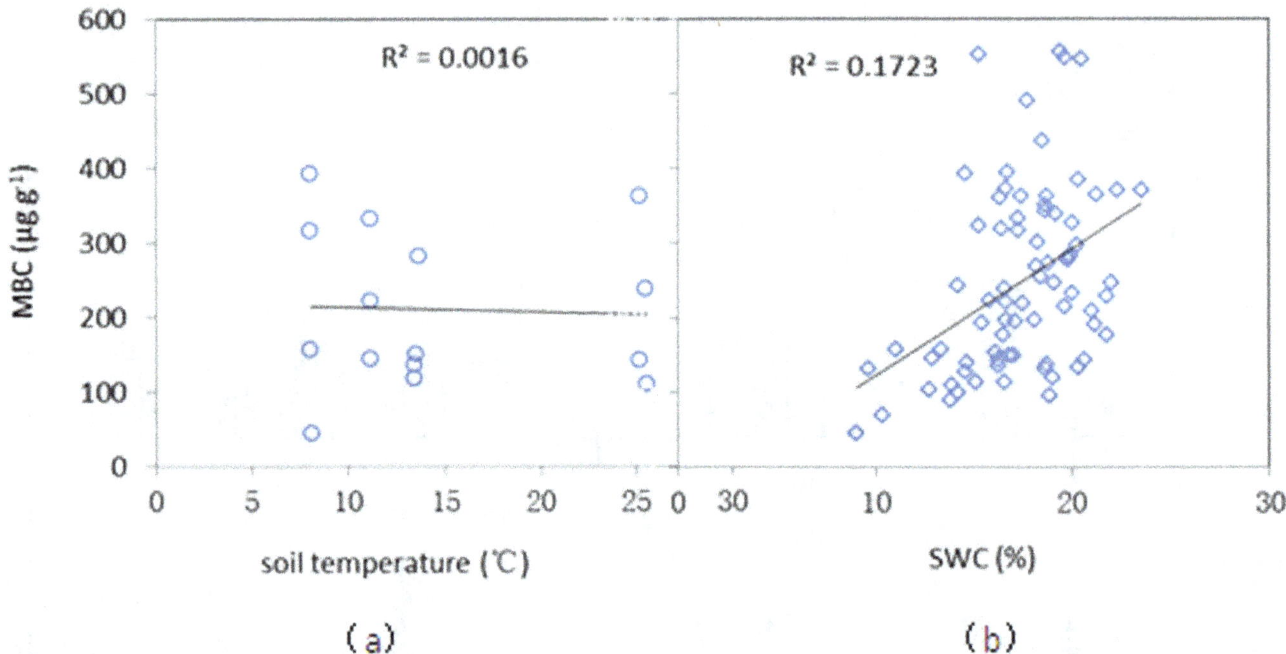


**Figure 4. The correlations between soil microbial biomass carbon (MBC) and soil temperature (a), and between MBC and soil water content (SWC, b).**

## Conclusion

After 4 consecutive years of application, biochar increased soil MBC significantly compared to CK. The biochar application at a rate of 9.0 t $ha^{-1}$ $yr^{-1}$ reached and even exceeded the effect of wheat residue return, whereas smaller differences in soil MBN were found among treatments. Soil MBC for biochar treatments showed less seasonal fluctuation compared to CK and SR treatments, suggesting that biochar provided a more suitable habitat for soil microorganisms. Biochar treatment showed the highest value of soil microbial biomass C:N ratio, and one possible reason might be that biochar could decease the fraction of biomass N mineralized.


## Acknowledgments

This work was supported by the S&T Innovation Program of Chinese Academy of Agricultural Sciences.


## Author Contributions

Conceived and designed the experiments: QZZ. Performed the experiments: JH NL QZZ. Analyzed the data: QZZ FAD XRL YDW. Contributed reagents/materials/analysis tools: QZZ FAD YDW XRL. Contributed to the writing of the manuscript: QZZ FAD XRL.

## References


1. Lehmann J, Rillig MC, Thies J, Masiello CA, Hockaday WC, et al. (2011) Biochar effects on soil biota – A review. Soil Biology and Biochemistry 43: 1812–1836.
2. Lehmann J (2007) A handful of carbon. Nature 447: 143–144.
3. Zimmerman AR, Gao B, Ahn M-Y (2011) Positive and negative carbon mineralization priming effects among a variety of biochar-amended soils. Soil Biology and Biochemistry 43: 1169–1179.
4. Luo Y, Durenkamp M, De Nobili M, Lin Q, Brookes PC (2011) Short term soil priming effects and the mineralisation of biochar following its incorporation to soils of different pH. Soil Biology and Biochemistry 43: 2304–2314.
5. Cross A, Sohi SP (2011) The priming potential of biochar products in relation to labile carbon contents and soil organic matter status. Soil Biology and Biochemistry 43: 2127–2134.
6. Keith A, Singh B, Singh BP (2011) Interactive priming of biochar and labile organic matter mineralization in a smectite-rich soil. Environmental Science & Technology 45: 9611–9618.
7. Thiessen S, Gleixner G, Wutzler T, Reichstein M (2013) Both priming and temperature sensitivity of soil organic matter decomposition depend on microbial biomass – An incubation study. Soil Biology & Biochemistry 57: 739–748.
8. Zavalloni C, Alberti G, Biasiol S, Vedove GD, Fornasier F, et al. (2011) Microbial mineralization of biochar and wheat straw mixture in soil: A short-term study. Applied Soil Ecology 50: 45–51.
9. Steinbeiss S, Gleixner G, Antonietti M (2009) Effect of biochar amendment on soil carbon balance and soil microbial activity. Soil Biology and Biochemistry 41: 1301–1310.
10. Luo Y, Durenkamp M, De Nobili M, Lin Q, Devonshire BJ, et al. (2013) Microbial biomass growth, following incorporation of biochars produced at 350°C or 700°C, in a silty-clay loam soil of high and low pH. Soil Biology and Biochemistry 57: 513–523.
11. Castaldi S, Riondino M, Baronti S, Esposito FR, Marzaioli R, et al. (2011) Impact of biochar application to a Mediterranean wheat crop on soil microbial activity and greenhouse gas fluxes. Chemosphere 85: 1464–1471.
12. Dempster DN, Gleeson DB, Solaiman ZM, Jones DL, Murphy DV (2012) Decreased soil microbial biomass and nitrogen mineralisation with Eucalyptus biochar addition to a coarse textured soil. Plant and Soil 354: 311–324.
13. Zhang QZ, Wang XH, Du ZL, Liu XR, Wang YD (2013) Impact of biochar on nitrate accumulation in an alkaline soil. Soil Research 51: 521–528.
14. Zhang QZ, Wang YD, Wu YF, Wang XH, Du ZL, et al. (2013) Effects of biochar ammendment on soil thermal conductivity, reflectance, and temperature. soil Science Society of America Journal 77: 1478–1487.
15. Brookes PC, Landman A, Pruden G, Jenkinson DS (1985) Chloroform fumigation and the release of soil nitrogen: A rapid direct extraction method to measure microbial biomass nitrogen in soil. Soil Biology and Biochemistry 17: 837–842.
16. Wu J, Joergensen RG, Pommerening B, Chaussod R, Brookes PC (1990) Measurement of soil microbial biomass C by fumigation-extraction—an automated procedure. Soil Biology and Biochemistry 22: 1167–1169.
17. Kolb SE, Fermanich KJ, Dornbush ME (2008) Effect of charcoal quantity on microbial biomass and activity in temperate soils. Soil Science Society of America Journal 73: 1173–1181.

18. O' Neill B, Grossman J, Tsai MT, Gomes JE, Lehmann J, et al. (2009) Bacterial community composition in Brazilian Anthrosols and adjacent soils characterized using culturing and molecular identification. Microbial Ecology 58: 23–35.
19. Liang B, Lehmann J, Sohi SP, Thies JE, O'Neill B, et al. (2010) Black carbon affects the cycling of non-black carbon in soil. Organic Geochemistry 41: 206–213.
20. Steiner C, Das KC, Garcia M, Förster B, Zech W (2008) Charcoal and smoke extract stimulate the soil microbial community in a highly weathered xanthic Ferralsol. Pedobiologia 51: 359–366.
21. Lehmann J, Joseph S (2009) Biochar for Environmental Management: An Introduction. In: Lehmann J, Joseph S, editors. Biochar for Environmental Management: Science and Technology. London: Earthscan. pp. 1–12.
22. Guo W, Chen HX, Zhang QZ, Wang YD (2011) Effects of biochar application on total nitrogen and alkali-hydrolyzable nitrogen content in the topsoil of the high-yield cropland in north China Plain. Ecology and Environmental Sciences 20: 425–428.
23. Nicolardot B, Recous S, Mary B (2001) Simulation of C and N mineralisation during crop residue decomposition: a simple dynamic model based on the C: N ratio of the residues. Plant and Soil 228: 83–103.
24. Kushwaha CP, Tripathi SK, Singh KP (2000) Variations in soil microbial biomass and N availability due to residue and tillage management in a dryland rice agroecosystem. Soil and Tillage Research 56: 153–166.
25. Kallenbach C, Grandy AS (2011) Controls over soil microbial biomass responses to carbon amendments in agricultural systems: A meta-analysis. Agriculture, Ecosystems and Environment 144: 241–252.
26. Joergensen RG (1996) The fumigation-extraction method to estimate soil microbial biomass: Calibration of the kEC value. Soil Biology and Biochemistry 28: 25–31.
27. Greenfield LG (1995) Release of microbial cell N during chloroform fumigation. Soil Biology and Biochemistry 27: 1235–1236.
28. Sparling G, Zhu C (1993) Evaluation and calibration of biochemical methods to measure microbial biomass C and N in soils from western australia. Soil Biology and Biochemistry 25: 1793–1801.
29. Chen Y, Shinogi Y, Taira M (2010) Influence of biochar use on sugarcane growth, soil parameters, and groundwater quality. Australian Journal of Soil Research 48: 526–530.
30. Piccolo A, Pietramellara G, Mbagwu JSC (1996) Effects of coal derived humic substances on water retention and structural stability of Mediterranean soils. Soil Use and Management 12: 209–213.
31. Joergensen RG, Brookes PC, Jenkinson DS (1990) Survival of the soil microbial biomass at elevated temperatures. Soil Biology and Biochemistry 22: 1129–1136.
32. Contin M, Corcimaru S, De Nobili M, Brookes PC (2000) Temperature changes and the ATP concentration of the soil microbial biomass. Soil Biology and Biochemistry 32: 1219–1225.
33. Wardle DA (1992) A comparative assessment of factors which influence microbial biomass carbon and nitrogen levels in soil. Biological Reviews of the Cambridge Philosophical Society 67: 321–358.
34. Kumar K, Goh KM (1999) Crop residues and management practices: effects of soil quality, soil nitrogen dynamics, crop yield, and nitrogen recovery. Advances In Agronomy 68: 197–319.
35. Lou Y, Li Z, Zhang T, Liang Y (2004) $CO_2$ emissions from subtropical arable soils of China. Soil Biology & Biochemistry 36: 1835–1842.

22

# Carbon, Nitrogen and Phosphorus Accumulation and Partitioning, and C:N:P Stoichiometry in Late-Season Rice under Different Water and Nitrogen Managements

**Yushi Ye[1], Xinqiang Liang[1]*, Yingxu Chen[2], Liang Li[1], Yuanjing Ji[1], Chunyan Zhu[2]**

**1** Institute of Environmental Science and Technology, College of Environmental and Resource Sciences, Zhejiang University, Hangzhou, China, **2** Zhejiang Province Key Laboratory for Water Pollution Control and Environmental Safety, Hangzhou, China

## Abstract

Water and nitrogen availability plays an important role in the biogeochemical cycles of essential elements, such as carbon (C), nitrogen (N) and phosphorus (P), in agricultural ecosystems. In this study, we investigated the seasonal changes of C, N and P concentrations, accumulation, partitioning, and C:N:P stoichiometric ratios in different plant tissues (root, stem-leaf, and panicle) of late-season rice under two irrigation regimes (continuous flooding, CF; alternate wetting and drying, AWD) and four N managements (control, N0; conventional urea at 240 kg N $ha^{-1}$, UREA; controlled-release bulk blending fertilizer at 240 kg N $ha^{-1}$, BBF; polymer-coated urea at 240 kg N $ha^{-1}$, PCU). We found that water and N treatments had remarkable effects on the measured parameters in different plant tissues after transplanting, but the water and N interactions had insignificant effects. Tissue C:N, N:P and C:P ratios ranged from 14.6 to 52.1, 3.1 to 7.8, and 76.9 to 254.3 over the rice growing seasons, respectively. The root and stem-leaf C:N:P and panicle C:N ratios showed overall uptrends with a peak at harvest whereas the panicle N:P and C:P ratios decreased from filling to harvest. The AWD treatment did not affect the concentrations and accumulation of tissue C and N, but greatly decreased those of P, resulting in enhanced N:P and C:P ratios. N fertilization significantly increased tissue N concentration, slightly enhanced tissue P concentration, but did not affect tissue C concentration, leading to a significant increase in tissue N:P ratio but a decrease in C:N and C:P ratios. Our results suggested that the growth of rice in the Taihu Lake region was co-limited by N and P. These findings broadened our understanding of the responses of plant C:N:P stoichiometry to simultaneous water and N managements in subtropical high-yielding rice systems.

**Editor:** Xiujun Wang, University of Maryland, United States of America

**Funding:** This work was funded by the National Natural Science Foundation of China (No. 41271314, 21077088) and the National Key Science and Technology Project: Water Pollution Control and Treatment (2014ZX07101-012). The funders had no role in study design, data collection and analysis, decision to publish, or preparation of the manuscript.

**Competing Interests:** The authors have declared that no competing interests exist.

* Email: liang410@zju.edu.cn

## Introduction

Nitrogen (N) is one of the most important mineral nutrients. It promotes large leaf area index [1], long duration of photosynthesis [2], high nutrient uptake [3], and ultimately high crop productivity [4,5]. Along with N, phosphorous (P) is another vital mineral nutrient influencing plant photosynthesis assimilation and biomass production [2,6,7]. To meet the challenge of food security, a large amount of chemical fertilizers has been applied to the rice cropping systems, particularly in China [8]. Long-term use of high rates of fertilization with improper water and nutrient managements has resulted in low water and nutrient use efficiencies, leading to detrimental effects on ecology, environment and human health [4,8]. The seasonal absorption, accumulation and allocation of N and P in rice may deserve special attention for implementing sound water and nutrient management practices in sustainable rice production systems.

Carbon (C), which provides the structural basis and constitutes a fairly stable 50% of a plant's dry mass, can also act as a limiting element for plant [9]. Rice crop may play an important role in terrestrial C cycle by both C sequestrations through photosynthesis and C releases through residues decomposition and/or root respiration [10,11]. Management practices such as irrigation and fertilization can influence crop physio-ecological activities [1], hence affect the sequestration and emission of C in paddy fields, which presumably have an effect on the mitigation of global warming and the stability of food security [12]. Most previous studies have focused on the accumulation and partitioning of dry matter of rice in response to elevated [$CO_2$] (free-air $CO_2$ enrichment) [5,7,13,14]. However, little has been done to evaluate the effects of water and N managements on the assimilation, accumulation and distribution of C in rice plant.

Since C, N and P are strongly coupled in their biochemical functioning [9,15] and their balance generally affects crop production and food-web dynamics [16], the C:N:P stoichiometry is the most investigated factor in ecological interactions. To date, C:N:P stoichiometry has been widely applied in diverse ecological processes, and successfully incorporated into explain many phenomena at all levels of biology, from genes and molecules to whole organisms and even to ecosystems and the biosphere [16–19]. Some measurements have already been made on the spatiotemporal variations, biological regulation mechanisms and the ecological implications of C:N:P stoichiometric ratios in soil, plant and litter at different trophic levels on a regional, national and even global scale [9,17–20]. Elucidating changes in C:N:P ratios during plant growth could be useful in calibrating plant mechanistic models and developing terrestrial biogeochemical models [21,22]. However, the seasonal changes of C:N:P ratios in rice plant, particularly with their responses to different water and N managements have not been well characterized.

In an earlier study, we investigated the effects of two controlled-release N fertilizers (CRNFs) (controlled-release bulk blending fertilizer and polymer-coated urea: BBF and PCU) under two irrigation regimes (continuous flooding and alternate wetting and drying: CF and AWD) in comparison with urea on the dry matter accumulation and partitioning, grain yield, and water and N use efficiencies in late-season rice in the Taihu Lake region of China, and found the agronomic performances played individually or jointly by the irrigation and N managements [4]. The objectives of this research were to: (1) investigate the seasonal changes of C, N and P concentrations, accumulation, allocation, and stoichiometric ratios in different plant tissues under different water and N managements, (2) get the relationships between tissue C:N:P ratios and rice grain yield, and (3) evaluate the limiting patterns of nutrients via C:N:P stoichiometry in rice production systems.

## Materials and Methods

(This work was unrelated to ethics issues, and no specific permit was required for the described field study, and we confirmed that the field study did not involve endangered or protected species).

### Site description

This study was conducted at Yuhang town of Zhejiang province, Taihu Lake region of China (30°21′50″ N, 119°53′17″ E) in 2010 and 2011. The study site has a subtropical monsoon climate with an average temperature of 16.2°C and an annual precipitation of 1290 mm. The soil type of the experimental field is hydromorphic paddy soil (Ferric-accumulic Stagnic Anthrosols). Initial soil properties of the plow layer (0–20 cm) were: pH 5.8 (1:5, soil/water), soil organic C 21.75 g $kg^{-1}$, total N 3.46 g $kg^{-1}$, mineral N 24.04 mg $kg^{-1}$, and total P 0.32 g $kg^{-1}$. Single cropping of late-season rice has been widely adopted in the region. The average routine rate of fertilization is 240 kg N $ha^{-1}$ (as urea, 46% N), 120 kg $P_2O_5$ $ha^{-1}$ (as superphosphate, 12% $P_2O_5$), and 120 kg $K_2O$ $ha^{-1}$ (as potassium chloride, 60% $K_2O$) for one rice season.

### Experimental design

The field experiment was arranged in a split-plot design with three replications in both years. Main plots consisted of two irrigation regimes: CF and AWD. All plots were flooded during the first 10–14 days after transplanting (DAT), allowing seedlings to recover from the shock of transplanting prior to the imposition of AWD treatments. Further details on the application of CF and AWD were described by Ye et al. [4]. The daily temperature, rainfall, irrigation, and field water depth from transplanting to harvest under CF and AWD irrigation in 2010 and 2011 were also reported in detail in our previous publication [4]. Subplots consisted of four N treatments: the control (N0), a urea application of 240 kg N $ha^{-1}$ (UREA), and two basal CRNF treatments both at the rate of 240 kg N $ha^{-1}$ (BBF and PCU). The PCU (42% N) is one of the most widely used coated granular CRNF fertilizers, and the BBF (24% N-12% $P_2O_5$-12% $K_2O$) is a compound CRNF product in which N source is made up of 70% controlled-release N and 30% ordinary quick-acting N. Both BBF and PCU (Kingenta Ecological Engineering Co., Ltd., Shandong, China) are 90 days of N release period and need only one-off fertilization.

Each plot was 6 m×3 m in size. All bunds were mulched with plastic film to minimize lateral seepage between adjacent plots. Individual inlet and outlet in the boundary side of the bunds were established in each plot for irrigation and drainage. A local late-season rice cultivar named Xiushui 134 (*Oryza sativa* L.) with high-yielding potential and pest resistance was used. Urea was applied in three splits for the late-season rice, 40% was basally applied (0 DAT), 40% was topdressed at the tillering stage (32 DAT), and the remaining 20% was topdressed at the panicle initiation stage (63 DAT). Full doses of superphosphate and potassium chloride in the N0, UERA and PCU treatments were applied basally at rates of 120 kg $P_2O_5$ $ha^{-1}$ and 120 kg $K_2O$ $ha^{-1}$, respectively. The 3-week-old seedlings were transplanted at a spacing of 25 cm×16 cm with three seedlings per hill on 23 June 2010 and 1 July 2011. Plots were regularly hand-weeded until canopy leaves were extremely crowded to prevent weed damage. The pests and diseases managements followed the local tradition. Final harvests were done on 9 November 2010 and 17 November 2011, and the growth duration was 140 days in all treatments for either year.

### Plant sampling and measurements

Five hills of plants (except border plants) in each plot were dug out by a shovel at seedling, tillering, booting, filling, and maturity stages (harvest), 1, 35, 56, 91, 140 DAT in both years. All visible root tissues were collected from the soil. Plants were washed free of soil and separated into two parts: root, stem-leaf before heading (75–80 DAT) and three parts: root, stem-leaf, and panicle after heading. All plant sub-samples were oven-dried to a constant weight at 70°C, weighted, and ground to sift through a 0.15 mm sieve for chemical measurements. Tissue C and N concentrations were determined with combustion technique on a Vario MAX CNS elemental analyzer (Elementar Analysensysteme GmbH, Hanau, Germany) [10]. Tissue P concentration was analyzed colorimetrically by the molybdenum blue method following digestion in concentrated $H_2SO_4$ and $H_2O_2$ [23].

### Calculations and statistical analysis

Accumulation of C, N and P in root, stem-leaf, and panicle at various growth stages was calculated from the element concentration multiplied by the dry matter. The element accumulation rate (kg $ha^{-1}$ $day^{-1}$) at each growth stage was obtained from the element accumulation divided by the number of days of the growth stage. Stoichiometric ratios of C:N, N:P and C:P in plant tissues were calculated on mass basis.

All statistical analyses were performed using PASW Statistics 18.0 (SPSS Inc. Chicago, USA). Combined analysis of variance using data from two years indicated that the interactions between years and irrigation regimes, and between years and N managements on the measured parameters (C, N and P concentrations, accumulation, partitioning, and stoichiometric ratios) were not significant in both seasons. Therefore, the data were arithmetic averaged across two years (a total of six replications) for further analyses [1]. Irrigation methods and N managements were considered as fixed factors. Two-way ANOVA were used to assess the effects of both water regimes (CF vs. AWD) and N managements (N0 vs. UREA vs. BBF vs. PCU) on the analysis of variance (*F*-value) of C, N and P concentrations, accumulation, partitioning, and C:N:P stoichiometric ratios in root, stem-leaf, and panicle at various growth stages, and also to test the interactions of water regimes × N managements. The least significant difference (LSD) test at the 0.05 probability level was used to compare significant differences among treatment means.

Spearman rank correlation was performed to reveal the interrelations between C:N:P ratios of plant tissues and the grain yield at harvest.

## Results

Water and N managements did alter the patterns of C, N and P concentrations, accumulation and partitioning in different plant tissues after transplanting, but no significant water by N interaction effect was found at any stage of sampling ($P>0.05$). Besides, the effects of different N treatments on tissue C, N and P concentrations, accumulation and partitioning were always more significant than those of different irrigation regimes (Tables S1 and S2 in File S1). Because irrigation and N treatments had remarkable effects on the assimilation, accumulation and allocation of C, N and P in different organs, the plant C:N:P stoichiometry was greatly affected by the water and N managements, particularly during the late-cultivation period. However, the water × N interactions on tissue C:N:P ratios were generally not significant (except root N:P ($F=3.08$, $P<0.05$) and stem-leaf N:P ($F=3.02$, $P<0.05$) at tillering stage) through the rice growing seasons (Table S3 in File S1).

### C, N and P concentrations

C concentration ranged from 295.3 to 343.2 g $kg^{-1}$ in root, 374.1 to 403.7 g $kg^{-1}$ in stem-leaf, and 401.9 to 414.2 g $kg^{-1}$ in panicle over the rice growing seasons (Fig. 1). Root C concentration showed a remarkable increase from seedling (299.0 g $kg^{-1}$) to filling (325.5 g $kg^{-1}$), and then decreased to maturity (312.2 g $kg^{-1}$). Stem-leaf C concentration increased only up to tillering (398.3 g $kg^{-1}$), and then decreased to harvest (378.0 g $kg^{-1}$). Panicle C concentration showed a slight decrease from filling (411.5 g $kg^{-1}$) to maturity (403.9 g $kg^{-1}$). Tissue C concentration was not obviously affected by the two irrigation regimes or by the three N-fertilized treatments, and UREA, BBF and PCU did not give consistently higher C concentrations than N0 from transplanting to harvest.

N concentration ranged from 6.6 to 16.3 g $kg^{-1}$ in root, 7.1 to 27.4 g $kg^{-1}$ in stem-leaf, and 11.7 to 19.5 g $kg^{-1}$ in panicle over the planting seasons (Fig. 2). Both root and stem-leaf N concentrations in N0 treatment showed clearly decreasing trends from seedling (12.9 and 18.6 g $kg^{-1}$) to harvest (6.7 and 7.2 g $kg^{-1}$), while those in N-fertilized treatments first increased from seedling (12.9 and 18.8 g $kg^{-1}$) to tillering (15.6 and 25.5 g $kg^{-1}$) and then decreased to harvest (10.8 and 11.7 g $kg^{-1}$). Panicle N concentration dropped from filling (17.0 g $kg^{-1}$) to maturity (13.4 g $kg^{-1}$) irrespective of N addition. There was no significant difference in tissue N concentration between CF and AWD at any stage of observation. As expected, N fertilization dramatically increased tissue N concentration after seedling, particularly during the late growth period. At harvest, N concentration was significantly increased by 64.7%, 61.2% and 59.0% in root, 62.6%, 63.9% and 65.7% in stem-leaf, and 20.4%, 22.2% and 11.9% in panicle in UREA, BBF and PCU compared with those of N0, respectively. However, no consistent significant differences among the three N-fertilized treatments were observed through the growing seasons.

P concentration ranged from 1.3 to 4.2 g $kg^{-1}$ in root, 1.5 to 4.3 g $kg^{-1}$ in stem-leaf, and 2.9 to 3.9 g $kg^{-1}$ in panicle over the planting seasons (Fig. 3). Both root and stem-leaf P concentrations displayed slight increases from seedling (3.4 and 3.6 g $kg^{-1}$) to tillering (3.8 and 4.1 g $kg^{-1}$), then gradual decreases until harvest (1.5 and 1.8 g $kg^{-1}$). Interestingly, panicle P concentration increased from filling (3.2 g $kg^{-1}$) to maturity (3.6 g $kg^{-1}$), exhibiting an opposite trend to those of panicle C and N concentrations. Water regimes did alter tissue P concentration, particularly during the late growth period. Compared with CF, AWD significantly decreased P concentration by 9.1%, 5.8%, and 5.6% at filling and 9.6%, 12.5%, and 7.8% at maturity in root, stem-leaf, and panicle, respectively. Tissue P concentration was not obviously affected by the N fertilization before heading, hereafter enhanced by 5.7–38.6% at filling and 2.6–30.4% at maturity in those N-fertilized treatments compared with N0, though the differences among the four N treatments were not always significant at the $P<0.05$ level.

### C, N and P accumulation

For the N0 control, both root and stem-leaf C accumulation peaked at booting stage (506 and 1824 kg $ha^{-1}$). However, root C accumulation peaked at filling stage (656–695 kg $ha^{-1}$) and stem-leaf C accumulation peaked at harvest (2904–3052 kg $ha^{-1}$) for the N-fertilized treatments (Fig. 4A). Panicle C accumulation peaked at harvest irrespective of N-fertilizer application (1996, 3120, 3132, and 3548 kg $ha^{-1}$ in N0, UREA, BBF, and PCU, respectively). The maximum rates of average C accumulation across all treatments were observed at booting in root and stem-leaf (13.8 and 78.1 kg $ha^{-1}$ $day^{-1}$) and at filling in panicle (46.8 kg $ha^{-1}$ $day^{-1}$). Tissue C accumulation did not differ remarkably between CF and AWD or among UREA, BBF and PCU at any stage of sampling, but increased significantly in the N-fertilized treatments compared with N0 after tillering. At harvest, C accumulation was enhanced by 59.1%, 71.3% and 62.5% in root, 69.5%, 62.1% and 70.4% in stem-leaf, and 56.3%, 56.9% and 77.8% in panicle in UREA, BBF and PCU compared with those of N0, respectively.

The maximum N accumulation in root and stem-leaf was observed at booting in N0 control (16.1 and 60.2 kg $ha^{-1}$), but at filling in treatments with N addition (25.5, 24.0 and 26.0 kg $ha^{-1}$ for root and 109.6, 109.5 and 118.9 kg $ha^{-1}$ for stem-leaf in UREA, BBF and PCU, respectively) (Fig. 4B). Panicle N accumulation showed a clear uptrend from filling (68.4 kg $ha^{-1}$) to maturity (98.9 kg $ha^{-1}$) regardless of N addition. The maximum rates of average N accumulation across all treatments were observed at tillering in root (0.39 kg $ha^{-1}$ $day^{-1}$), booting in stem-leaf (3.03 kg $ha^{-1}$ $day^{-1}$), and filling in panicle (1.96 kg $ha^{-1}$ $day^{-1}$). There was no significant difference in tissue N accumulation between CF and AWD at all observation stages. Obviously, N addition greatly improved plant N uptake after seedling, and the difference between fertilized and unfertilized plots expanded continuously with crop growth. At harvest, N accumulation was enhanced by 155.8%, 157.5% and 167.4% in root, 170.4%, 162.8% and 179.1% in stem-leaf, and 109.4%, 111.5% and 116.0% in panicle in UREA, BBF and PCU compared with those of N0, respectively. Besides, PCU obtained higher plant N accumulation than BBF and UREA after tillering, suggesting that PCU tended to be more effective in promoting the N transformation from soil to plant during rice growth period.

P accumulation increased remarkably up to booting in both root and stem-leaf (6.1 and 20.0 kg $ha^{-1}$), and up to harvest in panicle (26.5 kg $ha^{-1}$) (Fig. 4C). The maximum rates of average P accumulation across all treatments were observed at booting in root and stem-leaf (0.12 and 0.70 kg $ha^{-1}$ $day^{-1}$) and at filling in panicle (0.36 kg $ha^{-1}$ $day^{-1}$). AWD decreased tissue P accumulation after seedling, particularly at booting and filling stages ($P<0.05$). N fertilization had a positive effect on plant P uptake which increased with the increase of plant maturity. At harvest, P accumulation was significantly enhanced by 67.9%, 108.3% and 75.6% in root, 89.2%, 91.7% and 90.5% in stem-leaf, and 64.9%,

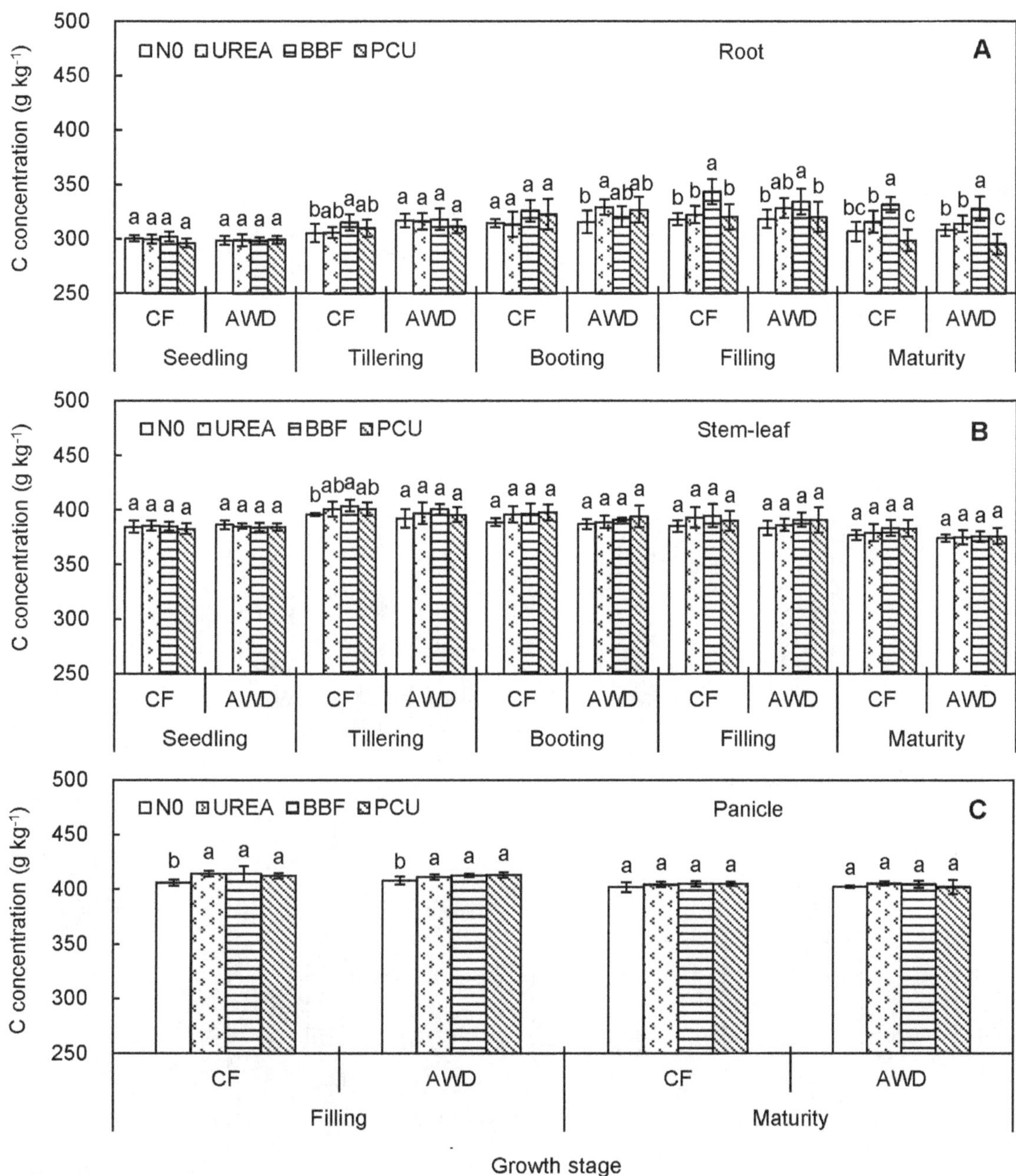


**Figure 1. Seasonal changes of carbon concentration in root (A), stem-leaf (B), and panicle (C) of late-season rice under different water and N managements (2-year average).** Vertical bars represent ± standard deviation of the mean (n = 6). The different letters listed above bars represent significant differences at $P<0.05$.

68.3% and 81.6% in panicle in UREA, BBF and PCU compared with those of N0, respectively. However, no consistent significant differences among the three N-fertilized treatments were observed through the whole rice seasons.

## C, N and P partitioning

Crop element composition depends on the process of accumulation, translocation and allocation of the elements. The seasonal variations of tissue C, N and P partitioning were shown in Table 1. C partitioning displayed a sharp decrease from transplanting (37.6%) to harvest (8.3%) in root but a remarkable increase from filling (34.8%) to maturity (47.9%) in panicle. Stem-leaf C partitioning initially increased up to booting (78.5%) and then decreased with the increase of plant maturity (43.8% at harvest). Tissue C partitioning was comparable between the two irrigation treatments at seedling, tillering and maturity stages, but increased significantly in root (7.7% and 7.9%) vs. decreased in stem-leaf (2.0% and 2.1%) in AWD compared with CF at booting and filling stages. No significant differences in tissue C partitioning among the four N treatments were found before ripening (110–130 DAT), while the highest C partitioning of root, stem-leaf, and panicle was observed in BBF (8.9%), UREA (45.3%), and PCU (49.5%) at harvest, respectively.

Tissue N partitioning exhibited a similar seasonal pattern to C partitioning (Table 1). N partitioning was also significantly higher

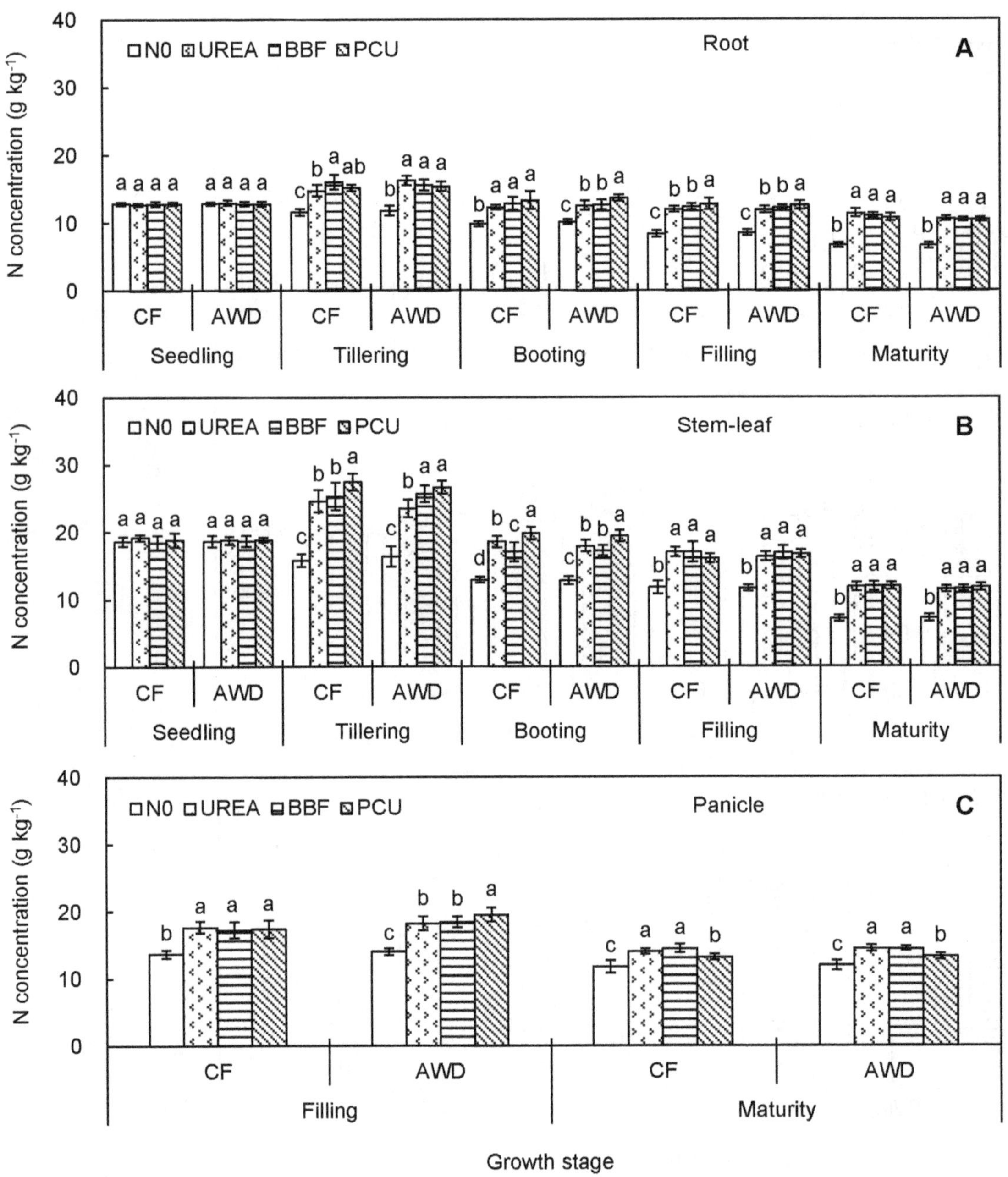


**Figure 2. Seasonal changes of nitrogen concentration in root (A), stem-leaf (B), and panicle (C) of late-season rice under different water and N managements (2-year average).** Vertical bars represent ± standard deviation of the mean (n = 6). The different letters listed above bars represent significant differences at $P<0.05$.

in root (8.1% and 5.3%) but lower in stem-leaf (2.0% and 3.9%) in AWD than CF at booting and filling stages. There were no clear differences in tissue N partitioning among the four N treatments before ripening. However, N0 resulted in the lowest N partitioning in root (7.5%) and stem-leaf (34.1%) but the highest one in panicle (58.4%) at harvest.

The seasonal changes of P partitioning in root, stem-leaf, and panicle were similar to those of C and N partitioning (Table 1). Tissue P partitioning was unaffected by the two water regimes at seedling, tillering and filling stages, but increased significantly by 6.9% in root at booting and 2.2% in panicle in AWD compared with CF at harvest. Tissue P partitioning was comparable among the four N treatments at all observation stages, indicating that N fertilization had no significant influence on P allocation.

## C:N:P stoichiometric ratios

C:N ratio ranged from 19.4 to 46.6 in root, 14.6 to 52.1 in stem-leaf, and 21.3 to 34.6 in panicle over the planting seasons (Fig. 5). Tissue C:N ratio increased gradually with crop growth after seedling, and peaked at harvest. Notably, C:N ratio was comparable between root and stem-leaf during the rice seasons, though C and N concentrations were much higher in stem-leaf than root (Figs. 1 and 2). Tissue C:N ratio did not differ significantly between CF and AWD or among UREA, BBF and

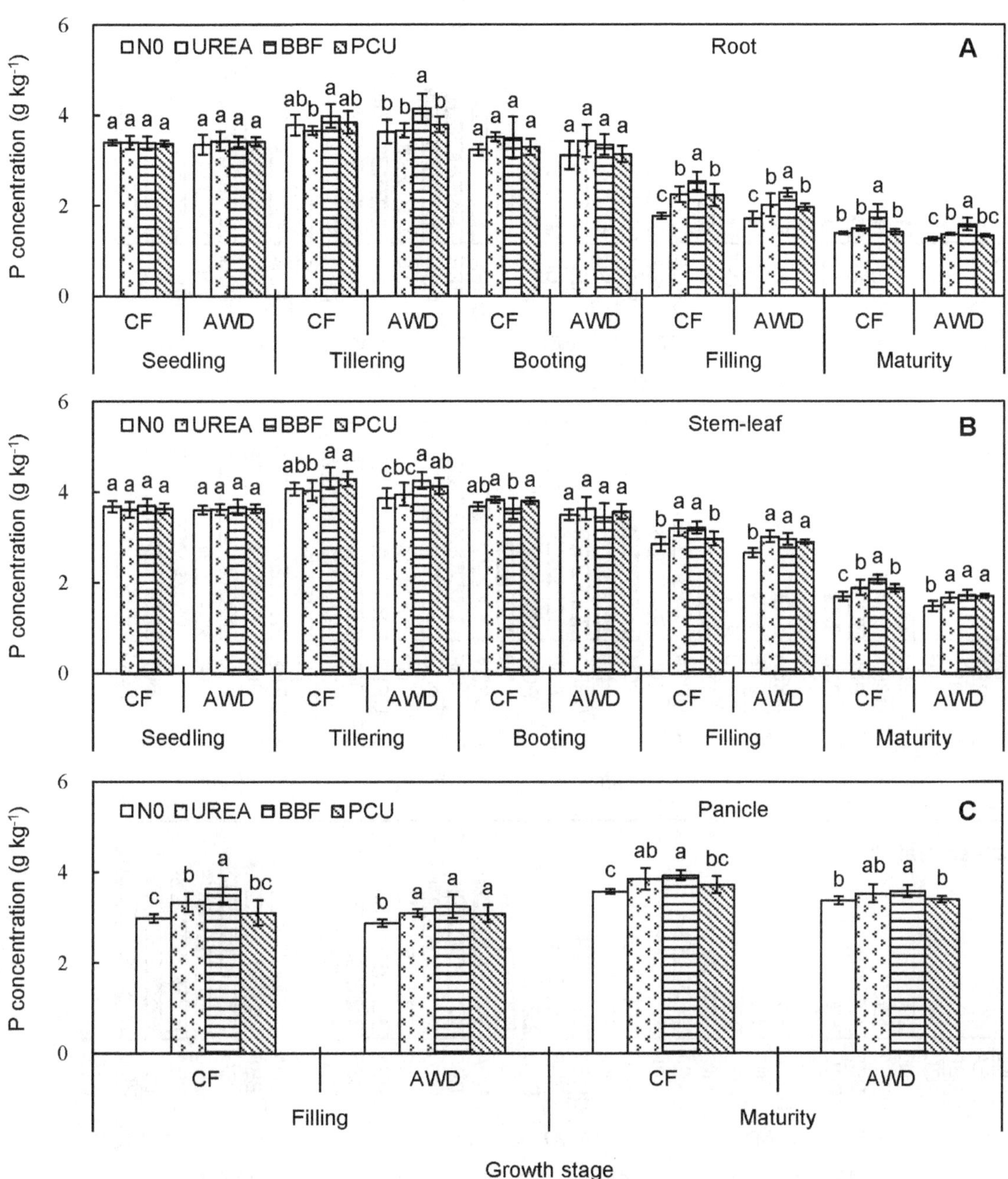


**Figure 3. Seasonal changes of phosphorus concentration in root (A), stem-leaf (B), and panicle (C) of late-season rice under different water and N managements (2-year average).** Vertical bars represent ± standard deviation of the mean (n = 6). The different letters listed above bars represent significant differences at $P<0.05$.

PCU at any stage of observation, but decreased significantly by 23.4–40.1%, 17.8–33.3%, 21.0–32.9%, and 10.6–39.3% in the N-fertilized treatments compared with those of N0 at tillering, booting, filling, and maturity stages, respectively.

N:P ratio ranged from 3.1 to 7.8 in root, 3.5 to 6.9 in stem-leaf, and 3.3 to 6.3 in panicle over the planting seasons (Fig. 6). Root N:P ratio showed no obvious difference from seedling (3.8) to tillering (3.7), then increased remarkably up to maturity (6.7). Stem-leaf N:P ratio exhibited a slight increase from booting (4.7) to maturity (6.0). Panicle N:P ratio displayed a 31.0% average reduction from filling (5.4) to maturity (3.7). The N:P ratio of panicle was significantly lower than those of root and stem-leaf at harvest (81.0% and 62.1%), though N and P concentrations of panicle were the highest among the three plant parts (Figs. 2 and 3). Tissue N:P ratio was comparable between the two irrigation regimes before heading, but enhanced significantly by 9.5% and 4.9% in root, 5.8% and 12.1% in stem-leaf, and 12.1% and 10.2% in panicle in AWD compared with CF at filling and maturity stages, respectively. N fertilization significantly increased tissue N:P ratio after seedling. The average N:P ratio was 52.2%, 24.6% and 52.5% higher in root, 44.6%, 37.3% and 46.0% higher in stem-leaf, and 13.7%, 13.2% and 9.2% higher in panicle in UREA, BBF and PCU than those of N0 at harvest, respectively. No consistent significant differences in N:P ratios of stem-leaf and panicle among the three N-fertilized treatments were observed through the rice growing seasons, except that UREA and PCU

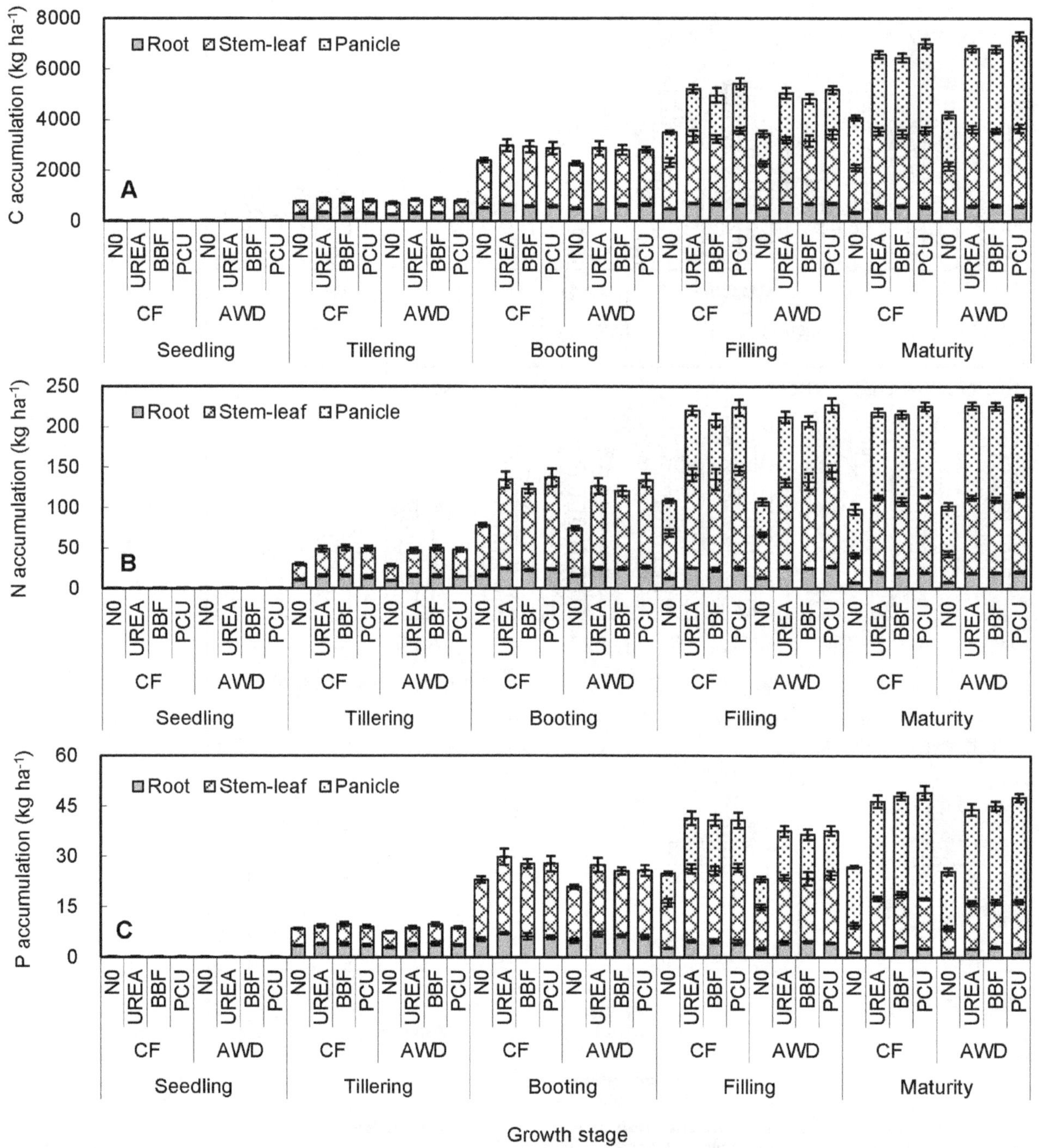


**Figure 4. Seasonal changes of carbon (A), nitrogen (B) and phosphorus (C) accumulation in root, stem-leaf, and panicle of late-season rice under different water and N managements (2-year average).** Vertical bars represent ± standard deviation of the mean (n = 6).

obtained significantly greater root N:P ratios (22.2% and 22.4%) than BBF at harvest.

C:P ratio ranged from 76.9 to 244.1 in root, 94.0 to 254.3 in stem-leaf, and 103.1 to 142.0 in panicle over the growing seasons (Fig. 7). Both root and stem-leaf C:P ratios fluctuated at around 100 before booting and then increased to 215.4 and 216.6 at harvest. As with panicle N:P ratio, panicle C:P ratio also decreased from filling (130.7) to maturity (111.9). The C:P ratio of panicle was significantly lower than those of root and stem-leaf at harvest (92.5% and 93.6%), though panicle possessed the highest C and P concentrations (Figs. 1 and 3). Tissue C:P ratio did not change significantly in response to irrigation nor to N fertilization before heading, but decreased by 5.3–9.4% at filling and 8.4–12.6% at maturity in CF compared with AWD, and reduced by 3.8–28.6% at filling and 2.0–17.1% at maturity in the N-fertilized treatments compared with N0 (although not in all cases significantly).

## Correlation studies

Tissue C:N ratio displayed significantly positive correlation with tissue C:P ratio but negative correlation with tissue N:P ratio (Table 2). Meanwhile, C:N ratios of root, stem-leaf, and panicle were significantly and positively correlated with each other.

**Table 1.** Seasonal changes of carbon, nitrogen and phosphorus partitioning in root, stem-leaf, and panicle of late-season rice under different water and N managements (2-year average).

| Growth stage | Water regime | Nitrogen treatment | C partitioning (%) | | | N partitioning (%) | | | P partitioning (%) | | |
|---|---|---|---|---|---|---|---|---|---|---|---|
| | | | Root | Stem-leaf | Panicle | Root | Stem-leaf | Panicle | Root | Stem-leaf | Panicle |
| Seedling | CF | N0 | 38.0 a | 62.0 a | - | 35.2 a | 64.8 a | - | 42.2 a | 57.8 a | - |
| | | UREA | 37.1 a | 62.9 a | - | 33.6 a | 66.4 a | - | 42.0 a | 58.0 a | - |
| | | BBF | 36.3 a | 63.7 a | - | 33.7 a | 66.3 a | - | 40.1 a | 59.9 a | - |
| | | PCU | 38.5 a | 61.5 a | - | 35.7 a | 64.3 a | - | 43.0 a | 57.0 a | - |
| | | Avg. | 37.5 A | 62.5 A | - | 34.6 A | 65.4 A | - | 41.7 A | 58.3 A | - |
| | AWD | N0 | 37.3 a | 62.7 a | - | 34.8 a | 65.2 a | - | 41.8 a | 58.2 a | - |
| | | UREA | 37.2 a | 62.8 a | - | 34.7 a | 65.3 a | - | 42.0 a | 58.0 a | - |
| | | BBF | 37.9 a | 62.1 a | - | 35.1 a | 64.9 a | - | 42.2 a | 57.8 a | - |
| | | PCU | 38.7 a | 61.3 a | - | 35.6 a | 64.4 a | - | 43.3 a | 56.7 a | - |
| | | Avg. | 37.8 A | 62.2 A | - | 35.1 A | 64.9 A | - | 42.3 A | 57.7 A | - |
| Tillering | CF | N0 | 36.4 a | 63.6 a | - | 35.5 a | 64.5 c | | 40.9 a | 59.1 a | |
| | | UREA | 37.9 a | 62.1 a | - | 32.5 b | 67.5 b | - | 42.0 a | 58.0 a | - |
| | | BBF | 35.9 a | 64.1 a | - | 31.3 bc | 68.7 ab | - | 39.9 a | 60.1 a | - |
| | | PCU | 36.4 a | 63.6 a | - | 29.1 c | 70.9 a | - | 39.8 a | 60.2 a | - |
| | | Avg. | 36.6 A | 63.4 A | - | 32.1 A | 67.9 A | - | 40.7 A | 59.3 A | - |
| | AWD | N0 | 36.3 a | 63.7 a | - | 33.8 a | 66.2 b | - | 39.9 a | 60.1 a | - |
| | | UREA | 37.5 a | 62.5 a | - | 34.4 a | 65.6 b | - | 41.1 a | 58.9 a | - |
| | | BBF | 36.4 a | 63.6 a | - | 30.5 b | 69.5 a | - | 41.3 a | 58.7 a | - |
| | | PCU | 37.4 a | 62.6 a | - | 30.5 b | 69.5 a | - | 41.1 a | 58.9 a | - |
| | | Avg. | 36.9 A | 63.1 A | - | 32.3 A | 67.7 A | - | 40.8 A | 59.2 A | - |
| Booting | CF | N0 | 21.6 a | 78.4 a | - | 20.8 a | 79.2 b | - | 23.1 a | 76.9 a | - |
| | | UREA | 21.4 a | 78.6 a | - | 18.6 b | 81.4 a | | 24.0 a | 76.0 a | |
| | | BBF | 19.6 a | 80.4 a | - | 18.3 b | 81.7 a | - | 22.3 a | 77.7 a | - |
| | | PCU | 20.3 a | 79.7 a | - | 17.5 b | 82.5 a | - | 21.5 a | 78.5 a | - |
| | | Avg. | 20.7 B | 79.3 A | - | 18.8 B | 81.2 A | - | 22.7 B | 77.3 A | - |
| | AWD | N0 | 21.8 a | 78.2 a | - | 21.4 a | 78.6 a | - | 23.4 a | 76.6 a | - |
| | | UREA | 22.9 a | 77.1 a | - | 19.8 a | 80.2 a | - | 24.9 a | 75.1 a | - |
| | | BBF | 22.1 a | 77.9 a | - | 20.4 a | 79.6 a | - | 25.2 a | 74.8 a | - |
| | | PCU | 22.4 a | 77.6 a | - | 19.7 a | 80.3 a | - | 23.5 a | 76.5 a | - |
| | | Avg. | 22.3 A | 77.7 B | - | 20.3 A | 79.7 B | - | 24.3 A | 75.7 B | - |
| Filling | CF | N0 | 13.6 a | 52.3 a | 34.1 a | 11.6 a | 51.3 a | 37.1 a | 10.6 b | 54.3 a | 35.1 a |
| | | UREA | 13.1 a | 51.1 a | 35.8 a | 11.6 a | 52.8 a | 35.6 a | 11.5 a | 52.3 ab | 36.2 a |
| | | BBF | 13.2 a | 52.3 a | 34.7 a | 11.2 a | 53.7 a | 35.1 a | 11.7 a | 51.6 b | 36.7 a |

**Table 1.** Cont.

| Growth stage | Water regime | Nitrogen treatment | C partitioning (%) | | | N partitioning (%) | | | P partitioning (%) | | |
|---|---|---|---|---|---|---|---|---|---|---|---|
| | | | Root | Stem-leaf | Panicle | Root | Stem-leaf | Panicle | Root | Stem-leaf | Panicle |
| | | PCU | 11.6 b | 54.1 a | 34.3 a | 11.2 a | 53.9 a | 34.9 a | 10.8 b | 54.9 a | 34.3 a |
| | | Avg. | 12.8 B | 52.5 A | 34.7 A | 11.4 B | 52.9 A | 35.7 A | 11.1 A | 53.2 A | 35.7 A |
| | AWD | N0 | 14.2 a | 51.4 ab | 34.4 a | 12.1 a | 50.3 a | 37.6 a | 11.2 b | 52.8 a | 36.0 a |
| | | UREA | 14.1 a | 49.4 b | 36.5 a | 12.2 a | 49.6 a | 38.2 a | 11.6 b | 51.5 a | 36.9 a |
| | | BBF | 14.0 a | 51.5 ab | 34.5 a | 11.9 a | 52.1 a | 36.0 a | 12.6 a | 51.5 a | 35.9 a |
| | | PCU | 13.2 a | 52.9 a | 33.9 a | 11.8 a | 51.7 a | 36.5 a | 11.2 b | 53.9 a | 34.9 a |
| | | Avg. | 13.9 A | 51.3 B | 34.8 A | 12.0 A | 50.9 B | 37.1 A | 11.6 A | 52.5 A | 35.9 A |
| Maturity | CF | N0 | 8.2 b | 43.4 b | 48.4 a | 7.4 b | 34.2 c | 58.4 a | 5.6 b | 29.4 b | 65.0 a |
| | | UREA | 8.1 b | 45.7 a | 46.2 b | 8.9 a | 42.9 a | 48.2 b | 5.4 b | 32.2 a | 62.4 b |
| | | BBF | 9.0 a | 44.2 b | 46.8 b | 8.9 a | 41.1 b | 50.0 b | 6.7 a | 32.2 a | 61.1 b |
| | | PCU | 7.7 b | 43.3 b | 49.0 a | 8.7 a | 41.9 ab | 49.4 b | 5.3 b | 30.3 ab | 64.4 a |
| | | Avg. | 8.3 A | 44.1 A | 47.6 A | 8.5 A | 40.0 A | 51.5 A | 5.8 A | 31.0 A | 63.2 B |
| | AWD | N0 | 8.5 b | 43.2 b | 48.3 b | 7.5 b | 34.0 b | 58.5 a | 5.7 b | 27.9 b | 66.4 a |
| | | UREA | 8.3 b | 44.8 a | 46.9 c | 8.4 a | 41.3 a | 50.3 b | 5.6 b | 31.0 a | 63.4 b |
| | | BBF | 8.9 a | 43.4 b | 47.7 bc | 8.6 a | 40.2 a | 51.2 b | 6.5 a | 29.9 a | 63.6 b |
| | | PCU | 7.9 c | 42.1 c | 50.0 a | 8.6 a | 40.5 a | 50.9 b | 5.5 b | 29.3 ab | 65.2 ab |
| | | Avg. | 8.4 A | 43.4 A | 48.2 A | 8.3 A | 39.0 A | 52.7 A | 5.8 A | 29.6 B | 64.6 A |

Within a column for each growth stage, means followed by the same letter are not significantly different at $P<0.05$ by LSD test. Lowercase and uppercase letters indicate comparisons among four N treatments and between two irrigation regimes, respectively.

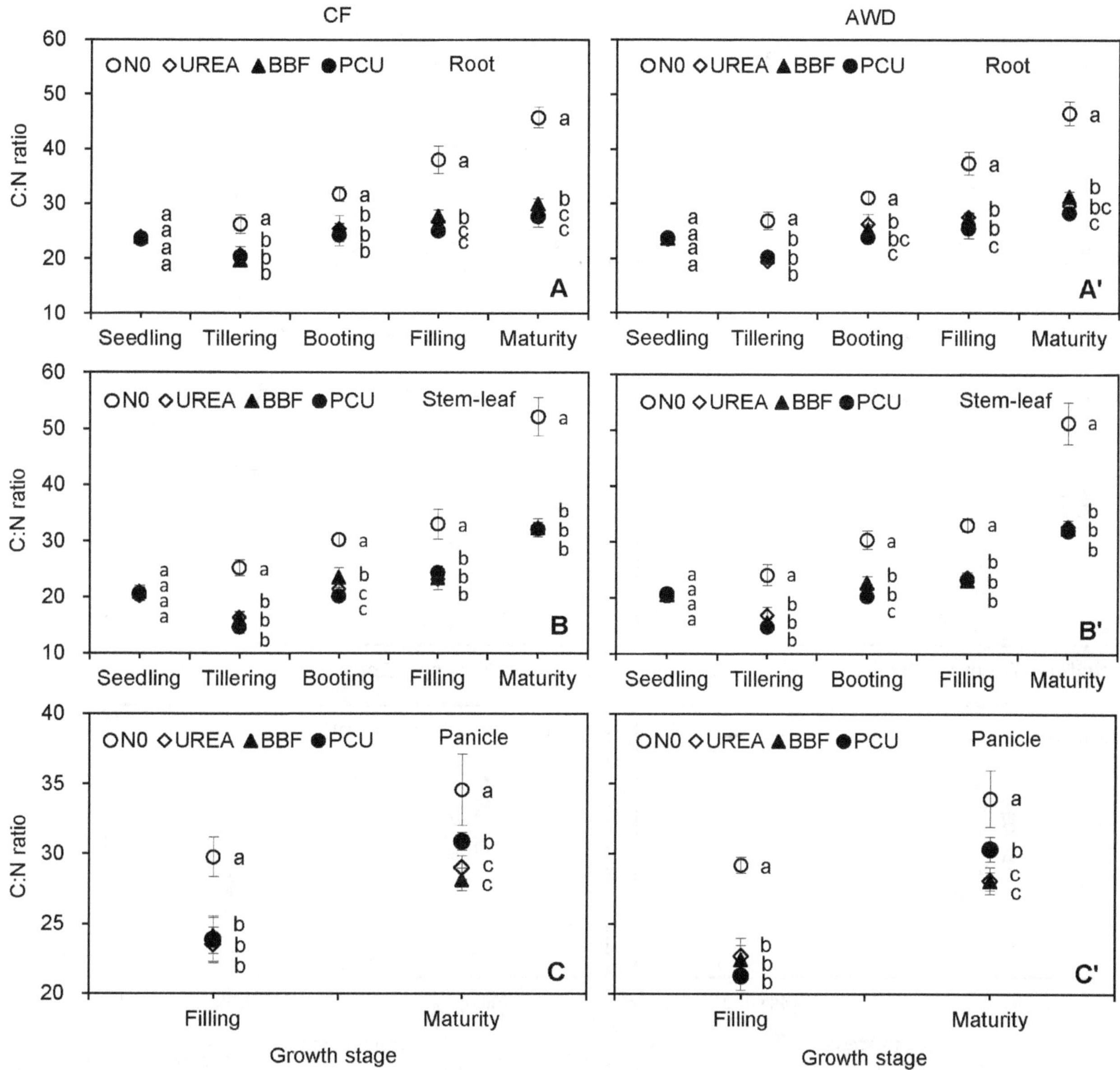


**Figure 5. Seasonal changes of C:N ratio in root (A, A′), stem-leaf (B, B′), and panicle (C, C′) of late-season rice under different water and N managements (2-year average).** Vertical bars represent ± standard deviation of the mean (n = 6). The different letters listed around bars represent significant differences at $P<0.05$.

Similar patterns of correlation were also found for N:P and C:P ratios. These results indicated that C:N:P stoichiometric ratios were highly correlated both within and across plant tissues.

Tissue N:P ratio was significantly and positively correlated with grain yield, while tissue C:N ratio got the opposite correlations. No obvious correlations were found between tissue C:P ratio and grain yield. Herein, tissue C:N and N:P ratios could have much more important implications than C:P ratio for expressing ecological stoichiometric relations in rice crop.

## Discussion

### Seasonal changes of C, N and P concentrations under different water and N managements

The C, N and P concentrations differed remarkably with plant tissues and crop growth stages. The overall increases of N and P concentrations in root and stem-leaf from seedling to tillering likely reflected the rapid soil exploration and high nutrient uptake by the crop roots. Besides, N and P concentrations in root increased only up to tillering while C concentration increased up to filling (Fig. 1), indicating that root C concentration was less affected by crop senescence during the late growth period. For stem-leaf, there exhibited systematic reductions in C, N and P concentrations from tillering to harvest (Figs. 1, 2 and 3). Decreases of nutrient concentration in vegetative parts with ontogenetic development of individual plants were documented not only in rice [5,7], but also in corn, wheat, barley, and soybean [3,24], resulting from an analogous dilution effect caused by increased plant size and biomass [13,25,26,27]. Notably, P concentration in panicle increased from filling to maturity (Figs. 1, 2 and 3), showing an opposite trend to those of C and N concentrations. This could be explained by the fact that rapidly growing organ needs relatively

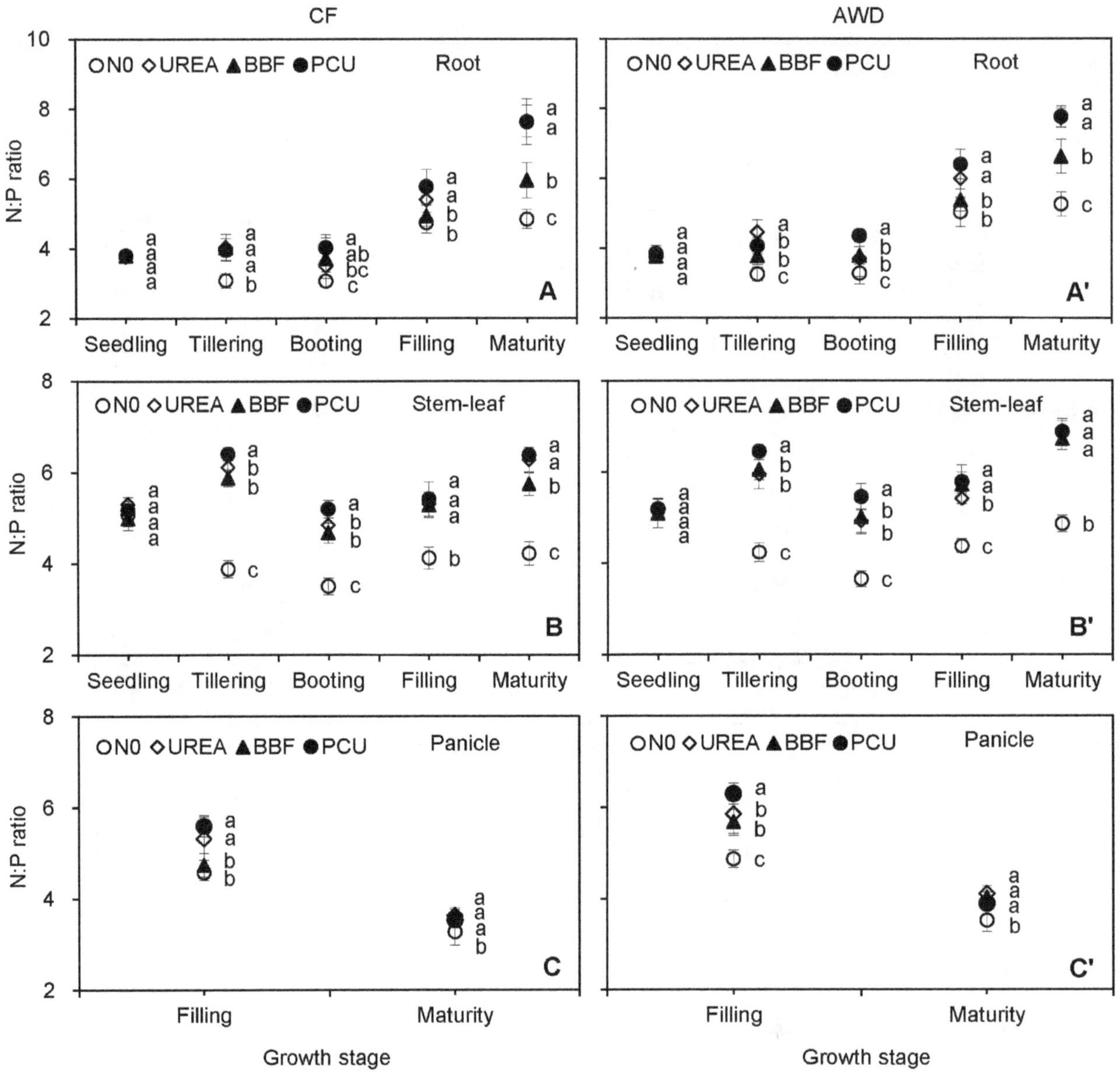


**Figure 6. Seasonal changes of N:P ratio in root (A, A'), stem-leaf (B, B'), and panicle (C, C') of late-season rice under different water and N managements (2-year average).** Vertical bars represent ± standard deviation of the mean (n = 6). The different letters listed around bars represent significant differences at $P<0.05$.

more P-rich ribosomal RNA (approx. 9% by mass) to maintain rapid rate of protein synthesis [28,29].

We noted different responses of C, N and P concentrations to CF and AWD in different plant tissues and at different growth stages, confirming that water availability played a key role in mediating element status in rice plant. Tissue C and N concentrations were comparable between CF and AWD (Figs. 1 and 2), implying that the water-saving irrigation had only a marginal effect on crop C and N assimilation. However, tissue P concentration was significantly reduced by AWD after heading (Fig. 3), probably because soil drying increased soil P sorption and led to less P availability for plant [30]. These results implied that the process governing tissue P concentration might be independent of those determining C and N concentrations. Indeed, in terms of plant, P is mainly derived from soil whereas C and N have diversified sources (e.g. $CO_2$ enrichment, SOC mineralization, N deposition and N fixation). Therefore, soil water conditions could have more influence on P compared with C and N.

Tissue C, N and P concentrations responded differently to N fertilization. C concentration was almost unaffected by N fertilization (Fig. 1), in line with previous researches [18,31], mainly due to the stable plant carbon composition and structural basis [9]. N and P concentrations were positively affected by the N application for most stages after tillering (Figs. 2 and 3), consistent with the reports of Bélanger et al. [25] who stated that increases in N concentration from increased N fertilization resulted in increased P concentration. This phenomenon was partly attributed to the increased capacity of roots to absorb more nutrients [24], and partly because of the stimulated P mobilization by the enhanced extracellular phosphatase activity [6,32] and/or mycorrhizal colonization [27,33].

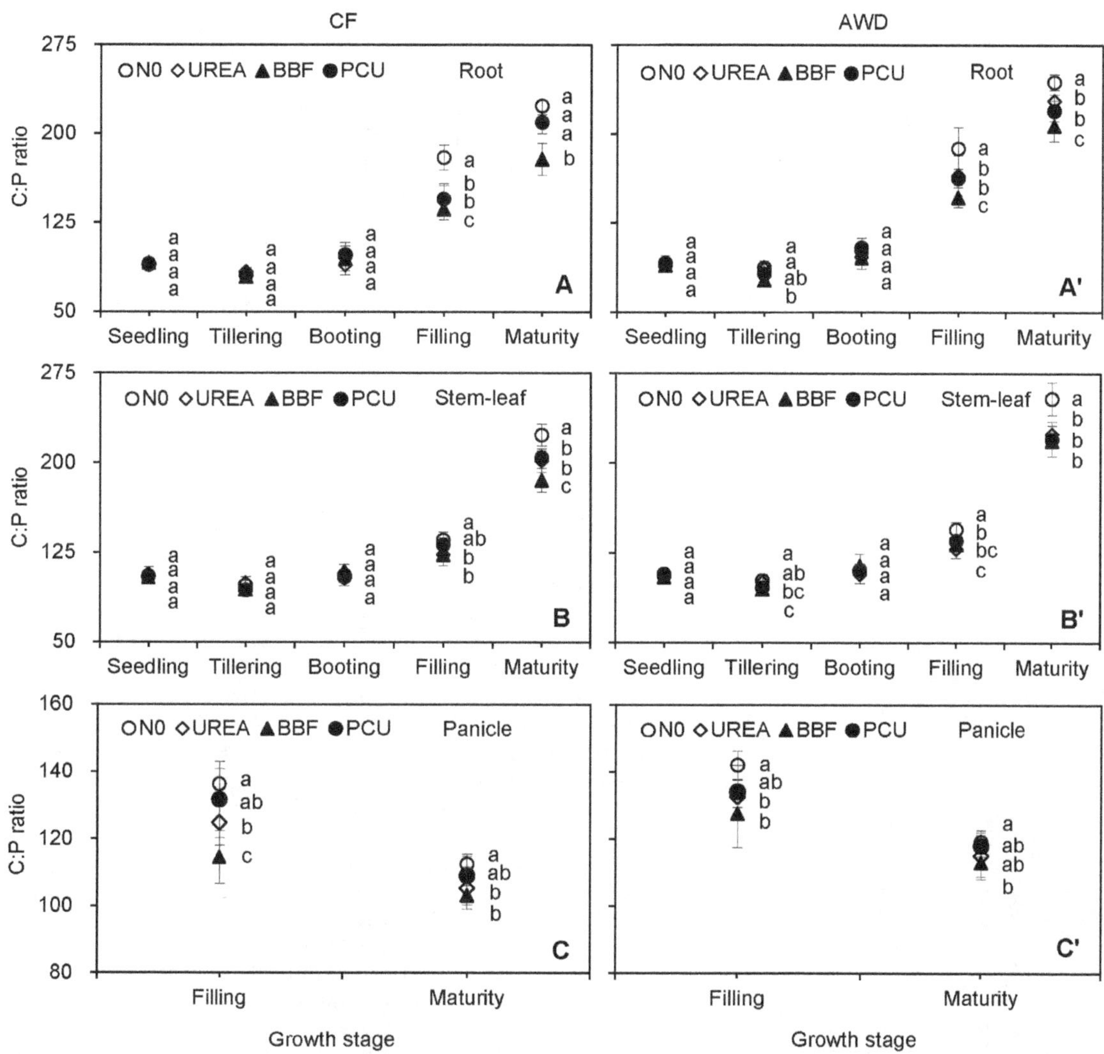


**Figure 7. Seasonal changes of C:P ratio in root (A, A′), stem-leaf (B, B′), and panicle (C, C′) of late-season rice under different water and N managements (2-year average).** Vertical bars represent ± standard deviation of the mean (n = 6). The different letters listed around bars represent significant differences at $P<0.05$.

## Seasonal changes of C, N and P accumulation and partitioning under different water and N managements

An unsynchronized tissue C, N and P accumulation was observed during rice growth. The highest accumulation of C, N and P in root, stem-leaf, and panicle emerged at different growth stages. N and P uptakes in root and stem-leaf were much lower at maturity than those at filling or even booting stage, while C, N and P accumulation in panicle all peaked at final harvest (Fig. 4). These results illustrated that carbohydrate provided by photosynthesis and nutrient derived by N and P remobilization were translocated from the senesced vegetative parts to the newly generated reproductive parts during the reproductive period [3]. The different behaviors among plant C, N and P accumulation were possibly due to the high flexibility of element composition and migration as a result of the trade-off between elements uptake and storage efficiency by the plant [34].

The complete knowledge of element allocation among plant organs is critical to evolutionary explanations of plant functional diversity [35]. As for the C, N and P partitioning of late-season rice in this study (Table 1), seasonal decreasing trends in root vs. increasing trends in panicle were exhibited. Stem-leaf was more like a transmission part between root and panicle, as its element partitioning initially increased during the early vegetative period and then decreased during the late reproductive period. Such seasonal allocation patterns of C, N and P were closely linked to (1) the obvious remobilization and retranslocation of C, N and P from root to aboveground parts before heading, and from stem-leaf to panicle hereafter, (2) the corresponding seasonal changes of biomass partitioning in different organs [4], and (3) the combined changes in plant nutrient absorbability, leaf phenology, and soil nutrient availability [32,33]. Besides, C, N and P proportions in panicle increased substantially from 34.8%, 36.5% and 35.7% at filling to 47.9%, 52.1% and 63.9% at maturity (Table 1). The rapid increases in panicle N and P fractions were in line with the data reported by Yang et al. [5,7] who found that the average N and P proportions were 10% and 7% at heading and 61% and 56% at maturity in panicle to the total aboveground rice plant. A similar pattern of element allocation was also found in maize [2] and wheat [3], demonstrating that the grain rather than stem-leaf was the major and most active sink for assimilated C, N and P in field crops. Furthermore, the reproductive allocation of N (48.2–

**Table 2.** Spearman rank correlations among tissue C:N:P stoichiometric ratios and grain yield of late-season rice at harvest.

| Trait[a] | $C:N_R$ | $C:N_{SL}$ | $C:N_P$ | $N:P_R$ | $N:P_{SL}$ | $N:P_P$ | $C:P_R$ | $C:P_{SL}$ | $C:P_P$ | GY |
|---|---|---|---|---|---|---|---|---|---|---|
| $C:N_R$ | 1 | | | | | | | | | |
| $C:N_{SL}$ | **0.648**** | 1 | | | | | | | | |
| $C:N_P$ | **0.336*** | **0.509**** | 1 | | | | | | | |
| $N:P_R$ | **−0.870**** | **−0.618**** | **−0.390**** | 1 | | | | | | |
| $N:P_{SL}$ | **−0517**** | **−0.559**** | **−0.532**** | **0.667**** | 1 | | | | | |
| $N:P_P$ | **−0.315*** | **−0.332*** | **−0.661**** | **0.385**** | **0.677**** | 1 | | | | |
| $C:P_R$ | **0.327*** | **0.365*** | **0.366*** | $0.031^{ns}$ | $-0.073^{ns}$ | $-0.010^{ns}$ | 1 | | | |
| $C:P_{SL}$ | **0.494**** | **0.605**** | **0.374**** | $-0.281^{ns}$ | $0.076^{ns}$ | $0.072^{ns}$ | **0.637**** | 1 | | |
| $C:P_P$ | **0.306*** | **0.302*** | **0.307*** | $-0.114^{ns}$ | $0.163^{ns}$ | **0.437**** | **0.504**** | **0.651**** | 1 | |
| GY | **−0.699**** | **−0.633**** | **−0.302*** | **0.852**** | **0.832**** | **0.458**** | $0.200^{ns}$ | $-0.158^{ns}$ | $0.094^{ns}$ | 1 |

Data from both CF and AWD treatments in both years were included (n = 48).
[a] $C:N_i$, $N:P_i$ and $C:P_i$: C:N, N:P and C:P ratios in different plant tissues, *i* refers to root (R), stem-leaf (SL), and panicle (P), respectively. *GY*: grain yield (data was drawn from Ye et al. [4]).
*Significant at $P<0.05$.
**Significant at $P<0.01$.
[ns] No significant.

58.5%) and P (61.1–66.4%) to the panicle exceeded that of C (46.2–48.4%) at harvest (Table 1), implying that the grain reproduction required higher fractions of N and P than C in rice.

Since intermittent irrigation has certain influence on the microclimatic conditions of paddy fields [1,36], AWD would affect crop C assimilation and nutrient uptake. In fact, plant C and N accumulation was not remarkably affected by irrigation regimes through the rice growing seasons (Figs. 4A and 4B), whereas plant P accumulation was decreased by AWD compared with CF after seedling (Fig. 4C), resulting from the decreased available P in soil under the intermittent irrigation. This phenomenon implied that the AWD irrigation influenced P accumulation in a different way from its effect on the C and N accumulation. Water regimes affected tissue C, N and P partitioning after tillering (Table 1). In general, AWD obtained lower proportions of C, N and P in stem-leaf but higher ones in root and panicle (although not in all cases significantly) especially where N supply was abundant, mainly due to the functional shifts of plant organs and the distributional differences of biomass and nutrients [33] in response to the water-saving irrigation.

N fertilization greatly enhanced tissue C, N and P accumulation after transplanting (Fig. 4), resulting from both increased tissue element concentration (Figs. 1, 2 and 3) and biomass [4]. Meanwhile, N fertilizer (especially the PCU) application delayed the emerging of peak values of C and N accumulation in root and stem-leaf, owing to the extended soil N availability. In addition, the maximum rates of average C, N and P accumulation always appeared at the booting stage, mostly attributing to the fact that fast-growing plant tissues need relatively more elements to support rapid rate of cell proliferation. In order to take full advantage of the high uptake rate of N during the middle growth period and facilitate N harvesting during the late growth period, the delay of N release was necessary, which could be realized by the CRNFs [5]. In fact, the PCU enhanced not only N uptake but also C and P accumulation in panicle, shoot and whole plant compared with BBF and UREA at an equivalent N rate (Fig. 4). As a compound CRNF product, BBF failed to achieve a comparative effect on the increment of element accumulation as the PCU, owing to the lower components of controlled-release N source (70%) and the less effectiveness in controlling N release than the PCU [4]. N fertilization also altered N partitioning in different plant tissues. Zero N addition resulted in the lowest N partitioning in root (7.5%) and stem-leaf (34.1%) but the highest one in panicle (58.4%) at harvest (Table 1), which was in agreement with Yang et al. [5], suggesting that rice has evolved some internal regulation and conveyance strategies [37] to promote the absorbed N preferentially transferred from vegetative part to the reproductive organ when confronted with soil N deficiency.

## Seasonal variations of C:N:P ratios in different tissues under different water and N managements

C:N:P stoichiometry is an important and sensitive index reflecting diverse ecological processes. Knowledge on the C:N ratio of crop residues is of great importance for modelling C and N dynamics in agricultural systems [38]. We found that the stem-leaf C:N ratio showed an overall uptrend after seedling (Fig. 5), agreeing with the results of Ruan et al. [39] who observed that leaf C:N ratio increased from 20.8 at heading to 30.6 at filling and then to 32.0 at maturity in hybrid rice varieties. The stem-leaf C:P ratio exhibited a similar increasing trend as stem-leaf C:N ratio, owing to the reduced allocation of N and P to the senesced leaves. Panicle, which had higher C, N and P concentrations than root and stem-leaf, presented the lowest C:N, N:P and C:P ratios at harvest (Figs. 5, 6 and 7). Gan et al. [38] found that C:N ratio of seed (6–17) was significantly lower than those of straw (14–55) and root (17–75) at maturity in oilseed and pulse crops. Bélanger et al. [25] reported that grain N:P ratio (2.6–7.4) was lower and less variable than those of whole shoot (3.6–12.9) and the uppermost collared leaf (6.8–16.6) in maize. These results revealed that the storage-related tissues (panicle/seed) which optimized N and P use relative to C assimilation for grain production possessed lower C:N:P ratios than the growth-related tissues (stem/leaf).

Water availability that influences leaf phenology and photosynthesis rate could affect plant growth, nutritional status, and ultimately the stoichiometric ratios [32]. In this study, AWD irrigation did not significantly affect the tissue C:N ratio (Fig. 5), because neither tissue C nor N concentrations varied significantly under water-saving irrigation (Figs. 1 and 2). Besides, rice was unlikely to be water-limited under current experimental conditions because the AWD irrigation experienced in this study was within the 'Safe AWD' threshold (water level at 150 mm below the soil surface), and the soil in AWD plots was kept relatively wet and saturated during the non-submerged periods [4]. Plant N:P and C:P ratios were significantly enhanced by AWD irrigation at filling and maturity stages (Figs. 6 and 7), resulting from the reduced tissue P concentration under the alternate submergence-nonsubmergence (Fig. 3). These results indicated that the C:N ratio was much less sensitive to water conditions than the other two stoichiometric ratios, and confirmed that the AWD irrigation had important implications for plant-mediated C, N and P biogeochemical cycling in rice production systems. Lü et al. [32] noted that when water availability was enhanced, higher tissue P and unaltered N concentrations resulted in lower foliar N:P ratio in grassland plant species. Sardans & Peñuelas [11] pointed out that plants under drought tended to have high C:P ratio in leaves and litters.

Terrestrial plants change their C:N and N:P ratios in response to changes in N availability [20,32]. N dynamics have been proved to drive stoichiometric shifts in both plant tissues and mineral soils [31]. It has, in fact, become a widely stated view that increasing soil N availability through N fertilization or atmospheric N deposition would increase N:P but decrease C:N ratios in plants [20,25,32,39]. For instance, a meta-analysis demonstrated that N addition significantly increased the N:P but reduced the C:N ratios of the photosynthetic tissues of woody and herbaceous plants [20]. Our results were in accordance with these studies because N fertilization obtained higher increases in tissue N concentration than P concentration but had no visible effect on tissue C concentration (Figs. 1, 2 and 3). Although the responses of plant C:N and N:P ratios to N addition have been widely investigated, the response of C:P ratio to N addition has received less attention. We found that N application decreased tissue C:P ratio in the late growth period (Fig. 7), reflecting that N availability had significant impacts on P uptake (Fig. 3) but little effects on C assimilation (Fig. 1).

### Implication of the C:N:P stoichiometry for evaluating nutrient limitation in rice production systems

Plant C:N:P stoichiometry determines plant community composition and structure (resource allocation), trophic dynamics, and nutrient limitation [19,20,33]. Among those stoichiometric ratios, N:P ratio is the most widely investigated because it reflects important biochemical constraints on relative investments in N-rich proteins (approx. 16% by mass) and the P-rich ribosomal RNA used to generate them [19]. Though fluctuating across a broad range (approx. 1–100) in individual measurements [15], N:P ratio, with high diagnostic value for nutrient limitation [22], offers a simple and cost-effective tool for evaluating the limiting patterns of nutrient from individual plant species to terrestrial ecosystems [6,19,40,41]. Based on theoretical considerations and laboratory results, the 'Liebig's law of the minimum' for N and P suggests that there is a 'critical (also called ideal or optimal) N:P ratio' below which plant growth is limited only by N and above which growth is limited only by P, and within which growth is co-limited by both elements [15,22,40,42]. Tremendous researches were carried out on this critical N:P ratio in natural vegetations. Koerselman & Meuleman [40] suggested an optimal aboveground biomass N:P ratio of 14–16 in wetland plant species by a review of 40 fertilization studies. Güsewell [15] proposed a broader range of 10–20 as the ideal N:P ratio in terrestrial plants from short-term fertilization experiments. Knecht & Göransson [42] sorted out an average optimal N:P ratio of about 10.0 in field-grown terrestrial plants from published data. Sadras et al. [22] reported that the critical N:P ratio was 4.5 for oilseed crops (n = 81), 5.6 for cereals (n = 134) and 8.7 for legumes (n = 52), and stated that over 40% of cereal and oilseed crops attaining maximum yield had N:P ratios in a relatively narrow range between 4 and 6. Aulakh & Malhi [43] summarized that the main cereal crops, such as rice, wheat, and corn, typically had optimal N:P ratios in both grain and straw in a fairly narrow range of 4.2–6.7. As for rice, Witt et al. [44] simulated balanced nutrient uptakes of 14.7 kg N and 2.6 kg P per ton of grain with a corresponding N:P ratio of 5.7. In this study, the rice aboveground plant N:P ratio (excluding N0 treatment) ranged from 4.4 to 6.4 through the growing seasons, which was somewhat lower than those critical N:P ratios in other plant species mentioned above (10–20) but still remained within the normal range (4.2–6.7), suggesting that the growth of rice in this region was co-limited by N and P. Collectively, these findings provided valuable insights into the mechanisms underpinning plant essential elements cycling in response to simultaneous changes in water and N availability, and broadened the knowledge of the C:N:P stoichiometry in subtropical high-yield rice systems.

## Supporting Information

**File S1 Table S1,** Combined analysis of variance (*F* values) for C, N and P concentrations in root, stem-leaf, and panicle of late-season rice at various growth stages under different water and N managements in 2010–2011. **Table S2,** Combined analysis of variance (*F* values) for C, N and P accumulation and partitioning in root, stem-leaf, and panicle of late-season rice at various growth stages under different water and N managements in 2010–2011. **Table S3,** Combined analysis of variance (*F* values) for C:N:P stoichiometric ratios in root, stem-leaf, and panicle of late-season rice at various growth stages under different water and N managements in 2010–2011.

## Author Contributions

Conceived and designed the experiments: XL. Performed the experiments: YY. Analyzed the data: YY. Contributed reagents/materials/analysis tools: YC YJ LL CZ. Wrote the paper: YY.

## References

1. Mahajan G, Chauhan BS, Timsina J, Singh PP, Singh K (2012) Crop performance and water- and nitrogen-use efficiencies in dry-seeded rice in response to irrigation and fertilizer amounts in northwest India. Field Crops Research 134: 59–70.

2. Ning P, Li S, Yu P, Zhang Y, Li C (2013) Post-silking accumulation and partitioning of dry matter, nitrogen, phosphorus and potassium in maize varieties differing in leaf longevity. Field Crops Research 144: 19–27.
3. Dordas C (2009) Dry matter, nitrogen and phosphorus accumulation, partitioning and remobilization as affected by N and P fertilization and source-sink relations. European Journal of Agronomy 30: 129–139.
4. Ye YS, Liang XQ, Chen YX, Liu J, Gu JT, et al. (2013) Alternate wetting and drying irrigation and controlled-release nitrogen fertilizer in late-season rice. Effects on dry matter accumulation, yield, water and nitrogen use. Field Crops Research 144: 212–224.
5. Yang LX, Huang HY, Yang HJ, Dong GC, Liu HJ, et al. (2007a) Seasonal changes in the effects of free-air $CO_2$ enrichment (FACE) on nitrogen (N) uptake and utilization of rice at three levels of N fertilization. Field Crops Research 100: 189–199.
6. Ågren GI, Martin Wetterstedt JÅ, Billberger MFK (2012) Nutrient limitation on terrestrial plant growth - modeling the interaction between nitrogen and phosphorus. New Phytologist 194: 953–960.
7. Yang LX, Wang YL, Huang JY, Zhu JG, Yang HJ, et al. (2007b) Seasonal changes in the effects of free-air $CO_2$ enrichment (FACE) on phosphorus uptake and utilization of rice at three levels of nitrogen fertilization. Field Crops Research 102: 141–150.
8. Ju XT, Xing GX, Chen XP, Zhang SL, Zhang LJ, et al. (2009) Reducing environmental risk by improving N management in intensive Chinese agricultural systems. Proceedings of the National Academy of Sciences of the United States of America 106: 3041–3046.
9. Ågren GI (2008) Stoichiometry and nutrition of plant growth in natural communities. Annual Review of Ecology, Evolution, and Systematics 39: 153–170.
10. Pampolino MF, Laureles EV, Gines HC, Buresh RJ (2008) Soil carbon and nitrogen changes in long-term continuous lowland rice cropping. Soil Science Society of America Journal 72: 798–807.
11. Sardans J, Peñuelas J (2012) The role of plants in the effects of global change on nutrient availability and stoichiometry in the plant-soil system. Plant Physiology 160: 1741–1761.
12. Lal R (2004) Soil carbon sequestration impacts on global climate change and food security. Science 304: 1623–1627.
13. Kim HY, Lim SS, Kwak JH, Lee DS, Lee SM, et al. (2011) Dry matter and nitrogen accumulation and partitioning in rice (*Oryza sativa* L.) exposed to experimental warming with elevated $CO_2$. Plant and Soil 342: 59–71.
14. Seneweera S (2011) Effects of elevated $CO_2$ on plant growth and nutrient partitioning of rice (*Oryza sativa* L.) at rapid tillering and physiological maturity. Journal of Plant Interactions 6: 35–42.
15. Güsewell S (2004) N:P ratios in terrestrial plants: variation and functional significance. New Phytologist 164: 243–266.
16. Elser JJ, Fagan WF, Denno RF, Dobberfuhl DR, Folarin A, et al. (2000) Nutritional constraints in terrestrial and freshwater food webs. Nature 408: 578–580.
17. Ågren GI, Weih M (2012) Plant stoichiometry at different scales: element concentration patterns reflect environment more than genotype. New Phytologist 194: 944–952.
18. Elser JJ, Fagan WF, Kerkhoff AJ, Swenson NG, Enquist BJ (2010) Biological stoichiometry of plant production: metabolism, scaling and ecological response to global change. New Phytologist 186: 593–608.
19. Sterner RW, Elser JJ (2002) Ecological Stoichiometry: the Biology of Elements from Molecules to the Biosphere. Princeton University Press, Princeton.
20. Sardans J, Rivas-Ubach A, Peñuelas J (2012) The C:N:P stoichiometry of organisms and ecosystems in a changing world: A review and perspectives. Perspectives in Plant Ecology, Evolution and Systematics 14: 33–47.
21. Greenwood DJ, Karpinets TV, Zhang K, Bosh-Serra A, Boldrini A, et al. (2008) A unifying concept for the dependence of whole-crop N:P ratio on biomass: theory and experiment. Annals of Botany 102: 967–977.
22. Sadras VO (2006) The N:P stoichiometry of cereal, grain legume and oilseed crops. Field Crops Research 95: 13–29.
23. Bao SD (2000) Soil and Agricultural Chemistry Analysis, 3rd ed. China Agriculture Press, Beijing, 268–270. (in Chinese).
24. Ziadi N, Bélanger G, Cambouris AN, Tremblay N, Nolin MC, et al. (2007) Relationship between P and N concentrations in corn. Agronomy Journal 99: 833–841.
25. Bélanger G, Claessens A, Ziadi N (2012) Grain N and P relationships in maize. Field Crops Research 126: 1–7.
26. Zhang HY, Wu HH, Yu Q, Wang ZW, Wei CZ, et al. (2013) Sampling date, leaf age and root size: implications for the study of plant C:N:P stoichiometry. Plos One 8: e60360.
27. Gifford RM, Barrett DJ, Lutze JL (2000) The effects of elevated [$CO_2$] on the C:N and C:P mass ratios of plant tissues. Plant and Soil 224: 1–14.
28. Yu Q, Wu HH, He NP, Lü XT, Wang ZP, et al. (2012) Testing the growth rate hypothesis in vascular plants with above- and below-ground biomass. Plos One 7: e32162.
29. Matzek V, Vitousek PM (2009) N:P stoichiometry and protein:RNA ratios in vascular plants: an evaluation of the growth-rate hypothesis. Ecology Letters 12: 765–771.
30. Haynes RJ, Swift RS (1989) The effects of pH and drying on adsorption of phosphate by aluminium-organic matter associations. Journal of Soil Science 40: 773–781.
31. Yang YH, Luo YQ, Lu M, Schadel C, Han WX (2011) Terrestrial C:N stoichiometry in response to elevated $CO_2$ and N addition: a synthesis of two meta-analyses. Plant and Soil 343: 393–400.
32. Lü XT, Kong DL, Pan QM, Simmons ME, Han XG (2012) Nitrogen and water availability interact to affect leaf stoichiometry in a semi-arid grassland. Oecologia 168: 301–310.
33. Zheng SX, Ren HY, Li WH, Lan ZC (2012) Scale-dependent effects of grazing on plant C:N:P stoichiometry and linkages to ecosystem functioning in the Inner Mongolia grassland. Plos One 7: e51750.
34. Abbas M, Ebeling A, Oelmann Y, Ptacnik R, Roscher C, et al. (2013) Biodiversity effects on plant stoichiometry. Plos One 8: e58179.
35. Kerkhoff AJ, Fagan WF, Elser JJ, Enquist BJ (2006) Phylogenetic and growth form variation in the scaling of nitrogen and phosphorus in the seed plants. The American Naturalist 168: E103–E122.
36. Mao Z (2001) Water efficient irrigation and environmentally sustainable irrigated rice production in China. ICID website. Available: http://www.icid.org/wat_mao.pdf. Accessed 2014 Jun 12.
37. Rentsch D, Schmidt S, Tegeder M (2007) Transporters for uptake and allocation of organic nitrogen compounds in plants. FEBS Letters 581: 2281–2289.
38. Gan YT, Liang BC, Liu LP, Wang XY, McDonald CL (2011) C:N ratios and carbon distribution profile across rooting zones in oilseed and pulse crops. Crop & Pasture Science 62: 496–503.
39. Ruan XM, Shi FZ, Luo ZX (2011) Effects of nitrogen application on the leaf C:N and nitrogen uptake and utilization at later developmental stages in different high yield hybrid rice varieties. Soil and Fertilizer Sciences in China (2): 35–38. (in Chinese).
40. Koerselman W, Meuleman AFM (1996) The vegetation N:P ratio: a new tool to detect the nature of nutrient limitation. Journal of Applied Ecology 33: 1441–1450.
41. Tessier JT, Raynal DJ (2003) Use of nitrogen to phosphorus ratios in plant tissue as an indicator of nutrient limitation and nitrogen saturation. Journal of Applied Ecology 40: 523–534.
42. Knecht MR, Göransson A (2004) Terrestrial plants require nutrients in similar proportions. Tree Physiology 24: 447–460.
43. Aulakh MS, Malhi SS (2005) Interactions of nitrogen with other nutrients and water: effect on crop yield and quality, nutrient use efficiency, carbon sequestration, and environmental pollution. Advances in Agronomy 86: 341–409.
44. Witt C, Dobermann A, Abdulrachman S, Gines HC, Guanghuo W, et al. (1999) Internal nutrient efficiencies of irrigated lowland rice in tropical and subtropical Asia. Field Crops Research 63: 113–138.

# PERMISSIONS

All chapters in this book were first published in PLOS ONE, by The Public Library of Science; hereby published with permission under the Creative Commons Attribution License or equivalent. Every chapter published in this book has been scrutinized by our experts. Their significance has been extensively debated. The topics covered herein carry significant findings which will fuel the growth of the discipline. They may even be implemented as practical applications or may be referred to as a beginning point for another development.

The contributors of this book come from diverse backgrounds, making this book a truly international effort. This book will bring forth new frontiers with its revolutionizing research information and detailed analysis of the nascent developments around the world.

We would like to thank all the contributing authors for lending their expertise to make the book truly unique. They have played a crucial role in the development of this book. Without their invaluable contributions this book wouldn't have been possible. They have made vital efforts to compile up to date information on the varied aspects of this subject to make this book a valuable addition to the collection of many professionals and students.

This book was conceptualized with the vision of imparting up-to-date information and advanced data in this field. To ensure the same, a matchless editorial board was set up. Every individual on the board went through rigorous rounds of assessment to prove their worth. After which they invested a large part of their time researching and compiling the most relevant data for our readers.

The editorial board has been involved in producing this book since its inception. They have spent rigorous hours researching and exploring the diverse topics which have resulted in the successful publishing of this book. They have passed on their knowledge of decades through this book. To expedite this challenging task, the publisher supported the team at every step. A small team of assistant editors was also appointed to further simplify the editing procedure and attain best results for the readers.

Apart from the editorial board, the designing team has also invested a significant amount of their time in understanding the subject and creating the most relevant covers. They scrutinized every image to scout for the most suitable representation of the subject and create an appropriate cover for the book.

The publishing team has been an ardent support to the editorial, designing and production team. Their endless efforts to recruit the best for this project, has resulted in the accomplishment of this book. They are a veteran in the field of academics and their pool of knowledge is as vast as their experience in printing. Their expertise and guidance has proved useful at every step. Their uncompromising quality standards have made this book an exceptional effort. Their encouragement from time to time has been an inspiration for everyone.

The publisher and the editorial board hope that this book will prove to be a valuable piece of knowledge for researchers, students, practitioners and scholars across the globe.

# LIST OF CONTRIBUTORS

**Syed Tahir Ata-Ul-Karim, Xia Yao, Xiaojun Liu, Weixing Cao and Yan Zhu**
National Engineering and Technology Center for Information Agriculture, Jiangsu Key Laboratory for Information Agriculture, Nanjing Agricultural University, Nanjing, Jiangsu, P. R. China

**Martin Lechenet, Sandrine Petit and Nicolas M. Munier-Jolain**
Institut National de la Recherche Agronomique, Unité Mixte de Recherche 1347 Agroécologie, Dijon, Côte d'Or, France

**Vincent Bretagnolle**
Centre d'Etudes Biologiques de Chizé – Centre National de Recherche Scientifique, Beauvoir sur Niort, Deux-Sévres, France

**Christian Bockstaller**
Institut National de la Recherche Agronomique, Unité de Recherche 1121 Agronomie et Environnement, Colmar, Haut-Rhin, France
Université de Lorraine, Vandoeuvre-lés-Nancy, Meurthe-et-Moselle, France

**François Boissinot**
Chambre d'Agriculture des Pays de la Loire, Angers, Maine-et-Loire, France

**Marie-Sophie Petit**
Chambre Régionale d'Agriculture de Bourgogne, Quetigny, Côte d'Or, France

**Keiko Midorikawa, Kaede Terauchi, Masako Hoshi, Yoshiro Ishimaru and Tomiko Asakura**
Department of Applied Biological Chemistry, Graduate School of Agricultural and Life Sciences, The University of Tokyo, Bunkyo-ku, Tokyo, Japan

**Masaharu Kuroda**
Crop Development Division, NARO Agricultural Research Center, Inada, Joetsu, Niigata, Japan

**Sachiko Ikenaga**
Field Crop and Horticulture Research Division, NARO Tohoku Agricultural Research Center, Morioka, Iwate, Japan

**Keiko Abe**
Food Safety and Reliability Project, Kanagawa Academy of Science and Technology, Takatsu-ku, Kawasaki, Kanagawa, Japan

**Tao Ren**
College of Resources and Environmental Science, China Agricultural University, Beijing, China
College of Resources and Environment, Huazhong Agricultural University, Wuhan, China

**Jingguo Wang, Qing Chen and Fusuo Zhang**
College of Resources and Environmental Science, China Agricultural University, Beijing, China

**Shuchang Lu**
Department of Agronomy, Tianjin Agricultural University, Tianjin, China

**Forough Aghili, Hannes A. Gamper and Emmanuel Frossard**
Institute of Agricultural Sciences, Department of Environmental Systems Science, Swiss Federal Institute of Technology (ETH) Zürich, Switzerland

**Jost Eikenberg**
Paul Scherrer Institute (PSI), Radioanalytics Laboratory, Villigen, Switzerland

**Amir H. Khoshgoftarmanesh and Majid Afyuni**
College of Agriculture, Department of Soil Sciences, Isfahan University of Technology, Isfahan, Iran

**Rainer Schulin**
Institute of Terrestrial Ecosystems, Department of Environmental Systems Science, Swiss Federal Institute of Technology (ETH) Zürich, Switzerland

**Jan Jansa**
Institute of Microbiology, Academy of Sciences of the Czech Republic, Prague, Czech Republic

**Wei Ouyang, Guanqing Cai and Fanghua Hao**
School of Environment, State Key Laboratory of Water Environment Simulation, Beijing Normal University, Beijing, China

**Siyang Chen**
Marine Monitoring and Forecasting Center of Zhejiang, Hangzhou, China

**Zhimin Du, Yan Xie, Longxing Hu and Jinmin Fu**
Key Laboratory of Plant Germplasm Enhancement and Specialty Agriculture, Wuhan Botanical Garden, Chinese Academy of Sciences, Wuhan, Hubei, China
Graduate University of Chinese Academy of Sciences, Beijing, Hebei, China

**Liqun Hu**
Key Laboratory of Plant Germplasm Enhancement and Specialty Agriculture, Wuhan Botanical Garden, Chinese Academy of Sciences, Wuhan, Hubei, China

**Shendong Xu, Daoxin Li and Gongfang Wang**
National Dalaoling Forest Park, Yichang, Hubei, China

**Yangquanwei Zhong and Zhouping Shangguan**
State Key Laboratory of Soil Erosion and Dryland Farming on the Loess Plateau, Northwest A & F University, Yangling, Shaanxi, P.R. China

**Magalie Canuel**
Institut national de santé publique du Québec (INSPQ), Québec City, Canada

**Belkacem Abdous**
Centre de recherche du Centre hospitalier universitaire de Québec, Québec City, Canada
Département de médecine sociale et préventive de l9Université Laval, Québec City, Canada

**Diane Bélanger**
Centre de recherche du Centre hospitalier universitaire de Québec, Québec City, Canada
Institut national de la recherche scientifique, Centre Eau Terre Environnement, Québec City, Canada

**Pierre Gosselin**
Institut national de santé publique du Québec (INSPQ), Québec City, Canada
Centre de recherche du Centre hospitalier universitaire de Québec, Québec City, Canada
Institut national de la recherche scientifique, Centre Eau Terre Environnement, Québec City, Canada

**Jean A. Hall and William R. Vorachek**
Department of Biomedical Sciences, College of Veterinary Medicine, Oregon State University, Corvallis, Oregon, United States of America

**Gerd Bobe**
Department of Animal and Rangeland Sciences, College of Agricultural Sciences, Oregon State University, Corvallis, Oregon, United States of America
Linus Pauling Institute, Oregon State University, Corvallis, Oregon, United States of America

**Janice K. Hunter**
Department of Animal and Rangeland Sciences, College of Agricultural Sciences, Oregon State University, Corvallis, Oregon, United States of America
Deep Springs College, Dyer, Nevada, United States of America

**Whitney C. Stewart**
Department of Animal and Rangeland Sciences, College of Agricultural Sciences, Oregon State University, Corvallis, Oregon, United States of America
Texas AgriLife Research, San Angelo, Texas, United States of America

**Jorge A. Vanegas**
Department of Clinical Sciences, College of Veterinary Medicine, Oregon State University, Corvallis, Oregon, United States of America

**Charles T. Estill**
Department of Animal and Rangeland Sciences, College of Agricultural Sciences, Oregon State University, Corvallis, Oregon, United States of America
Department of Clinical Sciences, College of Veterinary Medicine, Oregon State University, Corvallis, Oregon, United States of America

**Wayne D. Mosher and Gene J. Pirelli**
Department of Animal and Rangeland Sciences, College of Agricultural Sciences, Oregon State University, Corvallis, Oregon, United States of America

**Liang Wu, Xinping Chen, Zhenling Cui, Weifeng Zhang and Fusuo Zhang**
Center for Resources, Environment and Food Security, China Agricultural University, Beijing, People's Republic of China

**Dejun Li**
Department of Microbiology and Plant Biology, University of Oklahoma, Norman, Oklahoma, United States of America

**Marijn van der Velde, Linda See, Juraj Balkovič Steffen Fritz, Nikolay Khabarov and Michael Obersteiner**
International Institute for Applied Systems Analysis (IIASA), Ecosystem Services and Management Program, Laxenburg, Austria

**Liangzhi You**
International Food Policy Research Institute (IFPRI), Washington D.C., United States of America
College of Economics and Management, Huazhong Agricultural University, Wuhan, China

**Stanley Wood**
Global Development Program, Bill & Melinda Gates Foundation, Seattle, Washington, United States of America

**John K. Senior**
School of Plant Science, University of Tasmania, Hobart, TAS, Australia

**Jennifer A. Schweitzer and Joseph K. Bailey**
School of Plant Science, University of Tasmania, Hobart, TAS, Australia
Department of Ecology and Evolutionary Biology, University of Tennessee, Knoxville, Tennessee, United States of America

**Julianne O'Reilly-Wapstra**
School of Plant Science and National Centre for Future Forest Industries, University of Tasmania, Hobart, TAS, Australia

**Samantha K. Chapman and Adam Langley**
Department of Biology, Villanova University, Villanova, Pennsylvania, United States of America

**Dorothy Steane**
School of Plant Science, University of Tasmania, Hobart, TAS, Australia
University of the Sunshine Coast, Sippy Downs, Queensland, Australia

**Xia Zhu**
Chengdu Institute of Biology, Chinese Academy of Sciences, Chengdu, China
University of Chinese Academy of Sciences, Beijing, China
Department of Land, Air, and Water Resources, University of California Davis, Davis, California, United States of America

**Lucas C. R. Silva, Timothy A. Doane and William R. Horwath**
Department of Land, Air, and Water Resources, University of California Davis, Davis, California, United States of America

**Avat Shekoofa**
Department of Crop Science, North Carolina State University, Raleigh, North Carolina, United States of America

**Yahya Emam**
Department of Crop Production and Plant Breeding, Shiraz University, Shiraz, Iran

**Navid Shekoufa**
Department of Computer Engineering and Information Technology, Amirkabir University of Technology, Tehran, Iran

**Mansour Ebrahimi**
Department of Biology, School of Basic Sciences, University of Qom, Qom, Iran

**Esmaeil Ebrahimie**
Department of Crop Production and Plant Breeding, Shiraz University, Shiraz, Iran
School of Molecular and Biomedical Science, The University of Adelaide, Adelaide, Australia

**Fanqiao Meng, Xiangping Sun and Wenliang Wu**
College of Resources and Environmental Sciences, China Agricultural University, Beijing, China

**Jørgen E. Olesen**
Department of Agroecology and Environment, Faculty of Agricultural Sciences, Aarhus University, Tjele, Denmark

**Yuying Wang, Chunsheng Hu, Hua Ming, Wenxu Dong, Yuming Zhang and Xiaoxin Li**
Key Laboratory of Agricultural Water Resources, Center for Agricultural Resources Research, Institute of Genetics and Developmental Biology, Chinese Academy of Sciences, Shijiazhuang, Hebei, China

**Oene Oenema**
Department of Soil Quality, Wageningen University, Alterra, Wageningen, The Netherlands

**Douglas A. Schaefer**
Key Lab of Tropical Forest Ecology, Xishuangbanna Tropical Botanical Garden, Chinese Academy of Sciences, Menglun, Yunnan, China

**Belén Carbonetto, Nicolás Rascovan and Martin P. Vázquez**
Instituto de Agrobiotecnología de Rosario (INDEAR), Predio CCT Rosario, Santa Fe, Argentina

**Roberto Álvarez**
Facultad de Agronomía, Universidad de Buenos Aires, Buenos Aires, Argentina

**Alejandro Mentaberry**
Departamento de Fisiología y Biología Molecular y Celular, Facultad de Ciencias Exactas y Naturales, Universidad de Buenos Aires, Buenos Aires, Argentina

**Zongzhuan Shen, Chao Xue, Jian Zhang, Rong Li and Qirong Shen**
National Engineering Research Center for Organic-based Fertilizers, Key Laboratory of Plant Nutrition and Fertilization in Low-Middle Reaches of the Yangtze River, Ministry of Agriculture, Jiangsu Key Lab and Engineering Center for Solid Organic Waste Utilization, Jiangsu Collaborative Innovation Center for Solid Organic Waste Resource Utilization, Nanjing Agricultural University, Nanjing, China

**Dongsheng Wang**
Nanjing Institute of Vegetable Science, Nanjing, China

**Yunze Ruan**
Hainan key Laboratory for Sustainable Utilization of Tropical Bio-resources, College of Agriculture, Hainan University, Haikou, China

**Qing-zhong Zhang, Xing-ren Liu, Yi-ding Wang, Jian Huang and Ning Lu**
Key Laboratory of Agricultural Environment, Ministry of Agriculture, Sino-Australian Joint Laboratory For Sustainable Agro-Ecosystems, Institute of Environment and Sustainable Development in Agriculture, Chinese Academy of Agricultural Sciences, Beijing, China

**Feike A. Dijkstra**
Centre for Carbon, Water and Food, Department of Environmental Sciences, The University of Sydney, Camden, New South Wales, Australia

**Yushi Ye, Xinqiang Liang, Liang Li and Yuanjing Ji**
Institute of Environmental Science and Technology, College of Environmental and Resource Sciences, Zhejiang University, Hangzhou, China

**Yingxu Chen and Chunyan Zhu**
Zhejiang Province Key Laboratory for Water Pollution Control and Environmental Safety, Hangzhou, China

# Index